ANNUAL REVIEW OF NEUROSCIENCE

ANNUAL REVIEWS INC.
Palo Alto, California, USA

International Standard Serial Number: 0147–006X
International Standard Book Number: 0–8243-2407-2

Typesetting by Kachina Typesetting Inc., Tempe, Arizona; John Olson, President Typesetting coordinator, Jeannie Kaarle

PRINTED AND BOUND IN THE UNITED STATES OF AMERICA

ANNUAL REVIEW OF NEUROSCIENCE

VOLUME 7, 1984

W. MAXWELL COWAN, *Editor*

Salk Institute for Biological Studies

ERIC M. SHOOTER, *Associate Editor*

Stanford University School of Medicine

CHARLES F. STEVENS, *Associate Editor*

Yale University School of Medicine

RICHARD F. THOMPSON, *Associate Editor*

Stanford University

ANNUAL REVIEWS INC.   4139 EL CAMINO WAY   PALO ALTO, CALIFORNIA 94306 USA

EDITORIAL COMMITTEE (1984)

Annual Review of Neuroscience
Volume 7, 1984

CONTENTS

(Continued)   v

vi   CONTENTS   (*Continued*)

SOME RELATED ARTICLES IN OTHER *ANNUAL REVIEWS*

From the *Annual Review of Biochemistry*, Volume 52, 1983

Adenylate Cyclase–Coupled Beta-Adrenergic Receptors: Structure and Mechanisms
of Activation and Desensitization, R. J. Lefkowitz, J. M. Stadel, and M. G. Caron

Cell Surface Interactions with Extracellular Materials, K. M. Yamada

A Molecular Description of Nerve Terminal Function, L. F. Reichardt and
R. B. Kelly

From the *Annual Review of Biophysics and Bioengineering*, Volume 12, 1983

Intracellular Measurements of Ion Activities, R. Y. Tsien

Calcium Transport and Regulation in Nerve Fibers, J. Requena

Sodium Channel Gating: Models, Mimics, and Modifiers, R. J. French and R. Horn

Acetylcholine Receptor-Controlled Ion Translocation: Chemical Kinetic Investigations of the Mechanism, G. P. Hess, D. J. Cash, and H. Aoshima

From the *Annual Review of Medicine*, Volume 34, 1983

Sensory Evoked Potentials: Clinical Applications in Medicine, L. J. Dorfman

From the *Annual Review of Pharmacology and Toxicology*, Volume 23, 1983

Structure-Activity Relationships of Dopamine Agonists, J. G. Cannon

Nonopioid Neuropeptides in Mammalian CNS, L. L. Iversen

The Endorphins: A Growing Family of Pharmacologically Pertinent Peptides,
F. E. Bloom

From the *Annual Review of Physiology*, Volume 46, 1984

Neural Mechanisms for Sound Localization, R. B. Masterton

Neural Coding of Complex Sounds: Speech, M. B. Sachs

Peripheral and Central Chemoreceptors in the Fetus and Newborn, D. Walker

Classical Conditioning of Cardiovascular Responses, D. H. Cohen

Patch Clamp Techniques for Studing Ionic Channels in Excitable Membranes,
B. Sakmann and E. Neher

The Electric Sense of Weakly Electric Fish, W. Heiligenberg

Magnetic Field Sensitivity in Animals, J. L. Gould

Physiological Mechanisms for Spatial Filtering and Image Enhancement in the Sonar
of Bats, J. Simmons and S. A. Kick

From the *Annual Review of Psychology*, Volume 35, 1984

Brain Function: Neural Adaptations and Recovery from Injury, J. F. Marshall

Hormones and Sexual Behavior, H. H. Feder

Spatial Vision, G. Westheimer

Human Learning and Memory, D. L. Horton and C. B. Mills

Ann. Rev. Neurosci. 1984. 7:1–11

DOWNWARD CAUSATION?

János Szentágothai

Department of Anatomy, Semmelweis University Medical School, H-1450, Budapest, Hungary

Introduction

Following the lead of R. W. Sperry in his 1981 prefatory chapter of the *Annual Review of Neuroscience,* I shall avoid biographical aspects. I shall not even allow myself the tempting opportunity to look back on the neurosciences, even to give a perspective of the neuroanatomy of the mid-thirties from a contemporary vantage point—a view embracing entirely new techniques that bring together the tools of the micro-anatomist, the physiologist, and the neurochemist. A good reason for leaving biographical aspects aside is that I have been given ample opportunity over the last few years to indulge in some psychic exhibitionism, by having had the honor to be invited to write three autobiographical notes of my scientific activities and views (Szentágothai 1975b,c, 1982a). I do hope that I have succeeded in confirming my image as a "crazy Hungarian" and an impossible romantic adventurer, because I would be unhappy if the real flavor of my life became lost on my fellow neuroscientists.

I have to put the misgivings of the reader quickly to rest by promising that the parallel with Sperry's paper shall end here. Not because I would not agree almost completely with everything he expressed so beautifully about the importance of developing new priorities and values for mankind in which brain research may have a crucial word to say. In addition to being utterly unable to express such noble thoughts so articulately, even in my mother language, I am afraid that I lack the crucial stamina of a prophet, and particularly the zeal to speak up for something that I believe to be the right solution for mankind's predicament. (As a practicing Christian I believe that everything relevant was said close to 2000 years ago. There would be no problem if we would simply follow those words with the amendments made necessary by the changes that have occurred—I am thinking particularly of the need to control population growth—but I think even the amendments can be derived from the same pronouncements.) I am not too hopeful, though, if we continue mutually to

1

0147–006X/84/0301–0001$02.00

paint exaggerated images of our opponents and contribute to the real danger by adding fuel to the distrust instead of trying to end the deadly spiral.

What I shall try to do instead is to concentrate on a single problem, which I shall term "downward causation," although this issue is of general philosophical relevance, it gains particular significance in the problem of brain-mind relationship. My own "legitimate" field of studies—attempting to unravel the blueprint of neuronal and synaptic arrangements in various systems of the CNS, venturing into some speculations on the operative principles of neuron networks—was rather removed from this problem. However, I became fascinated with the gallant undertaking by R. W. Sperry (1965, 1969, 1972, 1977, 1980) to come to grips with the problem of consciousness or mental processes in general as an "operative force" intervening with the purely physiological mechanisms of the brain. That was exactly what I had always felt intuitively to be the crucial problem of the brain-mind relationship. But how to envisage such an interference within the framework of the second law of thermodynamics? "Ay there's the rub." As much as one might look, there is in Sperry's papers no solution or even a hint of how to get out of the dilemma. Sperry rejects both the remarkable arguments of MacKay (1980) and the dualistic-interactionalist philosophy offered by Popper & Eccles (1977). But if we were to accept neither, we would have to abandon the concept of "downward causation" entirely and accept one of the many current variations of the psycho-neural identity theories, or admit defeat in the form of a "non sequitur."

Is there really no way out of this dilemma? Intuitively I feel very strongly that within the framework of the natural sciences there is a solution, or at least a direction from which a solution might be sought, because without some kind of "downward causation" things simply make no sense. Since I want to keep away, as far as possible, from philosophical arguments[1], the best I can try is to take stock of some aspects of problems that have largely been neglected, or their significance underestimated by most recent authors. All of these issues have been treated separately by various authors in sufficient depth; what is lacking is an attempt to discuss them in terms of neurobiological realities and the problem of mind-brain relationship. These issues are (a) the consideration of neuron networks as self-organizing systems, (b) the possible functioning of self-organizing neuron networks as carriers of information, and (c) the question of how the principles of organization of neural centers, as we begin to understand them, might be integrated with (a) and (b).

---

[1]What I would resent particularly would be to be labeled as a something or other *ist*. I am simply a humble neurobiologist who wants to understand how the nervous system operates. In doing so I feel committed to stay within the strictest rules of natural science, though admitting that these are in continuous change and development. I have indicated my personal stand in the ultimate questions of existence, but this is beyond scientific reasoning. I would agree verbatim with the last paragraph of D. M. MacKay (1980). I do not think, however, that such questions need be discussed here because they have no influence upon my scientific reasoning.

There is obviously no room in this chapter for dealing with such issues in depth, so I shall only indicate the main lines along which I think some progress can be made. Lacking the appropriate mathematical insight into thermodynamics, information theory, and engineering for the proper treatment of self-organizing systems, I have joined forces with Peter Érdi for the elaboration of a "general brain theory" in which the necessary details can be defined and stated—not, of course, with the objective of trying to explain consciousness, but with the more modest aim of outlining a concept that, in principle, can lead to an explanation of a two-way relationship between conventional neural functions and the higher, more global events encompassed by the term "mind."

Neuron Networks and Self-Organization

The best experimental models for self-organizing neuron networks are tissue cultures, in which the activities of individual neurons can be recorded by microelectrodes (Crain 1973a,b). When I first became interested in this problem these techniques were not available, so we chose the developmental tissue recombination technique, devised by Paul Weiss (1950), of deplantation of neural tissue fragments into the dorsal fin of amphibian larvae. Our technique was essentially the same, but with more emphasis on a very careful histological analysis of the preparations, consisting of very small central tissue fragments from the medullary tube deplanted close to a limb bud (Székely & Szentágothai 1962). We observed exactly the same results as those described by Weiss (1950): the limb buds became innervated by the prospective motoneurons of the medullary tube implants; when the buds developed into supernumerary limbs, these became active, showing the characteristic spontaneous movements, generally in unpredictable bursts, termed by Weiss "epileptiform." These activities could not be influenced by local mechanical stimulation of the fin, so they were assumed to be spontaneous; they could be greatly enhanced by local application of strychnine or tetanus toxin. True reflex preparations could be produced easily by implanting an additional fragment of a dorsal root ganglion, the cells of which became connected with the medullary tube fragments and with the skin covering the graft. In such cases many spots developed from which characteristic movements of the supernumerary limb could be elicited by light mechanical stimulation. We concluded upon careful histological analysis that 10–20 scattered cells, without even a hint of any organized arrangement, were sufficient to produce spontaneous activity in the supernumerary limbs. There was no indication of a relationship among the size of the fragment, the number of surviving nerve cells, and the frequency of the movements. Because there was no neural input into these fragments it was assumed that the movements observed originated from the highly randomized neuronal network of the fragment and that the bursts of activity were transmitted to the supernumerary limb muscles by the few motoneurons in the fragment. The effect of strychnine could be explained by assuming the presence of inhibitory interneurons in the neural deplants. Although the activities of these preparations were absolutely unpredictable for any limited period of time, series of bursts, or of single twitch-like movements, were interrupted occasionally by periods of motionlessness lasting from a few seconds to a few minutes; each preparation had some distinctive general activity pattern, which would have been difficult to characterize more precisely due to differences in the muscles that happened to be innervated and the joints that were moved in the several specimens.

This may not seem terribly exciting, and from all that we now know about synapse formation and individual neuronal activities in tissue cultures, it is only as one might expect. Spontaneous activity of excitable tissues is a well-known phenomenon; isolation of such tissues enhances the spontaneous activity. But it is also quite generally assumed that many elements of higher centers show a great deal of spontaneous activity in the intact nervous system too. It is, of course, very difficult to assess how much of the neural activity that occurs during complete rest, for example in sleep, is really spontaneous and how much is dependent on a background of afferent inflow, on pacemaker sites or systems, and possibly on many other influences. It is obviously impossible to discuss even marginally the huge body of knowledge that has accumulated around this problem over the last half century. It seems remarkable, though, that so little thought has been given to the role of spontaneous activity in recent neuroscience. That apparently spontaneous activity and self-organization can go pretty far in producing natural-like outputs in isolated neural centers has been shown by Székely & Czéh (1971), who, in continuing our earlier experiments, have shown that pieces of neural tube tissue deplanted into the dorsal fin of larval amphibians can generate "walking-type" movements in supernumerary limbs, provided that the fragments are taken from limb-innervating segments of the neural tube and that some organized structure (including the central canal) is retained in the deplant. Preservation of dorsal horn tissue is not necessary, and the preparations need no neural input. Alternating movements of two supernumerary limbs can be observed (Brändle & Székely 1973) in similar experiments if a few segments rostral to the limb-bearing region are included in the model. This indicates that although "stepping" movements can be generated by the limb segment itself, the alternation of two limbs needs an additional "programmer" and that this is normally situated in segments rostral to the "stepping" generator.

How much neural activity is normally based on spontaneous activity and self-organization in various programmer networks? This problem has been studied extensively in invertebrates—albeit usually in connection with various kinds of pacemaker neurons—but recently has been almost completely neglected in the literature on the vertebrate nervous system. If isolated neural fragments that lack all neural inputs are capable of such sophisticated functions as indicated in the preceding paragraphs, the assumption that spontaneous activity and self-organization in neuronal networks, either with randomized or with more ordered internal connectivity, is by no means far fetched. The theoretical background for the treatment of such problems is available in the form elaborated by Wilson & Cowan (1972, 1973), or in the dissipative structural approach of Prigogine (1969), in the conceptual framework of synergetics (Haken 1978), or in catastrophe theories (Thom 1972). What I propose, therefore, is that the role of spontaneous activity be included in future considerations of neural organization.

Information Flow Aspects

D. M. MacKay's patient, unrelenting efforts over the last thirty-odd years to come to grips with the information flow aspects of "mind-like" behavior have fascinated me since the early 1960s. (I came to it relatively late because of my isolation—for various reasons—from 1939 to 1960.) I have often wondered why MacKay's gallant efforts, which have gained him world-wide recognition and have been repeatedly reprinted in subsequent "collections," have had so little impact on the neurosciences in the narrow sense. His information flow models (e.g. MacKay 1951, 1953, 1954, 1956, 1966, 1980) deserve much more serious attention from the neuroscience community and should be integrated into the framework of our concepts about neural organization.

Philosophically I feel comfortable with the argument of MacKay (1953a,b, 1980) that in order to denote the interdependence of mind and matter, terms other than "causality" and "interaction" ought to be used; these terms should be reserved "for the links between events or entities at the same categorical level, whether mental or physical." If we want to retain Aristotelian logic we have to label the "causality" between mental activity and the correlated brain activity as *"formal."* However, I also concur (with some reservations) with MacKay's (1978) argument that "purely physical determinancy" is "fully compatible with the determination of the informational function of the CNS by the mental activity it embodies," because the two belong to different categories.

So much for the philosophical side. Being a pedestrian neurobiologist, however, I would feel more comfortable with a more concrete explanation of how to envisage, for example, the influence that World 3 "objects" (in Popper's sense) exercise upon our brain processes. The analogy of the relationship between the physical functioning of a computer and the equation that determines its behavior cannot be transferred directly to this influence because there is no programmer who feeds "World 3 information" into the machine. Manfred Eigen (1982) expresses it beautifully when he states that the nervous system, like all living matter, is open not only in the sense that it feeds on extraneous sources of energy, but also in the informational sense. But I still wonder whether it would not be possible to approach the problem from the controversial relationship between entropy and information. Granted that the thermodynamic concept of entropy and "informational entropy" may have a purely "formal" correspondence, this seems to be true also of the "causal" relationship between mind and matter. Many physicists would agree, though not in the direct sense assumed by Brillouin (1962), that there is a genuine relationship between information and order and that the two may be linked with one another, not only formally. I have played with this thought rather intuitively over the last few years [and have published a few thoughts about this problem in Hungarian (Szentágothai 1979a, 1982b)], but I have not dared to publish anything as yet for an international readership. With the aid of Peter Érdi, I am now prepared to stick my neck out with a detailed account of what are still some very preliminary thoughts. We propose that if the carriers of information (about, say, the "objects" of Popper's World 3) are neuron networks endowed with the property of self-organization, some of the inconsistencies around the mind-matter problem might be removed without having to turn to either a radically monistic or explicitly dualistic philosophy.

If self-organization in neural networks is based to a considerable extent on the spontaneous activity of individual neurons (as discussed above), we are dealing essentially with what is considered "noise" in information theory and engineering. So we have here an example of "noise" being put to some useful purpose. But this should immediately make us aware of the danger of smuggling in some notion of "indeterminancy" through the back door. Physical indeterminancy in cerebral information processing has already been clearly ruled out by MacKay (1953b), and it would be misguided if this were to be reintroduced at the higher level of spontaneous neuronal activity.

Organization Principle of the CNS

J. C. Eccles first spelled out the crucial significance of the modular architectonic principle of the cerebral cortex (Popper & Eccles 1977, Eccles 1978, 1979, 1980; see especially his Heisenberg Lecture 1981) as the possible material basis of mental events. The modular concept had its antecedents in many illustrations of Ramón y Cajal (1909, 1911); it was explicitly stated first by Scheibel & Scheibel (1958) as a purely architectural notion. Eventually it became generalized under a different name by myself (Szentágothai 1967), and formulated as a general principle of neural organization by Szentágothai & Arbib (1974). For obvious reasons it was primarily the modular architectural principle of the cerebral cortex that lent itself to speculations about the mind-brain relationship. So it was quite logical that my earlier attempts to give an anatomical explanation of the "columnar arrangement" (Mountcastle 1957, Hubel & Wiesel 1959) in the sensory cortical areas (Szentágothai 1969, 1970, 1973, 1975a) were not really useful for the development of a coherent concept of the higher brain (or mind) functions. It was not until the discovery by

Goldman & Nauta (1977) of the anatomical reality of cortico-cortical columns that an overall structural framework of the cortical module was developed (Szentágothai 1978a,b).

The modular architectonic concept was used by Mountcastle (1978), with less ambitious goals in mind, to define an organizing principle for cortical function. A crucial new element was introduced by calling attention to the massive reentrant circuitry, which, along with the very complex and very high ordering of the extraneous connections, provides the essential background for "distributed systems." Although we do not have any firm data about quantitative relations, I would deem it a fair estimate that 80–90% of all connections (neglecting the local intracortical connections) of the cerebral cortex are cortico-cortical and only 10–20% are truly extrinsic. (It may be debatable whether cortico-striate connections should be labeled extrinsic or intrinsic.) What is of special importance in Mountcastle's concept is the assumption of phasic cycling of internally generated activity and its far-reaching functional consequences. I have tried in some of my more recent circuit diagrams (Szentágothai 1981, 1983) to illustrate the morphological substrates of this principle. Another important notion that Mountcastle has introduced is that of "developmental minicolumns" that are linked together mainly by extraneous inputs into much larger macrocolumns, 500–1000 μm in diameter. Although there is little anatomical evidence for such minicolumns the notion that they may be the smallest processing units of the cortex is convenient. Certainly the cortico-cortical column with its four to five thousand neurons contains a number of subsets that may be coupled in various combinations, which may not necessarily be fixed, but may very well be functional units within a dynamic continuum of combinations, changing as the moment may require.

But what is really essential in the "organizing principle" of Mountcastle is the phasic cycling of activities made possible by the massive reentrant connectivity. This would strongly support the basic notion mentioned above that neural functions be envisaged as dynamic patterns of neuronal network activities as in dissipative structures (Prigogine 1969). These could be imagined as being superimposed upon the conventional physiological chains of events elicited by stimuli reaching the different sensors as some kind of hierarchically coupled dissipative structure (Szentágothai 1978b, 1979a). This concept of neural organization was originally proposed by Aharon Katchalsky (Katchalsky et al 1974), but his important suggestions had little immediate influence upon the neurosciences. The basic notion of reentrance has been developed by Edelman (1978) into a very elegant theory of higher brain function of "group selection and phasic reentrant signaling." The hypothesis is attractive for several reasons: (a) it is in overall agreement with known anatomical facts (with one notable exception to be indicated below), (b) it is expressed without having to include mentalistic categories and depends on a set of mechanisms that are either verifiable or refutable, rather than on philosophical alliances, (c) it makes a coherent hypothesis that includes evolution, ontogeny, and experience throughout the entire lifetime, (d) it makes certain clear and testable propositions, including the very likely assumption that cortical, limbic-reticular, and thalamocortical relations are probably those primarily responsible for higher neural functions, and finally (e) it specifies the conditions that would refute the hypothesis. In spite of its many advantages this theory still suffers from being too abstract: What do the terms "recognizers" and "recognizers of recognizers" mean in realistic neurophysiological terms? What drives the "clockwork" for phasic reentrance? Other similar questions might be raised, but this is a general feature of all theories of this kind.[2] There is one aspect of the theory that is almost certainly not true: this is the suggestion that neocortical development is paralleled by a large increase in the number of local circuit neurons (LCN), in the sense of Rakić (1975) and Schmitt et al (1976). If anything the reverse is true: Dendro-dendritic synapses (many mutual and of reversed polarity) are very abundant in the olfactory bulb, in all thalamic nuclei, and even in some lower brainstem structures, but they are found only very occasionally in the amygdaloid nuclei, and they are virtually absent in the hippocampal formation and the neocortex. Neocortical development is characterized by an abundance and density of interneurons with locally ramifying axons. All of these interneurons fit into the classical concept of histodynamic polarity of Ramón y Cajal. So the LCNs can hardly have the significance attributed to them in Edelman's theory; their role is probably much more modest, as suggested by some theoretical considerations of Lábos (1977). However, these discrepancies are relatively minor, and the majority of anatomical assumptions in the theory I consider to be correct (although this opinion is not shared by many neuroanatomists). I include among these the assumption that most interneuronal connections are not rigidly determined genetically or developmentally. I have argued this view on purely anatomical grounds (with, I must admit, little success) for many years, but without attributing any special significance to the notion. It was only relatively recently that I became conscious of its possible significance (Szentágothai 1978a–c; see also 1981). What I had in mind was the occurrence of "dissipative structures" superimposed upon, or coexistent with, the conventional (reflex) functions of the neural organs.

All this may seem to have been dealt with in unnecessary detail, but it shows that the organizing principles of such widely divergent, yet "nonabstract," theories of higher brain functions as those of Eccles, Mountcastle, and Edelman have two major features in common: (a) all of them rely on some kind of modular architectonics principle (or at least on groups of neurons, in Edelman's theory); (b) all recognize *reentrance* as one of the crucial features. This is

---

[2]Edelman & Finkel (1983) have considerably elaborated on the group selection theory using the somatosensory cortex as a paradigm. Unfortunately, this as yet unpublished manuscript became known to me only after I had written this paper. In spite of the difference in the mental strategies applied, Edelman's concept has gained much in attraction in this new, greatly improved, version.

explicitly stated as the decisive element of the whole design in Mountcastle's (1978) theory and especially in Edelman's (1978) theory, but it is also implicit in Eccles' concept (see the diagram in Figure 4, Eccles 1981).

"Downward Causation?"

It appears then that the concept of "downward causation," though simple at first glance and intuitively so appealing, is frought with serious inconsistencies. According to Popper (Popper & Eccles 1977, p. 207) it would be the causal closeness of the physical world that is intuitively appealing to the majority of the scientific community, but what is intuitively appealing to me, rather, is the beautiful three world concept of Popper, although I cannot go along with its ultimate consequence of philosophical dualism (or trialism). Neither does MacKay, with whose philosophical positions I can identify myself without reservation. Although they have been expressed only very occasionally and with little reference to the nervous system, I feel closest to the views of Manfred Eigen (see the remarkable book *Das Spiel,* Eigen & Winkler 1975, and somewhat casually more recently, Eigen 1982) in considering the nervous system as "dissipative structures" open to the outside both for energy flow and for information flow.

"Downward causation," however appealing, cannot be expressed in physical terms, because the category difference mentioned above makes such a relationship meaningless. But does this compel us to accept a dualistic view? I do not think so; there has to be an acceptable third way between the essentially reductionist and the dualist interpretations. This is what I intend to present soon with Peter Érdi in appropriate detail and formalism for an outline of general brain theory. We shall try to incorporate into such a theory (or more correctly the outlines of a theory) the already partially cited definition restated by MacKay (1980) "that purely physical determinacy of the CNS is fully compatible with the determination of the informational function of the CNS by the mental activity it embodies, and has no need to be *refuted* in order to make room for the causal efficacy of human thinking and deciding."

What I am arguing is that the real principles of neural organization include: (a) the self-organizing capacity of neural networks as dissipative structures, (b) openness for information, including conventional sensory information, stored information (even genetically inbuilt programs), and information belonging to World 3 of Popper (used here only in the metaphorical sense to denote "things" or concepts that belong to mental categories), and (c) the anatomical and functional organization of neural matter into modules (or groups, having a high degree of degeneracy in the sense of Edelman) connected with one another in a way that allows for reentrance of activity patterns in an infinite number and variety of loops. Collectively these make it possible to elaborate a theory to account for the basic requirements of higher mental activities, including self-awareness. It may be debatable whether this is a realistic goal for the neurosciences, given the present state of knowledge; however, I believed with Sperry (1981) that this is the kind of contribution that mankind is entitled to expect from the brain sciences.

Literature Cited

Brändle, K., Székely, Gy. 1973. The control of alternating coordination of limb pairs in the newt *(Triturus vulgaris)*. *Brain Behav. Evol.* 8:366–85

Brillouin, L. 1962. *Science and Information Theory*. New York: Academic. 2nd ed.

Crain, S. M. 1973a. Microelectrode recording in brain and tissue cultures. In *Methods in Physiological Psychology*, Vol. 1, *Bioelectric Recording Techniques*, Pt. A, ed. R. F. Thompson, M. M. Patterson, pp. 39–75. New York: Academic

Crain, S. M. 1973b. Tissue culture models of developing brain functions. In *Developmental Studies of Behavior and the Nervous System*, Vol. 2, *Aspects of Neurogenesis*, pp. 69–114. New York: Academic

Eccles, J. C. 1978. An instruction-selection hypothesis of cerebral learning. In *Cerebral Correlates of Conscious Experience*, ed. P. A. Buser, A. Rougeul-Buser, pp. 155–75, Amsterdam: North-Holland 364 pp.

Eccles, J. C. 1979. *The Human Mystery*. Berlin: Springer. 255 pp.

Eccles, J. C. 1980. *The Human Psyche*. Berlin: Springer

Eccles, J. C. 1981. The modular operation of the cerebral neocortex considered as the material basis of mental events. *Neuroscience* 6:1839–56

Edelman, G. M., Finkel, L. H. 1983. Neuronal group selection in the cerebral cortex. In *Dynamic Aspects of the Neocortical Function. 1st Symp. Neurosci. Inst., La Jolla, Calif., Oct. 3–8, 1982*, ed. G. M. Edelman, W. M. Cowan, W. E. Gall. New York: Wiley. In press

Edelman, G. N., Mountcastle, V. B. 1978. *The Mindful Brain*. Cambridge, Mass./London: MIT Press. 100 pp.

Eigen, M., Winkler, R. 1975. *Das Spiel*. Munich: Piper. 426 pp.

Eigen, M. 1982. Vom Problem, das Genie zu erkennen. *Neue Zürcher Zeitung* 298:45

Goldman, P. S., Nauta, W. J. H. 1977. Columnar distribution of cortico-cortical fibers in the frontal association, limbic, and motor cortex of the developing Rhesus monkey. *Brain Res.* 122:393–413

Haken, H. 1978. *Synergetics. An Introduction*. Berlin: Springer-Verlag. 349 pp. 2nd ed.

Hubel, D. H., Wiesel, T. N. 1959. Receptive fields of single neurones in the cat's striate cortex. *J. Physiol. London* 148:574–91

Katchalsky, A. K., Rowland, V., Blumenthal, R. 1974. Dynamic patterns of brain cell assemblies. *Neurosci. Res. Progr. Bull.* 12(1): 1–187

Lábos, E. 1977. Theoretical considerations of local neuron circuits and their triadic synaptic arrangements (TSA) in subcortical nuclei. *J. Neurosci. Res.* 3:1–10

MacKay, D. M. 1951. Mindlike behaviour in artefacts. *Br. J. Philos. Sci.* 2:105–21

MacKay, D. M. 1953. Mindlike behaviour in artefacts. *Br. J. Philos. Sci.* 3:352–53

MacKay, D. M. 1954. Supralogical behaviour in automata. *Proc. 14th Int. Congr. Psychol.* Amsterdam: North Holland. 204 pp.

MacKay, D. M. 1956. Towards an information-flow model of human behaviour. *Br. J. Psychol.* 47:30–43

MacKay, D. M. 1966. Cerebral organization and the conscious control of action. In *Brain and Conscious Experience*, ed. J. C. Eccles, pp. 422–45, and 566–74. New York: Springer. 591 pp.

MacKay, D. M. 1978. Selves and brains. *Neuroscience* 3:599–606

MacKay, D. M. 1980. The interdependence of mind and brain. *Neuroscience* 5:1389–91

Mountcastle, V. B. 1957. Modalities and topographic properties of single neurons of cat's sensory cortex. *J. Neurophysiol.* 20:408–34

Mountcastle, V. B. 1978. The organizing principle for cerebral function: The unit module and the distributed system. See Edelman & Mountcastle 1978, pp. 7–50

Popper, K. R., Eccles, J. C. 1977. *The Self and Its Brain—An Argument for Interactionism*. Berlin: Springer Intl.

Prigogine, I. 1969. Structure, dissipation and life. In *Theoretical Physics and Biology*, ed. M. Marois, pp. 23–52. Amsterdam: North Holland

Rakić, P. 1975. Local circuit neurons. *Neurosciences Res. Progr. Bull.* 13:295–446

Ramón y Cajal, S. 1909. *Histologie du système nerveux de l'homme et des vertébrés*, Vol. 1. Paris: Maloine

Ramón y Cajal, S. 1911. *Histologie du système nerveux de l'homme et des vertébrés*, Vol. 2. Paris: Maloine

Scheibel, M. E., Scheibel A. B. 1958. Structural substrates for integrative patterns in the brain stem reticular core. In *Reticular*

Formation of the Brain, ed. H. H. Jasper, L. D. Proctor, R. S. Knighton, W. C. Noshay, R. T. Ostello, pp. 31–55. Boston: Little Brown

Schmitt, F. O., Dev, P., Smith, B. H. 1976. Electronic processing of information by brain cells. *Science* 193:114–20 (9 July)

Sperry R. W. 1965. Mind, brain and humanist values. In *New Views of the Nature of Man*, ed. J. R. Platt, pp. 71–92. Chicago: Univ. Chicago Press

Sperry, R. W. 1969. A modified concept of consciousness. *Psychol. Rev.* 76:532–36

Sperry, R. W. 1972. Science and the problem of values. *Perspect. Biol. Med.* 16:115–30

Sperry, R. W. 1977. Bridging science and values: A unifying view of mind and brain. *Am. Psychol.* 32:237–45

Sperry, R. W. 1980. Mind-brain interaction: Mentalism yes; dualism, no. *Neuroscience* 5:195–206

Sperry, R. W. 1981. Changing priorities. *Ann. Rev. Neurosci.* 4:1–15

Székely, G., Czéh, G. 1971. Activity of spinal cord fragments and limbs deplanted in the dorsal fin of *Urodele Larvae*. *Acta Physiol. Acad. Sci. Hung.* 40:303–12

Székely, G., Szentágothai, J. 1962. Experiments with "model nervous systems." *Acta Biol. Acad. Sci. Hung.* 12:253–69

Szentágothai, J. 1967. The anatomy of complex integrative units in the nervous system. In *Recent Development in Neurobiology in Hungary*, ed. K. Lissák, 1:9–45. Budapest: Akad. Kiadó

Szentágothai, J. 1969. Architecture of the cerebral cortex. In *Basic Mechanisms of the Epilepsies*, ed. H. H. Jasper, A. A. Ward, A. Pope, pp. 13–28. Boston: Little, Brown. 835 pp.

Szentágothai, J. 1970. Les circuits neuronaux de l'écorce cérébrale. *Bull. Acad. R. Med. Belgique* (Ser. 7) 10:475–92

Szentágothai, J. 1973. Synaptology of the visual cortex. *Handb. Sensory Physiol.* 7(3b):269–324

Szentágothai, J. 1975a. The "module concept" in cerebral cortex architecture. *Brain Res.* 95:475–96

Szentágothai, J. 1975b. From the last skirmishes around the neuron theory to the functional anatomy of neuron networks. In *The Neurosciences: Paths of Discovery*, ed. F. G. Worden, J. P. Swazey, G. Edelman, pp. 103–20. Cambridge, Mass.: MIT Press 622 pp.

Szentágothai, J. 1975c. Under the spell of hypothalamic feedback. In *Pioneers in Neuroendocrinology*, ed. J. Meites, B. T. Donovan, pp. 297–311. New York/ London: Plenum. 327 pp.

Szentágothai, J. 1978a. The neuron network of the cerebral cortex: A functional interpretation. The Ferrier Lecture. *Proc. R. Soc. London. Ser. B* 201:219–48

Szentágothai, J. 1978b. The local neuronal apparatus of the cerebral cortex. In *Cerebral Correlates of Conscious Experience*, ed. P. A. Buser, A. Rougeul-Buser, pp. 131–38. Amsterdam: North Holland. 364 pp.

Szentágothai, J. 1978c. Specificity versus (quasi-) randomness in cortical connectivity. In *Architectonics of the Cerebral Cortex*, ed. M. A. B. Brazier, H. Petsche, pp. 77–97. New York: Raven. 486 pp.

Szentágothai, J. 1979a. Egységes agyelmélet: Utopia vagy realitás (A general brain theory: Utopia or reality.) In Hungarian. *Magyar Tudomány* 8/9: 601–16

Szentágothai, J. 1979b. Local neuron circuits of the neocortex. In *The Neurosciences Fourth Study Program*, ed. F. O. Schmitt and F. G. Worden, pp. 399–415. Cambridge, Mass./ London: MIT Press 1185 pp.

Szentágothai, J. 1981. Principles of neural organization. In *Advances in Physiological Sciences*, Vol. 1, *Regulatory Functions of the CNS Principles of Motion and Organization*, ed. J. Szentágothai, M. Palkovits, J. Hámori, pp. 1–16. Oxford-Budapest: Pergamon Press-Akadémiai Kiadó. 335 pp.

Szentágothai, J. 1982a. Too "much" and too "soon." A lifetime inquiry into the functional organization of the nervous system. *Acta Biol. Acad. Sci. Hung.* 33:107–26

Szentágothai, J. 1982b. Elmélkedés egy általános agyelméletről (Thoughts on a general brain theory.) In Hungarian. *Valóság* 9:1–9

Szentágothai, J. 1983. The "modular" architectonic principle of neural centers. In *Reviews of Physiology, Biochemistry, and Pharmacology*. Heidelberg: Springer Verlag. In press

Szentágothai, J., Arbib, M. A. 1974. Conceptual models of neural organization. *Neurosciences Res. Progr. Bull.* 12:313–510

Thom, R. 1972. *Stabilité structurelle et morphogenèse*. New York: Benjamin

Weiss, P. 1950. The deplantation of fragments of nervous system in amphibians. I. Central reorganization and the formation of nerves. *J. Exp. Zool.* 113:397–461

Wilson, H. R., Cowan, J. D. 1972. Excitatory and inhibitory interactions in localized populations of model neurons. *Biophys. J.* 12:1–24

Wilson, H. R. Cowan, J. D. 1973. A mathematical theory of the functional dynamics of cortical and thalamic nervous tissue. *Kybernetic* 13:55–80

Ann. Rev. Neurosci. 1984. 7:13–41

THE ACCESSORY OPTIC SYSTEM

John I. Simpson

Department of Physiology and Biophysics, New York University Medical Center, New York, New York 10016

INTRODUCTION

Despite its name, the accessory optic system (AOS) is a primary visual system. Not only do retinal ganglion cell axons constitute a major input to AOS nuclei, but the AOS contributes fundamentally to retinal image stabilization, which is important for the proper functioning of other visual systems. Although the AOS has been recognized anatomically for well over 100 years and is present in all vertebrate classes (Mai 1978), the first physiological studies were not reported until 1959 (Marg et al 1959). Until quite recently, progress in understanding AOS physiology was limited. Indeed, in the most recent review of the physiology of the AOS, Marg (1973) expressed concern as to whether the AOS even had a normal function. In the intervening years substantial progress has been made in revealing both the anatomy and physiology of the AOS, and these advances are the focus of this review. After a brief survey of the gross anatomical features of the AOS, the following topics are treated: retinal ganglion cell morphology and distribution, efferents and nonvisual afferents of the AOS, visual response properties of AOS neurons and their interpretation, behavioral consequences of AOS lesions, and relations and comparisons between the AOS and an allied pretectal nucleus.

ANATOMICAL DESCRIPTION AND SOME HISTORY

What is the Accessory Optic System?

Because the AOS is not a well-known entity, even to some visual specialists, it is appropriate to begin by addressing the question, "What is the AOS?" The

13

0147–006X/84/0301–0013$02.00

terminology and general topographical plan proposed by Hayhow & collaborators (Hayhow 1959, 1966, Hayhow et al 1960) provides the most widely accepted answer for the mammalian AOS. In Hayhow's schema the AOS consists of two sets of optic fibers of contralateral retinal origin, the inferior and superior accessory optic fasciculi, and three target nuclei in the anterior midbrain, the medial, lateral, and dorsal terminal nuclei (Figure 1). An introduction to the course and termination of these fasciculi is afforded by the initial description presented by Hayhow et al (1960) for the rat. The inferior fasciculus of the accessory optic tract (AOT-IF) is formed by a group of late crossing optic fibers that are medially displaced from the main optic tract by the fibers that make up the supraoptic commissures of Gudden and Meynert. At the medial edge of the cerebral peduncle the AOT-IF deviates from the course of the main optic tract and proceeds posteriorly, some fibers passing through the lateral hypothalamic and adjacent subthalamic regions and some remaining in relation to the medial edge of the cerebral peduncle, to terminate in a prominent nucleus in the ventromedial midbrain tegmentum (Figure 2). This nucleus, the medial terminal nucleus, is bounded laterally by the cerebral peduncle and the substantia nigra and medially by the mammillary peduncle and the medial lemniscus. The superior fasciculus of the accessory optic tract (AOT-SF) is formed by a more or less distributed set of optic fibers that proceed from the posterior margin of the optic tract and the ventral margin of the brachium of the superior colliculus to run superficially across the cerebral peduncle to its medial edge and ultimately to the medial terminal nucleus. The most posterior bundle of the superior fasciculus contains fibers that terminate in two superficially located neuronal aggregations termed the dorsal and lateral terminal nuclei. The dorsal terminal nucleus lies in the groove between the ventral edge of the brachium of the superior colliculus and the dorsoposterior border of the medial geniculate body. It forms a small superficial elevation and is encapsulated by the posterior fibers of the AOT-SF. The lateral terminal nucleus is situated immediately ventral to the medial geniculate body in apposition to the dorsolateral edge of the cerebral peduncle. The AOT-SF was subsequently divided more formally into three components termed the anterior, middle, and posterior fibers (Hayhow 1966). The anterior fibers leave the ventroposterior margin of the main optic tract anterior to where this line of departure comes into close relation with the lateral geniculate body. The anterior fibers run superficially in a posteromedial direction to the medial edge of the cerebral peduncle and then proceed posteriorly close to, but more or less distinguishable from, the AOT-IF to attain the medial terminal nucleus. The middle fibers of the AOT-SF leave the main optic tract in relation to the ventroposterior pole of the lateral geniculate body. They descend superficially across the anterolateral surface of the medial geniculate body and some of them probably send collaterals to the lateral terminal nucleus while proceeding across the surface of the peduncle to

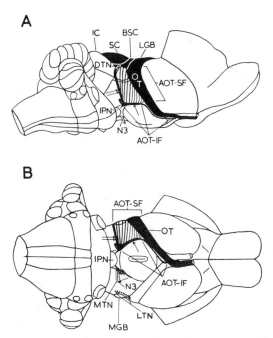

Figure 1   A right ventrolateral view (*A*) and a direct ventral view (*B*) of the rat brainstem, showing the course taken by degenerating contralateral optic axons *(shown in black)* resulting from section of the left optic nerve. The course of the inferior fasciculus (AOT-IF) is indicated by *fine stipple*. The course of the superior fasciculus (AOT-SF) is indicated semidiagrammatically by *unbroken lines*. The *single arrow* indicates the termination in the anterior extremity of the MTN of the slender fiber bundle formed by the most anterior fibers of the AOT-SF. The *double arrows* indicate the course of the more posterior fibers of the AOT-SF (the classical tractus peduncularis transversus of von Gudden). The location of the DTN is depicted by *white dots* in *A*, while the location of the LTN and MTN is shown in *B* by *black dots*. **Abbreviations for *Figures 1* and *2:*** AOT-IF, accessory optic tract - inferior fasciculus; AOT-SF, accessory optic tract - superior fasciculus; BP, basis pedunculi; BSC, brachium of superior colliculus; DTN, dorsal terminal nucleus; HPT, habenulo-peduncular tract; IC, internal capsule; IPN, interpeduncular nucleus; LGB$_d$, lateral geniculate body, pars dorsalis; LGB$_v$, lateral geniculate body, pars ventralis; LTN, lateral terminal nucleus; MB, mammillary body; MGB, medial geniculate body; MP, mammillary peduncle; MTN, medial terminal nucleus; MTT, mammillo-thalamic tract; N3, oculomotor nerve; OT, optic tract; PC, posterior commissure; PR, pretectal region; SC, superior colliculus; SN, subthalamic nucleus; SNI, substantia nigra; SOC, supraoptic commissure. (From Hayhow et al 1960.)

the medial terminal nucleus. The posterior fibers of the AOT-SF arise largely from the ventrolateral margin of the brachium of the superior colliculus. Some of these fibers synapse in the dorsal terminal nucleus; those that proceed farther ventrally do so in relation to the posterolateral surface of the medial geniculate body to attain the lateral or the medial terminal nuclei. Neurons are also located outside of these nuclei, within the posterior fiber bundle of the AOT-SF.

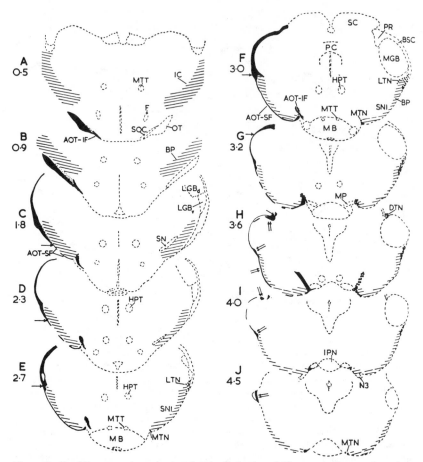

Figure 2 Semidiagrammatic projection drawings of selected Nauta-stained coronal sections through the diencephalic and midbrain regions of the rat following section of the right optic nerve. The drawings *A-J* form an anteroposterior series. The degeneration patterns presented by the main and accessory optic axons are shown on the left-hand side. The degenerating AOT-IF is indicated by *stipple;* the degenerating axons of the main optic tract and AOT-SF are shown by *solid lines.* The *single arrow* shows the point of departure of the fibers of the AOT-SF from the main optic tract. The *double arrows* indicate the various portions of the most posterior fibers of the AOT-SF that correspond with the classical tractus peduncularis transversus of von Gudden. The location of the terminal cells of the AOS has been indicated by *dots* on the right-hand side. The degeneration patterns in the LGB, pretectal region, and SC have not been indicated. The numbers following the alphabetical figure designations indicate the approximate distance in millimeters of each section from the posterior edge of the optic chiasm. (From Hayhow et al 1960.)

The AOS of some mammalian species is not complete, in the sense that one or more components that compose the general plan have been reported missing. Later studies using more sensitive anatomical methods, however, have often revealed some of the previously undetected components (e.g. cat, Lin & Ingram 1972; opossum, Lent et al 1976; ferret, Thorpe & Herbert 1976a; golden hamster, Pickard & Silverman 1981). In addition to their use in revealing previously undetected crossed components of the AOS, anatomical methods based on axonal transport have consistently shown minor uncrossed components that usually escaped detection with degeneration techniques (e.g. Takahashi et al 1977, Winfield et al 1978, Erickson & Cotter 1983). An AOS has not been found in echolocating bats, but has been found in non-echolocating bats (Cotter & Pierson Pentney 1979).

A striking departure from a "complete" AOS is found in primates. Giolli (1963) first reported for cynomolgus monkey that both the AOT-IF and, more surprisingly, the medial terminal nucleus are absent. Subsequent investigations of both prosimian and anthropoid primates have consistently failed to find a direct retinal projection to the ventral tegmental region where the medial terminal nucleus is customarily found in other mammals (Campos-Ortega & Glees 1967, Campos-Ortega & Cluver 1968, Campbell 1969, Tigges & Tigges 1969a,b, Giolli & Tigges 1970, Hendrickson et al 1970, Tigges & O'Steen 1974, Tigges et al 1977, Lin & Giolli 1979). The reduced AOS characteristic of primates consists of the posterior fibers of the AOT-SF, a dorsal terminal nucleus, and a comparatively substantial lateral terminal nucleus. In the cynomolgus monkey, Giolli (1963) described a minute superficial pathway that was free of degenerating optic fibers and extended from the lateral terminal nucleus to a nucleus located in the customary position of the medial terminal nucleus. In the rhesus monkey, such a pathway and nucleus could not be clearly seen (Lin & Giolli 1979). Interestingly, the tree shrew, which has been considered an osculant between insectivores and primates, has a fully developed AOS with both a medial terminal nucleus and an AOT-IF (Tigges 1966, Campbell et al 1967, Campbell 1969).

The layout of the AOS in nonmammals is less complicated than in mammals. In birds, amphibians, and reptiles, the AOS is classically considered to consist of a single discrete fasciculus that separates from the main optic tract just after crossing in the chiasm, and a single terminal nucleus located at the ventral margin of the midbrain tegmentum, slightly rostral to the exit of the third cranial nerve (Shanklin 1933, Ariens-Kappers et al 1936, Herrick 1948, Mai 1978). This fasciculus and its terminal nucleus have been given many names, but those most widely used at present are *basal optic root* (BOR) and *nucleus of the basal optic root* (nBOR). In bird, the nBOR has been divided on morphological grounds into three parts: the nBOR proper, nBOR pars lateralis and nBOR pars dorsalis, which together make up the nBOR complex (Brecha et al 1980).

The AOS has been consistently and readily recognized in mammals, birds, reptiles, and amphibians, but this was not the case for fish prior to the use of autoradiographic methods to study retinofugal pathways (Ariens-Kappers et al 1936, Ebbesson 1972, but see Campbell & Ebbesson 1969). When autoradiographic methods have been used in fish, an AOS has usually been revealed, and now species differences are apparent with regard to location, size, and number of presumed AOS nuclei (e.g. Northcutt & Butler 1976, Reperant et al 1976, Finger & Karten 1978, Northcutt & Wathey 1980, Smeets 1981). Whether structures labeled as AOS nuclei in fish are all properly so described has been questioned, raising the issue of what criteria should be met in assigning a structure to the AOS (Finger & Karten 1978, Prasada Rao & Sharma 1982).

How the Accessory Optic System Got Its Name

During the long anatomical history of the AOS, an extensive and somewhat bewildering body of names for its optic pathways and terminal nuclei has accumulated. Access to the history of AOS nomenclature is readily available (Hamasaki & Marg 1960a, Giolli 1963, Marg 1964, Mai 1978), but a few points here on terminology may be helpful to those not wishing to make their own historical journey. The first informative description of an AOS component derived from von Gudden's (1870) observations of the surface anatomy of the rabbit brainstem. He described a macroscopically visible tract that arises from the brachium of the superior colliculus, courses along the anterior edge of the superior colliculus in the groove between the colliculus and the posterior part of the medial geniculate body, descends superficially over the brachium of the inferior colliculus and the cerebral peduncle, and then courses medially over the surface of the cerebral peduncle to disappear just rostral to the exit of the third nerve. Von Gudden called this fiber bundle the *tractus peduncularis transversus,* which has been anglicized to *transpeduncular tract* (Giolli 1961). This tract, which is poorly developed after removal of the retina in newborn rabbit (von Gudden 1870, 1881), is identical to the compact bundle of posterior fibers of the AOT-SF. In 1903 Marburg applied the name tractus peduncularis transversus to a set of optic fibers that ran with the main optic tract up to the lateral geniculate body and then coursed ventrocaudally around the cerebral peduncle to enter the midbrain medial to the cerebral peduncle. This tractus peduncularis transversus corresponds most closely to the middle fibers of the AOT-SF. These two different descriptions probably came about because Marburg worked largely with guinea pigs and these animals do not have a compact bundle of the type described by von Gudden (Giolli & Creel 1973). Subsequent investigators (e.g. Tsai 1925, Le Gros Clark 1932, Gillilan 1941) were not careful to distinguish the transpeduncular tract as described by von Gudden from the transpeduncular tract as described by Marburg. Considerable confusion, of course, arose from applying the identical name to two tracts having substantially different courses, but each terminating in the same nucleus, the medial terminal nucleus.

In 1908 Bochenek described in the rabbit a collection of optic fibers that were separated, just posterior to the chiasm, from the main body of optic fibers by a collection of nonoptic fibers. Bochenek called this tract the *fasciculus accessorius optici anterior,* and it is to him that we can trace the demeaning label "accessory" as applied to the presently recognized AOS. Although Bochenek did not explicitly propose that the transpeduncular tract(s) be called the *fasciculus accessorius optici posterior,* it was natural for others to make this association. In the course of time, the label *posterior accessory optic tract* came to be used indiscriminantly as an alternative name for both the transpeduncular tract of Marburg and the transpeduncular tract of von Gudden. The anterior accessory optic tract of Bochenek corresponds to the AOT-IF only up to a point, for Bockenek and others (e.g. Loepp 1911, Tsai 1925, Gillilan 1941) believed that the anterior accessory optic tract went posteriorly only as far as the subthalamic nucleus of Luys. Later it was recognized that these optic fibers passed through the subthalamus to terminate in the medial terminal nucleus (e.g. Lashley 1934, Hayhow et al 1960, Giolli 1961).

RETINAL GANGLION CELL MORPHOLOGY AND DISTRIBUTION

The morphology and retinal distribution of AOS ganglion cells have been studied in different species by using retrograde horseradish peroxidase (HRP) methods (pigeon, Karten et al 1977, Fite et al 1981; chicken, Reiner et al 1979; chinchilla, Kimm et al 1979; rabbit, Oyster et al 1980; frog, Montgomery et al 1981; turtle, Reiner 1981; cat, Farmer & Rodieck 1982). The initial experiments of this type were performed in pigeon (Karten et al 1977, Fite et al 1981) and revealed that the retinal projection to the nBOR complex originated almost exclusively from the displaced ganglion cells of Dogiel. In chinchilla, turtle, and frog only a small portion of the retinal projection to accessory optic nuclei was found to originate from displaced ganglion cells; in rabbit and cat no labeled displaced ganglion cells were found after HRP injections into the medial terminal nucleus.

Even though a large interspecies difference is present with regard to the relation of displaced ganglion cells to the AOS, in some species AOS ganglion cells do have characteristics in common beyond their projection to AOS nuclei. In rabbit, the soma size of the ganglion cells projecting to the contralateral medial terminal nucleus places them among the 20% largest ganglion cells, and their anatomically determined partitioning between the visual streak and the rest of the retina closely corresponds to the physiologically determined partitioning of only the on-direction selective class of ganglion cells. Although the

AOS ganglion cells of rabbit and cat have about the same average absolute soma size, the relatively small to medium soma size of cat AOS ganglion cells contrasts with what pertains in the rabbit. The particular dendritic pattern of cat AOS ganglion cells suggests that they constitute a distinct morphological type within the "gamma class" of cat ganglion cells. In turtle, the AOS ganglion cell somas are located in different layers within the retina (80% orthotopic in the ganglion cell layer, 6% in the inner plexiform layer, 14% at the inner margin of the inner nuclear layer), but even so, they have in common the features of large soma size and wide dendritic ramifications in the outer portion of the inner plexiform layer. The most heterogeneous population of AOS ganglion cells has been found in the frog. Six morphological types were identified: two types of displaced cells and four types of orthotopic ganglion cells. From the above, it can be seen that the answer to the question, "Do AOS ganglion cells form a single morphological type common to all species?" is "no." When the question is restricted to one particular species, the answer would appear to be "yes" for the pigeon, chicken, rabbit, and cat, but "no" for the frog.

The above question can be turned around: "Do ganglion cells that project to the AOS nuclei do so exclusively?" For pigeon and chicken the answer is a qualified "yes." HRP injections into the pretectum and optic tectum have not resulted in labeled displaced ganglion cells, at least of the type that projects to the nBOR complex, but the remaining primary visual receiving areas have not yet been so examined. For the other species studied, it is not yet clear whether some ganglion cells project exclusively to the AOS nuclei.

The number of AOS ganglion cells could be determined in some of the studies referenced above. In pigeon, approximately 3700 cells were labeled with a large HRP injection into the nBOR complex, in reasonable agreement with an estimate of 4300 BOR fibers. In the adult chicken up to 7700 cells were similarly labeled; the increase over that found in pigeon has been attributed to the chicken's larger eye and to differences in the retinal distribution pattern, as discussed below. A large HRP injection into rabbit and cat medial terminal nuclei resulted in about 2000 and 1000 labeled cells in the contralateral retina, respectively. In rabbit, the average number of retinal fibers projecting to the medial terminal nucleus in the posterior bundle of the AOT-SF was found to be 1900 (Giolli 1961). In comparing the two figures, it should be noted that the count of labeled ganglion cells should reflect projections to the medial terminal nucleus via all components of the accessory optic tract. In turtle, the number of fibers in the basal optic root is estimated to be 1500. In the frog about 850 large, medium and displaced ganglion cells were labeled following HRP application to the transected BOR. This figure is close to the number of myelinated fibers (700) found in the frog BOR using electron microscopic analysis (Vinogradova & Manteifel 1976). The many small labeled ganglion cells found throughout the frog retina after application of HRP to the cut BOR are presumed to give rise to the 1800 unmyelinated fibers also present in the frog BOR.

A quantitative measure of the uncrossed AOS projection to the medial terminal nucleus is available from the HRP studies in rabbit and cat. In cat, the ipsilateral retina contained no more than 3% of the number of labeled cells found in the contralateral retina; in one rabbit the corresponding figure was 0.3%. The ipsilateral:contralateral ratio of grain counts obtained from autoradiographic studies in chinchilla (Winfield et al 1978) and rabbit (E. Takahashi, personal communication) suggests a higher percentage for the uncrossed input to the medial terminal nucleus (up to 7%) than found in the retrograde labeling studies. Whatever technical limitations underlie these differences, both assessments point to a small uncrossed component. In frog the uncrossed component in the basal optic root appears to be made up of unmyelinated fibers.

A detailed picture of the retinal distribution of AOS ganglion cells is available from HRP studies performed in rabbit, cat, and pigeon (Figure 3). In the rabbit and cat, the density distribution roughly mirrors that of the overall ganglion cell population. In both rabbit and cat there is a paucity of AOS ganglion cells in the far retinal periphery. Several other features are noteworthy. In the rabbit, AOS ganglion cells projecting to the medial terminal nucleus are absent in a horizontally extended region in which the ganglion cell axons are myelinated and in which the overall ganglion cell density is very low. In the cat, there is no evidence of a nasotemporal division for AOS ganglion cells. The distribution of chinchilla AOS ganglion cells is qualitatively similar to that found in cat and rabbit, in that labeled ganglion cells were located chiefly in the central retina and were absent from the peripheral retina. The distribution present in pigeon contrasts with that in mammals insofar as very few labeled cells were found in regions of visual specialization (central fovea and red field). Outside of these areas of specialization labeled cells are found throughout the retina, with a moderate tendency to concentrate more peripherally in the temporoinferior and nasosuperior quadrants. In chickens, the AOS ganglion cells are also more concentrated toward the periphery, but unlike the pigeon there is no clear evidence of a differential distribution among retinal quadrants. Presumably the relative absence in the chicken of specialized retinal regions, which correlate with a low AOS ganglion cell density in pigeon, results in not only a greater number of AOS ganglion cells but also in a more uniform distribution.

EFFERENTS AND NONRETINAL AFFERENTS

For many years, knowledge of the connectivity of the mammalian AOS was limited, in essence, to a catalog of species variations in the course and

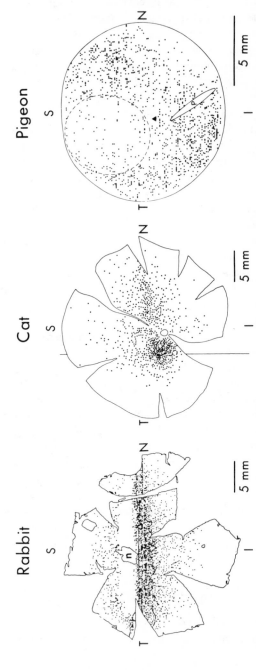

Figure 3 Retinal distribution of ganglion cells retrogradely labeled following HRP injection into the contralateral MTN of rabbit and cat and the contralateral nBOR complex of pigeon. Whole-mount preparations were used for rabbit and cat and each dot represents one labeled ganglion cell. The pigeon distribution is based on a reconstruction from horizontal serial sections and shows the location of the labeled cells found in every third section. The *countour line* on the rabbit retina is the 2000 ganglion cell/mm² isodensity line, which is taken as the boundary of the visual streak. The *vertical marks* near the cat retina indicate the zero vertical meridian. The *area within the broken line* on the pigeon retina indicates the approximate location of the red field. Abbreviations: T, temporal; N, nasal; I, inferior; S, superior; n, optic nerve head; P, pecten; ▲, approximate location of the fovea. (From Oyster et al 1980, Farmer & Rodieck 1982, and Fite et al 1981.)

terminations of the retinal axons and to the frequent observation that the posterior fiber bundle of the AOT-SF is not made up only of optic fibers. This situation began to change when Giolli et al (1968) revealed, in a Golgi study, some of the organization of the rabbit medial terminal nucleus in relation to its immediate neighborhood, including a reciprocal connection with the substantia nigra. The efferent connections of the AOS with more distant specific structures began to be revealed by Maekawa & Simpson (1972, 1973), who suggested, on the basis of an electroanatomical mapping study in rabbit, that a direct connection from the lateral pretectum to the ipsilateral inferior olive was a component in the pathway carrying visual signals from the AOS to the cerebellar flocculo-nodular lobe.

The Efferent Connections

The efferent connections from the AOS to the inferior olive are now known in some detail, particularly in the rabbit. The pathway originally described by Maekawa & Simpson was shown to go to the dorsal cap of Kooy of the inferior olive (Mizuno et al 1973, 1974), which gives rise to the major climbing fiber projection to the flocculo-nodular lobe (Alley et al 1975). Subsequently, Takeda & Maekawa (1976; Maekawa & Takeda 1979) used retrograde HRP techniques to reveal a finer structure in the relation of the AOS to the dorsal cap. The caudal half of the dorsal cap was found to receive a projection from both the ipsilateral dorsal terminal nucleus and the immediately adjacent pretectal nucleus known, in mammals, as the nucleus of the optic tract (NOT). In addition, Takeda & Maekawa (1976) stated that the ipsilateral lateral terminal nucleus projected to the dorsal cap, but judging from their figures, the labeled cells said to be in the lateral terminal nucleus are actually within the posterior fiber bundle of the AOT-SF, between the dorsal and the lateral terminal nuclei. Such an arrangement was found by J. I. Simpson and R. E. Soodak (unpublished observations) after HRP injections into the rabbit dorsal cap; no labeled cells were found in the region that is customarily defined as the lateral terminal nucleus. Neurons interstitial to the posterior fibers of the AOT-SF have often been described, and those that project to the caudal half of the dorsal cap are presumably part of the dorsal terminal nucleus. The midbrain projections to the rabbit rostral dorsal cap contrast with those described above for the caudal dorsal cap in that few, if any, cells whose soma is located within AOS retinal terminal fields project to the rostral dorsal cap. Rather, an ipsilateral projection arises from two groups of neurons located in the ventral tegmentum, dorsal to the medial terminal nucleus and ventral to the medial lemniscus and red nucleus (Maekawa & Takeda 1979). An ipsilateral projection from the NOT to the dorsal cap has been reported in other mammals, but the likely contribution from the dorsal terminal nucleus has not been explicitly distinguished (e.g. Itoh 1977, Terasawa et al 1979, Weber & Harting 1980, Walberg et al 1981).

Recently, a small projection from the NOT to the contralateral dorsal cap was reported for rabbit (Holstege & Collewijn 1982). In pigeon, a bilateral, but ipsilaterally predominating, projection from the nBOR complex (nBOR proper and nBOR pars dorsalis) to the dorsal division of the medial accessory inferior olive has been found (Brecha et al 1980); whether this olivary region projects to the flocculus and nodulus as well as to the uvula and paraflocculus remains to be determined. The frog nBOR has a medullary projection that may correspond to the projection to the inferior olive described above (Montgomery et al 1981).

The first general survey of AOS efferents was performed in the pigeon (Brauth & Karten 1977, Brecha et al 1980). The nBOR complex was shown to project (a) bilaterally to the vestibulocerebellum, (b) bilaterally to the inferior olive (see above), (c) bilaterally to the oculomotor complex, (d) bilaterally to the interstitial nucleus of Cajal, (e) contralaterally to the nBOR complex, and (f) ipsilaterally to the pretectal nucleus lentiformis mesencephali, pars magnocellularis (LMmc). Some of these connections appear to be comparable to AOS efferent paths found in other species, but species differences are also apparent.

In pigeon, the AOS projects as mossy fibers to a restricted part of the vestibulocerebellum—namely, a portion of the uvula (folia IX c,d) and the paraflocculus. This projection arises from all parts of the nBOR complex and is about evenly distributed with regard to laterality. In turtle, the nBOR and adjacent nucleus give rise to a predominantly ipsilateral projection to the cerebellum (Reiner & Karten 1978), and in fish a midbrain nucleus, which may be part of the AOS, projects to the cerebellum (Finger & Karten 1978). In mammals, a direct AOS projection to the vestibulocerebellum has been reported only for the chinchilla (Winfield et al 1978); HRP injections into the flocculus resulted in labeled cells in the medial terminal nucleus, predominantly on the contralateral side, but not in the lateral or dorsal terminal nucleus or in the pretectum. In similar studies in other mammals, a projection from the medial terminal nucleus to the flocculus has been looked for, but not found (cat, Kawasaki & Sato 1980, R. Blanks and J. Blanks, personal communication; rat, Blanks et al 1983; rabbit, J. I. Simpson, unpublished observations, K. Maekawa, personal communication). The generality of a direct projection from the AOS to the vestibulocerebellum is also brought into question by the report of Montgomery et al (1981) that HRP injections into frog cerebellum did not result in labeled cells in the nBOR.

The pigeon nBOR complex projects upon specific divisions of the oculomotor nuclei, as described below in conjunction with the associated physiology. On the basis of the anatomical observations, a disynaptic connection from the retina to extraocular motoneurons is probably present in the bird. A similar arrangement has been advocated for the frog on electrophysiological grounds (Cochran et al 1981); anatomical studies of the frog indicate that the soma of the interposed neuron lies outside the nBOR neuropil, but has dendrites extending into it (Montgomery et al 1981). In a recent autoradiographic study of rat and rabbit medial terminal nucleus efferents, no projection to the oculomotor nuclei was detected (Blanks et al 1982b).

In the pigeon, some neurons in all three divisions of the nBOR complex project contralaterally through the posterior commissure to terminate within the contralateral nBOR pars dorsalis. A posterior commissural pathway from the medial terminal nucleus has also been found in the rat and rabbit to terminate contralaterally upon cells whose location in the ventral tegmentum is compatible with their projection to the dorsal cap of the inferior olive (Blanks et al 1982b: R. Blanks and J. I. Simpson, unpublished observations). A posterior commissural pathway has also been identified in relation to the frog AOS (Montgomery et al 1981).

A bilateral projection to the interstitial nucleus of Cajal has been described from the nBOR complex in the pigeon (Brecha et al 1980) and from the medial terminal nucleus in the rat and rabbit (Blanks et al 1982b). This path provides a rather short route whereby visual signals can influence neck muscle activity.

The prominent projection from the entire nBOR complex upon the ipsilateral pretectal nucleus LMmc completes the catalog of nBOR efferents in pigeon. The medial terminal nucleus in rabbit (Simpson & Soodak 1978) and rat (Blanks et al 1982a) also projects to the NOT. The similarity between mammals and birds with regard to this projection serves as one criterion for the commonly proposed homology between NOT and LMmc. A comparable pathway also exists in the frog (Montgomery et al 1981). The course of the projections from the medial terminal nucleus to the NOT has been determined in some detail for rat (Blanks et al 1982a); one route is superficial within the posterior fiber bundle of the AOT-SF, whereas two other routes are internal. The termination of these fibers in rat and rabbit is not restricted to the NOT but also includes the ipsilateral dorsal and lateral terminal nuclei (Blanks et al 1982a; R. McCrea et al, unpublished observations).

A recent study of medial terminal nucleus efferents in rat and rabbit revealed several projections that were not described for pigeon, including a predominantly contralateral projection to the superior and lateral vestibular nuclei (Blanks et al 1982b).

The Nonretinal Afferents

A comprehensive search for the nonretinal afferents to the AOS nuclei has not yet been undertaken. Therefore, our knowledge of these inputs is probably quite incomplete, for it has been obtained mostly from studies focused elsewhere. Whether the ipsilateral projection from the NOT to the medial and lateral terminal nuclei [recently found in several mammals (tree shrew, Weber & Harting 1980; cat, Berson & Graybiel 1980; rabbit, Holstege &

Collewijn 1982)] should be regarded as an afferent projection to the AOS or as an internal connection of the AOS depends upon one's view of the NOT vis-à-vis the dorsal terminal nucleus. Nevertheless, its existence, along with the reciprocally directed pathway discussed above, indicates the presence of communication between nuclei whose neurons share the feature of direction selectivity but are distinguishable by their direction preferences (see below). These interconnecting pathways are possibly involved in direction-selectivity tuning. In addition to the ipsilateral projections between the NOT and AOS nuclei, certain of the possible combinations of commissural projections between the dorsal terminal nucleus and the NOT have been reported in rabbit (Holstege & Collewijn 1982).

A projection from the ventral lateral geniculate body to the lateral or medial terminal nuclei has been found in the rat and cat (Graybiel 1974, Swanson et al 1974, Edwards et al 1974). A possibly comparable secondary visual input from the thalamus to the nBOR has been reported in the frog (Montgomery et al 1981). Other nonretinal but visually related inputs to AOS nuclei in the cat arise from some, but not all, parts of the ipsilateral visual neocortex (Berson & Graybiel 1980, Marcotte & Updyke 1982). The medial Clare-Bishop area and area 21 are the cortical regions primarily responsible for the projection to each of the three terminal nuclei in the cat. A projection from visual cortical areas I and II to the medial terminal nucleus has not been found in the rabbit (Giolli & Guthrie 1971, Giolli et al 1978), but these visual cortical areas may not be comparable to the ones in cat that give rise to AOS afferents. Finally, a projection to AOS nuclei in the cat from an imprecisely known part of the midbrain reticular formation has been described (Berson & Graybiel 1980).

VISUAL RESPONSE PROPERTIES

Neurons in the Nuclei of the Accessory Optic System

Physiological investigations of the AOS were first conducted in the rabbit; this seems only fitting since the rabbit featured prominently in the anatomical history of the AOS. The earliest electrophysiological experiments (Hamasaki & Marg 1960b, 1962, Hill & Marg 1963) served mainly to rule out certain previously proposed functions for the AOS. Some of the visual response properties characteristic of AOS neurons in general were initially observed by Walley (1967) in a study of the medial terminal nucleus in the rabbit. Over half the neurons showed a type of direction selectivity for vertical movement, and the vast majority of them preferred upward movement within their quite large receptive fields. Walley concluded that the AOS is specialized for detecting movement in the vertical plane and may be concerned with changes in eye position relative to the horizon (Marg 1964).

Identification and interpretation of the visual response properties of AOS neurons was advanced by Simpson et al (1979), who recorded from each of the three terminal nuclei in the rabbit. Their studies revealed that AOS neurons process signals about both the speed and direction of movement of large parts of the visual world (Figure 4). Neuronal activity was well modulated by moving large (30° × 30°) textured patterns slowly in certain directions within the even larger contralateral receptive fields. The greatest modulation occurred at quite slow speeds, ranging from 0.1 to 1.0°/sec. For movement in the preferred excitatory direction, neural activity increased in a sustained manner up to two to three times over a typical background activity of 25–50 spikes/sec. Movement in the preferred inhibitory direction could silence the neural activity. This type of direction selectivity, with both incremental and decremental responses, contrasts with the purely incremental type of direction selectivity reported by Walley (1967). Although the preferred inhibitory direction has been commonly referred to as the "null direction," this label should be abandoned, for it is not in keeping with the meaning of the word "null." The preferred directions of medial and lateral terminal nuclei neurons were vertical but with a posterior (temporal) component, and therefore, the preferred excitatory direction and the preferred inhibitory direction of individual neurons typically were not 180° apart. The proposed significance of this non-collinear organization is treated below. With the visual stimulus restricted to a 70° × 70° region of visual space centered on the optic axis, medial terminal nucleus neurons excited by upward and somewhat posterior movement were twice as numerous as those excited by downward and somewhat posterior movement. Lateral terminal nucleus neurons also responded best to off-vertical movement, but for most of them the preferred excitatory direction was downward and somewhat posterior. In contrast to medial and lateral terminal nuclei neurons, dorsal terminal nucleus neurons were best modulated by horizontal movement, and posterior to anterior (temporanasal) movement in the visual field was the preferred excitatory direction.

Electrophysiological studies of the AOS of other animals (frog, Kondrashev & Orlov 1976, Cochran et al 1980, 1981, Gruberg & Grasse 1980; chicken, Burns & Wallman 1981; pigeon, Morgan & Frost 1981; cat, Grasse & Cynader 1982a,b) have revealed visual response properties that are qualitatively similar to those of the rabbit. Among the various animals studied, the shared characteristics of AOS neurons are speed- and direction-selective responses to movement of large, textured visual patterns. The optimal speed for exciting AOS neurons ranges from .5 to 10°/sec, depending partly on the species. Neurons of the lateral and medial terminal nuclei and the nBOR have in common vertical or off-vertical direction preferences, and the preferred excitatory direction and the preferred inhibitory direction are often non-collinear. There are, of course exceptions to the above generalizations, particularly for the frog.

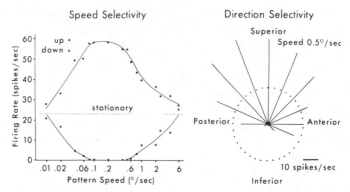

Figure 4 Speed- and direction-selective characteristics of a rabbit MTN neuron, illustrating the typical preference for slow speeds and for vertical movements with a posterior component. The position and effective size of the visual stimulus used in obtaining these data are illustrated in *Figure 7*. The speed selectivity curve was obtained for vertical movement. The direction selectivity polar plot was obtained by moving the stimulus at 0.5°/sec in each of the 16 directions indicated by the *solid lines*. The length of these lines represents the firing rate for each particular direction. The *dotted circle* represents the level of spontaneous activity when the stimulus was stationary. For some neurons, the preferred excitatory direction was up and somewhat posterior in the visual world (as illustrated), but for others the preferred excitatory direction was down and somewhat posterior. (From Simpson et al 1979.)

Recordings from neurons located in the monkey ventral tegmentum and purported to be part of the AOS revealed pursuit eye movement signals in addition to direction selective retinal slip signals (Westheimer & Blair 1974). However, as pointed out above, primates lack a medial terminal nucleus, and the relation of these neurons to the AOS is unclear.

Ganglion Cells of the Accessory Optic System

A detailed comparison between the response properties of retinal ganglion cells and central neurons in the AOS can be made at present only for the rabbit. The visual response properties of neurons in the rabbit's AOS nuclei are in several ways markedly similar to those of a type of direction-selective ganglion cell found in the rabbit retina (Barlow et al 1964, Oyster & Barlow 1967, Oyster 1968, Oyster et al 1972). Like the majority of rabbit medial terminal nucleus neurons (Hamasaki & Marg 1960b, 1962, Walley 1967, Simpson et al 1979), these ganglion cells are excited only at the onset of steady illumination and are thus called on-direction-selective ganglion cells. The speed selectivity of these ganglion cells is identical to that of neurons in the terminal nuclei of the AOS (Figure 5). The preferred directions of these ganglion cells have been determined for that portion of visual space within about 20° of the optic axis (Oyster 1968). For this region of visual space the preferred directions, taken collectively, define three directions: anterior, up with a posterior component, and down with a posterior component (Figure 5). When the preferred excitatory direc-

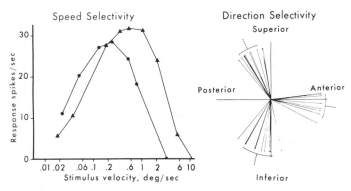

Figure 5   Speed and direction selectivity of the on-direction-selective class of retinal ganglion cells in the rabbit retina. The two speed selectivity curves represent the range of variation found for movement in the preferred excitatory direction. The spatial distribution of the preferred excitatory directions for a number of ganglion cells shows that collectively they establish three clusters of preferred directions for a certain portion of visual space (see text). (From Oyster et al 1972 and Oyster 1968.)

tions of the rabbit AOS neurons are compared with those of the on-direction-selective ganglion cells, a striking similarity is apparent. That is, the dorsal terminal nucleus cells have a preferred excitatory direction from posterior to anterior in the horizontal direction, and the lateral and medial terminal nuclei neurons have a preferred excitatory direction that is either up with a posterior component or down with a posterior component. The marked similarities of speed and direction selectivity indicate that the retinal input to the rabbit AOS is largely, if not exclusively, from the on-direction-selective ganglion cells. The preferred excitatory and inhibitory directions of rabbit lateral and medial terminal nuclei neurons are typically non-collinear; both have a posterior component along with the major vertical component. This arrangement suggests that both the preferred excitatory direction and the preferred inhibitory direction of lateral and medial terminal nuclei neurons are related in a simple way to the two groups of on-direction-selective ganglion cells that have a vertical preference. The decremental component of the modulation of AOS nuclei neurons is probably due to inhibitory interneurons within the AOS (Simpson et al 1979), because the on-direction-selective ganglion cells in the rabbit have virtually no spontaneous activity (Oyster 1968, Wyatt & Daw 1975); this precludes a disfacilitory origin for the decrement. A model suggesting how the directional tuning properties characteristic of rabbit medial and lateral terminal nuclei neurons may be synthesized is illustrated in Figure 6.

Functional Interpretations

On the basis of the response properties of AOS neurons, Simpson et al (1979) proposed a specific function for the AOS. They reasoned that the rather large visual stimulus required to modulate AOS neurons well would be present

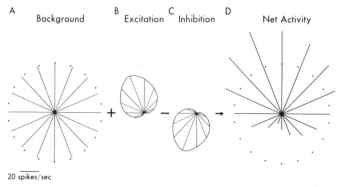

20 spikes/sec

Figure 6 Synthesis of direction-selective properties of rabbit medial and lateral terminal nuclei neurons from direction-selective properties of rabbit ganglion cells. The direction-selective tuning curves of rabbit direction-selective ganglion cells are cardioid-like when represented in polar form, as in *B* and *C* (Oyster 1968, Levick et al 1969). Tuning curves (e.g. D) highly similar to those found for rabbit MTN and LTN neurons can be constructed by summing a spontaneous activity (*A*) with excitatory (*B*) and inhibitory (*C*) inputs represented by the two ganglion cell tuning curves positioned with their characteristic off-vertical orientation.

outside the laboratory only as a result of an animal's own movement, and they concluded that the AOS serves to signal self-motion as reflected in the slip of the visual world image over the retina. Detection of self-motion is also a property of the vestibular system, and the fact that AOS neurons are selective to slow speeds of retinal image movement makes the AOS an ideal complement to the vestibular system. One of the possible behavioral consequences of self-motion detection is the reflex stabilization of the eyes and head. Investigators of the AOS generally agree that it is intimately involved in stabilizing the eyes and head in space. The efferent relations of the AOS, as presented above, are consistent with this view.

The proposition that the accessory optic and vestibular systems are complementary detectors of self-motion raised the question of how signals of self-motion from two different sensory modalities are represented so as to facilitate their interaction. The tripartite distribution of preferred directions of AOS ganglion cells and the non-collinearity of the preferred excitatory and inhibitory directions of AOS neurons were both important clues in arriving at the proposition that the AOS represents global retinal slip in a coordinate system like that of the semicircular canals (Simpson & Soodak 1978, Simpson et al 1979). As originally presented, this proposition related the directions of movement of the optic axis produced by rotation of the rabbit about the three principal semicircular canal axes, to the three preferred directions of the on-direction-selective ganglion cells. The movement of the optic axis was chosen as an indicator of the rotational movement because it approximately represents the visual world movement that would be seen by that population of retinal ganglion cells whose preferred direction distribution had been obtained by Oyster (1968). The general situation is pictured in Figure 7, along with the particular visual stimulus arrangement most often used by Simpson et al (1979) in determining the response properties of AOS neurons. Figure 7 illustrates how the optic flow field set up by rotation about each of the two vertical canal principal axes can be associated with the two groups of on-direction-selective ganglion cells that have off-vertical direction preferences in the neighborhood of the optic axis. Extension of this association to include horizontal movement is straightforward. The non-collinearity of the preferred excitatory and inhibitory directions of rabbit medial and lateral terminal nuclei neurons can be seen as the outcome of combining an excitation for rotation about one vertical canal principal axis with inhibition for rotation about the other vertical canal principal axis (Simpson et al 1982).

Burns & Wallman (1981) pooled direction preference measures from chicken nBOR neurons with those from the pigeon nBOR (Morgan & Frost 1981), and found that the neurons could be placed into one of two categories according to the organization of their preferred excitatory and inhibitory directions. One group of neurons had an upward preferred excitatory direction and a downward preferred inhibitory direction, separated on average by 185°. The other group had a downward and anterior preferred excitatory direction and an upward and anterior preferred inhibitory direction, separated on average by 145°. Burns & Wallman (1981) related these two groups to rotation about one or the other of the vertical canal principal axes, using as a reference the movement of the optic axes. In the chicken, the optic axis is angled about 65° to the sagittal plane and therefore is not symmetrically positioned with respect to the two principal axes of the vertical canals. Therefore, the curvature of the arc swung through by the optic axis is appreciably greater for rotation about one principal axis than the other. This difference was taken to be consistent with the difference in preferred direction organization found for the two groups of nBOR neurons, thus supporting the notion that the AOS is organized along the same motion axes as the vestibular system. The receptive field organization proposed for the chicken is somewhat different from that proposed for the rabbit, since in the chicken the inhibitory and excitatory contributions should arise from separate regions of the receptive field, one above the horizon and the other below the horizon.

The projection pattern of the pigeon nBOR complex upon specific divisions of the oculomotor nuclei has an attractive physiological correlate (Burns & Wallman 1981). The neurons of the dorsal part of the nBOR complex, which are excited by upwardly moving visual stimuli presented to the contralateral eye, project ipsilaterally to that division of the oculomotor complex that innervates the superior rectus muscle of the contralateral eye (Brecha et al 1980). The neurons of the ventral part of the nBOR complex, which are excited by downwardly moving visual stimuli presented to the contralateral eye,

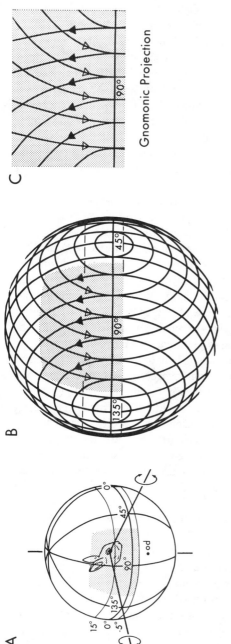

Orthographic Projection

Gnomonic Projection

Figure 7 Interpretation of AOS direction preferences in terms of tuning for detection of rotational optic flow. The physical relation between the rabbit and the surrounding visual world, considered as an enveloping spherical surface, are portrayed in *A*. The visual world is referenced to the rabbit using the coordinates of azimuth and elevation, with the origin at the anterior nodal point of the eye. The parallels at elevations 15° and −5° indicate the boundary of the visual streak (Hughes 1971) and *od* denotes the projection of the optic disc. The *stippled region* represents the region of effective visual stimulation transferred from a tangent plane, centered at azimuth 90° and elevation 0°, to the reference sphere representing the visual world. The stimulus in the tangent plane was 70° square and centered on the optic axis, but because of the retinal distribution of the AOS ganglion cells the effective visual stimulus extended below the horizon by only about 5°. The two axes of 45° and 135° indicate the principal axes of the vertical semicircular canals as well as axes of relative rotation between the rabbit and the visual world. A set of optic flow lines can be associated with rotation about each of the axes. The two sets of optic flow lines are depicted in *B*, using an orthographic projection. Within the area subtended by the effective visual stimulus, the set of flow lines associated with the 135° axis is marked by *solid triangles* while the set associated with the 45° axis is marked with *open triangles*. In each case, the directions of the arrows correspond to a particular cluster of preferred directions of the on-direction-selective ganglion cells, i.e. the *solid arrows* represent the up and posterior group while the *open arrows* represent the down and posterior group (see *Figure 5*). In *C*, the two sets of rotational optic flow lines are shown gnomonically projected onto the tangent screen. The flow line curves in the tangent screen are conic sections.

project contralaterally to that division of the oculomotor complex that inner-vates the inferior rectus muscle of the contralateral eye. There is also a weaker bilateral projection to the trochlear nucleus, suggesting that a projection to the inferior oblique motor nucleus also exists.

In the cat, the direction tuning profiles for medial terminal nucleus neurons show that the preferred excitatory and inhibitory directions are commonly not 180° apart (Grasse & Cynader 1982a). In an attempt to represent each neuron's direction selectivity by a single excitatory and a single inhibitory preferred direction, a vector summation was performed separately for the excitatory and inhibitory components of the direction tuning profile. The physiological mean-ing to be attached to the average preferred directions obtained by this manipula-tion is obscure. Because the excitation and inhibition are not symmetrically distributed with respect to their respective peak values, a vector summation yields average preferred inhibitory and excitatory directions that are more nearly collinear than are the directions corresponding to the peak values. To offer a proper explanation of the shapes of direction tuning profiles it is important that their essential geometrical properties not be lost to averaging. Aspects of directional response distributions that are lost to averaging may reflect variations in the position and weighting of each neuron's receptive field with respect to spatially nonuniform optic flow fields.

In the rabbit, some medial terminal nucleus neurons have the intriguing characteristic of distinctly different preferred direction organizations in sepa-rate parts of their monocular, contralateral receptive field (Leonard et al 1981). Typically, the receptive field extended mostly above the horizon, and in one part of the receptive field the preferred excitatory direction was up and some-what posterior, whereas in the other part of the receptive field the preferred excitatory direction was down and somewhat posterior. The direction selectiv-ity of these neurons can be characterized by a preferred axis of rotation. For example, when the excitatory preferred direction in the receptive field region extending from 0° (the nose) to 45° posterior is up and somewhat posterior, and the excitatory preferred direction in the region from 45° to greater than 160° posterior is down and somewhat posterior, rotation of the visual world about an axis at about 45° to the midline will best match the spatial organization of the preferred directions. Other neurons located in the ventral tegmentum just dorsal to the medial terminal nucleus, and antidromically identified as projecting to the inferior olive, were typically binocular, with one or the other eye dominant. These neurons have particular spatial organizations of preferred directions within their virtually global receptive fields that make them appropriate precur-sors for the visual response properties of rostral dorsal cap neurons. Neurons within the rostral dorsal cap respond preferentially to rotation of the visual world about axes that bear a striking geometrical resemblance to the principal axes of the vertical semicircular canals (Simpson et al 1981).

LESION STUDIES

As a result of lesion studies, three disparate functions have been proposed for the AOS: mediation of light-induced tonic changes in neuroendocrine activity; transmission of visual signals necessary for performing luminous flux discrimination in the absence of visual striate cortex; and mediation of optokinetic nystagmus.

At a time when the presence of a direct retinohypothalamic connection was debatable, Moore et al (1967) lesioned components of the rat visual system in various combinations and concluded that the AOT-IF transmits the photic signals controlling the activity in the pineal body of the melatonin-forming enzyme hydroxyindole-O-methyltransferase. Subsequently, relations between the AOS and other light-modulated neuroendocrine functions were sought but not found (e.g. Moore 1974, Thorpe & Herbert 1976b). Upon readdressing the issue of which visual pathway and central neural structures are involved in the photic regulation of pineal hydroxyindole-O-methyltransferase activity, Klein & Moore (1979) concluded from a lesion study in rat that the relevant pathway is the retinohypothalamic tract, not the AOT-IF.

From a variety of lesion studies, Pasik & Pasik (1971, 1973) concluded that the visual input through the AOS is essential for the ability of rhesus monkeys deprived of visual striate cortex to discriminate differences in total luminous flux. This conclusion should be viewed with reservation, if only for the reason that the nucleus described and pictured (Pasik & Pasik 1971, Pasik et al 1973) as the AOS structure targeted for lesioning does not correspond topographically to the description and location of monkey AOS nuclei, as consistently given by others (Giolli 1963, Campos-Ortega & Glees 1967, Tigges & Tigges 1969a, Hendrickson et al 1970, Tigges & O'Steen 1974, Lin & Giolli 1979). The nucleus described by the Pasiks is close to, but posterior to, the lateral terminal nucleus. The problems associated with trying to link the AOS to visual discrimination abilities are made evident by the studies of Legg (1979) in the rat, and by the finding that the performance of a variety of visual descrimination tasks by pigeons is not compromised by lesions of the nBOR (Hodos & Bonbright 1975).

The first lesion study that showed behavior altered in a way consistent with contemporary understanding of AOS function was performed by Lazar in the frog (1973). Interestingly, Lazar's specific findings that sectioning of the BOR or destruction of nBOR abolished horizontal optokinetic head nystagmus could not be replicated, but even so, his study marks the first clear association between the AOS and optokinetic nystagmus. In contrast to the findings of Lazar, bilateral destruction of nBOR in the turtle and pigeon resulted only in a reduction of horizontal nystagmus frequency at the higher range of effective stimulus velocities (Fite et al 1979). A similar change was recorded in the chinchilla following bilateral lesion of the medial terminal nucleus (Kimm et al 1979). The discrepancy between the outcome of the initial experiments of Lazar and attempts to replicate them have been largely resolved by the recent findings of Montgomery et al (1982) that, in frogs, only lesions located medial to nBOR, in the peri-nBOR, abolished horizontal optokinetic nystagmus. Lesions of BOR or nBOR only altered horizontal nystagmus frequency over different parts of the range of effective stimulus velocities. In adult chickens, nBOR lesions abolish vertical and torsional, but not horizontal, optokinetic nystagmus, a finding in concert with the directional preferences of chicken nBOR neurons (Wallman et al 1981).

A PRETECTAL ALLY OF THE ACCESSORY OPTIC SYSTEM

Both anatomical and physiological studies in mammals have shown that some neurons in the pretectal NOT are closely allied with those of the AOS. Anatomical relations between the AOS and the NOT have been reviewed above. The NOT and dorsal terminal nucleus have more than what may be a false border in common; the visual response properties of neurons in both nuclei are characterized by direction and speed selectivity with excitatory preference for horizontal, temporonasal movement of large patterned stimuli (rabbit, Collewijn 1975a, Simpson et al 1979; cat, Hoffmann & Schoppman 1975, 1981, Hoffmann et al 1976, Grasse & Cynader 1982b). In rabbit, some NOT neurons have a low speed selectivity identical to that of dorsal terminal nucleus neurons, while other NOT neurons exhibit an optimal response at speeds higher than those preferred by dorsal terminal nucleus neurons. In cat, the range of optimal speeds for NOT neurons (1–10°/sec) is comparable to that for dorsal terminal nucleus neurons (3–13°/sec). In rat, the speed selectivity for NOT or dorsal terminal nucleus neurons has not been specifically reported, but for pretectal direction-selective neurons as a group the optimal response is about 1°/sec (Cazin et al 1980), much the same as for the rabbit dorsal terminal nucleus. In rabbit and cat, the direction-selective modulation of NOT neurons to contralateral monocular stimulation typically consists of both incremental and decremental changes from the spontaneous discharge rate. That pattern was seldom found in the rat; rather, the most common form of direction-selective modulation consisted simply of an increase in activity for temporonasal movement, without a decrease below spontaneous activity for oppositely directed movement.

Collewijn (1975b) showed the NOT to be an essential structure in the horizontal optokinetic circuit of rabbit; after degeneration of the primary optic fibers, unilateral electrical stimulation of the NOT elicited horizontal optokinetic nystagmus with the slow phase toward the stimulated side, and small lesions in the lateral pretectum abolished horizontal optokinetic nystagmus. The possibility of a contribution from the dorsal terminal nucleus was not explicitly investigated, probably because preliminary stimulation experiments at the ventrolateral side of the mesodiencephalon, where the posterior fibers of the AOT-SF are coursing, never elicited any eye movement. This observation seems odd in light of the presently known physiological and anatomical similarities between the dorsal terminal nucleus and the NOT, but it is reminiscent of the finding of Maekawa & Simpson (1973) that stimulation of the ventral part of AOT-SF led to blockage, at the lateral pretectum, of transmission from the retina to the inferior olive. Hoffmann (1981, 1982) has related the visual response properties of direction-selective NOT cells that project to the inferior olive to the quality of optokinetic nystagmus in normal cats, in monocularly deprived cats, and in cats with large lesions of the visual cortex. The ocular dominance distribution and speed selectivity of these NOT neurons is dependent upon inputs from the visual cortex, and the changes seen in these characteristics of NOT neurons following monocular deprivation and visual cortex lesion correlate well with many of the concomitant changes in optokinetic nystagmus.

In pigmented rabbits, nearly all NOT neurons are preferentially excited by temporonasally moving stimuli presented to the contalateral eye, but in rabbits with ocular albinism, an inversion of this direction selectivity occurs in that part of the receptive field of NOT neurons that corresponds to the anterior visual field (Collewijn et al 1978, Winterson & Collewijn 1981). This arrangement is likely to be responsible for the inversion in horizontal optokinetic nystagmus elicited by stimulating the anterior visual field of albino rabbits.

The mammalian NOT and the pretectal nucleus LMmc of the bird have often been proposed, on anatomical grounds, to be homologous. The appropriateness of this proposal is supported by the outcome of a 2-deoxyglocose study identifying brain regions in chicken that responded to retinal slip in different directions (McKenna & Wallman 1981). Slow, horizontal, temporonasal visual world movement presented to one eye caused the contralateral LMmc to become heavily labeled, while nasotemporal movement resulted in little, if any, detectable label. In addition, the contralateral nBOR pars lateralis was differentially labeled in response to horizontal movement, whereas no differences in the labeling of the other regions of the nBOR complex on the two sides of the brain could be distinguished. This study also revealed that the nBOR pars lateralis and the LMmc merge with one another, and the homology of the nBOR pars lateralis and the dorsal terminal nucleus was proposed.

CLOSING REMARKS

Over the past decade our knowledge and understanding of the AOS has evolved rapidly. No longer are there any doubts as to whether the AOS has a normal function. Contemporary physiological and anatomical studies in a variety of vertebrates have consistently pointed to the importance of the AOS in processing and distributing visual signals subserving compensatory eye and head movement. Could it not, therefore, be argued that an "AOS" is also present in those invertebrates that show optokinetic behavior?

The AOS stands as a specific example of how vision is used for proprioception, and illustrates how messages about self-movement that are inherent in optic flow can be detected and analyzed. Studies of the AOS show clearly the importance of including geometrical considerations in attempts to understand sensorisensory integration as well as sensorimotor integration.

The AOS has at last come into its own as a primary visual system, but there is yet much to learn from it and about it. Among the topics for future investigation are elaboration of the anatomy and physiology of the nonretinal afferents to the AOS, and assessment of the contribution of the different cell types within the AOS nuclei to overall function.

ACKNOWLEDGMENTS

The author's research is supported in part by USPHS Grant NS-13742 and by the Irma T. Hirschl Trust.

Literature Cited

Alley, K. E., Baker, R., Simpson, J. I. 1975. Afferents to the vestibulo-cerebellum and the origin of the visual climbing fibers in the rabbit. *Brain Res.* 98:582–89

Ariens-Kappers, C. U., Huber, G. C., Crosby, E. C. 1936. *The Comparative Anatomy of the Nervous System in Vertebrates, Including Man*, Vol. 2. New York: MacMillan

Barlow, H. B., Hill, R. M., Levick, W. R. 1964. Retinal ganglion cell responding selectively to direction and speed of image motion in the rabbit. *J. Physiol.* 173:377–407

Berson, D. M., Graybiel, A. M. 1980. Some cortical and subcortical fiber projections to the accessory optic nuclei in the cat. *Neuroscience* 5:2203–17

Blanks, R. H. I., Giolli, R. A., Pham, S. V. 1982a. Projections of the medial terminal nucleus of the accessory optic system upon pretectal nuclei in the pigmented rat. *Exp. Brain Res.* 48:228–37

Blanks, R. H. I., Giolli, R. A., Torigoe, Y. 1982b. Descending projections of the medial terminal nucleus of the accessory optic system: A light autoradiographic study in rat and rabbit. *Neurosci. Abstr.* 8:204

Blanks, R. H. I., Precht, W., Torigoe, Y. 1983. Afferent projections to the cerebellar flocculus in the pigmented rat demonstrated

by retrograde transport of horseradish peroxidase. *Exp. Brain Res.* In press

Bochenek, A. 1908. Über zentrale Endigungun des Nervus opticus. *Anz. Akad. Wiss. Krak. No. 1. Bull. Internat. Acad. Sci. Cracovie Classe Sc. Math Nat.*, pp. 91–95

Brauth, S. E., Karten, H. J. 1977. Direct accessory optic projections to the vestibulo-cerebellum: A possible channel for oculomotor control systems. *Exp. Brain Res.* 28:73–84

Brecha, N., Karten, H. J., Hunt, S. P. 1980. Projections of the nucleus of the basal optic root in the pigeon: An autoradiographic and horseradish peroxidase study. *J. Comp. Neurol.* 189:615–70

Burns, S., Wallman, J. 1981. Relation of single unit properties to the oculomotor function of the nucleus of the basal optic root (accessory optic system) in chickens. *Exp. Brain Res.* 42:171–80

Campbell, C. B. G. 1969. The visual system of insectivores and primates. *Ann. NY Acad. Sci.* 167:388–403

Campbell, C. B. G., Ebbesson, S. O. E. 1969. The visual system of a teleost. *Holocentrus* re-examined. *Brain Behav. Evol.* 2:415–30

Campbell, C. B. G., Jane, J. A., Yashon, D. 1967. The retinal projections of the tree shrew and hedgehog. *Brain Res.* 5:406–18

Campos-Ortega, J. A., Cluver, P. F. de V. 1968. The distribution of retinal fibers in *Galago crassicaudatus*. *Brain Res.* 7:487–89

Campos-Ortega, J. A., Glees, P. 1967. The subcortical distribution of the optic fibers in *Saimiri sciureus* (squirrel monkey). *J. Comp. Neurol.* 131:131–42

Cazin, L., Precht, W., Lannou, J. 1980. Firing characteristics of neurons mediating opto-kinetic responses to rat's vestibular neurons. *Pflug. Arch.* 386:221–30

Cochran, S. L., Precht, W., Dieringer, N. 1980. Direction-selective neurons in the frog's visual system. *Neurosci. Abstr.* 6:121

Cochran, S. L., Precht, W., Dieringer, N. 1981. Response of sensory relay neurons and extraocular motoneurons to optokinetic stimulation in the frog. *Neurosci. Abstr.* 7:23

Collewijn, H. 1975a. Direction-selective units in the rabbit's nucleus of the optic tract. *Brain Res.* 100:489–508

Collewijn, H. 1975b. Oculomotor areas in the rabbit's midbrain and pretectum. *J. Neurobiol.* 6:3–22

Collewijn, H., Winterson, B. J., Dubois, M. F. W. 1978. Optokinetic eye movements in albino rabbits: Inversion in the anterior visual field. *Science* 199:1351–53

Cotter, J. R., Pierson Pentney, R. J. 1979. Retinofugal projections of nonecholocating *(Pteropus giganteus)* and echolocating *(Myotis lucifugus)* bats. *J. Comp. Neurol.* 184:381–400

Ebbesson, S. O. E. 1972. A proposal for a common nomenclature for some optic nuclei in vertebrates and the evidence for a common origin of two such cell groups. *Brain Behav. Evol.* 6:75–91

Edwards, S. B., Rosenquist, A. C., Palmer, L. A. 1974. An autoradiographic study of ventral lateral geniculate projections in the cat. *Brain Res.* 72:282–87

Erickson, R. G., Cotter, J. R. 1983. Uncrossed retinal projections to the accessory optic nuclei in rabbits and cats. *Exp. Brain Res.* 49:143–46

Farmer, S. G., Rodieck, R. W. 1982. Ganglion cells of the cat accessory optic system: Morphology and retinal topography. *J. Comp. Neurol.* 205:190–98

Finger, T. E., Karten, H. J. 1978. The accessory optic system in teleosts. *Brain Res.* 153:144–49

Fite, K. V., Brecha, N., Karten, H. J., Hunt, S. P. 1981. Displaced ganglion cells and the accessory optic system of pigeon. *J. Comp. Neurol.* 195:279–88

Fite, K. V., Reiner, A., Hunt, S. P. 1979. Optokinetic nystagmus and the accessory optic system of pigeon and turtle. *Brain Behav. Evol.* 16:192–202

Gillilan, L. A. 1941. The connections of the basal optic root (posterior accessory optic tract) and its nucleus in various mammals. *J. Comp. Neurol.* 74:367–408

Giolli, R. A. 1961. An experimental study of the accessory optic tracts (Transpeduncular tract and anterior accessory optic tracts) in the rabbit. *J. Comp. Neurol.* 117:77–95

Giolli, R. A. 1963. An experimental study of the accessory optic system in the cynomolgus monkey. *J. Comp. Neurol.* 121:89–107

Giolli, R. A., Braithwaite, J. R., Streeter, T. T. 1968. Golgi study of the nucleus of the transpeduncular tract in the rabbit. *J. Comp. Neurol.* 133:309–28

Giolli, R. A., Creel, D. J. 1973. The primary optic projections in pigmented and albino guinea pigs: An experimental degeneration study. *Brain Res.* 55:25–39

Giolli, R. A., Guthrie, M. D. 1971. Organization of subcortical projections of visual areas I and II in the rabbit. An experimental degeneration study. *J. Comp. Neurol.* 142:351–76

Giolli, R. A., Tigges, J. 1970. The primary optic pathways and nuclei of primates. In *The Primate Brain*, ed. C. R. Noback, W. Montagna, pp. 29–54. New York: Appleton-Century-Crofts

Giolli, R. A., Towns, L. C., Takahashi, T. T., Karamanlidis, A. N., Williams, D. D. 1978. An autoradiographic study of the projections of visual cortical area I to the thalamus, pretectum and superior colliculus of the rabbit. *J. Comp. Neurol.* 180:743–52

Grasse, K. L., Cynader, M. S. 1982a. Electrophysiology of medial terminal nucleus of accessory optic system in the cat. *J. Neurophysiol.* 48:490–504

Grasse, K. L., Cynader, M. S. 1982b. Distribution of direction selectivity in the medial, lateral and dorsal terminal nuclei of the cat accessory optic system. *Neurosci. Abstr.* 8:407

Graybiel, A. M. 1974. Visuo-cerebellar and cerebello-visual connections involving the ventral lateral geniculate nucleus. *Exp. Brain Res.* 20:303–6

Gruberg, E. R., Grasse, K. L. 1980. Basal optic projection in the frog *(Rana pipiens)*. *Neurosci. Abstr.* 6:121

Hamasaki, D., Marg, E. 1960a. An historical review of the accessory optic tracts. *Am. J. Optom.* 37:53–66

Hamasaki, D., Marg, E. 1960b. Electrophysiological study of the posterior accessory optic tract. *Am. J. Physiol.* 199:522–28

Hamasaki, D., Marg, E. 1962. Microelectrode study of accessory optic tract in the rabbit. *Am. J. Physiol.* 202:480–86

Hayhow, W. R. 1959. An experimental study of the accessory optic fiber system in the cat. *J. Comp. Neurol.* 113:281–314

Hayhow, W. R. 1966. The accessory optic system in the marsupial phalanger, *Trichosurus vulpecula*. An experimental degeneration study. *J. Comp. Neurol.* 126:653–72

Hayhow, W. R., Webb, C., Jervie, A. 1960. The accessory optic fiber system in the rat. *J. Comp. Neurol.* 115:187–215

Hendrickson, A., Wilson, M. E., Toyne, M. J. 1970. The distribution of optic nerve fibers in *Macaca mulatta*. *Brain Res.* 23:425–27

Herrick, C. J. 1948. *The Brain of the Tiger Salamander*. Chicago: Univ. Chicago Press

Hill, R. M., Marg, E. 1963. Single cell responses of the nucleus of the transpeduncular tract in rabbit to monochromatic light on the retina. *J. Neurophysiol.* 26:249–57

Hodos, W., Bonbright, J. C. Jr. 1975. Intensity and pattern discrimination after lesions of the pretectal complex, accessory optic nucleus and ventral geniculate in pigeons. *J. Comp. Neurol.* 161:1–18

Hoffmann, K.-P. 1981. Neuronal responses related to optokinetic nystagmus in the cat's nucleus of the optic tract. In *Progress in Oculomotor Research*, ed. A. F. Fuchs, W. Becker, *Dev. Neurosci* 12:443–54. Amsterdam/New York/Oxford: Elsevier/North Holland

Hoffmann, K.-P. 1982. Cortical versus subcortical contributions to the optokinetic reflex in the cat. In *Functional Basis of Ocular Motility Disorders*, ed. G. Lennerstrand et al, pp. 303–10. Oxford/New York: Pergamon

Hoffmann, K.-P., Behrend, K., Schoppmann, A. 1976. A direct afferent visual pathway from the nucleus of the optic tract to the inferior olive in the cat. *Brain Res.* 115:150–53

Hoffmann, K.-P., Schoppmann, A. 1975. Retinal input to direction selective cells in the nucleus tractus opticus of the cat. *Brain Res.* 99:359–66

Hoffmann, K.-P., Schoppmann, A. 1981. A quantitative analysis of the direction-specific response of neurons in the cat's nucleus of the otpic tract. *Exp. Brain Res.* 42:146–57

Holstege, G., Collewijn, H. 1982. The efferent connections of the nucleus of the optic tract and the superior colliculus in the rabbit. *J. Comp. Neurol.* 209:139–75

Hughes, A. 1971. Topographical relationships between anatomy and physiology of the rabbit visual system. *Doc. Ophthalmol.* 30:33–159

Itoh, K. 1977. Efferent projections of the pretectum in the cat. *Exp. Brian Res.* 50:89–105

Karten, H. J., Fite, K. V., Brecha, N. 1977. Specific projection of displaced retinal ganglion cells upon the accessory optic system in the pigeon *(Columbia livia)*. *Proc. Natl. Acad. Sci. USA* 74:1753–56

Kawasaki, T., Sato, Y. 1980. Afferent projection from the dorsal nucleus of the raphe to the flocculus in cats. *Brain Res.* 197:496–502

Kimm, J., Winfield, J. A., Hendrickson, A. E. 1979. Visual-vestibular interactions and the role of the flocculus in the vestibulo-ocular reflex. In *Reflex Control of Posture and Development*, ed. R. Granit, O. Pompeiano, *Prog. Brain Res.* 50:703–13. Amsterdam/New York/Oxford: Elsevier/North Holland

Klein, D. C., Moore, R. Y. 1979. Pineal *n*-acetyltransferase and hydroxyindole-O-methyltransferase: Control by the retinohypothalamic tract and the suprachiasmatic nucleus. *Brain Res.* 174:245–62

Kondrashev, S. L., Orlov, O. Y. 1976. Direction-sensitive neurons in the frog visual system. *Neurophysiologica* 8:196–98

Lashley, K. S. 1934. The mechanism of vision. VII. The projection of the retina upon the primary optic centers of the rat. *J. Comp. Neurol.* 59:341–73

Lazar, G. 1973. Role of the accessory optic system in the optokinetic nystagmus of the frog. *Brain Behav. Evol.* 5:443–60

Legg, C. R. 1979. Visual discrimination impairments after lesions in zona incerta or lateral terminal nucleus of accessory optic tract. *Brain Res.* 177:461–78

Le Gros Clark, W. E. 1932. The structure and connections of the thalamus. *Brain* 55:406–70

Lent, R., Cavalcante, L. A., Rocha-Miranda, C. E. 1976. Retinofugal projections in the opossum. An anterograde degeneration and radioautographic study. *Brain Res.* 107:9–26

Leonard, C. S., Simpson, J. I., Soodak, R. E. 1981. From local direction preferences to global rotation preferences in the accessory optic system. *Neurosci. Abstr.* 7:24

Levick, W. R., Oyster, C. W., Takahashi, E. 1969. Rabbit lateral geniculate nucleus: Sharpener of directional information. *Science* 165:712–14

Lin, H., Giolli, R. A. 1979. Accessory optic system of rhesus monkey. *Exp. Neurol.* 63:163–76

Lin, H., Ingram, W. R. 1972. An anterior component of the accessory optic system of the cat, with evidence for the absence of reticuloretinal fibers. *Exp. Neurol.* 37:37–49

Loepp, W. H. 1911. Ueber die zentralen Opticusendigungen beim Kaninchen. *Anat. Anz.* 40:309–23

Maekawa, K., Simpson, J. I. 1972. Climbing fiber activation of Purkinje cells in the flocculus by impulses transferred through the visual pathway. *Brain Res.* 39:245–51

Maekawa, K., Simpson, J. I. 1973. Climbing fiber responses evoked in the vestibulocerebellum of rabbit from visual system. *J. Neurophysiol.* 36:649–66

Maekawa, K., Takeda, T. 1979. Origin of descending afferents to the rostral part of dorsal cap of inferior olive which transfers contralateral optic activities to the flocculus. An HRP study. *Brain Res.* 172:393–405

Mai, J. K. 1978. The accessory optic system and the retino-hypothalamic system. A review. *J. Hirnforsch.* 19:213–88

Marburg, O. 1903. Basale Opticuswurzel und Tractus peduncularies transversus. *Arb. Neurol. Inst. Univ. Wien* 10:66–80

Marcotte, R. R., Updyke, B. V. 1982. Cortical visual areas of the cat project differentially onto the nuclei of the accessory optic system. *Brain Res.* 242:205–17

Marg, E. 1964. The accessory optic system. *Ann. NY Acad. Sci.* 117:35–51

Marg, E. 1973. Neurophysiology of the accessory optic system. *Hand. Sensory Physiol.* ed. R. Jung, 7/13:103–11. Heidelberg/New York: Springer Verlag

Marg, E., Hamasaki, D., Giolli, R. 1959. Responses of the posterior accessory optic tract to photic stimulation of the retina and electrical stimulation of the optic nerve in rabbit. *21st Intl. Congr. Physiol. Sci., Buenos Aires*, p. 176

McKenna, O. C., Wallman, J. 1981. Identification of avian brain regions responsive to retinal slip using 2-deoxyglucose. *Brain Res.* 210:455–60

Mizuno, N., Mochizuki, K., Akimoto, C., Matsushima, R. 1973. Pretectal projection to the inferior olive in the rabbit. *Exp. Neurol.* 39:498–506

Mizuno, N., Nakamura, Y., Iwahori, N. 1974. An electron microscope study of the dorsal cap of the inferior olive in the rabbit, with special reference to the pretecto-olivary fibers. *Brain Res.* 77:385–95

Montgomery, N., Fite, K. V., Bengston, L. 1981. The accessory optic system of *Rana pipiens:* Neuroanatomical connections and intrinsic organization. *J. Comp. Neurol.* 203:595–612

Montgomery, N., Fite, K. V., Taylor, M., Bengston, L. 1982. Neural correlates of optokinetic nystagmus in the mesencephalon of *Rana pipiens:* A functional analysis. *Brain Behav. Evol.* 21:137–50

Moore, R. Y. 1974. Visual pathways and the central neural control of diurnal rhythms. In *The Neurosciences Third Study Program,* ed. F. O. Schmitt, F. G. Worden, pp. 537–42. Cambridge, Mass./London: MIT Press

Moore, R. Y., Heller, A., Wurtman, R. J., Axelrod, J. 1967. Visual pathway mediating pineal response to environmental light. *Science* 155:220–23

Morgan, B., Frost, B. J. 1981. Visual response characteristics of neurons in nucleus of basal optic root of pigeons. *Exp. Brain Res.* 42:181–88

Northcutt, R. G., Butler, A. B. 1976. Retinofugal pathways in the longnose gar *Lepisosteus osseus (Linnaeus).* *J. Comp. Neurol.* 166:1–16

Northcutt, R. G., Wathey, J. C. 1980. Guitarfish possess ipsilateral as well as contralateral retinofugal projections. *Neurosci. Lett.* 20:237–42

Oyster, C. W. 1968. The analysis of image motion by the rabbit retina. *J. Physiol.* 199:613–35

Oyster, C. W., Barlow, H. B. 1967. Direction-selective units in rabbit retina: Distribution of preferred directions. *Science* 155:841–42

Oyster, C. W., Simpson, J. I., Takahashi, E. S., Soodak, R. E. 1980. Retinal ganglion cells projecting to the rabbit accessory optic system. *J. Comp. Neurol.* 190:49–61

Oyster, C. W., Takahashi, E., Collewijn, H. 1972. Direction selective retinal ganglion cells and control of optokinetic nystagmus in the rabbit. *Vision Res.* 12:183–93

Pasik, P., Pasik, T. 1973. Extrageniculostriate vision in the monkey. V. Role of accessory optic system. *J. Neurophysiol.* 36:450–57

Pasik, T., Pasik, P. 1971. The visual world of monkeys deprived of striate cortex: Effective stimulus parameters and the importance of the accessory optic system. *Vision Res. Suppl.* 3:419–35

Pasik, T., Pasik, P., Hamori, J. 1973. Nucleus of the accessory optic tract. Light and electron microscopic study in normal monkeys and after eye enucleation. *Exp. Neurol.* 41:612–27

Pickard, G. E., Silverman, A. J. 1981. Direct retinal projections to the hypothalamus, piriform cortex, and accessory optic nuclei in the golden hamster as demonstrated by a sensitive anterograde horseradish peroxidase technique. *J. Comp. Neurol.* 196:155–72

Prasada Rao, P. D., Sharma, S. C. 1982. Retinofugal pathways in juvenile and adult channel catfish, *Ictalurus (Ameiurus) punctatus:* An HRP and autoradiographic study. *J. Comp. Neurol.* 210:37–48

Reiner, A. 1981. A projection of displaced ganglion cells and giant ganglion cells to the accessory optic nuclei in turtles. *Brain Res.* 204:403–9

Reiner, A., Brecha, N., Karten, H. J. 1979. A specific projection of retinal displaced ganglion cells to the nucleus of the basal optic root in the chicken. *Neuroscience* 4:1679–88

Reiner, A., Karten, H. J. 1978. A bisynaptic retinocerebellar pathway in the turtle. *Brain Res.* 150:163–69

Reperant, J., Lemire, M., Miceli, D., Peyrichoux, J. 1976. A radioautographic study of the visual system in fresh water teleosts following intraocular injection of tritiated fucose and proline. *Brain Res.* 118:123–31

Shanklin, W. M. 1933. The comparative neurology of the nucleus opticus tegmenti with special reference to *Chameleon vulgaris*. *Acta Zool.* 14:163–84

Simpson, J. I., Graf, W., Leonard, C. 1981. The coordinate system of visual climbing fibers to the flocculus. In *Progress in Oculomotor Research*, ed. A. F. Fuchs, W. Becker, *Dev. Neurosci.* 12:475–84. Amsterdam/New York/Oxford: Elsevier/North Holland

Simpson, J. I., Soodak, R. E. 1978. The accessory optic system: A visual system in vestibular coordinates. *Neurosci. Abstr.* 4:645

Simpson, J. I., Soodak, R. E., Hess, R. 1979. The accessory optic system and its relation to the vestibulocerebellum. In *Reflex Control of Posture and Movement*, ed. R. Granit, O. Pompeiano, *Prog. Brain Res.* 50:715–24. Amsterdam/New York/Oxford: Elsevier/North Holland

Simpson, J. I., Soodak, R. E., Leonard, C. S. 1982. Synthesis of an analyzer of self-motion: Modeling the accessory optic system. *Neurosci. Abstr.* 8:407

Smeets, W. J. A. J. 1981. Retinofugal pathways in two chondrichthyans, the shark *Scyliorhinus canicula* and the ray *Raja clavata*. *J. Comp. Neurol.* 195:1–11

Swanson, L. W., Cowan, W. M., Jones, E. G. 1974. An autoradiographic study of the efferent connections of the ventral lateral geniculate nucleus of the albino rat and the cat. *J. Comp. Neurol.* 156:143–63

Takahasi, E. S., Hickey, T. L., Oyster, C. W. 1977. Retinogeniculate projections in the rabbit: An autoradiographic study. *J. Comp. Neurol.* 175:1–12

Takeda, T., Maekawa, K. 1976. The origin of the pretecto-olivary tract. A study using the horseradish peroxidase method. *Brain Res.* 117:319–25

Terasawa, K., Otani, K., Yamada, J. 1979. Descending pathways of the nucleus of the optic tract in the rat. *Brain Res.* 173:405–17

Thorpe, P. A., Herbert, J. 1976a. The accessory optic system of the ferret. *J. Comp. Neurol.* 170:295–310

Thorpe, P. A., Herbert, J. 1976b. The effect of lesions of the accessory optic tract terminal nuclei on the gonadal response to light in ferrets. *Neuroendocrinology* 22:250–58

Tigges, J. 1966. Ein experimenteller Beitrag zum subkortikalen optischen System von *Tupaia glis*. *Flia Primat.* 4:103–23

Tigges, J., Bos, J., Tigges, M. 1977. An autoradiographic investigation of the subcortical visual system in chimpanzee. *J. Comp. Neurol.* 172:367–80

Tigges, J., O'Steen, W. K. 1974. Termination of retinofugal fibers in squirrel monkey: Re-investigation using autoradiographic methods. *Brain Res.* 79:489–95

Tigges, J., Tigges, M. 1969a. The accessory optic system and other optic fibers of the squirrel monkey. *Folia Primat.* 10:245–62

Tigges, J., Tigges, M. 1969b. The accessory optic system in *Erinaceus* (insectivora) and *Galago* (primates). *J. Comp. Neurol.* 137:59–70

Tsai, C. 1925. The optic tracts and centers of the opossum, *Didelphis virginiana*. *J. Comp. Neurol.* 39:173–216

Vinogradova, V. M., Manteifel, Y. B. 1976. Axonal composition of the marginal and basal optic tracts in frog. *Neurophysiologica* 8:54–61

von Gudden, B. 1870. Ueber einen bisher nicht beschrieben Nervenfasernstrang im Gehirne der Saugethiere und des Menschen. *Arch. Psychiatr. Nervenkr.* 2:364–66

von Gudden, B. 1881. Ueber den Tractus peduncularis transversus. *Arch. Psychiatr. Nervenkr.* 11:415–23

Walberg, F., Nordby, T., Hoffmann, K.-P., Hollander, H. 1981. Olivary afferents from the pretectal nuclei in the cat. *Anat. Embryol.* 161:291–304

Walley, R. E. 1967. Receptive fields in the accessory optic system of the rabbit. *Exp. Neurol.* 17:27–43

Wallman, J., McKenna, O. C., Burns, S., Velez, J., Weinstein, B. 1981. Relation of the accessory optic system and pretectum to optokinetic responses in chickens. In *Progress in Oculomotor Research*, ed. A. F. Fuchs, W. Becker, *Dev. Neurosci.* 12:435–42. Amsterdam/New York/Oxford: Elsevier/North Holland

Weber, J. T., Harting, J. K. 1980. The efferent projections of the pretectal complex: An autoradiographic and horseradish peroxidase analysis. *Brain Res.* 194:1–28

Westheimer, G., Blair, S. M. 1974. Unit activity in accessory optic system in alert monkeys. *Invest. Ophthalmol.* 13:533–34

Winfield, J. A., Hendrickson, A., Kimm, J. 1978. Anatomical evidence that the medial terminal nucleus of the accessory optic tract in mammals provides a visual mossy fiber input to the flocculus. *Brain Res.* 151:175–82

Winterson, B. J., Collewijn, H. 1981. Inversion of direction-selectivity to anterior fields in neurons of nucleus of the optic tract in rabbits with ocular albinism. *Brain Res.* 220:31–49

Wyatt, H. J., Daw, N. W. 1975. Directionally selective ganglion cells in the rabbit retina: Specificity for stimulus direction, size and speed. *J. Neurophysiol.* 38:613–26

Ann. Rev. Neurosci. 1984. 7:43–65
Copyright © 1984 by Annual Reviews Inc. All rights reserved

CENTRAL NEURAL INTEGRATION FOR THE CONTROL OF AUTONOMIC RESPONSES ASSOCIATED WITH EMOTION

Orville A. Smith and June L. DeVito

Regional Primate Research Center and Departments of Physiology & Biophysics and Biological Structure, University of Washington, Seattle, Washington 98195

INTRODUCTION

The continual search for the neural substrates underlying integrated behavior has been aided dramatically by the development of tracer techniques based on axoplasmic transport, cytochemistry, and metabolic markers. Over the past ten years the major use of these approaches has been to revamp our ideas of how the nervous system is interconnected and, to the delight of publishers, to revise the textbooks in the field. The race to establish what is connected to what has peaked and there are now more attempts to use the knowledge and techniques to study the neural basis of integrated functions. Recent particularly fine examples are those of Swanson and colleagues (Swanson & Mogenson 1981, Swanson & Sawchenko 1983).

In this regard there are few areas of research that need elucidation regarding the underlying neural basis more than that of emotion and its relation to the activity of the autonomic nervous system. For several reasons this area has been particularly refractory to experimental analysis. The first problem lies in the specification of what is meant by emotion and how to measure it. Psychologists have criticized the concept to the point where many refuse even to use the term; it is clearly a situation that calls for the use of a stringent operational definition,

43

which is rarely done. The other major drawback, on the biological side, is that autonomic responses are difficult to measure in the awake animal experiencing an emotional situation.

To complicate matters even further, until recently it has been impossible to gain any reasonable picture of the central neural connections involved in autonomic control. Probably because of the small diameter of the fibers involved, or some peculiarity of the chemistry of the cells involved, staining with reduced silver techniques, and particularly the very old techniques for staining degenerated myelin (Marchi), failed to demonstrate convincingly how the higher neural levels connect with the known autonomic cells in the cord, the intermediolateral cell column. In one of the earliest studies of descending autonomic connections from the diencephalon, Beattie et al (1930) concluded that the route of central autonomic fibers was related to the central gray matter of the aqueduct and the surrounding medial longitudinal fasciculus and dorsal tegmentum. In contradiction, Magoun et al (1938), using vasomotor, respiratory, and urinary bladder contractions produced by hypothalamic stimulation before and after transections of various portions of the brain stem, concluded that "the descending connections from the hypothalamus are diffusely and widely represented through a large extent of the mesencephalic tegmentum." The implication here was that the system was multisynaptic. This view persisted until the advent of the Nauta technique, when the problem was readdressed. Smith (1965) first used the Nauta technique in an examination of the central neural control of cardiovascular function, and reported projections to the vicinity of the intermediolateral cell columns of the thoracic cord from upper brain stem areas that produced large cardiovascular changes when electrical stimulation was applied. However, Nauta & Haymaker (1969) and Enoch & Kerr (1967) did not find such projections, and there matters sat until the axoplasmic transport techniques arrived to provide the definitive answers.

In the past seven years, enough information has been accumulated on the direct inputs to the intermediolateral cells to allow at least a tentative look at the possible relations between the higher levels of the nervous system that have been shown to be involved in emotional behavior and the neuroanatomical connections that link them to the output cells of the autonomic nervous system. In this review we will determine the sources of the axons that have been demonstrated to project directly to the intermediolateral cells, and detail some of the major interconnections among those sources. Then we examine connections of the major CNS structures that have been shown to influence emotional behavior, with specific attention directed to those that also have a demonstrable influence on cardiovascular control, as an index of autonomic involvement. The relation between these structures and the sources of input to intermediolateral cells is then considered.

BRAIN STEM INPUTS TO THE INTERMEDIOLATERAL CELL COLUMN

Although Schramm (1982) has warned of the potential power exerted by intrinsic spinal systems and spinal afferents upon intermediolateral cell output, the paucity of information on these systems dictates that they be set aside for the purposes of this review.

Because of Alexander's (1946) physiological investigations, one clearly would expect to find that the "pressor" area of the "vasomotor center" in the medulla would be a major source of input to the intermediolateral cells. Alexander had located this center in an extensive area of the lateral reticular formation. This view had been accepted generally despite Dittmar's (1873) demonstration that vasomotor tone could be maintained with a lower ventrolateral quadrant of intact medulla. As Loewy & McKellar (1980) have pointed out, these views of a unitary medullary pressor center do not stand the test of recent anatomical analyses.

Amendt et al (1979) injected small quantities of horseradish peroxidase (HRP) into segments T3 and L1 of cats in which the intermediolateral cells had been localized by recording antidromic potentials elicited from stimulating the cut end of the white ramus. Their analysis showed accumulations of labeled cell bodies in (a) the medial nucleus of the nucleus of the solitary tract (NTS), (b) the postpyramidal and inferior central nuclei of the raphe, and (c) the ventrolateral reticular formation immediately below the surface of the medulla. Loewy & Burton (1978), in a previous investigation of morphology and connections of the NTS, had demonstrated solitario-spinal projections including input to the intermediolateral cell column.

The projection from the raphe nuclei (magnus, pallidus, and obscurus of rats) has been verified (Loewy et al 1979, Loewy 1981, Holstege & Kuypers 1982). The projection from the ventrolateral reticular formation has turned out to be intriguing because of the accumulating evidence that it may be cells in this vicinity that are responsible for the vasomotor tone that emanates from somewhere in the ventral medulla. The anatomical projections have been verified in the opossum (Martin et al 1979) and analyzed in the rabbit (Blessing et al 1981). The latter paper demonstrates a much more widespread distribution of labeled cells in the medulla, but basically supports the projections from the three general regions reported by Amendt et al (1979).

In considering the possible role of these areas in the control of the cardiovascular system (the measure selected for assessing influence on the autonomic nervous system), the best understood is that of the NTS and its focal position in the control of the baroreflex. It is interesting that even with this very strong projection to the thoracic cord and its well known function as the first relay in the baroreflex, there were no experimental demonstrations of a direct projection to the intermediolateral cell column before the report by Loewy & Burton (1978). Other than the role in the baroreflex modulation of sympathetic activity, no more specific function for these projections can be identified. Physiological studies of the functions of raphe cells have yielded conflicting results (Loewy & Neil 1981), but the data support the finding of a pressor response to electrical stimulation of some portions of the raphe and a depressor response to stimulation of other portions (Adair et al 1977). For our purposes, these studies at least have demonstrated a cardiovascular influence. In passing, one must lament that Bowker et al (1982) did not specifically inject the intermediolateral cells with HRP for use with their elegant double-label peroxidase antiperoxidase (PAP) method. It is clear, however, that these raphe cells react strongly with antisera raised to serotonin and are the cells described by Dahlstrom & Fuxe (1964) as B1 and B3.

To list the third major projection as being merely from ventrolateral reticular cells is an oversimplification. In Amendt's study (1979), the cells so designated encompassed about an 8-mm range from obex to lower pons. Other analyses have indicated that although a large number of A1 cells in the ventrolateral reticular formation project to the intermediolateral cells, few of these contain norepinephrine (NE) and may instead represent the source of cells carrying epinephrine to the intermediolateral cells (Loewy et al 1979, Loewy & Neil 1981, Blessing et al 1981). The NE cells of this ventrolateral group that project to the intermediolateral cells seem to be located in the A5 region, just lateral to the superior olive.

The physiological literature on this group of cells indicates that as far back as 1972, Nathan had postulated cells in the ventrolateral reticular formation as the final step in transmission to the intermediolateral cells because of the short latency of the evoked response in the splanchnic nerves. Feldberg & Guertzenstein (1972) showed that blood pressure could be influenced by chemical manipulations of the surface of the ventral medulla, and Wennergren & Öberg (1980) and McAllen et al (1982) have extended these observations. The potential importance of cells in this vicinity has been highlighted by Loewy et al (1979), who suggest that the A5 cells may be the site of the vasomotor center of Dittmar, and Dampney (1981) suggests that the ventrolateral cells may at least be an important contributor to vasomotor tone. A different hypothesis has been put forward that these cells represent a way station for the path of the "defense" reactions (Guertzenstein et al 1978), but others disagree (McAllen et al 1982).

Another major area of the brain stem recently implicated in autonomic control is the dorsal pons. While agreement is not unanimous, Westlund & Coulter (1980) believe that in the monkey, the locus ceruleus projects to the parasympathetic neurons of the sacral cord as well as to the dorsal motor nucleus of the vagus and nucleus ambiguus, while the nearby cells of the medial parabrachial nuclei and nucleus subceruleus project to the intermediolateral cells. Saper & Loewy (1980) reported direct projections to the intermediolateral cells from the closely associated Kolliker-Fuse nucleus in the rat, but did not support the other observations. In contrast, Holstege et al (1979) provided evidence for a projection from subceruleus to sacral parasympathetic cells.

The physiological effects of stimulating these groups of cells in the dorsal pons include striking effects on blood pressure (Ward & Gunn 1976, Mraovitch et al 1982a) and renal blood flow (Ward et al 1978). However, Crawley et al (1980) and Mraovitch et al (1982a) have presented strong evidence indicating that the locus ceruleus is not involved in the mediation of those responses and that the cardiovascular control resides within the parabrachial cells themselves. The latter paper presents particularly convincing evidence that these responses cannot be attributed to antidromic involvement of hypothalamus, amygdala, NTS, or fastigial nucleus.

DIENCEPHALIC PROJECTIONS TO THE INTERMEDIOLATERAL CELL COLUMN

The early divergent opinions on the existence of direct hypothalamic-intermediolateral cell connections are alluded to in the Introduction. The first definitive evidence of direct projections was provided by the retrograde study of Kuypers & Maisky (1975). In a landmark paper in 1976, Saper et al focused on these autonomic connections in three animal species. Using both HRP and tritiated amino acid techniques, they demonstrated substantial projections from the paraventricular nucleus of the hypothalamus (PVH), lateral and dorsal hypothalamic areas, the lateral portion of the dorsomedial nucleus, and the posterior hypothalamic area to the thoracic cord and the intermediolateral cell column. A few cells were also found in zona incerta, the central gray, and the Edinger-Westphal nucleus of the mesencephalon.

Swanson & Sawchenko (1983) have presented a truly elegant analysis of the connections between the PVH and the intermediolateral cells, dorsal vagal complex, the pituitary, and the median eminence. They demonstrate that what was at one time considered to be a single-function nucleus is in reality composed of "essentially separate groups of neurons (that) appear to project to the posterior pituitary, to the external lamina of the median eminence, and to autonomic centers in the brain and spinal cord."

While Saper et al (1979) did not verify the projection from lateral hypothalamus and zona incerta to thoracic cord, Swanson & Kuypers (1980) were able to do so. In addition, with the double labeling technique, they demonstrated that about 15% of these cells have collateral projections to both the thoracic cord and the dorsal vagal complex.

Physiologic demonstrations of the influence of the hypothalamus in general on autonomic function have been reported repeatedly since the time of Karplus & Kreidl (1910). Ciriello & Calaresu (1980) have demonstrated increased blood pressure and heart rate, and inhibition of reflex brachycardia of baroflex origin in response to stimulation of the PVH.

In summary, direct projections to intermediolateral cells have been shown to originate in the NTS, nuclei of the raphe, ventrolateral reticular formation (including A1 and A5), parabrachial nuclei, nucleus subceruleus, Kolliker-Fuse nucleus, lateral hypothalamus, posterior hypothalamus, dorsal hypothalamus, PVH, zona incerta, and central gray. It is significant, although possibly expected, that each of the structures listed above in addition to projecting to intermediolateral cells has been demonstrated to exert some degree of control on the cardiovascular system.

At this point a caveat must be enjoined. Much of the information on descending pathways to the intermediolateral cells has been garnered with HRP injections into the thoracic cord in the vicinity of the intermediolateral cells. The possibility of spurious uptake of HRP by injured fibers from electrode penetration is well recognized. Also, the amount of spread of the enzyme is not well controlled and therefore the exact uptake boundaries will be indistinct. Taking into account these possible sources of technical error, it is nonetheless very surprising to find that when HRP is placed in spinal segment C2–3, well away from any identified autonomic outflow cells, the pattern of HRP marking of cells in the brain stem is strikingly similar to that found after thoracic segment injections. Specifically, cells are found in the NTS, ventrolateral reticular formation, raphe, parabrachial nuclei, subceruleus, locus ceruleus, central gray, lateral hypothalamic area, posterior hypothalamic area, and zona incerta (Figures 8 and 9 in Huerta & Harting 1982). The significance of this overlap is completely unknown, but it represents a problem that should be addressed.

INTERCONNECTIONS OF MAJOR AREAS PROJECTING TO THE INTERMEDIOLATERAL CELL COLUMN

Nucleus Tractus Solitarius

Ricardo & Koh (1978) provided the first analysis of projections from the NTS by using the combined autoradiographic and retrograde HRP techniques. The projections to other structures of the brain stem included a major input to the parabrachial complex as well as the inferior olive and the reticular formation in general. Forebrain projections to the NTS include the PVH, the bed nucleus of the stria terminalis dorsomedial, perifornical (Figure 2 in Ricardo & Koh 1978) and arcuate nuclei of the hypothalamus, the central nucleus of the amygdala, the medial preoptic area, and the periventricular nucleus of the thalamus.

Loewy & Burton (1978) verified the projection to the parabrachial nuclei, the Kolliker-Fuse nucleus, the ventro-lateral reticular formation, and its closely associated ambiguus nuclei. They also found projections to the periaqueductal gray. Ross et al (1981), using the tetramethyl benzidine (TMB) technique, verified projections from the Kolliker-Fuse and the paraventricular and the posterolateral hypothalamic nuclei. Of great interest was the extensive labeling found in the fastigial nucleus of the cerebellum. The possible relation of this connection to the fastigial pressor response (Achari et al 1978, Miura & Reis 1969) is intriguing. The recent failure to demonstrate a fastigial projection to autonomic segments of the cord (Del Bo et al 1982) implies that this pressor response must be mediated in a polysynaptic fashion through other structures of the brain stem. Another unique finding by Ross et al (1981) was the projection from the insular cortex to the NTS. Stimulation of this cortical area produces a pressor response and tachycardia (Mraovitch et al 1982b).

Ventrolateral Medulla

Loewy et al (1981) analyzed the efferent projections from the ventral medulla and found projections to the dorsal vagal complex (including the NTS), the Kolliker-Fuse-parabrachial complex, the locus ceruleus, and subceruleus. In the midbrain there are inputs to the central gray, cuneiform, and restricted portions of the colliculi. The hypothalamic projections include paraventricular, perifornical, and dorsal areas, as well as the supraotic nucleus and median eminence. Fibers continue through the septal area to innervate the hippocampal formation.

The afferent projections to this region have not been determined by retrograde (HRP) approaches but include the lateral hypothalamus (Hosaya & Matsushita 1980, Berk & Finkelstein 1982), parabrachial nuclei (Saper & Loewy 1980), locus ceruleus (Loewy & Saper 1978), and subceruleus/medial parabrachial nuclei (Westlund & Coulter 1980).

Parabrachial Nuclei

King (1980) reported major projections of the brain stem to the parabrachial and Kolliker-Fuse areas from the NTS and medullary parvocellular reticular formation, while the major rostral projections originate from central nucleus of the amygdala (Takeuchi et al 1982) and the PVH (Swanson & Sawchenko 1983).

The efferents from this complex (Saper & Loewy 1980) are very extensive, including most of the intralaminar, midline, and ventro-basal thalamic nuclei; the hypothalamic nuclei, including the dorsomedial, ventromedial, paraventricular, lateral hypothalamic, and lateral preoptic; the bed nucleus of the stria terminalis; the anterior, central, medial, basomedial, and basolateral nuclei of the amygdala; and areas of the brain stem including the ventrolateral medulla, ventrolateral reticular formation, and NTS. As indicated above, the medial parabrachial and subceruleus nuclei have been shown to project to intermediolateral cells in the monkey (Westlund & Coulter 1980).

Paraventricular Nucleus

In their article in the 1983 *Annual Review of Neuroscience,* Swanson & Sawchenko described the PVH projections to the posterior lobe of the pituitary, the median eminence, and the descending brain stem projections, and provided strong evidence for the proposition that the neurons concerned with autonomic control are anatomically separate from those involved with pituitary control. The dorsal, lateral, and ventral portions of the parvocellular division are the major sources of these descending fibers. In immunohistochemical studies, neurophysin-stained fibers were found to project to parabrachial areas, locus ceruleus, central gray, NTS, dorsal motor nucleus of the vagus, and thoracic cord.

The inputs to the PVH are extensive, including connections to the brain stem from the A1 group of ventral medulla, NTS, locus ceruleus, parabrachial nucleus, and laterodorsal tegmental nucleus. The hypothalamic inputs are preoptic, ventromedial nucleus, anterior and lateral hypothalamic areas, and the suprachiasmatic nucleus. Although amygdalar afferents to the PVH have been reported (Silverman et al 1981, Berk & Finkelstein 1981), Swanson & Sawchenko (1983) question their existence because autoradiographic studies (Krettek & Price 1978) have failed to show such a connection. They believe that projection sites from the amygdala lie immediately adjacent to the PVH so that injections of HRP into the area of the PVH may encroach on these other sites.

The details of the relations of these nuclei to the perifornical-lateral hypothalamus are given in Figures 1 to 3, and are discussed in the next section.

NEURAL STRUCTURES INVOLVED IN EMOTIONAL BEHAVIOR

Amygdala

The amygdala became a focus of attention for those interested in emotional behavior after Klüver & Bucy (1939) demonstrated the extreme placidity and other behavioral changes that resulted from bilateral temporal lobectomy. Kaada (1972) has reviewed the experimental literature and Gloor (1972) the data from human studies. While the data from many hundreds of studies on this topic leave little doubt about some kind of involvement of the amygdala in the many facets of emotional behavior, in general the area is clouded by contradictions in results and conflicts in functional localization associated with particular nuclear subgroups. This confusion must be due in part to the inadequate specification of the emotional behavior, which frequently has been purely an observationally based descriptive technique in which there is no possibility of quantification. Other reasons may be species difference, production of irritative foci, differing stimulation parameters, or simply the complexity of the system at this level. Regardless of the confusion, it is generally accepted that the amygdala operates as a modulator or regulator of the hypothalamus with regard to emotional behavior and associated autonomic responses (Isaacson 1982), although not uniquely.

The cardiovascular responses that may be elicited by electrical stimulation of the amygdala have suffered from the same ambiguities as the behavior. For many years it has been accepted that stimulation of the structure produces changes in respiration and heart rate and that changes in the latter may be caused by hypoxia brought about by the inhibition of the former (Reis & McHugh 1968). Recently, Kapp et al (1982a) demonstrated cardiac slowing without respiratory sequelae in response to electrical stimulation of the amygdala in the rabbit, and related this to previous work (Kapp et al 1979) in which they showed an alteration of a conditioned bradycardia response after lesions of the central nucleus. The cardiovascular changes of sympathetic excitation that would be expected to accompany severe emotional displays of rage, anger, or defense were, until recently, obtainable only through stimulation of the amygdala in unanesthetized preparations (Hilton & Zbrozyna 1963). However, Timms (1981) has described the use of a steroid anesthetic, Althesin, which apparently does not distort synaptic transmission along efferent pathways of the amygdala. The full cardiovascular response of tachycardia, elevated arterial pressure, vasoconstriction in kidneys, intestines and skin, and dilation in muscle results from stimulation of several amygdalar sites and along the path of the amygdalofugal pathway.

The efferent connections of the amygdala are classically described as following two major paths, the stria terminalis and the ventral amygdalofugal path. The stria originates from the medial, central, and basolateral subdivisions and projects to the preoptic area, the bed nucleus of the stria, the nucleus accumbens, the septal area, olfactory structures, and around the ventromedial nucleus of the hypothalamus. The ventral amygdalofugal path is much larger than the stria and arises from the periamygdaloid cortex as well as having a general contribution from most of the subnuclei. It projects in a fashion similar to the stria, i.e. to the bed nucleus of the stria, the preoptic areas, and the ventromedial nucleus itself.

While it is probably true that each of the subnuclei or divisions of the amygdala has unique projection targets (Kosel & Rosene 1982), for the purposes of this review we will focus on the output of the central nucleus. Hopkins (1975) first demonstrated the existence of a long, descending monosynaptic system originating in the central nucleus of the amygdala and projecting as far as the caudal medulla in the rat, cat, and monkey. Subsequent work showed that the central nucleus projects to the far lateral posterior hypothalamus, the central gray, the midbrain tegmentum lateral to the red nucleus, substantia nigra, the ventral tegmental area of Tsai, the lateral reticular formation of the pons, the parabrachial nuclei (Takeuchi et al 1982), the locus ceruleus and subceruleus, the NTS, and the dorsal motor nucleus of X (Hopkins & Holstege 1978). Krettek & Price (1978) and Schwaber et al (1980) essentially confirmed these anatomical results. Hopkins et al (1981) suggested that this nucleus may be the source of the fibers mediating the "defense" reaction throughout the brain stem.

The projections to prefrontal and cingulate cortices are discussed in the following section.

The inputs to the amygdala are just as varied as the outputs, and in general follow the reciprocal projections principle found to be so common in neurobiology. Nineteen-eighty was a bountiful year for studies dealing with afferents to the amygdala in monkeys (Mehler 1980, Aggleton et al 1980, Norita & Kawamura 1980) and in rats and cats (Ottersen 1980). In addition to the prefrontal connections detailed in the following section, cortical afferents arise from the subcallosal gyrus, temporal pole, and anterior insula. The subcortical afferents include, among others, hypothalamic sources from areas dorsal and medial to the PVH, ventromedial nucleus, premammillary nucleus, arcuate nucleus, and the lateral edge of the lateral hypothalamic area. Scattered cells are found in supramammillary, dorsomedial, and posterior nuclei (Ottersen 1980, Mehler 1980). The most common projections of the brain stem include the ventral tegmental area, lateral parabrachial nucleus (Mehler 1980, Norita & Kawamura 1980), periaqueductal gray, peripeduncular nucleus, substantia nigra, and substantia innominata (Aggleton et al 1980, Norita & Kawamura 1980). Ricardo & Koh (1978) had reported a projection to the central nucleus of the amygdala from the NTS; and while Mehler (1980) expressed some doubts about those cells, Hopkins & Holstege's (1978) clear demonstration of the efferents from the central nucleus to the NTS provides additional support on the basis of the high probability of reciprocal connections. This listing of reported subcortical projections to the amygdala is not exhaustive and it must be recognized that each subdivision of the amygdala probably receives a unique set of afferents. No attempt was made to detail these specific connections for the purposes of this review.

A major stumbling block to understanding the neural basis of emotional behavior is a fundamental one of knowing how a sensory input is translated into a perception and how that perception then is translated into a motor output—in this case, a pattern of cardiovascular changes. Turner et al (1980) attempted to bridge that process by tracing the pathways from the primary sensory receiving areas in the neocortex to the modality-specific association areas one or more steps removed. They succeeded in tracing specific modality pathways each to a specific pattern of afferents to the amygdaloid complex, and arrived at five basic conclusions: (a) all sensory systems contain areas that project to the amygdaloid complex, but not to more central limbic structures in the basal forebrain or hypothalamus (consequently, whatever influence the sensory systems have on emotional processes mediated by these more central limbic structures is likely to depend largely on relays through the amygdala); (b) except for the olfactory system, the amygdalopetal projections arise only from the later stages of cortical processing (i.e. not the primary sensory area); (c) a progressively greater influence on amygdaloid activity is expected by successively more highly processed sensory information; (d) each part of the amygdala is under the major influence of a particular sensory system; and (e) the same cortical efferents that give rise to separate sensory channels to the amygdala send efferents that converge upon the major sources of input to the hippocampus. This intriguing study is all the more remarkable for having been done with the Fink-Heimer technique instead of the *au courant* axoplasmic transport approaches. The deductions generated from these data will provide many testable hypotheses for years to come.

Prefrontal and Cingulate Cortex

Freeman & Watts (1950) quoted C. F. Jacobsen, who had been training an extremely emotional chimpanzee that had subsequently been given a prefrontal lobectomy, as saying that after the surgery the ape acted " . . . as if she had joined the happiness cult of Elder Michaux and placed her burdens on the Lord." This observation plus some very vague theoretical notions led to the rash of prefrontal lobotomies and what was probably an overemphasis on the role of the frontal lobes in emotional behavior. Subsequent research has made apparent the tremendous complexity of this area of association cortex and has decreased the focus on its role in emotional behavior. Anatomical analyses have revealed that there are at least two major categories of relationships in the prefrontal region—a region that has strong connections with the limbic system and the remainder which relates to the major sensory systems (Nauta 1971). Happily enough, the limbic portion (ventral-orbital) is also that which has been demonstrated to influence autonomic responses, blood pressure, and respiration (Kaada et al 1949).

Although the prefrontal cortex does project to several subcortical structures that then relay to the intermediolateral cells, such as the lateral hypothalamus, zona incerta, and central gray (DeVito & Smith 1964, Nauta 1964), there is apparently no evidence of projections either to other major nuclear groups providing input to the intermediolateral cells or directly to the intermediolateral cells themselves. Of potentially greater significance is the recent study by Porrino et al (1981), who described the details of the relations between frontal cortex and the amygdala. They showed projections from the basolateral, basomedial, and basal accessory nuclei of the amygdala to the orbital prefrontal cortex, the gyrus rectus, superior frontal gyrus, and the anterior cingulate gyrus. Aggleton et al (1980) provided evidence for most of the expected reciprocal projections; specifically, afferents to the amygdala originate from the orbital frontal cortex, anterior cingulate gyrus, and subcallosal gyrus, as well as temporal pole and insula. Porrino et al (1981) concluded that the frontal lobe of the monkey can be divided into (*a*) "a ventromedial region including the anterior cingulatae gyrus, which receives both direct (amygdalo-cortical) and indirect (amygdalo-thalamo-cortical) input from the amygdala," and (*b*) "a dorsolateral frontal region, which is essentially devoid of either direct or indirect amygdalofugal axons." They consider the ventromedial region to be the limbic part of the frontal lobe.

Conspicuous by its absence to this juncture is any mention of the "Papez cycle of emotion" (Papez 1937) with its heavy reliance on the hippocampus, fornix, mammillary body, anterior thalamic nuclei, and cingulate gyrus portions of the limbic system as the neural substrate of emotion. The above-mentioned involvement of the anterior cingulate gyrus and its major projection to the amygdala provides one point of overlap with the Papez cycle. The only other overlap is possibly the septal area, which is a major projection target of the hippocampus (Swanson & Cowan 1979) and has been implicated in emotional behavior (Brady & Nauta 1953). The remainder of the structures in the cycle have been refractory in demonstrating any significant influence on either autonomic activity or emotional behavior. The septal areas do relate directly to the lateral hypothalamus and via the stria terminalis to portions of the amygdala (Swanson & Cowan 1979, Krettek & Price 1978).

Hypothalamus

The hypothalamus has long been considered the site of neural integration for emotional behavior. The reasons for this idea are multiple and probably valid. Bard's (1928) classic demonstration that a decorticate (thalamic) animal is still capable of producing a complete display of emotional behavior (sham rage) is convincing in itself. The fact that stimulation of specific locations in the hypothalamus can produce a variety of integrated emotional behaviors, interpreted as flight, anger, or defense (Hess & Brügger 1943, Nakao 1958, DeMolina & Hunsperger 1959, Abrahams et al 1960, Kaada 1972), also supports the idea. Added to these reasons is the hypothalamic influence on autonomic responses similar to those that invariably accompany emotional behavior.

Abstracting from the numerous studies on behavioral changes induced by stimulation of the hypothalamus, the most reactive sites include the perifornical area, the lateral hypothalamic area, and the dorsal and posterior areas. The most dramatic effect of hypothalamic lesions is produced by ablation of the ventromedial nucleus (Wheatley 1944), which produces an explosive, biting attack in response to usually innocuous stimuli.

Over the past several years a series of related papers dealing with the neural substrate for cardiovascular responses accompanying emotional behavior have appeared. They have attempted to deal with the major problems listed in the Introduction by (a) using an operationally defined situation, the "conditioned emotional response" (CER) (Stebbins & Smith 1964), (b) using an array of cardiovascular measuring probes implanted so as to enable a pattern of responses associated with emotion to be measured in the unanesthetized animal, and (c) using axoplasmic transport techniques to determine neural connections functionally specified as central autonomic pathways. They first demonstrated that a particular pattern of cardiovascular responses could be reproduced consistently in animals trained in the CER paradigm (Smith et al 1979) and that this pattern was different from cardiovascular patterns associated with other behaviors (Smith et al 1980a). A particular area of the hypothalamus was found that, when stimulated, reproduced the pattern of cardiovascular responses during the CER (Smith et al 1980b). Ablation of this same area, designated the hypothalamic area controlling emotional responses (HACER), led to the complete disappearance of the cardiovascular responses when the animal was retested in the CER situation (Smith et al 1980b).

The major anatomical loci of the functionally defined HACER include the perifornical region and the most medial portion of the lateral hypothalamus. The afferent projections to the HACER region determined by the HRP technique (DeVito & Smith 1982) are shown in Figures 1 and 2, and the efferents, determined with tritiated amino acids, are presented in Figure 3. Comparison of these figures reveals the striking degree of reciprocal connection between the afferents and efferents of the HACER.[1]

[1]Abbreviations used in Figures 1–3: Ab, nucleus ambiguus; AB, basal amygdaloid nucleus; Acc, nucleus accumbens; ACo, cortical amygdaloid nucleus; AM, medial amygdaloid nucleus; AV3V, anteroventral third ventricle; Cd, caudate nucleus; CeM, central medial nucleus; CM, centromedian nucleus; CT, central tegmentum; Dm, dorsomedial nucleus of hypothalamus; DM, dorsomedial nucleus of thalamus; DR, dorsal raphe nucleus; DSCP, decussation of superior cerebellar peduncle; DT, dorsal tegmentum; DV, dorsal motor nucleus of vagus; F, fornix; GP, globus pallidus; Hb, habenula; Hp, posterior hypothalamic area; In, infundibular nucleus; Ip, interpeduncular nucleus; LC, locus ceruleus; LM, medial lemniscus; LS, lateral septal nucleus; MLF, medial longitudinal fasciculus; MM, medial mammillary nucleus; NCA, bed nucleus of anterior commissure; NCS, superior central nucleus; NDBB, nucleus of diagonal band of Broca; NRa, nucleus of the raphe; NTS, nucleus of the solitary tract; n III, third nerve; N IV, fourth nerve nucleus; n VII, facial nerve; N VII, facial nerve nucleus; Pa, paraventricular nucleus of thalamus; PAG, periaqueductal gray; Pbl, lateral parabrachial nucleus; PbM, medial parabrachial nucleus; PL, lateral preoptic area; Pm, median preoptic nucleus; PM, medial preoptic area; Pp, peripeduncular nucleus; Put, putamen; Pv, paraventricular nucleus of hypothalamus; PVG, periventricular gray; Re, nucleus reuniens; RF, reticular formation; RL, lateral reticular nucleus; RN, red nucleus; SCP, superior cerebellar peduncle; SFO, subfornical organ; SNc, substantia nigra, pars compacta; SNr, substantia nigra, pars reticulata; SI, substantia innominata; So, supraoptic nucleus; Sub, subiculum; TM, tuberomammilary nucleus; TS, solitary tract; TSV, spinal tract of trigeminal nerve; VA, anteroventral nucleus; Ves, vestibular nucleus; Vm, ventromedial nucleus of hypothalamus; VTA, ventral tegmental area.

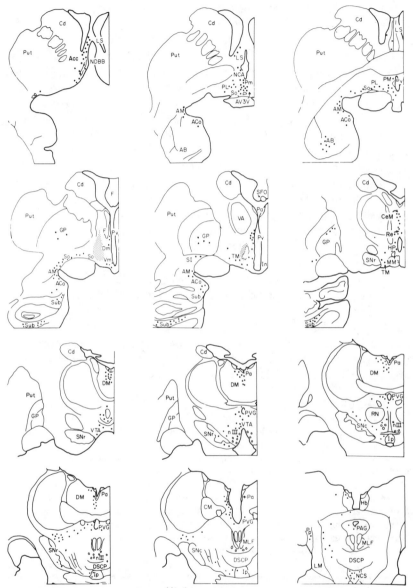

Figure 1   **HRP** injection into the perifornical-lateral hypothalamic region *(shaded area)*. Sites of retrograde transport are shown as *filled circles* in tracings of coronal sections of rostral subcortical regions. (From DeVito & Smith 1982, with permission from Elsevier Biomedical Press.)

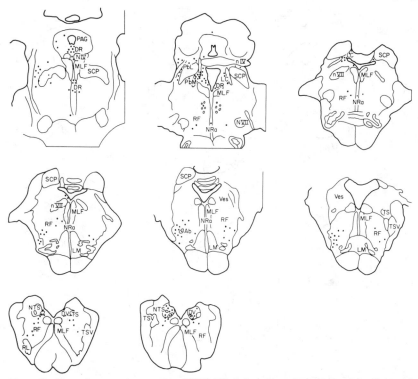

Figure 2 Sites of retrograde transport of HRP *(filled circles)* are shown on tracings of coronal sections of brain stem. (From DeVito & Smith 1982, with permission from Elsevier Biomedical Press.)

The connections of the HACER include several areas of the amygdala—the septal nuclei, the preoptic area, and the diagonal band of Broca—all of which have been implicated as being involved in emotional behavior. There are also connections with the central gray, the zona incerta, PVH, medial and lateral parabrachial nuclei, the locus ceruleus, the raphe nuclei, the NTS, and the vagal complex. All of these have been demonstrated repeatedly to have cardiovascular control functions and projections to the autonomic outflow in the cord.

The most striking efferent projections are illustrated in Figure 4, which depicts all of the structures of the CNS that have been shown to project directly to the intermediolateral cells. As can be seen, the HACER area itself projects directly to the intermediolateral cells and to every other structure of the CNS that also projects directly to the intermediolateral cells. This degree of control over the output of the sympathetic division of the autonomic nervous system provides concrete backing for the idea that the hypothalamus is the "head

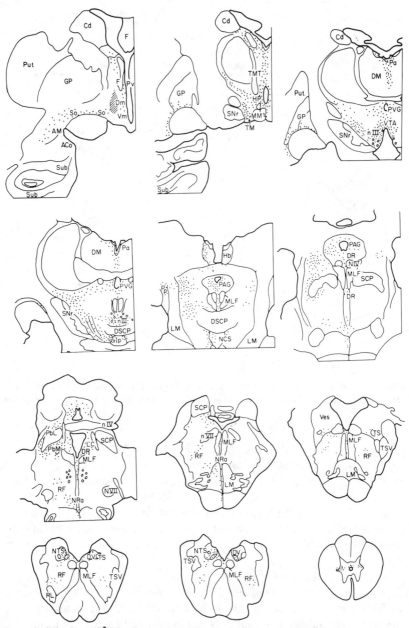

Figure 3 Injections of ³H-leucine and proline into the perifornical-lateral hypothalamic region *(shaded area)*. Descending projections are shown as *black dots* on tracings of coronal sections.

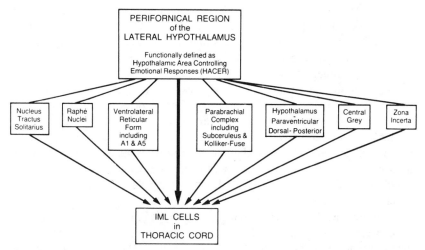

Figure 4   Anatomical basis for the integrative role of the HACER. Projections are shown to the intermediolateral cells and also to every other region that projects to them.

ganglion" of the autonomic nervous system. It also provides one more piece of evidence that this HACER region fits the definition of a "center" controlling the cardiovascular responses accompanying emotion, as has been suggested previously (Smith et al 1982a,b).

These connections are supported by Berk & Finkelstein (1982). Their concern was an analysis of the lateral hypothalamic area in general, but their injections of ³H leucine included the HACER region. They did not report the direct projection to the thoracic cord.

CONCLUDING REMARKS

The enigma of how neural information is communicated from the higher levels of the nervous system to the preganglionic cells of the autonomic nervous system has been at least partially resolved. It has been possible to identify several nuclear groups that have direct, ostensibly monosynaptic (although this is largely assumed rather than proven), connections with the intermediolateral cells of the thoracic cord. There continues to be the question of the existence and relative importance of polysynaptic routes, which will have to be investigated largely through electrophysiological approaches.

The several nuclear groups defined extend from the lower medulla through the brain stem as far rostral as the PVH. In each case there is evidence that these nuclei serve a functional role in autonomic regulation in that manipulation of these structures by electrical, chemical, or destructive techniques produces changes in the cardiovascular system.

There are several salient features of these cellular groups that must be of biologic importance. One is the high degree of interconnectivity among these nuclei. Even though these groups were identified primarily because they showed the common feature of directly innervating the intermediolateral cells, analysis of the connections of the groups themselves shows a remarkable degree of communication from one group to the other. The principle of reciprocal connections is readily demonstrated in this system. The significance of this network will be exciting to unravel. The obvious possibility of simple redundancy—a safety factor—should be assessed with relative ease; the more probable role of integrative function such as matching autonomic responses to somatic adjustments will be much more difficult. In either case, these tests will ultimately have to be made in the unanesthetized state under conditions of strict environmental and behavioral control, where multiple, precise measurements of autonomic and somatic responses can be made.

In discovering these multiple direct inputs to the intermediolateral cells, we have inferential evidence of a system that is capable of very precise, specific output that is dependent on higher neural demands. The direct monosynaptic input provides the highest degree of control over the individual neurons, allowing for both rapidity of action and specificity. A necessary corollary for this specificity would be evidence of differential input to the various thoracic cord segments, which has been supplied recently by Saper (1982). He used retrograde fluorescence double labels at different cord levels and found in the hypothalamus "small groups of neurons projecting to one labeled spinal level . . . interspersed with small groups projecting to the second labeled level, but no double labeled neurons were seen unless the injections were placed into adjacent spinal levels." These results provide the anatomical basis for the modern view of autonomic action as opposed to the older "mass action" concept of undifferentiated autonomic output.

In relating this information on the central neural control of the autonomic nervous system to complex psychological phenomena like emotion, we are still faced with the nearly philosophical issues of how and where the learning occurs that links an innocuous perception with an affective situation that changes the originally neutral response to that perception into an emotional response. Just how that newly acquired information is transferred to the brain stem nuclear groups and thence to the intermediolateral cells to be reflected as cardiovascular or other autonomic responses remains a major question. Anatomically speaking, one immediately looks to an association cortex where multiple sensory modalities have the capability of interacting. Both the temporal lobe cortex and the prefrontal cortex are candidates in this regard, particularly because of the demonstrated influence on emotional or autonomic responses from these sites. The recent evidence indicating how both of these cortices project to and receive from the amygdala in very specific patterns presents new, exciting possibilities for parcelling out identifiable functions related to complex psychological processes and their relation to emotional processes.

The point of transfer from psychological processes to central autonomic control still seems to be at the level of the hypothalamus—probably via the amygdala—although there are some direct inputs to the hypothalamus from portions of prefrontal cortex. The hypothalamus receives inputs from the amygdala in a patterned fashion and the major afferent projections into the ventromedial nucleus and lateral hypothalamus provide direct contact with structures known to be important in emotional behavior and cardiovascular control.

A critical linkage point has been specified in defining the HACER. The dramatic results showing that the complete cardiovascular response pattern accompanying emotional behavior can be eliminated by electrolytic lesion pinpoints a crucial neural structure in this complicated system. There is, of course, concern that the lesion is merely destroying fibers of passage rather than cells, which would diminish the potential importance of this area as an integrative structure. However, the fact that the tritiated amino acids are taken up only by cell bodies means that the efferent projections shown in Figure 3 must originate in the HACER region. These projections provide ample basis for believing that the HACER is a major autonomic control system during emotional behavior. Indeed, the relation of this region to the total control of input to the intermediolateral cells has to be of substantial biologic significance. This area apparently influences every other region projecting to the intermediolateral cell column as well as providing a direct input itself.

This influence can be interpreted in one of two ways. First, it is just this arrangement that one would hope to find in order to stipulate that a particular group of neurons is acting in an integrative fashion. One would desire a direct input to the critical elements and, additionally, some measure of control over the other inputs to those critical elements. Second, taking an entirely different view, it is obvious that the autonomic activity during emotional behavior comes as close to complete activation of the system as can be imagined; it may well be, therefore, that the output of the HACER is designed to activate intermediolateral cells directly and, in addition, to activate every other nuclear group that provides excitatory input to the intermediolateral cells, thus guaranteeing a nearly total sympathetic discharge.

If either of the above hypotheses is correct, the HACER region stands at that critical junction between the psychological processes involved in emotion and the resultant autonomic responses.

ACKNOWLEDGMENTS

Preparation of this review was supported in part by National Institutes of Health grants RR00166, HL16910, and HL28525. We gratefully acknowledge the assistance of C. A. Astley, D. J. Taylor, and D. A. Baldwin in much of our own research reported here.

Literature Cited

Abrahams, V. C., Hilton, S. M., Zbrozyna, A. W. 1960. Active muscle vasodilation by stimulation of the brain stem: Its significance in the defense reaction. *J. Physiol. London* 154:491–513

Achari, N. K., Al-Ubraidy, S. S., Downman, C. B. B. 1978. Spinal sympathoexcitatory pathways activated by stimulating fastigial nuclei, hypothalamus, and lower brain stem in cats. *Exp. Neurol.* 62:230–40

Adair, J. R., Hamilton, B. C., Scappaticci, K. A., Helke, C. J., Gillis, R. A. 1977. Cardiovascular responses to electrical stimulation of the medullary raphe area of the cat. *Brain Res.* 128:141–45

Aggleton, J., Burton, M. J., Passingham, R. E. 1980. Cortical and subcortical afferents to the amygdala of the rhesus monkey *(Macaca mulatta)*. *Brain Res.* 190:347–68

Alexander, R. S. 1946. Tonic and reflex functions of medullary sympathetic cardiovascular centers. *J. Neurophysiol.* 9:205–17

Amendt, K., Czachurski, J., Dembowsky, K., Seller, H. 1979. Bulbospinal projections to the intermediolateral cell column: A neuroanatomical study. *J. Autonom. Nerv. Syst.* 1:103–17

Bard, P. 1928. A diencephalic mechanism for the expression of rage with special reference to the sympathetic nervous system. *Am. J. Physiol.* 84:490–515

Beattie, J., Brow, G. R., Long, C. N. H. 1930. Physiological and anatomical evidence for the existence of nerve tracts connecting the hypothalamus with spinal sympathetic centres. *Proc. R. Soc. B London Ser.* 105:253–75

Berk, M. L., Finkelstein, J. A. 1981. Afferent projections to the preoptic area and hypothalamic regions in the rat brain. *Neuroscience* 6:1601–24

Berk, M. L., Finkelstein, J. A. 1982. Efferent connections of the lateral hypothalamic area of the rat: An autoradiographic investigation. *Brain Res. Bull.* 8:511–26

Blessing, W. W., Goodchild, A. K., Dampney, R. A. L., Chalmers, J. P. 1981. Cell groups in the lower brain stem of the rabbit projecting to the spinal cord with special reference to catecholamine-containing neurons. *Brain Res.* 221:35–55

Bowker, R. M., Westlund, K. N., Sullivan, M. C., Coulter, J. D. 1982. Organization of descending serotinergic projections to the spinal cord. In *Descending Pathways to the Spinal Cord*, ed. H. G. J. M. Kuypers, G. F. Martin. New York: Elsevier

Brady, J. V., Nauta, W. J. H. 1953. Subcortical mechanisms in emotional behavior: Affective changes following septal forebrain lesions in the albino rat. *J. Comp. Physiol. Psychol.* 46:339–46

Ciriello, J., Calaresu, F. R. 1980. Role of paraventricular and supraoptic nuclei in central cardiovascular regulation in the cat. *Am. J. Physiol.* 8:R137–42

Crawley, J. N., Maas, J. W., Roth, R. H. 1980. Evidence against specificity of electrical stimulation of the nucleus locus coeruleus in activating the sympathetic nervous system in the rat. *Brain Res.* 183:301–11

Dahlstrom, A., Fuxe, K. 1964. Evidence for the existence of monoamine-containing neurones in the central nervous system. I. Demonstration of monoamines in the cell bodies of brain stem neurones. *Acta Physiol. Scand.* 62 (Suppl. 232):1–55

Dampney, R. A. L. 1981. Functional organization of central cardiovascular pathways. *Clin. Exp. Pharmacol. Physiol.* 8:241–60

Del Bo, A., Ruggiero, D. A., Ross, C. A., Wiley, R., Reis, D. J. 1982. Pathways from the fastigial pressor area in rat. *Soc. Neurosci. Abstr.* 8:76

DeMolina, A. F., Hunsperger, R. W. 1959. Central representation of affective reactions in forebrain and brain stem: Electrical stimulation of amygdala, stria terminalis, and adjacent structures. *J. Physiol. London* 145:251–65

DeVito, J. L., Smith, O. A. 1964. Subcortical projections of the prefrontal lobe of the monkey. *J. Comp. Neurol.* 123:413–24

DeVito, J. L., Smith, O. A. 1982. Afferent projections to the hypothalamic area controlling emotional responses (HACER). *Brain Res.* 252:213–26

Dittmar, C. 1873. Ueber die Lage des sogenannten Gefasszentrums in der Medulla oblongata. *Ber. Verh. Sachs Ges. Wiss. Leipzig* 25:449–69

Enoch, D. M., Kerr, F. W. L. 1967. Hypothalamic vasopressor and vesicopressor pathways. II. Anatomic study of their course and connections. *Arch. Neurol.* 16:307–20

Feldberg, W., Guertzenstein, P. G. 1972. A vasopressor effect of pentobarbitone sodium. *J. Physiol. London* 224:83–103

Freeman, W., Watts, J. W. 1950. *Psychosurgery*. Springfield, Ill: Thomas. 2nd ed.

Gloor, P. 1972. Temporal lobe epilepsy: Its possible contribution to the understanding of the functional significance of the amygdala and of its interaction with neocortical-temporal mechanisms. *Adv. Behav. Biol.* 2:423–57

Guertzenstein, P. G., Hilton, S. M., Marshall, J. M., Timms, R. J. 1978. Experiments on the origin of vasomotor tone. *J. Physiol. London* 275:78P–79P

Hess, W. R., Brügger, M. 1943. Das subkortikale Zentrum der affektiven Abwehrreaction. *Helv. Physiol. Acta* 1:37–52

Hilton, S. M., Zbrozyna, A. W. 1963. Amygdaloid region for defense reactions and its efferent pathway to the brain stem. *J. Physiol. London* 165:160–73

Holstege, G., Kuypers, H. G. J. M. 1982. The anatomy of brain stem pathways to the spinal cord in cat. A labeled amino acid tracing study. *Progr. Brain Res.* 54:145–75

Holstege, G., Kuypers, H. G. J. M., Boer, R. C. 1979. Anatomical evidence for direct brain stem projections to the somatic motoneuronal cell groups and autonomic preganglionic cell groups in cat spinal cord. *Brain Res.* 171:329–33

Hopkins, D. A. 1975. Amygdalotegmental projections in the rat, cat and rhesus monkey. *Neurosci. Lett.* 1:263–70

Hopkins, D. A., Holstege, G. 1978. Amygdaloid projections to the mesencephalon, pons and medulla oblongata in the cat. *Exp. Brain Res.* 32:529–47

Hopkins, D. A., McLean, J. H., Takeuchi, Y. 1981. Amygdalotegmental projections: Light and electron microscopic studies utilizing anterograde degeneration and the anterograde and retrograde transport of horseradish peroxidase (HRP). In *The Amygdaloid Complex*, INSERM Symp. No. 20, ed. Y. Ben-Ari. Amsterdam: Elsevier

Hosaya, T., Matsushita, M. 1980. Cells of origin of the descending afferents to the lateral hypothalamic area in the rat, studied with the horseradish peroxidase method. *Neurosci. Lett.* 18:231–36

Huerta, M. F., Harting, V. K. 1982. Tectal control of spinal cord activity: Neuroanatomical demonstration of pathways connecting the superior colliculus with the cervical spinal cord grey. *Progr. Brain Res.* 57:294–328

Isaacson, R. L. 1982. *The Limbic System*. New York: Plenum. 327 pp.

Kaada, B. R. 1972. Stimulation and regional ablation of the amygdaloid complex with reference to functional representations. *Adv. Behav. Biol.* 2:205–81

Kaada, B. R., Pribram, K. H., Epstein, J. A. 1949. Respiratory and vascular responses in monkeys from temporal pole, visula, orbital surface and cingulate gyrus. A preliminary report. *J. Neurophysiol.* 12:347–56

Kapp, B. J., Frysinger, R. C., Gallagher, M., Haselton, J. 1979. Amygdala central nucleus lesions: Effects on heart rate conditioning in the rabbit. *Physiol. Behav.* 23:1109–17

Kapp, B. S., Gallagher, M., Underwood, M. D., McNall, C. L., Whitehorn, D. 1982a. Cardiovascular responses elicited by electrical stimulation of the amygdala central nucleus in the rabbit. *Brain Res.* 234:251–62

Karplus, J. P., Kreidl, A. 1910. Gehirn und Sympathicus. II. Mitteilung. Ein Sympathicuszentrum in Zwischenburn. *Pflügers Arch. Physiol.* 135:401–16

King, G. W. 1980. Topology of ascending brainstem projections to nucleus parabrachialis in the cat. *J. Comp. Neurol.* 191:615–38

Klüver, H., Bucy, P. C. 1939. Preliminary analysis of functions of the temporal lobes in monkeys. *Arch. Neurol. Psychiatr.* 42:979–1000

Kosel, K. C., Rosene, D. L. 1982. Subcortical efferents of the amygdaloid complex in the monkey. *Soc. Neurosci. Abstr.* 8:217

Krettek, J. E., Price, J. L. 1978. Amygdaloid projections to subcortical structures within the basal forebrain and the brainstem of the rat and cat. *J. Comp. Neurol.* 178:255–80

Kuypers, H. G. J. M., Maisky, V. A. 1975. Retrograde axonal transport of horseradish peroxidase from spinal cord to brainstem cell groups in the cat. *Neurosci. Lett.* 1:9–14

Loewy, A. D. 1981. Raphe pallidus and raphe obscurus projections to the intermediolateral cell column in the rat. *Brain Res.* 222:129–33

Loewy, A. D., Burton, H. 1978. Nuclei of the solitary tract: Efferent projections to the lower brain stem and spinal cord of the cat. *J. Comp. Neurol.* 181:421–50

Loewy, A. D., McKellar, S., Saper, C. B. 1979. Direct projections from the A5 catecholamine cell group to the intermediolateral cell column. *Brain Res.* 174:309–14

Loewy, A. D., McKellar, S. 1980. The neuroanatomical basis of central cardiovascular control. *Fed. Proc.* 39:2495–2503

Loewy, A. D., Neil, J. J. 1981. The role of descending monoaminergic systems in central control of blood pressure. *Fed. Proc.* 40:2778–85

Loewy, A. D., Saper, C. B. 1978. Efferent projections of the locus coeruleus. *Neurosci. Abstr.* 4:277

Loewy, A. D., Wallach, J. H., McKellar, S. 1981. Efferent connections of the ventral medulla oblongata in the rat. *Brain Res. Rev.* 3:63–80

Magoun, H. W., Ranson, S. W., Heatherington, A. 1938. Descending connections from the hypothalamus. *Arch. Neurol. Psychiatr.* 39:1127–49

Martin, G. F., Humbertson, A. O., Laxson, C., Panneton, W. M. 1979. Evidence for direct bulbospinal projections to laminae IX, X and the intermediolateral cell column. Studies using axonal transport techniques in the

North American opossum. *Brain Res.* 170:165–71

McAllen, R. M., Neil, J. J., Loewy, A. D. 1982. Effects of kainic acid applied to the ventral surface of the medulla oblongata on vasomotor tone, the baroreceptor reflex and hypothalamic autonomic responses. *Brain Res.* 238:65–76

Mehler, W. R. 1980. Subcortical afferent connections of the amygdala in the monkey. *J. Comp. Neurol.* 190:733–62

Miura, M., Reis, D. J. 1969. Cerebellum: A pressor response elicited from the fastigial nucleus and its efferent pathway in brain stem. *Brain Res.* 13:595–99

Mraovitch, S., Kumada, M., Reis, D. 1982a. Role of the nucleus parabrachialis in cardiovascular regulation in the cat. *Brain Res.* 232:57–75

Mraovitch, W., Ruggiero, D. A., Ross, C. A., Reis, D. J. 1982b. Direct projections to autonomic centers of forebrain and brainstem from a cortical vasopressor area in rat. *Soc. Neurosci. Abstr.* 8:77

Nakao, H. 1958. Emotional behavior produced by hypothalamic stimulation. *Am. J. Physiol.* 194:411–18

Nathan, M. A. 1972. Pathways in medulla oblongata mediating splanchnic nerve activity. Electrophysiological and anatomical evidence. *Brain Res.* 45:115–26

Nauta, W. J. H. 1964. Some efferent connections of the prefrontal cortex in the monkey. In *Frontal Granular Cortex*, ed. J. M. Warren, K. Akert, pp. 397–409. New York: McGraw-Hill

Nauta, W. J. H. 1971. The problem of the frontal lobe: A reinterpretation. *J. Psychiatr. Res.* 8:167–87

Nauta, W. J. H., Haymaker, W. 1969. Hypothalamic nuclei and fiber connections. In *The Hypothalamus*, ed. W. Haymaker, E. Anderson, W. J. H. Nauta, pp. 136–209. Springfield, Ill: Thomas

Norita, M., Kawamura, K. 1980. Subcortical afferents to the monkey amygdala: An HRP study. *Brain Res.* 190:225–30

Ottersen, O. P. 1980. Afferent connections to the amygdaloid complex of the rat and cat: II. Afferents from the hypothalamus and the basal telencephalon. *J. Comp. Neurol.* 194:267–89

Papez, J. W. 1937. A proposed mechanism of emotion. *Arch. Neurol. Psychiatr. Chicago* 38:725–43

Porrino, L. J., Crane, A. M., Goldman-Rakic, P. S. 1981. Direct and indirect pathways from the amygdala to the frontal lobe in rhesus monkeys. *J. Comp. Neurol.* 198:121–36

Reis, D. J., McHugh, P. R. 1968. Hypoxia as a cause of bradycardia during amygdala sti-

mulation in the monkey. *Am. J. Physiol.* 214:601–10

Ricardo, J., Koh, E. T. 1978. Anatomical evidence of direct projections from the nucleus of the solitary tract to the hypothalamus, amygdala, and other forebrain structures in the rat. *Brain Res.* 153:1–26

Ross, C. A., Ruggiero, D. A., Reis, D. J. 1981. Afferent projections to cardiovascular portions of the nucleus of the tractus solitarius in the rat. *Brain Res.* 223:402–8

Saper, C. B. 1982. *Soc. Neurosci. Abstr.* 8:76

Saper, C. B., Loewy, A. D. 1980. Efferent connections of the parabrachial nucleus in the rat. *Brain Res.* 197:291–317

Saper, C. B., Loewy, A. D., Swanson, L. W., Cowan, W. M. 1976. Direct hypothalamo-autonomic connections. *Brain Res.* 117:305–12

Saper, C. B., Swanson, L. W., Cowan, W. M. 1979. An autoradiographic study of the efferent connections of the lateral hypothalamic area in the rat. *J. Comp. Neurol.* 183:689–706

Schramm, L. P. 1982. Ganglionic, spinal and medullary substances for functional specificity in circulatory regulation. In *Circulation, Neurobiology and Behavior*, ed. O. A. Smith, R. A. Galosy, S. M. Weiss. New York: Elsevier

Schwaber, J. S., Kapp, B. S., Higgins, G. 1980. The origin and extent of direct amygdala projections to the region of the dorsal motor nucleus of the vagus and the nucleus of the solitary tract. *Neurosci. Lett.* 20:15–20

Silverman, A. J., Hoffman, D. L., Zimmerman, E. A. 1981. The descending afferent connections of the paraventricular nucleus of the hypothalamus (PVN). *Brain Res. Bull.* 6:47–61

Smith, O. A. 1965. Anatomy of central neural pathways mediating cardiovascular functions. In *Nervous Control of the Heart*, ed. W. C. Randall, pp. 34–53. Baltimore: Williams & Wilkins

Smith, O. A., Astley, C. A., DeVito, J. L., Stein, J. M., Walsh, K. E. 1980a. Functional analysis of hypothalamic control of the cardiovascular responses accompanying emotional behavior. *Fed. Proc.* 39:2487–94

Smith, O. A., Astley, C. A., Hohimer, A. R., Stephenson, R. B. 1980b. Behavioral and cerebral control of cardiovascular function. In *Neural Control of Circulation*, ed. M. J. Hughes, C. D. Barnes, pp. 1–21. New York: Academic

Smith, O. A., DeVito, J. L., Astley, C. A. 1982a. The hypothalamus in emotional behavior and associated cardiovascular correlates. In *Changing Concepts of the Nervous System*, ed. A. R. Morrison, P. L. Strick, pp. 569–84. New York: Academic

Smith, O. A., DeVito, J. L., Astley, C. A. 1982b. Cardiovascular control centers in the brain: One more look. In *Circulation, Neurobiology and Behavior*, ed. O. A. Smith, R. A. Galosy, S. M. Weiss, pp. 233–46. New York: Elsevier

Smith, O. A., Hohimer, A. R., Astley, C. A., Taylor, D. J. 1979. Renal and hindlimb vascular control during acute emotion in the baboon. *Am. J. Physiol.* 236:R193–R205

Stebbins, W. C., Smith, O. A. 1964. Cardiovascular concomitants of the conditioned emotional response in the monkey. *Science* 144:881–83

Swanson, L. W., Cowan, W. M. 1979. The connections of the septal region in the rat. *J. Comp. Neurol.* 186:621–56

Swanson, L. W., Kuypers, H. G. J. M. 1980. The paraventricular nucleus of the hypothalamus: Cytoarchitectonic subdivisions and organization of projections to the pituitary, dorsal vagal complex, and spinal cord as demonstrated by retrograde fluorescence double-labeling methods. *J. Comp. Neurol.* 194:555–70

Swanson, L. W., Mogenson, G. J. 1981. Neural mechanisms for the functional coupling of autonomic, endocrine and somatomotor responses in adaptive behavior. *Brain Res. Rev.* 3:1–34

Swanson, L. W., Sawchenko, P. E. 1983. Hypothalamic integration: Organization of the paraventricular and supraoptic nuclei. *Ann. Rev. Neurosci.* 6:269–324

Takeuchi, Y., McLean, J. H., Hopkins, D. A. 1982. Reciprocal connections between the amygdala and parabrachial nuclei: Ultra-structural demonstration by degeneration and axonal transport of horseradish peroxidase in the cat. *Brain Res.* 239:583–88

Timms, R. J. 1981. A study of the amygdaloid defence reaction showing the value of althesin anaesthesia in studies of the functions of the fore-brain in cats. *Plügers Arch.* 391:49–56

Turner, B. H., Mishkin, M., Knapp, M. 1980. Organization of the amygdalopetal projections from modality-specific cortical association areas in the monkey. *J. Comp. Neurol.* 191:515–43

Ward, D. G., Adair, J. R., Schramm, L. P., Gann, D. J. 1978. Parabrachial pole mediates hypothalamically induced vasoconstriction. *Am. J. Physiol.* 234:R223–28

Ward, D. G., Gunn, C. G. 1976. Locus coeruleus complex: Elicitation of a pressor response and a brain stem region necessary for its occurrence. *Brain Res.* 107:401–6

Wennergren, G., Öberg, B. 1980. Cardiovascular effects elicited from the ventral surface of medulla oblongata in the cat. *Plügers Arch.* 387:189–95

Westlund, K. N., Coulter, J. D. 1980. Descending projections of the locus coeruleus and subcoeruleus/medial parabrachial nuclei in monkey: Axonal transport studies and dopamine-β-hydroxylase immunocytochemistry. *Brain Res. Rev.* 2:235–64

Wheatley, M. D. 1944. The hypothalamus and affective behavior in cats: A study of the effects of experimental lesions, with anatomic correlations. *Arch. Neurol. Psychiatr.* 52:296–316

Ann. Rev. Neurosci. 1984. 7:67–93

LEARNING AND COURTSHIP IN *DROSOPHILA*: Two Stories with Mutants

William G. Quinn and Ralph J. Greenspan

Department of Biology, Princeton University, Princeton, New Jersey 08544

INTRODUCTION

Victorian ladies and gentlemen cultivated waltzing mice and tumbling pigeons as parlor pets. They derived mild social amusement from the inbred animals' curious behavior, unmindful of the hereditary ataxias and seizures that caused it. Present day fanciers of behavioral mutants continue to be amused by their organisms, but as biologists they are concerned above all with the etiology of curious behavior. Other things have changed as well. Simple organisms bearing single gene mutations have replaced complex, inbred strains, and the emphasis is on using these new mutants to dissociate complex behaviors and the neuronal processes that underlie them.

Recently, the cumulative effect of many experiments with many mutations, brought to bear on a few well-defined questions, has helped promote some of these studies from promising curiosities to coherent stories. Here we concentrate on two of these, in the organism we know best. In one case, mutations that perturb learning and memory in well-defined ways have led directly to the underlying biochemistry. In another, mutations and mosaics in combination with behavioral studies have resolved an ethologically complex behavior in unprecedented detail, showing the brain regions, sensory cues, and experiential determinants of many of its steps. For more comprehensive reviews, see Hall and colleagues (1982) and Hall (1982).

67

LEARNING AND MEMORY

> When from a long distant past nothing subsists, after the people are dead, after the things are
> broken and scattered, . . . the smell and taste of things remain poised a long time, . . . like
> souls, ready to remind us, . . . and bear unfaltering, in the tiny and almost impalpable drop of
> their essence, the vast structure of recollection.
>
> (Proust 1913)

In addition to its humanistic interest, learning has seemed "special" among behaviors in the apparent simplicity and generality of its basic phenomenology. With learning there has always been the hope that the underlying physiological change might be a simple one, conserved, with variations, across species. Convergent evidence, particularly from *Aplysia* and *Drosophila*, suggests that this may be true and outlines the principal features of the biochemical machinery involved. Information from both systems implicates cyclic AMP and monoamine transmitters as central to nonassociative and associative learning, and suggests that the chemical change underlying short-term memory may be simply an increase in cyclic AMP concentration in the relevant neurons.

The possibility of applying *Drosophila*'s convenient genetics to learning occurred to workers long ago (e.g. Thorpe 1939). However, lack of adequate experimental controls, together with *Drosophila's* perceived scholastic ineptitude, led to a lapse of years before this strategy could be realized. Some behavioral aspects of the work remain controversial, and this review represents our version.

Training Populations and Selecting Mutants

The first convincing report of learning in flies was by Nelson (1971), who classically conditioned individual blowflies, to respond differentially to water or taste cues. Her method, while carefully controlled and convincing, was not directly applicable to genetics, because training flies one by one was tedious, because the behavior of individuals was variable, and because her species, *Phormia regina*, had cumbersome genetics. Quinn, Harris & Benzer (1974) set out to condition populations of *Drosophila*, circumventing some of the problems of individual variability by training and testing flies *en masse*. Preliminary observations suggested that flies were particularly attentive to odor cues. Experimenters used discriminative (differential) training procedures because these provided rigorous internal controls for associative learning. Their method is described here because many of the more recent fly-training experiments are variations on this procedure.

About 40 flies are placed in an apparatus with tubes and trained by exposing them alternately to two chemical odorants (denoted "A" and "B"), one of which is coupled to a 90 V electric shock. They are then tested most simply (Dudai et al 1976) by transporting them to a choice point between tubes with odors A and B and seeing which way they run. Odor concentrations are arranged so that naive flies distribute themselves 50–50 between the two odorants. If the flies were shocked during training in the presence of odor A they now tend to run toward odor B (35%–65%). On the other hand, if they were shocked on odor B they now run toward A (65%–35%). The relevant results of this type of experiment can be expressed as a simple number, i.e. the fraction of the population avoiding the shock-associated odor minus the fraction avoiding the control odor (averaged for groups of flies trained in opposite directions). This numerical training index Λ normally can vary between 1 (perfect learning) and 0 (no learning). In the fly learning experiment outlined above Λ would be 0.30. A typical set of ten real experiments gives $\Lambda = 0.34 \pm 0.03$.

Recently, Jellies (1981) and also T. Tully (unpublished) have improved this learning procedure to get Λ values near 0.80 (90% correct odor choice by trained flies). Their major improvements were presenting odor stimuli at carefully controlled concentrations in air currents, and training flies classically in a quiet chamber, with exposure to odorants and shock made inevitable on several trials.

Given a reliable "group" learning test, selecting mutants is simple in principle, though brutally tedious in practice. Male flies from an inbred wild-type stock (C-S) are mutagenized, and their progeny are mated in appropriate genetic crosses to produce many populations, each with identical mutagenized X chromosomes (Lewis & Bacher 1968, Dudai et al 1976). Most mutations do not seriously alter learning behavior, but if a relevant one is present, it will affect all the flies of a given population. One selects learning mutants by training each population in the olfactory learning procedure above, and retaining those which show little learning, i.e. which give Λ values less than 0.05. Any such suspect population is analyzed in further behavioral experiments, to be sure that behavioral alteration in the mutants is genuinely interesting, not simply a defect in olfaction, locomotion, or general activity.

The first such mutant, isolated by D. Byers at the California Institute of Technology, was *dunce* (Dudai et al 1976). Five other mutations in the *dunce* gene have since been found. Three more mutations, *cabbage, turnip,* and *rutabaga,* all affecting different genes, were isolated at Princeton by Patricia Sziber (see Aceves-Pina & Quinn 1979, Quinn et al 1979). A fourth mutant, *amnesiac,* isolated by Sziber, learned normally but forgot after 1 hr compared to 4–6 hr for normal flies (Quinn et al 1979). All these mutants could sense electric shock and the odorants used to train them. All showed the normal tendency to migrate toward light, although *cabbage* and *turnip* did so more slowly than wild-type flies. Because of the breeding techniques used to isolate them, all the mutations were X-linked: *dunce, rutabaga,* and *amnesiac* have since been carefully mapped (see Duerr & Quinn 1982). *Dunce, turnip, cabbage,* and *rutabaga* showed virtually no learning ($\Lambda = 0.00$–0.05) on the olfactory discrimination learning test (Aceves-Pina & Quinn 1979). It remained to be seen whether they were "shallow" learning mutations affecting only the tasks for which they were isolated, or "deep" ones, which affected learning in a variety of situations using different cues, reinforcements, and responses. The answer seems to be that the mutations are "deep," though not perfectly so.

Several learning tests were devised, mostly variations on the basic discriminative scheme above, with different cues or reinforcements. Wild-type *Drosophila* larvae turned out to be nearly as acute at olfactory learning ($\Lambda = 0.26$) as adult flies, and larvae of the learning mutants failed to learn (Aceves-Pina & Quinn 1979). Menne & Spatz (1977) at Freiburg developed an elegant, automated procedure, and trained flies to discriminate between different colored lights, avoiding a color that was associated with severe mechanical shaking. Dudai & Bicker (1978) and Folkers (1982) tested the available learning mutants in this visual test and found that although they could eventually learn this task, it took them many more trials to do so. However, of all the tests developed so far, visual learning is the test on which the mutants' performance is the least altered.

Tempel et al (1983) trained flies to discriminate between odors as above, but substituted reward for punishment. They found that hungry flies would specifically migrate toward odorants previously associated with the opportunity to feed on sucrose. The magnitude of the learning effect was similar after training with sucrose ($\Lambda = 0.36$) to training with electric shock ($\Lambda = 0.34$), if measured immediately. Memory after reward persisted much longer—for days rather than 4–6 hours (Quinn et al 1974, Dudai 1977). Substituting reward for punishment also lengthened the memory span for *amnesiac* flies (3 hr vs 1 hr). The most surprising results came with the *dunce* and *rutabaga* mutants, which had showed virtually no learning in the shock-avoidance test. *Dunce* flies, tested with reward, learned normally but forgot within an hour. *Rutabaga* flies learned fairly well ($\Lambda = 0.16$), but forgot even more rapidly (Tempel et al 1983). The simplest explanation for these findings is that *dunce* and *rutabaga* are cryptic memory mutants—they can learn even in shock-avoidance training, but their memory span under those conditions is so short and labile that it is virtually undetectable (see also Dudai 1979). This finding is considered more fully below.

Training Individual Flies

In all the learning experiments above, flies were trained and tested *en masse.* This is a real experimental convenience; insofar as it circumvents the problem of variability in movement among genetically similar individuals, it is what has enabled us to select the learning mutants. Nevertheless, there are situations in which one wants to know whether a given individual can or cannot learn. This is particularly true of work using genetic mosaics, where no two individuals are alike.

The earliest reports of individual training, following on Nelson's research with *Phormia,* worked with the flies' reflex proboscis extension in response to stimulation with sugar solution. In one endearing variation, individual flies, tethered by their backs, were constrained to walk round and round the surface of little turning wheels that had patches of sucrose solution situated near patches of electrified grid. After several cycles of sucrose paired with shock, the flies learned to suppress their normal proboscis extension reflex to the sugar (Médioni & Vaysse 1977, Médioni et al 1978). These workers also selected for several generations, obtaining lines that were relatively better or worse than normal flies at this behavior modification. Although these lines are interesting as exemplars of the potential for variability latent in natural populations, they are not directly useful in understanding the mechanisms underlying learning, because they differ in many genes at once.

A very different method of training larger insects was also scaled down for *Drosophila.* If a tethered cockroach (Horridge 1962) is shocked every time it extends one leg, it rapidly comes to maintain that leg in a flexed position. This postural change is due to associative learning because it occurs only if shock is made contingent on leg position. Cockroaches show less extraneous movement and perform this task even better if they have heads removed, a not too surprising observation as insects have about a third of their CNS in thoracic ganglia. R. Booker adapted this procedure to *Drosophila* by becoming dexterous in handling tiny flies. Wild-type flies learned about as well as larger insects. The behavioral change was associative, and flies could also be trained to extend their legs by shocking them when their legs were flexed. Perhaps because leg flexion is a simple task and training is intensely repetitive, wild-type flies, particularly headless ones, performed well: 92% learned to a set criterion keeping their legs flexed over 90% of the time in a given ten minute period. The learning mutants did worse: only 20% of *turnip* flies, 25% of *dunce* flies, and 45% of *cabbage* flies met the criterion. Scores on the leg-extension test were comparable (Booker & Quinn 1981). Note, however, that some individuals of all genotypes did learn. As in other tests, the learning disabilities in the mutants appear to be relative, not absolute.

The behavioral separation between mutant and wild-type individuals, though imperfect with this test, was good enough for a start at mosaic mapping. Booker (1982) made mosaics that had *dunce* and *dunce/+* (essentially normal) tissue patches, together with markers that allowed him to identify the mutant tissue (see Kankel & Hall 1976). He tested these flies in the leg-flexion test, then sectioned them and scored the tissue distribution of mosaics that had passed and those that had failed the test. The imperfect reliability of the behavioral test and other complications severely lowered the resolution of the mosaic map. Nevertheless one tendency was clear. The anatomical site that showed the closest correlation to learning lay anterior to the thorax, in the region of neural tissue in the head. Apparently flies need normal brains to learn. This result, seemingly obvious, becomes paradoxical when one remembers that these were headless flies, with brain tissue destroyed before testing. One can make several rationalizations for this puzzlement. For example, mutant neurons from the head might make processes with functional synapses in the thoracic ganglion, processes which survive beheading and continue to function critically during the testing period. Any detailed understanding of what is going on will have to await a better behavioral assay.

On the subject of anatomical localization of learning circuits, it is interesting to note that the circuitry for olfactory learning is probably entirely in the brain, because Fukushi (1973) conditioned severed heads of houseflies (*Musca domestica*) to extend their proboscises to specific odors. Within the brain, the mushroom bodies are probably most important. Heisenberg has isolated two mutants, *mushroom bodies deranged* and *mushroom body miniature* (Heisenberg 1980; M. Heisenberg, unpublished) with severe, specific abnormalities in this structure. Both mutants fail to learn the olfactory discrimination task, whether positively or negatively reinforced (Borst, Byers & Heisenberg, submitted). Apparently they can, however, learn the visual discrimination task of Menne & Spatz (1977) (W. Roos, unpublished).

An instance of modifiable behavior that may have relevance to the lives of flies in the wild is the courtship-depression effect of Siegel & Hall (1979). Male flies placed with sexually unreceptive females become, as it were, discouraged. For about three hours afterwards they show markedly less ardor in courtship, even with receptive virgin females. As to whether this behavioral change represents learning, the best evidence is that the learning mutants show less courtship depression than wild-type flies, and that the depression effect is present in *amnesiac* males, but lasts less than an hour (Siegel & Hall 1979). This phenomenon is considered in detail below, along with other aspects of courtship.

In all the fly-learning tests described above, the experimenters have gone to some pains to ensure that the behavioral changes were *associative*, i.e. that the changes occurred only if the animal had experienced two environmental stimuli close enough together in time to suggest that one stimulus predicted the other. Animals also show simpler, nonassociative forms of learning: *habituation,* a decrease in responsiveness to a stimulus presented repeatedly, and *sensitization,* an increase in general responsiveness after a strong or noxious stimulus. Habituation and sensitization may seem less interesting than associative learning, but they are much better understood—their underlying mechanisms have been worked out in considerable detail in *Aplysia* (reviewed by Kandel &

Schwartz 1982). We wondered whether the (largely unknown) mechanism for associative learning had points of similarity with the (largely known) mechanism for, say, sensitization, at least in depending on the same gene products. To test this possibility, J. Duerr examined habituation and sensitization in *Drosophila*'s proboscis extension reflex. As expected (see Dethier 1976), normal flies showed both forms of plasticity. More to the point, the associative learning mutants showed abnormalities: *dunce* and *turnip* had low habituation; *dunce, rutabaga,* and *amnesiac* showed unusually brief sensitization (Duerr & Quinn 1982). The results suggested, albeit indirectly, that associative and nonassociative learning do have underlying mechanistic similarities, an idea that is strengthened by biochemical analysis of the mutants (see below) and, recently, by direct physiological analysis in *Aplysia* (Hawkins et al 1983, Walters & Byrne 1983).

Biochemical Lesions in Learning Mutants

The fly-learning project, spawned in psychology, has rapidly turned molecular. Several of the learning mutations turn out to produce well-defined metabolic defects once one knows where to look for them. The first to be understood in this way was *dunce*. Kiger & Golanty were interested for purely biochemical reasons in cyclic nucleotide metabolism, and they decided to map the genes coding for the relevant enzymes (e.g. guanyl cyclase, cyclic AMP-dependent protein kinase) as a prelude to further analysis. Techniques are available in *Drosophila* that allow one to do such mapping, even before mutants are isolated (reviewed by O'Brien & MacIntyre 1978). In this indirect way Kiger & Golanty (1977, 1979) demonstrated that cyclic AMP (cAMP) was hydrolyzed by at least two phosphodiesterases in flies, and they localized one phosphodiesterase enzyme, PdE II, to a well-defined genetic region.

D. Byers, at the California Institute of Technology, sensitized by other work to a possible role of cAMP in learning, noticed that the genetic locus for the *dunce* region mapped very near the published site for PdE II. He did genetic complementation tests, then collaborated with R. Davis and J. Kiger to do direct biochemical tests on *dunce* flies, which showed that they had low levels of PdE II enzyme and high levels of cAMP (Byers et al 1981). Continuing this line of work, Kauvar (1982) obtained good evidence that *dunce* is the structural gene for PdE II enzyme. Among the several different mutations isolated in this gene, one, *dunce*[1], alters the enzyme's Km; *dunce*[2] increases its thermolability measured in vitro.

A second learning mutation, *rutabaga,* alters another enzyme in this pathway: adenylate cyclase. This abnormality—a three-fold increase in Km, i.e. a *decrease* in the enzyme's apparent affinity (Km) for its substrate, ATP, measured in crude extracts—was discovered by M. S. Livingstone. Earlier work with the mutants had given inconclusive results (Uzzan & Dudai 1982) and the mutant difference became reliable only when Livingstone examined different tissues. At first, she could detect the mutation-induced abnormality only in enzyme from the abdomen, with brain enzyme apparently normal (Livingstone et al 1982). This posed at least a philosophical problem, since previous work as well as common sense suggests that learning takes place in the brain. New evidence by Livingstone has eased this difficulty somewhat. The *rutabaga* mutation also causes a three-fold increase in the thermolability, measured in vitro, of adenylate cyclase from abdominal tissue. With this difference, she was able to show a similar abnormality in 10–20% of enzyme from brain (Livingstone et al 1982). Therefore, the most likely explanation for *rutabaga*'s effect on learning is that it alters a form of adenyl cyclase that constitutes a minority of brain enzyme but that is critically important in some neurons and synapses involved in plasticity.

In the case of *rutabaga*, as with *dunce*, the mutant enzyme's thermolability measured in vitro suggests that the mutation lies in the structural gene for the catalytic enzyme, as does the mutant enzyme's altered Km. Other observations that support this idea are the finding that abdominal enzymatic activity from heterozygotes has two distinct thermal decay rates and two Km's (Livingstone et al 1982) and the finding that cyclase catalytic activity correlates well with the number of gene copies of the *rutabaga*$^+$ gene (M. S. Livingstone, unpublished). These results do, however, lead to a complication. If the *rutabaga* $^+$ gene does indeed code for an adenylate cyclase, then this conclusion, taken together with the finding that most brain cyclase activity is unaffected by the mutation, implies the presence of different isozyme species of catalytic cyclase. If true, it would be the first solid evidence for multiple adenylate cyclases. In fact, experiments with mammalian cell hybrids suggest a different picture, indicating that one type of cyclase enzyme functions interchangeably with different hormone receptors. The straightforward interpretation of the findings with *rutabaga* suggest that the adenylate cyclase biochemistry may have hidden complexities.

At present Livingstone cannot strictly rule out the possibility that the *rutabaga* mutation affects some tightly bound, stoichiometrically limiting regulator of adenyl cyclase rather than the catalytic subunit itself. She does have evidence that the traditional GTP binding regulatory subunit, called N or G (Ross et al 1978), is not involved. Cyclase activity in abdominal tissues continues to depend on the number of *rutabaga* gene copies even in the presence of forskolin (Seamon et al 1981) or Mn^{2+} ions, agents that circumvent this regulatory protein (Livingstone et al 1982).

The fact that two learning mutants affect cAMP metabolism is striking, and argues for a central role for this compound in learning. However, there is an apparent paradox. The *dunce* mutations, which decrease phosphodiesterase activity, should increase cAMP concentrations in cells. The *rutabaga* mutation, which makes adenylate cyclase less active, should have the opposite effect. Direct measurement of cAMP levels in flies (Byers et al 1981, Livingstone et al 1982) confirms these predictions. Why then do both mutations interfere with learning? Our working model is this: cAMP concentrations appear to be "buffered" in healthy cells, and a small change in concentration can be sufficient to produce a maximal cellular response (see Rasmussen et al 1972, Nathanson 1977). A large perturbation of cAMP levels in either direction to beyond its normal range should disrupt the signalling properties of the system. (To take an analogy from neurophysiology, either a transmitter or its antagonist, applied in excess, will disrupt the signalling properties of neurons.) The *dunce* and *rutabaga* mutants may well induce such perturbations.

In vertebrate and invertebrate brains, adenylate cyclase is most frequently coupled to receptors for monoamine transmitters (reviewed by Bloom 1976, Nathanson 1977). If cAMP is involved in learning, then probably one or more monoamine transmitters also are. For synaptic systems in which the role of monoamines has been carefully studied, these transmitters often have a modulatory role; that is, they act as accessory transmitters, increasing or decreasing the efficacy of synaptic signalling via another, principal neurotransmitter (reviewed by Kupfermann 1981). This observation ties in well with ideas of learning as a change in the strength of connections in neural circuits. Moreover, the neuroanatomy of monoamine cells in vertebrates is consistent with a role in learning; the widely dispersed "sprinkler system" arrangements of monoaminergic systems seem particularly suited to alter the attentiveness or the retentiveness of large areas of the brain at once. Pharmacological studies in vertebrates, although intrinsically limited by drug side effects, cumulatively argue for a role of monoamines in learning, particularly for a dopaminergic role in positive reinforcement (see Wise 1978). Finally, in *Aplysia* one monoamine, serotonin, was known to function in nonassociative learning (sensitization) (Kandel & Schwartz 1982). For these reasons, many of us were eager to examine the fly learning mutants for abnormalities in monoamine transmitter synthesis. Livingstone did so, using HPLC-electrochemistry (see Kissinger et al 1981), and found all the learning mutants to be disappointingly normal (M. S. Livingstone, unpublished). However, she and Tempel, examining the large published library of *Drosophila* stocks, found existing mutations (Wright et al 1981) that made it possible to attack the problem from a different standpoint— temperature-sensitive mutations and deficiencies in the structural gene, *Ddc*, for dopa decarboxylase. This enzyme is very abundant in peripheral tissue, where it is vitally necessary for outside hardening. Wright (1977), who isolated the mutants, and also Dewhurst et al (1972), had already adduced evidence for its presence in brain tissue. Livingstone & Tempel decided to try *Ddc* mutants in learning studies. First they circumvented the usual lethal effect of the mutations by raising temperature sensitive flies at permissive temperature

(18°C) through development, eclosion, and cuticle hardening. Then they shifted the flies to restrictive temperature (29°C) for three days, which caused dopa decarboxylase activity to fall to undetectable levels but caused no loss of viability or overt behavioral abnormalities. Using radioactive precursors, they measured transmitter synthesis in isolated fly brains or in brain homogenates and found that Ddc mutants, in contrast to normal flies, were unable to carry out two relevant decarboxylation reactions: from L-dopa to dopamine and from 5-hydroxytryptophan to serotonin (Livingstone 1981, Livingstone & Tempel 1983). Synthesis of the third important monoamine transmitter, octopamine, was virtually normal in Ddc mutants, but the decarboxylation step here was partially blocked (Livingstone & Tempel 1983) by another mutation, per^o, which abolishes circadian rhythms (Konopka & Benzer 1971).

With this partial dissection of monoamine transmitter synthesis in hand, Tempel & Livingstone set out to measure learning in the relevant mutants. They found that per^o mutants learned as well as normal flies in olfactory discrimination tests, and they remembered as long afterwards, with either electric shock or sucrose as reinforcement. Severely affected Ddc mutants, on the other hand, showed no detectable learning with either reinforcement. Their behavior in other respects was nearly normal. Temperature-sensitive Ddc^{ts1} mutants learned reasonably well ($\Lambda = 0.12$), provided they had been raised at permissive temperature (18°C) but not ($\Lambda = 0.00$) if shifted to 29° as adults. Tempel & Livingstone then constructed a number of Ddc stocks, with different combinations of alleles, kept at different temperatures as adults, which gave a range of dopa decarboxylase enzyme activity levels from wild-type to undetectable. In all stocks the flies' learning performance in both olfactory tests correlated well with their dopa decarboxylase activities (Tempel & Livingstone 1981). This correlation seems surprising, because the effects of synthetic enzyme levels on transmitter pools, transmitter secreted, and so on to overt behavior are unknown and extremely indirect. Still, we are left with robust phenomenology. With Ddc mutants the fact that one can modulate the severity of the behavioral defect has made it possible to measure what aspect of learning is affected. Tempel constructed stocks (e.g. $Ddc^{ts1}/+$ raised at 29°) with a partial enzymatic block and tested them behaviorally. Although, as expected, their ability to learn was somewhat reduced ($\Lambda = 0.22$), their memory span was normal, after both positively and negatively reinforced training (Tempel & Quinn 1982). Ddc lesions appear to decrease learning with no measurable affect on memory. This is in contrast to $rutabaga$ and particularly to $amnesiac$ and $dunce$ mutations, which, in the most sensitive tests, appear to abbreviate memory retention without affecting acquisition.

Biochemical studies with Ddc indicated that it blocked synthesis of two neurotransmitters, dopamine and serotonin. Which one is more critically involved with learning? We may be able to answer this question, if we can ameliorate the mutant's learning deficit by feeding it precursors to one or the other transmitter, by analogy with L-dopa therapy for Parkinsonian patients. Recent results suggest another way to identify the guilty transmitter. Neurotransmitters produce their effect by binding receptors on the postsynaptic membrane. In many cases the binding affinity of receptors for transmitters is high enough and specific enough to be detectable in filter assays using crude membrane fractions, as most dramatically exemplified by the identification of opiate receptors in mammalian brain (Pert & Snyder 1973). R. Smith at Princeton is currently investigating the serotonin-binding activity of homogenates from wild-type and mutant flies. Confirming previous reports (Dudai & Zvi 1982), he finds specific, high-affinity binding in normal flies and most of the learning mutants, with at least two dissociation constants for serotonin under his conditions ($Kd_1 = 0.05$ nM; $Kd_2 = 0.5$ nM). His exciting result is that the *turnip* mutation dramatically lowers these affinities ($Kd_1 = 0.2$ nM; $Kd_2 = 1.1$ nM) without greatly affecting the total amount, Bmax, of specific binding activity (R. Smith, unpublished). In the simplest picture, this argues for a structural alteration in serotonin receptors with no reduction in their numbers. At present we have no idea of the nature of the change, nor even whether other receptor types are affected. Nevertheless, Smith's result gives a well-defined clue to the biochemical deficit in another learning mutant. It also dovetails nicely with the defects in other *Drosophila* mutants, which all affect sequential steps on one biochemical signalling pathway.

Since four of the learning mutations affect important transmitters or metabolites, it is reasonable to ask why the mutants are not dead, let alone severely affected in nonlearning behavior. For *dunce, rutabaga,* and *turnip* the answer may be that the metabolic effects are small, affecting only one isozyme, in the case of *dunce,* subtly altering an enzyme's affinity for its substrate in the other three cases, without abolishing activity. This excuse does not hold with *Ddc;* the metabolic lesion is drastic. We can only surmise that in *Drosophila* the monoamine systems are purely modulatory, and do not disrupt vital circuits, an idea consistent with the small amounts of the relevant enzymes present in the fly's brain (Dewhurst et al 1972).

Conclusions and Speculations

This is where fly learning genetics stands at the moment. Single-gene changes can interfere with learning in several tests, using olfactory, taste, visual, or proprioceptive cues, with positive or negative reinforcement, while leaving other behaviors substantially unaltered. Behavioral tests with the mutants have suggested experience-dependent components in behavioral patterns, such as courtship, which had appeared to be hard-wired. Work with mutants has provided evidence that associative and nonassociative learning are mechanistically related, a conclusion now confirmed by direct evidence from another system (Hawkins et al 1983, Walters & Byrne 1983).

Because genes often specify enzymes, one can, with mutants and luck, jump directly from a behavior to a molecule. Luck has been with us all. Work with the mutants directly implicates monoamines and cyclic AMP as central to associative and nonassociative learning. The finding that *dunce* is a phosphodiesterase mutant was the first hard evidence for a role for cyclic nucleotides in associative learning. Behavioral examination of the mutants suggests a more detailed picture. Two, *dunce* and *rutabaga*, affect the kinetics of cAMP metabolism; these appear to affect short-term memory decay rates. A third mutation, *Ddc*, blocks a step before the AMP response; this one primarily affects learning acquisition. These results tie in nicely with the suggestion of Castellucci et al (1982) that the chemical change corresponding to a short-term memory may be simply an increase in cAMP concentration in the relevant neurons.

Could storing a memory be so simple as a hormone response? This seems implausibly simple. On the other hand, deep problems in science often have simple answers, once the smoke has cleared. Learning may derive its phenomenological richness from the infinitely varied geometry of circuits in the brain, with the underlying mechanism kept simple. Nonassociative learning (in its short-term form) might be like a hormone response. Short-term associative learning could be nearly this simple, with enough metabolic embroidery to make the system responsive to a second stimulus, along lines suggested by recent work in *Aplysia* (Hawkins et al 1983, Walters & Byrne 1983). Things may be more complex, but they don't have to be.

The repeated links to *Aplysia* illustrates a limitation as well as a strength of the work with mutants. Mutations can provide perturbations of living systems. Work with them, properly interpreted, can lead directly from an animal's behavior to its molecular heart of hearts. Nevertheless, the search for the salient biochemical lesion in a given mutant is a needle-in-a-haystack enterprise unless there are clues on where to look. Such clues often come from other, nongenetic work. Since the search for mutant abnormalities then becomes highly directed, there is a good chance that someone will jump on the first promising-looking difference he finds and call it important. Fortunately for the work at hand, there is good biochemical evidence that in *dunce, rutabaga,* and *Ddc,* mutations cause their primary lesions in the enzymes they are supposed to. At present *turnip* seems half understood. Mutations like *amnesiac,* which may be relevant to the different problem of long-term memory storage, are presently not understood at all.

COURTSHIP AND MATING BEHAVIOR

The wren goes to't, and the small gilded fly
Does lecher in my sight.
Let copulation thrive. . . .

(Shakespeare 1608)

Courtship and mating behavior seem to be what *Drosophila* do best. The behavior consists of a sequence of stereotypical actions whose normal performance requires an exchange of defined signals—olfactory, visual, and auditory—between males and females. Each signal, in its turn, increases the ardor of the other participant, and elicits a patterned behavior from sex-specific portions of the central nervous system. The current detailed picture, emerging during the last seven years, has depended largely on experiments with mutants and mosaics. While nothing in the description transcends common-sense expectations, it differs greatly from early formulations of cause and effect in the observed behavioral sequence.

To the outsider, the most interesting aspects of courtship in *Drosophila* are the movements and displays made by the male. At the outset, he faces the female's abdomen, standing within a few millimeters of her, and if she moves away he will follow her and continue to stare at her abdomen. (This stage of courtship is known as "orientation".) Next, the male will reach out and touch the female's abdomen with his foreleg ("tapping"). Following this, he will extend one of his wings perpendicularly out from his body and vibrate it to produce the temporally patterned "courtship song" of his species, an alternating series of hums and bumping sounds.

At this point, the male moves in and extends his mouthparts in order to lick the female's genitals, then, if he is so inclined, he will attempt to mount her ("attempted copulation"), curling his abdomen under hers and making genital contact. Copulation, once started, will normally go on for 20 min before disengagement. During courtship, the female appears to do much less. She must stand still while the male licks and attempts copulation, and, finally, she must spread her ovipositor to permit penetration by the male.

The foregoing scenario occurs when normal adult males court virgin females. If the female has already mated, then she will not permit the male to copulate and will flat out reject him, extruding her ovipositor in his face (see Spieth 1952, 1974 for full review of behaviors).

Studies of *Drosophila* mating behavior have always had a genetic leaning, reflecting the long history of genetic work in this organism. Many of these studies were undertaken primarily from an evolutionary point of view (e.g. Sturtevant 1915, Manning 1961, Parsons 1974) and they consisted mainly of species comparisons (e.g. Spieth 1952), multigene selections for differences in mating performance (e.g. Manning 1961), or tests of mutants with altered body color and eye color (e.g. Bastock 1956, Grossfield 1975). More recently, interest in the neurobiology of the behavior has motivated much of the genetic work. An early assumption about fly courtship was that it approximated a classical Lorenz-Tinbergen "fixed action pattern" or patterns (e.g. Benzer 1973). The working model was that male courtship, like some other insect behaviors, represented the playing out of a neural program that somewhat resembled a tape recording (Hoyle 1964) and that, because of its invariant sequence, might yield to a genetic analysis of steps in the pathway by analogy to genetic studies of metabolic pathways (e.g. Yanofsky 1971). This assumption, though wrong, was valuable because it provided a clear intellectual framework in which to design experiments with mutants and mosaics.

Sex-Specific Parts of the Central Nervous System

Male fly brains are different from female fly brains. This unstartling conclusion comes from functional tests on sex mosaics, composite animals in which some cells are genetically male, others genetically female. By testing many such mosaics and analyzing the arrangements of male and female cells in the nervous system with an appropriate enzymatic marker of sexual genotype (Kankel & Hall 1976), the sites which must be male to bring about male-specific behavioral components (and female to bring about female-specific components) can be mapped. The approach gives quite different information from anatomical studies that find sex-specific morphological differences (e.g. MacLusky & Naftolin 1981, Mobbs 1982). Mosaic analysis indicates which parts of the brain specify the particular behavioral patterns of each sex.

With this method, Hall (1977) mapped three early stages of male courtship to a site in the superior protocerebrum (the most dorsal part of the brain). That is, mosaics with male cells in this portion of the brain, in either of the two bilaterally symmetrical sides, showed orientation, following, and wing extension toward a virgin female. The unitary nature of the site for all three of these actions was reflected in the fact that most mosaics that performed any one of them went on to perform all three. In the few individuals that did not perform all three, no consistent anatomical pattern emerged for any one relative to the others. G. Technau (manuscript in preparation) has recently found a subtle anatomical dimorphism near this brain region: female flies have more nerve fibers than males in their mushroom bodies.

Although a mosaic with male tissue in the dorsal brain will extend its wings at a female, it will not sing a proper courtship song unless it also has male cells in part of the thoracic ganglion (von Schilcher & Hall 1979). The region in the thoracic ganglion was not as clearly defined as the one for early courtship actions, but the singing mosaics did show a greater tendency to have male cells in the ventral mesothoracic neuromere. The muscles whose activity precedes and correlates closely with song production are located near this part of the mesothorax (Ewing 1979). Perhaps the male-specific cells responsible for the song are their motorneurons or interneurons presynaptic to them. As with the site for wing extension, male tissue on one side of the ganglion was sufficient for a normal song; and in most of these cases, the fly was competent with either wing.

Hall (1979) also mapped the later steps in the courtship sequence, licking and attempted copulation. These turn out to have quite different kinds of sites. The one for licking behavior is not anatomically separable from the site previously described for orientation, following, and wing extension, but it is functionally distinct in that it requires male cells to be present on both sides of the brain. Whether licking actually maps to different groups of cells in these regions than early courtship is an open question. All of the mosaics that performed licking had previously gone through the early stages of courtship (cf Hotta & Benzer 1976, Cook 1978). This may indicate a hierarchical relationship. On the other hand, there may be two discrete sites so close together that the chance of finding a mosaic with a distribution of male cells that could resolve the issue is vanishingly small.

Of all the steps in courtship that have been mapped with mosaics, attempted copulation has given the least satisfactory answer. Likelihood of attempting copulation is associated with male cells in the thoracic ganglion. Moreover, the more male tissue in this ganglion, the more likely the attempt. In keeping with previous suggestions of an obligatory sequence of steps in courtship, no mosaics attempted copulation who had not done all of the preceding actions, through licking. However, no clearly defined site in the ganglion was consistently correlated with the behavior. A plausible explanation for this outcome is that male cells in any of several places are sufficient to produce the behavior (Hall 1979).

Tompkins & Hall (1983) made a mosaic study of a female-specific behavior, receptivity to copulation. In their experiments, sex mosaics were placed with normal males and observed for their willingness to copulate. As with the later stages in male courtship, this study necessarily used a subset of mosaics, those which had female genitalia and which also elicited attempted copulation by males. [The ability of a mosaic to stimulate courtship by a male has not been accurately mapped, but it depends on female cells in the abdomen or posterior thorax (Hall 1977, Jallon & Hotta 1979). This separation of "sex appeal" from the brain site for male behavior occasionally produces mosaics that court virgins and are themselves courted by other males (Hall 1977, Szabad & Fajszi 1982).]

Among mosaics meeting these criteria, most were not receptive to copulation, implying a separate site outside of the abdomen for behavioral receptivity.

This site, well defined, lay in the anterior portion of the medial inferior protocerebrum, apart from all sites for male courtship. In this case, bilaterally symmetric sites had to be female for the mosaic to be receptive to copulation.

Functional information of this sort, gained with sex mosaics, is valuable, detailed, and different in kind from what can be found out by other methods. One drawback, however, is that mosaic patch sizes tend to be large—usually 1/3 of the organism in these studies. In practice, this means that behaviors can be mapped to defined subdivisions of the brain, but not to individual cells or even groups of a few cells. Nor can a single site controlling two behaviors always be distinguished from two very close sites, as discussed above. A related drawback of the technique is the difficulty of obtaining multiple, discontinuous patches in the same animal for accurate identification of interacting sites, a problem that arose in connection with the apparent diffuseness of the male "site" for attempted copulation in the male thoracic ganglion. Dividing lines in mosaics are determined by patterns of cell division in the embryo and some orientations are more favored than others. In principle, this could be solved by scanning a sufficient number of mosaics, but a lifetime, particularly a research lifetime, is finite.

Intersex mosaics have confirmed our suspicions that fly brains differ between the sexes. However, they say nothing about when sex-specific differences actually affect function, because the cells in mosaics are sexually determined long before any differentiation begins in the nervous system (see Hall et al 1976 for discussion of time of mosaic induction). To distinguish between temporal, as opposed to spatial, determinants of sexual behavior, one needs a temperature-sensitive mutation of a sex-transforming gene.

The mutation is in the gene called *transformer-2*. It converts flies that should be female into phenotypic males in a temperature-dependent manner (Belote & Baker 1982). That is, when mutants with two X chromosomes are raised at 29°C, they look like males, whereas if they are raised at 16°C they look more like females (though they do have some intersexual external features). More relevant to our problem is the fact that mutants raised at 29°C will court just like males, whereas the mutants raised at 16°C are fertile, behaviorally competent females, with no detectable male behavior. J. M. Belote and B. S. Baker (unpublished) asked whether mutants, if raised at the temperature 16°C, which makes them look and act female, can be induced to behave like males if shifted to 29°C as adults after most of development is complete. The shifted mutants did change from female to male behavior, in several cases even attempting copulation, no small task for a fly with a female abdomen (but not unprecedented, cf Hall 1979).

This result is surprising. If the CNS of these mutants was authentically female prior to raising the temperature, then it suggests that a significant portion of the apparatus for male-specific behavior is not dependent on pre-adult, dimorphic development of the nervous system. Perhaps male circuits are present and ready to go in females, but are kept inactive by sex-specific gene action (see below). Alternatively, the *transformer-2* flies might be imperfect models of normal females, as suggested by their intermediate morphology. Still, interesting acute effects must be taking place to transform the behavior so rapidly after the temperature shift.

Another type of genetically induced sexual transformation creates phenotypic males that exhibit behaviors of both sexes. Appropriate combinations of alleles of the *Sex-lethal* gene (Cline 1978) will transform files with two X chromosomes into phenotypic males. One such combination, $Sxl^{M\#1,F\#3}/Sxl^{F\#7,M\#1}$ (Cline 1981), makes a morphological transformation so complete that the flies have all of the male-specific internal organs for sperm production, storage, and transmission (T. W. Cline, unpublished). These flies belie their appearance, however, exhibiting female-like rejection behaviors toward normal males who try to court them, including the unprecedented one of extruding their genitals as if they were ovipositors (L. Tompkins and T. W. Cline, unpublished). Since they lack female genitalia, these flies could not be scored for the most obvious female-specific behavior, receptivity (cf Tompkins & Hall 1983). When tested with normal virgin females, a few of these transformed mutants also performed the full repertoire of male courtship behaviors, more in line with their morphology. A different allele of this gene, Sxl^{F2593}/Sxl^{F2593} (Marshall & Whittle 1978), also produces a nearly complete morphological transformation of flies with two X chromosomes into males. In this case, a larger proportion of the transformed flies behave both as males and as females (L. Tompkins and T. W. Cline, unpublished).

Mutations in the *Sex-lethal* gene give no simple answers to the question of how the CNS is affected by sexual differentiation. The finding that a single fly can exhibit specific behaviors of either sex is particularly hard to deal with. The lack of correlation between sexual morphology and behavior suggests that cells in different parts of the organism will respond differently to a common genetic background, i.e. the very male-like cells of the cuticle and gonads in the mutant $Sxl^{M\#1,F\#3}/Sxl^{F\#7,M\#1}$ may have a different threshold than do cells of the CNS for intracellular signals of sex determination (cf Cline 1978). If it turns out that cells *within* the CNS can differ from each other in their responses, then it may also explain the bisexual behavior of many of the mutants.

One possible explanation for sex specificity in the CNS envisions the behavior of one sex as being the absence of a function present in the other sex (Hall et al 1979). This line of reasoning evolved from the classical notion of the fixed action pattern and its innate releasing mechanism. It postulates sex-specific behaviors as being restrained by an inhibitory switch, as copulatory behavior in the courting praying mantis is restrained until decapitation (reviewed by Roeder 1967). In this model, every *Drosophila* brain contains circuits that specify both male and female behaviors, but the circuits for the inappropriate sex are perpetually restrained by sex-specific mechanisms.

Another possibility is that there are sex-specific circuits in the brain whose developmental potential is present in both sexes, though latent in one. A situation like this exists in canaries, in which male hormones, if given to an adult female, will induce the development of male-like anatomical characteristics in specific regions of her brain (Nottebohm 1980).

Nonautonomy of the Behavioral "Program"

Male behavior patterns in *Drosophila* courtship follow an invariant sequence in the sense that a male will never omit steps: there is no attempt at copulation without a proper courtship first. Even the sex mosaics and genetic intersexes described above never omitted steps. A given mosaic might terminate the sequence at a specific point, such as wing extension, but the order to that point was conventional. The invariance of this sequence lends plausibility to the notion that courtship behavior is an autonomous fixed action pattern. However, this idea breaks down in the face of a growing body of new evidence for a continuous exchange of cues between male and female flies. Much of this evidence comes from studies of sensory, motor, and biochemical mutants.

In order for courtship to begin at all, a male must find a female "attractive." This attractiveness depends at least in part on olfactory and visual information. The olfactory cue is a volatile pheromone whose existence was initially surmised by watching the behavior of males kept "downwind" from females (Shorey & Bartell 1970). An organic extract from virgin females will stimulate male courtship in the absence of a female (Tompkins et al 1980, Venard & Jallon 1980); its most active component appears to be a heptacosadiene (Antony & Jallon 1982). Males that are mutant for the gene *smell-blind* (*sbl*, Aceves-Pina & Quinn 1979), and thus unable to respond to olfactory stimuli, will not respond to the extract. Moreover, *sbl* males will court a real virgin female less actively than will normal males (Tompkins et al 1980).

At first, visual information was assumed to be unimportant in courtship because flies mated successfully in the dark (e.g. Grossfield 1966), and stocks of blind mutants could be maintained without difficulty. However, blind males showed deficiencies in quantitative tests of courtship (Crossley & Zuill 1970, Siegel & Hall 1979, Tompkins et al 1980). Clues to the kind of visual information involved came from experiments with other mutants, ones with more specific visual defects.

A mutant, *no-receptor-potential (norpA),* which lacks all photoreceptor responses (Pak et al 1970), courted less actively than a normal male and copulated with one-third less probability in a 10 min test period (Siegel & Hall 1979, Tompkins et al 1980). A second mutant, *no-on-transient (non A),* can see but fails to respond to moving patterns that require high visual resolution (Heisenberg 1972). Cook (1980) measured male flies' accuracy in orienting toward females, calculating a males's "error angle" with respect to a female who is rotated around him in a chamber. Mutant *non A* flies oriented far more poorly than wild-type flies, suggesting, not surprisingly, that the fly uses the same visual circuitry to follow moving females as to follow moving stripes. In fact, at close range the male is probably following the stripes of pigment on the female's abdomen (Cook 1979).

Another mutant, *optomotor-blind* (omb^{H31}), with an even subtler visual impairment, also does poorly on this test (Cook 1980). These flies have high resolution vision, but they fail to respond to almost all moving patterns (Heisenberg et al 1978). Tompkins tested the effectiveness of omb^{H31} males in an actual courtship situation (Tompkins et al 1982). She found that the mutant males could orient if the female happened to be standing still but tended to "lose" her if she moved away. This was reflected quantitatively in the shorter "courtship bouts" performed by mutant males as compared to wild-type. Motion detection turns out to be important not only in following females, but also because female movement *per se* stimulates males to court more vigorously. Tompkins et al (1982) concluded this after observing that normal males spend less time courting females immobilized by a temperature-sensitive, paralytic mutation.

The results with omb^{H31} are particularly informative because of the anatomical defect caused by this mutation: it deletes several of the giant neurons in the proximal optic lobe, the lobula, which are thought to be essential for motion detection over large areas of the fly's visual field (Heisenberg et al 1978, Blondeau & Heisenberg 1982).

The results above can be assembled into a provisional scenario for early courtship. Olfactory cues (pheromones) will induce a male to start courting, but he will not be able to keep it up if he fails to see movement by the female. Olfactory cues, while sufficient, are not absolutely necessary since *sbl* males can still court a little bit if they can see. Deprive males of sight and smell, however, and they will not court at all unless they collide with a female (Tompkins et al 1980; L. Tompkins, personal communication).

As the female's pheromones and movements stimulate the male to court, he transmits information back to her via auditory, visual, and possibly olfactory stimuli which increase her receptivity. The evidence for auditory cues comes mainly from behavioral experiments with mutilated wild-type flies. Females "deafened" by removal of their auditory organs (aristae), or males "muted" by removal of their wings take longer to copulate (e.g. Manning 1967, von Schilcher 1976a). Moreover, the disadvantage of wingless males in courting normal females can be partially overcome by playing electronically produced, artificial courtship songs while the flies are mating (von Schilcher 1976a). The most visible effect of courtship song on a female is to decrease her tendency to run away (von Schilcher 1976a).

One indication that the interactions between male and female are not simply a series of triggered reflexes is that the effect of a male's singing on a female lasts for several minutes, and her "memory" of the song is controlled by genes that are involved in learning. Playing an electronically produced courtship song to a female will increase her subsequent mating speed when she is placed with a male (von Schilcher 1976b; C. P. Kyriacou and J. C. Hall, unpublished). Normally, this priming effect lasts 5 min. With mutant *amnesiac* females, however, the priming lasts only about 1 min, and *dunce* females show no priming at all (C. P. Kyriacou and J. C. Hall, unpublished).

Kyriacou & Hall (1980, 1982), in analyzing the fine structure of the courtship song, found a piece of information in the song that was particularly important to the listening female. Typically, the intervals between pulses in the song are variable and oscillate sinusoidally with a period of 55 sec. This periodic variation in interpulse intervals is as critical to a female's receptivity as is the interpulse interval itself (Kyriacou & Hall 1982). The period of the courtship song oscillation was influenced by mutations at the *period* (*per*) locus, previously identified for its effects on the fly's circadian rhythms (Konopka & Benzer 1971). Moreover, the effects of various *per* mutations had similar effects on these two, very different periods. The *per*[1] mutation, which lengthened circadian rhythms to 28 hr, also lengthened the song period to 82 sec; the *per*[s] mutation shortened the circadian period to 19 hr and the song period to 42 sec; and the *per*[o] mutation abolished both rhythms (Kyriacou & Hall 1980). In courtship tests, male *per* mutants that are either arrhythmic or short period are slower to achieve copulation than normal males (Kyriacou & Hall 1982; J. C. Hall and A. C. Gross, unpublished), a finding that underscores the importance of the song periodicity in courtship. Very interestingly, a female's ability to comprehend the song may also be under the control of the *per* locus. The poor performance of *per*[s] males with normal females is far less pronounced when these males are courting *per*[s] females (J. C. Hall, unpublished). This may turn out to be a clear example of matching oscillators in signal generation and reception, an idea with a long history in studies of animal communication.

The influence of the *per* gene on courtship song rhythms, dramatically different in period and expression from circadian rhythms, was very unexpected. It illustrates the fact that one gene product can influence several different behavior patterns, apparently by becoming incorporated into several systems and acting at different times and places. Support for this idea comes from analyses of the *per* mutation with tissue grafts and mosaics. The critical cells for circadian rhythms appear to be in the brain (Handler & Konopka 1979;

J. C. Hall, C. P. Kyriacou, R. J. Konopka, unpublished). In contrast, the critical cells for courtship song periodicity appear to be in the thoracic ganglion (J. C. Hall et al, unpublished), a finding that ties in well with the localization of song production to that tissue (von Schilcher & Hall 1979). Direct evidence for the anatomical separation of functions comes from the behavior of selected *per* mosaics. Some of them had normal circadian rhythms and mutant songs, while others were mutant for circadian rhythms and normal for song (J. C. Hall et al, unpublished).

A female's receptivity is affected by olfactory and visual stimuli as well as by the courtship song; all three modalities seem to work on her tendency either to move around or to come to a stop. The importance of olfactory cues in female receptivity (originally suggested by Ehrman 1969) emerged fortuitously during the course of experiments with blind flies, which required unusually long times to achieve copulation. Tompkins et al (1982) noticed that if courtship lasted for more than two minutes, the majority of the females would come to a complete stop for one minute or more, whereas females that copulated during the first couple of minutes of courtship did not. In contrast, mutant *smell-blind* females rarely stopped for more than a minute, even when courtship was prolonged for up to 15 min. This suggests that female locomotion during courtship is influenced by an olfactory cue, perhaps emanating from the male, which becomes apparent after the first 2–3 min of courtship.

The same female locomotory behavior is also affected by visual stimuli. However, in this case, the female's ability to see the male seems to increase her tendency to move, making her less receptive. The initial clue in this story came from placing omb^{H31} males with normal females. Usually these mutants were abnormally slow in copulating, but their relative inadequacies disappeared when the lights were turned off. Similarly, blind (*norpA*) males courted blind females as successfully as normal males courted normal females in the dark (Tompkins et al 1982, cf Markow & Manning 1982). The interaction of female coyness as influenced by sight, with female acceptance as influenced by sound and smell, has yet to be elaborated, but it may be at least as complex as in higher organisms (Marvell 1681).

The clearest deviation from fixed action patterns to be found in *Drosophila* courtship is the long-lived effect on a male of his frustrated attempts to court an unreceptive, previously mated female. All of the discussion so far has focused on the interactions between males and virgin females. If a male attempts to court a mated female, she will refuse to copulate with him, extruding her ovipostor at him (Connolly & Cook 1973), and the male will shortly give up the chase. If the same male is then placed with a virgin female, he will attempt to court her only about 25% as much as a naive male would. This "depression" of courtship lasts over 2 hr (Siegel & Hall 1979).

The message of scorn from mated females is not visual in nature, since blind *norpA* males become just as "depressed" as wild-type males (Siegel & Hall 1979). On the other hand, mutant *smell-blind* males show no "depression" effect (Tompkins et al 1983). This is consistent with the observation that extracts from virgin females contain attractive pheromones, whereas extracts from mated females contain inhibitory pheromones (Tompkins & Hall 1981). Nevertheless, extracts alone are not sufficient to "depress" a male. He must be in the presence of another fly while he is being exposed to extract in order to experience "depression." However, this other fly can be a virgin female or even a paralyzed male (Tompkins et al 1983).

The requirement that mated female extract be presented together with a fly to produce its depressive effect takes on more significance in light of the obervation that the learning mutants *dunce, cabbage, rutabega,* and *turnip* (Gailey & Siegel, submitted) and *Dopa decarboxylase* (Tempel, Livingston & Quinn, submitted) are all deficient in their ability to be conditioned, i.e. depressed, by mated females. The memory mutant *amnesiac* recovers from its depression much faster than wild-type, as if he is unable to remember the conditioning experience (Siegel & Hall 1979). The obligatory presence of a fly with the stimulus raises the possibility that this little drama of *Drosophila* courtship represents *bona fide* associative conditioning. If so, it argues even more strongly in favor of plasticity as an integral component of courtship behavior.

A female becomes behaviorally "mated," i.e. capable of depressing subsequent male courtship, after only a couple of minutes of copulation, before any sperm have actually been transferred (Tompkins & Hall 1981). Other components of seminal fluid, which are transferred earlier than sperm, include *cis*-vaccenylacetate, an esterified fatty acid, and also esterase-6, a carboxylesterase (Brieger & Butterworth 1970, Richmond & Senior 1981). *Cis*-vaccenylacetate was suspected at first as the "antiaphrodisiac" of mated females (Jallon et al 1981), but it was found to disappear too rapidly from inseminated females to account for the long-lasting behavioral effect (Jallon et al 1983, Tompkins & Hall 1981). Nevertheless, the picture may be nearly this simple. The fact that the ester and the esterase occur together in seminal fluid suggests that the enzyme might metabolize the ester in vivo into a longer lasting antiaphrodisiac. This idea is supported by experiments with mutants lacking esterase-6 activity (*Est 6°*). If mutant *Est 6°* males are mated to wild-type females, the females are able to inhibit subsequent courtship for only a brief time—about 10 min rather than the usual 6 hr (Mane et al 1983). The simplest explanation is that *cis*-vaccenylacetate can act as a short-term inhibitor, but that esterase-6 must act by cleaving the ester to *cis*-vaccenyl alcohol, and that this product is, makes, or induces the long-term inhibition.

A further test of this idea used an olfactory mutant, *olfC*, which is specifically defective in smelling acetates (Rodrigues & Siddiqi 1978). Mutant *olfC* males were not inhibited by females taken 10 min after copulating with a normal male (when *cis*-vaccenylacetate would be active), but they were inhibited by the same females 6 hr after they had copulated, as if a different chemical inhibitor was in effect (Mane et al 1983). Mutant *olfC* males, tested with females who had previously mated with *Est 6°* males, showed neither short-term nor long-term inhibition as expected. One additional chemical result makes this story very neat. Mane et al demonstrated that esterase-6 is capable of hydrolyzing *cis*-vaccenylacetate in vitro and that the product, *cis*-vaccenyl alcohol, is also capable of acting as an inhibitor of courtship.

Conclusion

The information on *Drosophila* courtship is different from that on learning. Courtship is more ethological than learning. Understanding a complex behavior pattern requires a different strategy from understanding physiological process. Here the direction has been more formal and conceptual, resolving a behavioral sequence into divisible components, indicating the types of sensory cues relayed from individual to individual during the courtship *pas de deux*. Traditional ethology has proceeded by observation, intuited experiments, or sometimes interspecies comparisons. The systematic use of single-gene mutants and mosaics adds a new tool and a new possibility for technical refinement, and the results show it.

A good deal has been written for and against the neurogenetic approach, much of it silly. Sometimes the approach works. We have picked two stories with happy outcomes, yielding information that is relevant in other fields. In each case, the experimental strategy has been devoutly eclectic—making mosaics, selecting mutants, adopting mutants, using these in conjunction with traditional ethological, psychological, or biochemical methods, testing home-grown ideas or models, doing what seems promising. The chief advantage of the neurogenetic approach, aside from the intrinsic cleanliness of mutational lesions, is that the field was founded by individuals who believed in young scientists and in improvisation. Given this tradition, and the rich circuitry of human and fly brains, we are confident of advances we cannot predict.

ACKNOWLEDGMENTS

We thank Joan Nielsen and Karin O'Hara for typing this manuscript. The work on learning from W. G. Q.'s laboratory cited here was supported by NIH Grant GM25578. R. J. G. is a Searle Scholar, a McKnight Scholar, and an Alfred P. Sloan Research Fellow.

Literature Cited

Aceves-Pina, E. O., Quinn, W. G. 1979. Learning in normal and mutant *Drosophila* larvae. *Science* 206:93–96

Antony, C., Jallon, J. M. 1982. The chemical basis for sex recognition in *Drosophila melanogaster*. *J. Insect Physiol.* 28:873–80

Bastock, M. 1956. A gene mutation which changes a behavior pattern. *Evolution* 10:421–39

Belote, J. M., Baker, B. S. 1982. Sex determination in *Drosophila melanogaster:* Analysis of transformer-2, a sex-transforming locus. *Proc. Natl. Acad. Sci. USA* 79:1568–72

Benzer, S. 1973. Genetic dissection of behavior. *Sci. Am.* 229 (6):24–37

Blondeau, J., Heisenberg, M. 1982. The three dimensional optomotor torque system of *Drosophila melanogaster*. Studies on wild-type and the mutant *optomotor-blind H31*. *J. Comp. Physiol.* 145:321–29

Bloom, F. E. 1976. The role of cyclic nucleotides in central synaptic function. *Adv. Biochem. Psychopharmacol.* 15:273–82

Booker, R. 1982. *A behavioral-genetic analysis of learning in Drosophila melanogaster*. PhD thesis. Princeton Univ., Princeton, NJ

Booker, R., Quinn, W. G. 1981. Conditioning of leg position in normal and mutant *Drosophila*. *Proc. Natl. Acad. Sci. USA* 78:3940–44

Borst, A., Byers, D., Heisenberg, M. 1983. Submitted for publication

Brieger, G., Butterworth, F. M. 1970. *Drosophila melanogaster:* Identity of male lipid in reproductive system. *Science* 167:1262

Byers, D., Davis, R. L., Kiger, J. A. 1981. Defect in cyclic AMP phosphodiesterase due to the *dunce* mutation of learning in *Drosophila melanogaster*. *Nature* 289:79–81

Castellucci, V. F., Nairn, A., Greengard, P., Schwartz, J. H., Kandel, E. R. 1982. Inhibitor of adenosine 3'5' monophosphate-dependent protein kinase blocks presynaptic facilitation in *Aplysia*. *J. Neurosci.* 2:1673–81

Cline, T. W. 1978. Two closely linked mutations in *Drosophila melanogaster* that are lethal to opposite sexes and interact with *daughterless*. *Genetics* 90:683–98

Cline, T. W. 1981. Positive selection methods for the isolation and fine-structure mapping of *cis*-acting, homeotic mutations at the *Sex-lethal* (S x l) locus of *D. melanogaster*. *Genetics* 97:S23

Connolly, K., Cook, R. 1973. Rejection responses by female *Drosophila melanogaster:* Their ontogeny, causality and effects upon the behavior of the courting male. *Behavior* 44:142–66

Cook, R. 1978. The reproductive behavior of gynandromorphic *Drosophila melanogaster*. *Z. Naturforsch. Teil C* 33:744–54

Cook, R. M. 1979. The courtship tracking of *Drosophila melanogaster*. *Biol. Cybernet.* 34:91–106

Cook, R. M. 1980. The extent of visual control in the courtship of tracking *D. melanogaster*. *Biol. Cybernet.* 37:41–51

Crossley, S., Zuill, E. 1970. Courtship behavior of some *Drosophila melanogaster* mutants. *Nature* 225:1064–65

Dethier, V. G. 1976. *The Hungry Fly*. Cambridge, Mass: Harvard Univ. Press. 489 pp.

Dewhurst, S. A., Croker, S. G., Ikeda, K. McCaman, R. E. 1972. Metabolism of biogenic amines in *Drosophila* nervous tissue. *Comp. Biochem. Physiol. B* 43:975–81

Dudai, Y. 1977. Properties of learning and memory in *Drosophila melanogaster*. *J. Comp. Physiol.* 114:69–89

Dudai, Y. 1979. Behavioral plasticity in a *Drosophila* mutant, dunce. *J. Comp. Physiol.* 130:271–75

Dudai, Y. 1981. Olfactory choice behavior of normal and mutant *Drosophila* in a conflict situation in a successive conditioning paradigm. *Soc. Neurosci. Abstr.* 7:643

Dudai, Y., Bicker, G. 1978. Comparison of visual and olfactory learning in *Drosophila*. *Naturwissenschaften* 65:495–96

Dudai, Y., Jan, Y. N., Byers, D., Quinn, W. G., Benzer, S. 1976. dunce, a mutant of *Drosophila* deficient in learning. *Proc. Natl. Acad. Sci. USA* 73:1684–88

Dudai, Y., Zvi, S. 1982. Heterogeneity of serotonin receptors in *Drosophila melanogaster*. *Soc. Neurosci. Abstr.* 8:989

Duerr, J. S., Quinn, W. G. 1982. Three *Drosophila* mutations that block associative learning also affect habituation and sensitization. *Proc. Natl. Acad. Sci. USA* 79:3646–50

Ehrman, L. 1969. The sensory basis of mate selection in *Drosophila*. *Evolution* 23:59–64

Ewing, A. W. 1979. The neuromuscular basis of courtship song in *Drosophila:* The role of the direct and axillary wing muscles. *J. Comp. Physiol.* 130:87–93

Folkers, E. 1982. Visual learning and memory of *Drosophila melanogaster* wild-type C-S and the mutants dunce, amnesiac, turnip and rutabaga. *J. Insect Physiol.* 28:535–39

Fukushi, T. 1973. Olfactory conditioning in the housefly *Musca domestica*. *Annot. Zool. Jpn.* 46:135–43

Gailey, D. A., Siegel, R. W. 1983. Submitted for publication

Grossfield, J. 1966. The influence of light on

the mating behavior of *Drosophila. Univ. Texas Publ.* No. 6615:147–76

Grossfield, J. 1975. Behavioral mutants of *Drosophila. Handb. Genet.* 3:679–702

Hall, J. C. 1977. Portions of the central nervous system controlling reproductive behavior in *Drosophila melanogaster. Behav. Genet.* 7:291–312

Hall, J. C. 1979. Control of male reproductive behavior by the central nervous system of *Drosophila:* Dissection of a courtship pathway by genetic mosaics. *Genetics* 92:437–57

Hall, J. C. 1982. Genetics of the nervous system in *Drosophila. Q. Rev. Biol.* 15:223–479

Hall, J. C., Gelbart, W. M., Kankel, D. R. 1976. Mosaic systems. In *Genetics and Biology of Drosophila,* ed. M. Ashburner, E. Novitski, 1A:265–314. New York:Academic. 486 pp.

Hall, J. C., Greenspan, R. J., Harris, W. A. 1982. *Genetic Neurobiology.* Cambridge, Mass: MIT Press. 284 pp.

Hall, J. C., Greenspan, R. J., Kankel, D. R. 1979. Neural defects induced by genetic manipulation of acetycholine metabolism in *Drosophila. Soc. Neurosci. Symp.* 4:1–42

Handler, A., Konopka, R. J. 1979. Transplantation of a circadian pacemaker in *Drosophila.* Nature 279:236–38

Hawkins, R. D., Abrams, T. W., Carew, T. J., Kandel, E. R. 1983. A cellular mechanism of classical conditioning in *Aplysia:* Activity-dependent amplification of presynaptic facilitation. *Science* 219:400–5

Heisenberg, M. 1972. Comparative behavioral studies on two visual mutants of *Drosophila. J. Comp. Physiol.* 80:119–36

Heisenberg, M. 1980. Mutants of brain structure and function: What is the significance of the mushroom bodies for behavior? In *Development and Neurobiology of Drosophila,* ed. O. Siddiqi, P. Babu, L. M. Hall, J. C. Hall, pp. 373–90. New York: Plenum. 496 pp.

Heisenberg, M., Wonneberger, R., Wolf, R. 1978. *Optomotor-blind H31—a Drosophila* mutant of the lobula plate giant neurons. *J. Comp. Physiol.* 124:287–96

Horridge, G. A. 1962. Learning of leg position by headless insects. *Nature* 193:697–98

Hotta, Y., Benzer, S. 1976. Courtship in *Drosophila* mosaics: Sex-specific foci for sequential action patterns. *Proc. Natl. Acad. Sci. USA* 73:4154–58

Hoyle, G. 1964. Exploration of neural mechanisms underlying behavior in insects. In *Neural Theory and Modelling,* ed. R. F. Reiss, pp. 346–76. Stanford: Stanford Univ. Press. 427 pp.

Jallon, J.-M., Antony, C., Benamar, O. 1981. Un anti-aphrodisiaque produit par les males de *Drosophila melanogaster* et transere aux femelles lors de la copulation. *CR Acad. Sci. III* 292:1147–49

Jallon, J.-M., Antony, C., Iwatsubo, T. 1983. Elements of chemical communication between Drosophilids and their modulation. In *Proc. Taniguchi Symp. Biophys.* ed. Y. Hotta. In press

Jallon, J.-M., Hotta, Y. 1979. Genetic and behavioral studies of female sex appeal in *Drosophila. Behav. Genet.* 9:257–75

Jellies, J. A. 1981. *Associative olfactory conditioning in Drosophila melanogaster and memory retention through metamorphosis.* Master's thesis. Illinois State Univ., Normal, Ill.

Kandel, E. R., Schwartz, J. H. 1982. Molecular biology of learning: Modulation of transmitter release. *Science* 218:433–43

Kankel, D. R., Hall, J. C. 1976. Fate mapping of nervous system and other internal tissues in genetic mosaics of *Drosophila melanogaster. Dev. Biol.* 48:1–24

Kauvar, L. M. 1982. Defective cyclic adenosine 3'5'-monophosphate phosphodiesterase in the *Drosophila* memory mutant *dunce. J. Neurosci.* 2:1347–58

Kiger, J. A., Golanty, E. 1977. A cytogenetic analysis of cyclic nucleotide phosphodiesterase activities in *Drosophila. Genetics* 85:609–22

Kiger, J. A., Golanty, E. 1979. A genetically distinct form of cyclic AMP phosphodiesterase associated with chromomere 3D4 in *Drosophila melanogaster. Genetics* 91:521–35

Kissinger, P. T., Bruntlett, C. S., Shoup, R. E. 1981. Neurochemical applications of liquid chromatography with electrochemical detection. *Life Sci.* 28:455–65

Konopka, R. J., Benzer, S. 1971. Clock mutants of *Drosophila melanogaster. Proc. Natl. Acad. Sci. USA* 68:2112–16

Kupfermann, I. 1981. Modulatory action of neurotransmitters. *Ann. Rev. Neurosci.* 2:447–65

Kyriacou, C. P., Hall, J. C. 1980. Circadian rhythm mutations in *Drosophila melanogaster* affect short-term fluctuations in the male's courtship song. *Proc. Natl. Acad. Sci. USA* 77:6729–33

Kyriacou, C. P., Hall, J. C. 1982. The function of courtship song rhythms in *Drosophila. Anim. Behav.* 30:794–801

Lewis, E. B., Bacher, F. 1968. Method of feeding ethyl methane sulfonate to *Drosophila* males. *Drosoph. Inform. Serv.* 43:193

Livingstone, M. S. 1981. Two mutations in *Drosophila* differentially affect the synthesis of octopamine, dopamine, and serotonin by altering the activities of two different amino acid decarboxylases. *Soc. Neurosci. Abstr.* 7:351

Livingstone, M. S., Sziber, P. P., Quinn, W.

G. 1982. Defective adenylate cyclase in the *Drosophila* learning mutant *rutabaga*. *Soc. Neurosci. Abstr.* 8:384

Livingstone, M. S., Tempel, B. L. 1983. Genetic dissection of monoamine neurotransmitter synthesis in *Drosophila*. *Nature* 303:67–70

MacLusky, N. J., Naftolin, F. 1981. Sexual differentiation of the central nervous system. *Science* 211:1294–1303

Mane, S. D., Tompkins, L., Richmond, R. C. 1983. Male esterase 6 catalyzes the synthesis of a sex pheromone in *Drosophila melanogaster* females. *Science*. In press

Manning, A. 1961. The effects of artificial selection for mating speed in *Drosophila melanogaster*. *Anim. Behav.* 9:82–92

Manning, A. 1967. Genes and the evolution of insect behavior. In *Behavior-Genetic Analysis*, ed. J. Hirsch, pp. 44–60. New York: McGraw-Hill. 522 pp.

Markow, T. A., Manning, M. 1982. Female olfaction, vision and reproductive isolation between *D. melanogaster* females and *D. simulans* males. *Drosoph. Inform. Serv.* 58:104–5

Marshall, T., Whittle, J. R. S. 1978. Genetic analysis of the mutation *Female-lethal* in *Drosophila melanogaster*. *Genet. Res.* 32:103–11

Marvell, A. 1681. To his coy mistress. In *Miscellaneous Poems* by A. Marvell. London: Robert Blouter

McGuire, T. R., Hirsch, J. 1978. Behavior-genetic analysis of *Phormia regina*: Conditioning, reliable individual differences, and selection. *Proc. Natl. Acad. Sci. USA* 74:5193–97

Médioni, J., Cadieu, N., Vaysse, G. 1978. Selection divergente pour le rapidite d'acquisition d'un conditionnement chez la Drosophile (*Drosophila melanogaster*.) *CR Soc. Biol.* 172:961–67

Médioni, J., Vaysse, G. 1975. Suppression of tarsal reflex in associative conditioning in *Drosophila melanogaster*—acquisition and extinction. *CR Soc. Biol.* 169:1386–91

Menne, D., Spatz, H. C. 1977. Color learning in *Drosophila*. *J. Comp. Physiol.* 114:301–12

Mobbs, P. G. 1982. The brain of the honeybee *Apis mellifera*. I. The connections and spatial organization of the mushroom bodies. *Philos. Trans. R. Soc. London Ser. B* 298:309–54

Nathanson, J. A. 1977. Cyclic nucleotides and nervous system function. *Physiol. Rev.* 57:157–256

Nelson, M. C. 1971. Classical conditioning in the blowfly (*Phormia regina*): Associative and excitatory factors. *J. Comp. Physiol. Psychol.* 77:353–68

Nottebohm, F. 1980. Testosterone triggers growth of brain vocal control nuclei in adult female canaries. *Brain Res.* 189:429–36

O'Brien, S. J., MacIntyre, R. J. 1978. Genetics and biochemistry of enzymes and specific proteins of *Drosophila*. In *Genetics and Biology of Drosophila*, ed. M. Ashburner, T. R. F. Wright, 2A:395–551. New York:Academic. 604 pp.

Pak, W. L., Grossfield, J., Arnold, K. S. 1970. Mutants of the visual pathway of *Drosophila melanogaster*. *Nature* 227:518–20

Parsons, P. A. 1974. Male mating speed as a component of fitness in *Drosophila*. *Behav. Genet.* 4:395–404

Pert, C. B., Snyder, S. H. 1973. Opiate receptor: Demonstration in nervous tissue. *Science* 179:1011–14

Proust, M. 1913. À la Recherche du Temps Perdu. Paris:Gallimard

Quinn, W. G., Harris, W. A., Benzer, S. 1974. Conditioned behavior in *Drosophila melanogaster*. *Proc. Natl. Acad. Sci. USA* 71:708–12

Quinn, W. G., Sziber, P. P., Booker, R. 1979. The *Drosophila* memory mutant *amnesiac*. *Nature* 277:212–14

Rasmussen, H., Goodman, D. B. P., Tenenhouse, A. 1972. The role of cyclic AMP and calcium in cell activation. *CRC Crit. Rev. Biochem.* 1:95–148

Richmond, R. C., Senior, A. 1981. Esterase 6 of *Drosophila melanogaster*: Kinetics of transfer to females, decay in females and male recovery. *J. Insect Physiol.* 27:849–53

Rodrigues, V., Siddiqi, O. 1978. Genetic analysis of chemosensory pathway. *Proc. Indian Acad. Sci. Sect. B* 87:147

Roeder, K. D. 1967. *Nerve Cells and Insect Behavior*. Cambridge, Mass: Harvard Univ. Press. 238 pp.

Ross, E. M., Howlett, A. C., Ferguson, K. M., Gilman, A. G. 1978. Reconstitution of hormone-sensitive adenylate cyclase activity with resolved components of the enzyme. *J. Biol. Chem.* 253:6401–12

Seamon, K. B., Padgett, W., Daly, J. W. 1981. Forskolin: Unique diterpene activator of adenylate cyclase in membranes and in intact cells. *Proc. Natl. Acad. Sci. USA* 78:3363–67

Shakespeare, W. 1608. *True Chronicle History of the Life and Death of King Lear and His Three Daughters*. London: Nathaniel Butter

Shorey, H. H., Bartell, R. J. 1970. Role of a volatile female sex pheromone in stimulating male courtship behavior in *Drosophila melanogaster*. *Anim. Behav.* 18:159–64

Siegel, R. W., Hall, J. C. 1979. Conditioned responses in courtship behavior of normal and mutant *Drosophila*. *Proc. Natl. Acad. Sci. USA* 76:3430–34

Spieth, H. T. 1952. Mating behavior within the genus *Drosophila* (Diptera). *Bull. Am. Mus. Nat. Hist.* 99:399–474

Spieth, H. T. 1974. Courtship behavior in *Drosophila. Ann. Rev. Entomol.* 19:385–405

Sturtevant, A. H. 1915. Experiments on sex recognition and the problem of sexual selection in *Drosophila. J. Anim. Behav.* 5:351–66

Szabad, J., Fajszi, C. 1982. Control of female reproduction in *Drosophila:* Genetic dissection using gynandromorphs. *Genentics* 100:61–78

Tempel, B. L., Bonini, N., Dawson, D. R., Quinn, W. G. 1983. Reward learning in normal and mutant *Drosophila. Proc. Natl. Acad. Sci. USA* 80:1482–86

Tempel, B. L., Livingstone, M. S. 1981. A mutation in *Drosophila* that reduces dopamine and serotonin synthesis abolishes associative learning. *Soc. Neurosci.* 7:351 (Abstr.)

Tempel, B. L., Livingstone, M. S., Quinn, W. G. 1983. Submitted for publication

Tempel, B. L., Quinn, W. G. 1982. Mutations in the dopa-decarboxylase gene affect learning but not memory in *Drosophila. Soc. Neurosci. Abstr.* 8:385

Thorpe, W. H. 1939. Further studies on preimaginal conditioning in insects. *Proc. R. Soc. London Ser. B* 127:424–33

Tompkins, L., Gross, A. C., Hall, J. C., Gailey, D. A., Siegel, R. W. 1982. The role of female movement in the sexual behavior of *Drosophila melanogaster. Behav. Genet.* 12:295–307

Tompkins, L., Hall, J. C. 1981 The different effects on courtship of volatile compounds from mated and virgin *Drosophila* females. *J. Insect Physiol.* 27:17–21

Tompkins, L., Hall, J. C. 1983. Identification of brain sites controlling female receptivity in mosaics of *Drosophila* melanogaster. *Genetics* 103:179–95

Tompkins, L., Hall, J. C., Hall, L. M. 1980. Courtship-stimulating volatile compounds from normal and mutant *Drosophila. J. Insect Physiol.* 26:689–97

Tompkins, L., Siegel, R. W., Gailey, D. A., Hall, J. C. 1983. Conditioned courtship in *Drosophila* and its mediation by association of chemical cues. *Behav. Genet.* In press

Uzzan, A., Dudai, Y. 1982. Aminergic receptors in *Drosophila melanogaster:* Responsiveness of adenylate cyclase to putative neurotransmitters. *J. Neurochem.* 38:1542–50

Venard, R., Jallon, J.-M. 1980. Evidence for an aphrodisiac pheromone of female *Drosophila. Experientia* 36:211–12

von Schilcher, F. 1976a. The role of auditory stimuli in the courtship of *Drosophila* melanogaster. *Anim. Behav.* 24:18–26

von Schilcher, F. 1976b. The function of pulse song and sine song in the courtship of *Drosophila melanogaster. Anim. Behav.* 24: 622–25

von Schilcher, F., Hall, J. C. 1979. Neural topography of courtship song in sex mosaics of *Drosophila melanogaster. J. Comp. Physiol. A* 129:85–95

Walters, E. T., Byrne, J. H. 1983. Associative conditioning of single sensory neurons suggests a cellular mechanism for learning. *Science* 219:405–8

Wise, R. A. 1978. Catecholamine theories of reward: A critical review. *Brain Res.* 152:215–47

Wright, T. R. F. 1977. The genetics of dopa decarboxylase and α-methyl dopa sensitivity in *Drosophila melanogaster. Am. Zool.* 17:707–21

Wright, T. R. F., Steward, R., Bentley, K. W., Adler, P. N. 1981. The genetics of dopa decarboxylase in *Drosophila melanogaster* III. Effects of a temperature sensitive dopa decarboxylase deficient mutation on female fertility. *Devel. Genet.* 2:223–35

Yanofsky, C. 1971. Tryptophan biosynthesis in *E. coli:* Genetic determination of the proteins involved. *J. Am. Med. Assoc.* 218:1026–35

Ann. Rev. Neurosci. 1984. 7:95–125
Copyright © 1984 by Annual Reviews Inc. All rights reserved

DEVELOPMENT OF THE SUPERIOR COLLICULUS

Barry E. Stein

Department of Physiology and Biophysics, Medical College of Virginia, Richmond, Virginia 23298

INTRODUCTION

Reference is often made to a study by Adamük (1872) to illustrate how long we have known that the superior colliculus (SC) plays a role in eye movements. The interest in the involvement of the SC in eye movements has not waned, but has actually increased as a result of recent work describing the activity of SC cells in behaving animals (for comprehensive discussions, see Sparks & Pollack 1977, Goldberg & Robinson 1978, and Wurtz & Albano 1980). In striking contrast to the decades of exploration of the role the SC plays in eye movements, and in vision in general, is the comparative recency of our appreciation of its multisensory roles. It is even difficult to point to a specific date when this began, because the view of the SC as a "visual" structure persisted even after anatomical studies demonstrated that it received nonvisual afferents (e.g. Marburg & Warner 1947, Poirier & Bertrand 1955, Anderson & Berry 1959, Mehler et al 1960). As late as 1962, the deep portion of the SC was described as containing " . . . cells that respond to extraoptic stimulation of unknown origin" (Altman & Malis 1962). Soon thereafter, however, demonstrations that SC cells could respond to nonvisual stimuli began to accumulate (Bell et al 1964, Jassik-Gerschenfeld 1965, Horn & Hill 1966), and now the involvement of the SC in auditory and somatosensory as well as visual functions is firmly established.

Many of the early studies that mention the SC do so in passing; seldom was it the major focus of attention. However, an upsurge of interest in the SC took place during the 1960s and has continued unabated since. Of considerable importance in initiating this interest was the description of the profound contralateral sensory neglect that occurred after destruction of the SC in cats

95

(Sprague & Meikle 1965). The most dramatic effect was a complete neglect of visual stimuli presented contralateral to the lesion despite the fact that the geniculocortical system was unimpaired. Sensorimotor dysfunctions of varying magnitudes have now been demonstrated in other species (Casagrande et al 1972, Goodale & Murison 1975, Schneider 1975, Albano et al 1982).

Since damage to the visual cortex and to the SC appeared to disrupt different visual capabilities, it was intimated or postulated that there were "two visual systems" (Sprague 1966, Diamond & Hall 1969, Schneider 1969). Although we now know that the SC and visual cortex are interrelated components of a complex whole, the concept of two visual systems was an impetus to compare the different properties of cells in both structures. Presumably, their different behavioral roles would be reflected in very different physiological properties. As a result, an enormous amount of information about the normal adult SC was generated, and its nonvisual as well as its visual properties became apparent. The magnitude of this interest is illustrated by the great number of review articles that have appeared since 1970 (e.g. Kruger 1970, McIlwain 1972, Palmer et al 1972, Schiller 1972, Sprague 1972, Gordon 1975, Schneider 1975, Sparks & Pollack 1977, Goldberg & Robinson 1978, Stein 1978a, Coulter et al 1979, Wurtz & Albano 1980, Huerta & Harting 1982).

Of primary concern in this discussion is how this adult SC develops. Data generated in developmental studies can be useful in relating the ontogeny of anatomy, physiology, and behavior, and may give us critical insight into how experience influences the development of the sensorimotor skills necessary for survival. Since the cat SC appears to be a reasonable model to use, and a good deal of the developmental work has been done in cat, this review leans heavily on data generated from this species. A brief description of each of the major features of the adult SC is provided first as a foundation for later discussions of development.

ORGANIZATION OF THE ADULT SUPERIOR COLLICULUS

Laminar Organization

This SC is composed of seven alternately fibrous and cellular laminae. From top to bottom, these are

1. stratum zonale
2. stratum griseum superficiale
3. stratum opticum
4. stratum griseum intermediale
5. stratum album intermediale
6. stratum griseum profundum
7. stratum album profundum.

Yet, despite the fact that even sublaminae can be distinguished (Kanaseki & Sprague 1974), the SC is often operationally divided into only two parts: superficial (I–III) and deep (IV–VII). This bipartite division is a result of differences between the superficial and deep laminae in cellular morphology, afferent-efferent connections, physiology, and behavioral involvement (for a more detailed discussion see Casagrande et al 1972, Harting et al 1973, Edwards 1980, Stein & Gordon 1981).

Deep Laminae and the Reticular Formation

Distinguishing the border between superficial and deep laminae is comparatively easy because of the relative paucity of cell bodies in the stratum opticum. However, drawing a boundary between the deepest laminae of the SC and the subjacent tegmentum (reticular formation) is a much more difficult task, and is sometimes estimated on the basis of a blood vessel that forms an "artificial" boundary between the stratum album profundum and the reticular formation (Edwards 1980).

The arbitrary separation of the deep laminae of the SC from the reticular formation has been emphasized by Edwards (1980). Because the deep SC and the reticular formation are similar in cellular morphology, physiological characteristics, and connections, Edwards (1980) suggested classifying the deep SC as a part of the reticular core. Certainly the observation that it is only when the deep layers of the SC are damaged that visuomotor deficits appear (Casagrande et al 1972) is a compelling reason to examine further the distinct properties of superficial and deep SC. However, the actual conceptual separation of the superficial and deep SC is not without serious problems. First, there are many physiological similarities between the visual cells above and below the stratum opticum, largely because both superficial and deep layer visual cell populations depend on corticotectal afferents for the same receptive field properties (see *Influence of Corticotectal Development*). As yet, equivalent dependencies have not been demonstrated for reticular cells. Second, the overall visual topography does not change below the stratum opticum as one might expect if these were unrelated structures. Third, the layering pattern of the SC as it is currently defined conforms more closely to its nonmammalian homologue, the optic tectum, than it would if the deep laminae were excluded. There is little doubt about the anatomical integrity of the optic tectum because it is clearly separated from underlying tissue by a large ventricle. Yet many of the reasons for distinguishing between superficial and deep SC are applicable to the optic tectum (see Stein & Gaither 1981, 1983). Thus, any review of the classification of the deep SC should be accompanied by a similar evaluation of the deep optic tectum.

Sensory Topographies

Although more is known about the visual than the nonvisual properties of the cat SC, this structure receives a substantial somatosensory and auditory input

(Garey et al 1968, Diamond et al 1969, Paula-Barbosa & Sousa-Pinto 1973, Stewart & King 1963, Tamai 1973, Baleydier & Mauguiere 1978, Blomqvist et al 1978, Edwards et al 1979, Kawamura & Konno 1979, Nagata & Kruger 1979, Tortelly et al 1980, Huerta et al 1981, Clemo & Stein 1981, Ogasawara & Kawamura 1982, Stein et al 1983). Whereas visual cells are most heavily represented in superficial laminae, many visual cells are also located in upper portions of the deep SC. In contrast, somatosensory and auditory cells are restricted to deep SC, and many cells here can respond to more than one sensory modality (Straschill & Hoffmann 1969, Gordon 1973, Stein & Arigbede 1972b, Abrahams & Rose 1975, Stein et al 1976b).

Despite this partial lamina segregation of the sensory representations, their organization is remarkably similar in that each representation is primarily contralateral and topographic. The topography results from converging ascending and descending (corticotectal) projections from several areas. Thus, the nervous system imposes a high degree of order on the manner in which afferents terminate in the SC. This has been demonstrated most clearly in the visual and somatosensory representations, where many SC cells receive convergent ascending and descending inputs from cells with similar receptive field positions (McIlwain & Fields 1971, Updyke 1977, Clemo & Stein 1981).

The individual sensory representations are in close topographic register with one another in cat (Gordon 1973, Stein et al 1975, 1976b, Harris et al 1980) and in rodents (Dräger & Hubel 1975, Chalupa & Rhoades 1977, Finlay et al 1978, Stein & Dixon 1979, Palmer & King 1982). The presence of this topographical register in the SC is a consequence of each of the individual maps having the same alignment of axes: superior-inferior axes are laid out along the medial-lateral SC, while their nasal-temporal (or rostral-caudal) axes are anterior-posterior. Figure 1 illustrates the topographical register of the visual and somatosenory representations in cat. As soon as one lowers a recording electrode into the lateral portion of the caudal SC (left), visual cells with receptive fields in right inferior temporal visual space will be found (Apter 1945, Feldon et al 1970, Kruger 1970, Berman & Cynader 1972, Lane et al 1974, McIlwain 1975). Similar results will be seen by sampling the visual cells somewhat deeper in the SC (though their receptive fields will be larger), but now somatic and auditory cells will also be encountered. The receptive fields of these nonvisual cells will be in corresponding zones: auditory receptive fields will be in the inferior and temporal aspect of auditory space, while somatic receptive fields will be located on lower portions of the contralateral (right) body, such as the limb.

It seems likely that this scheme of sensory register antedates the radiation of early mammals, as it has been demonstrated in the optic tectum of reptiles (Gaither & Stein 1979, Stein & Gaither 1981, Hartline 1983), amphibians

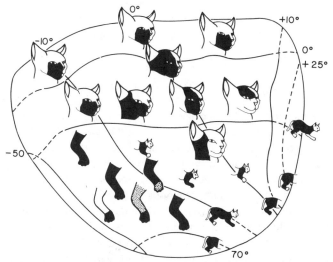

Figure 1   A schematic diagram of the dorsal surface of the left superior colliculus. The horizontal and vertical visual meridians are marked, respectively, by the vertical and horizontal degree lines, and the area anterior (rostral) to the vertical meridian illustrates the representation of the ipsilateral half-field. Note the nonlinearity of the visuotopic representation, with disproportionately large areas devoted to parafoveal and inferior temporal visual space. A summary of representative body sectors encountered in different areas of the superior colliculus is also presented. This figure minimizes the overlap of body sectors by the choice of body sectors most heavily represented in a given area. Note the nonlinearity of the somatotopy, with disproportionately large areas devoted to trigeminal and forelimb representations. These are the same areas devoted to parafoveal and inferior temporal visual space. (From Stein et al 1976b.)

(Harris 1982), fish (Bastian 1982), and birds (Cotter 1976, Ballam 1982, Knudsen 1982), and appears to be independent of phyletic level.

Motor Topographies

The deep laminae cells of many species send descending efferents to a variety of regions of brainstem and spinal cord involved in orienting the eyes, pinnae, vibrissae, head, and limbs (Martin 1969, Kawamura et al 1974, Graham 1977, Edwards & Henkel 1978, Henkel & Edwards 1978, Kawamura & Hashikawa 1978, Coulter et al 1979, Weber et al 1979, Holcombe & Hall 1981, Huerta & Harting 1982). This deep layer "motor" organization is also topographic and in register with the sensory representations in the SC. The functional relationship between the sensory and motor maps in the cat SC is apparent if one delivers an electrical stimulus through an electrode after first recording from it to determine its position in the sensory topographies of the SC. This focal electrical stimulation will produce eye movement that brings the area centralis to bear upon the point in space corresponding to the point in the visuotopic map that was stimulated (Syka & Radil-Weiss 1971, Stein et al 1976a, Harris 1980; but also see Straschill & Rieger 1973, Roucoux & Crommelinck 1976). Similarly, the contralateral pinna will move to "fixate" upon a presumptive auditory stimulus originating from the same location in space (Stein & Clamann 1981). Once again, this motor response to electrical stimulation of the SC occurs in other mammals (Schaefer 1970, Robinson 1972, McHaffie & Stein 1982).

The organizational covariance of sensory and motor aspects of the SC hardly seems coincidental. It is reasonable to suppose that the overlap of sensory and motor topographies is an adaptive characteristic, perhaps because it is the simplest and most economical way to integrate multiple sensory cues and produce the appropriate orientation of several receptor organs simultaneously. Yet retaining topographical register among maps could be a complex problem if a significant shift in gaze occurred and was maintained without a corresponding shift of the head and forequarters. However, cats tend not to maintain significant eccentric deviation of the eyes (perhaps for this reason) (Harris et al 1980). There is also recent evidence from work with monkeys that the maps may not be static, and that a shift in the position of the eyes produces a corresponding shift in the auditory map (Jay & Sparks 1982). The details of this "dynamic" mapping still must be worked out.

A topographical scheme such as the one in the superior colliculus may enable each sensory input to have access to the efferents controlling the various receptor organs directed by the SC. In this way, a unimodal (e.g. tactile) sensory cue can produce an orientation of the eyes, pinnae, head, and limbs. This serves the dual purpose of directing different receptor organs toward a stimulus to determine its nature, and of placing the organism in a better position to respond to it.

Visual Receptive Field Properties

Visual cells have been studied in the cat SC in many laboratories under a variety of conditions (e.g. Marchiafava & Pepeu 1966, Straschill & Taghavy 1967, McIlwain & Buser 1968, Sprague et al 1968, Sterling & Wickelgren 1969, Berman & Cynader 1972, Stein & Arigbede 1972a, Dreher & Hoffmann 1973, Harutiunian-Kozak et al 1973, Mandl 1974, Rizzolatti et al 1974, Mason 1979, Syka et al 1979, Pinter & Harris 1981), yet there is remarkable consistency in the descriptions of their physiological properties. The most striking characteristics of SC cells are that they

1. prefer moving stimuli as opposed to stationary flashed light;
2. are optimally activated by stimuli moved at comparatively low velocities;
3. prefer stimuli moved in specific directions across the receptive field (directionally selective);

4. receive inputs from both eyes (binocular);
5. have receptive fields that are not systematically divided into separate "on" and "off" regions;
6. have receptive fields that are larger than those found in the geniculocortical system;
7. exhibit suppressive flanks bordering the receptive field;
8. are optimally excited by stimuli much smaller than the borders of their excitatory receptive fields;
9. respond in transient fashion even to maintained stimuli;
10. have a tendency for "habituation" following repeated stimulus presentations even at long interstimulus intervals.

Of course, not every cell exhibits each of these features; some properties are more likely to be found in either superficial or deep layer cells. For example, cells with the largest receptive fields, and those most prone to habituation, are concentrated in the deeper layers.

One of the most intriguing mysteries in neurophysiology is how the integration of afferent inputs produces a "new" set of physiological characteristics. Many of the properties detailed above are the result of the convergence of retinal and corticotectal (as well as other) afferents, and it is now apparent that the presence of certain properties in SC cells is related to the influence of a given set of afferents. This is particularly evident in the case of the movement selectivity of visual SC cells and is dealt with in more detail below (see *Corticotectal Development*).

Nonvisual Receptive Field Properties

The characteristics of nonvisual cells in the SC have only recently begun to be studied in detail. This is due in part to the greater emphasis usually placed on the visual role of the SC. In addition, there are greater inherent difficulties in studying somatosensory and auditory cells of the SC with the same types of quantitative methods used to study its visual cells.

We do know, however, that somatosensory cells are primarily cutaneous (Dräger & Hubel 1975, Stein et al 1975, 1976b, Nagata & Kruger 1979, Stein & Dixon 1979) and respond best to rapid movements along particular axes (Clemo & Stein 1982a). Cells with nociceptive characteristics have been described in hamsters (Stein & Dixon 1978, Rhoades et al 1982), and, whereas it is likely that similar cells exist in other species, none has yet been described.

Auditory cells appear to be more sensitive to complex sounds than to pure tones (Stein & Arigbede 1972b, Gordon 1973, Wise & Irvine 1983), and species-specific sounds have been shown to be most effective in squirrel monkey (Allon & Wollberg 1978).

Both auditory and somatosensory cells resemble visual cells in that transient responses are evoked even when the stimulus is maintained. They seem most similar to their neighboring deep lamina visual cells in that their receptive fields are often large and they habituate rapidly.

The receptive field characteristics of these nonvisual cells are undoubtedly due to the integration of their various afferents. At present, though, we do not know which afferents are necessary for which properties. In fact, even the source of somatic corticotectal projections in cat has recently been an issue requiring clarification (Clemo & Stein 1981, 1982b, 1983). But despite these gaps in our knowledge, the existing information about afferent integration in the visual representation provides a workable conceptual framework for the investigation of nonvisual afferent integration.

Multimodal Integration

Most investigators studying the properties of deep SC or tectal cells have noted that many cells respond to more than one sensory modality, and there is now a growing interest in how such cells may facilitate the sensorimotor role of the SC (tectum). In cat, hamster (Meredith & Stein 1983), and rattlesnake (Newman & Hartline 1981), the simultaneous input from two modalities can either enhance or depress the activity of midbrain cells. Apparently, by pooling sensory inputs, many midbrain cells can amplify the effects of subtle environmental cues in certain conditions, whereas in others, reponses to normally effective stimuli can be blocked. These processes may in turn facilitate or inhibit overt responses, and they reflect the dynamic multisensory integrative nature of SC cells.

DEVELOPMENT OF THE SUPERIOR COLLICULUS

Development of Sensory Topographies

As noted above, the presence of a topographic organization of sensory and motor representations is an integral feature of the SC and is likely to be of critical importance in the ability of the structure to facilitate the remarkably accurate orientation and localization capabilities of many species. Yet we know little about its functional maturation. Fortunately, however, the advent of modern neuroanatomical techniques has led to a number of excellent and detailed anatomical descriptions of the pre- and postnatal maturation of developing sensory pathways, particularly of retinal projections (e.g. Currie & Cowan 1975, Rakic 1977, Cavalcante & Rocha-Miranda 1978, So & Schneider 1978, Frost et al 1979, Land & Lund 1979, Williams & Chalupa 1982, Shatz 1983).

Using these neuroanatomical methods in cat, Williams & Chalupa (1982) and Graybiel (1975) have shown that adult-like patterns of retinocollicular projections are already present several days prior to parturition (63–67 days after conception). These adult patterns include a segregation of contralateral and ipsilateral inputs into bands (Graybiel 1975, Harting & Guillery 1976) and the absence of ipsilateral inputs from the rostral and caudal collicular poles. Achieving this pattern involves a curious process of growth and retraction. By prenatal day 38 both retinas appear to project across the entire collicular mantle without any obvious ocular segregation; by prenatal day 56 the ipsilateral projection has retracted considerably, and by day 61 the organization of the ipsi- and contralateral inputs looks very much like that seen in the adult (Williams & Chalupa 1982). A similar process has been observed in the monkey (Rakic 1977), hamster (Frost et al 1979), rat (Land & Lund 1979), and opposum (Cavalcante & Rocha-Miranda 1978). Although it is not yet clear why this developmental process occurs, several possible operational factors have been pointed out (Williams & Chalupa 1982). These include maturational changes in axoplasmic transport, restriction of axonal terminals (see Hubel et al 1977, LeVay & Stryker 1979, Mason 1980), elimination of collaterals (see Land & Lund 1979, Innocenti 1981, Ivy & Killackey 1981), and degeneration of ganglion cells (see Kuwabara & Weidmann 1974, Frost et al 1979, Cunnigham et al 1981, Jeffery & Perry 1981, Sengelaub & Finlay 1981).

Even during embryological stages, the direct retinal input to the SC appears to be restricted to its superficial layers. Presumably, the deep lamina synaptic loci are occupied by developing somatosensory and auditory afferents, although little is known about nonvisual afferents to the SC during pre- or early postnatal maturation. In recent experiments in our laboratory, we have observed that trigeminocollicular (McHaffie et al 1983) and somatosensory corticotectal projections from the fourth somatosensory cortex (SIV) and its surround (Stein & Kruger 1983) are already present at birth, and that these afferents appear to have many of the same organizational features that characterize them in adulthood.

Just what factors account for this lamina segregation of developing afferents is not yet clear; however, it is known that at least one important factor is the competition between afferents for available synaptic space. Removal of one eye during pre- or early postnatal life alters the projection from the intact eye to the SC in a variety of species (Lund & Lund 1971, Frost & Schneider 1976, Land et al 1976, Sanderson et al 1978, Finlay et al 1979, Godement 1979, Land & Lund 1979, Lent & Mendez-Otero 1980, Rhoades & Chalupa 1980), producing an anomalous visual map in the SC. Enucleation also has been reported to increase the number of fibers maintained in the remaining optic nerve (Williams et al 1983). The afferent plasticity in visual projections led to the suggestion that removal of the eye might even alter the normal distribution and lamina segregation of nonvisual afferents (Chow et al 1973, Lund 1978). That suggestion was recently confirmed (see Dräger 1977, Rhoades et al 1981).

Unilateral enucleation in neonatal hamsters has been shown to produce an expansion of somatosensory afferents into the superficial laminae of the SC—an obvious departure from the normal terminal pattern of these inputs. However, the reorganization of somatosensory inputs did not violate the normal topographic pattern in the SC (Rhoades et al 1981) as might be expected on the basis of similar experiments detailing visual afferent reorganization. This is consistent with the observations of others that tectal maps with normal orientations can develop in the absence of a retinal projection (O'Leary & Cowan 1980, Constantine-Paton & Ferrari-Eastman 1981).

In the normal course of events, the prenatal maturation of visual inputs to the SC proceeds to the point where an adult-like organization is present even before birth. This is in striking contrast to the functional immaturity of SC visual cells at this time. Physiological studies have not yet adequately addressed the question of the functional maturation of the visuotopic map in the SC. However, behavioral experiments indicate that the orientation response to visual cues that occurs earliest in development is to stimuli falling on the central retina, followed by a gradual expansion of the effective visual field (Sireteanu & Maurer 1982). These behaviors are dependent on the SC and are eliminated by its destruction (Norton & Lindsley 1971). The assumption that these orientation responses are limited by the immaturity of the sensory properties of SC cells leads one to conclude that the visuotopic map develops in similar fashion. However, until there is more complete information about functional efferent maturation of SC cells, such a conclusion must remain tentative.

Data regarding the maturation of the nonvisual SC maps are even more scanty. However, the appearance of cells responsive to somatosensory and auditory stimuli (all of which are in deep laminae), coupled with the early functional maturation of at least some deep lamina efferents (see below), indicate that the deep laminae mature more rapidly than the superficial laminae.

Development of Motor Topographies

For early sensory cells in the SC to have any functional value for the neonate, there must be a way for them to influence other structures via efferents. In order to evaluate efferent maturation of the SC, my colleagues and I (Stein et al 1982) used autoradiographic and anterograde horseradish peroxidase tracing techniques in kittens six hours to five weeks of age. We found that even in the youngest animals studied, the major medial (predorsal bundle) and lateral projections from the SC were heavily labeled (Figure 2). In addition, terminal labeling was seen in each of the efferent targets of the SC thought to be involved in eye movements, including the ventral central gray matter overlying the oculomotor nucleus and those portions of the pontine and medullary reticular formation that provide excitatory and inhibitory inputs to the abducens nucleus. Further analysis of this tissue showed the presence in the neonates of the descending projection to areas of the reticular formation that project to the

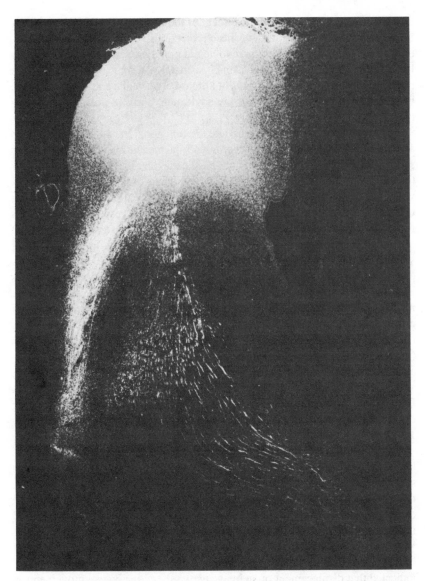

Figure 2   A low magnification darkfield photomicrograph of the caudal portion of the left superior colliculus and subjacent tegmentum. An injection of [³H] leucine was made in the colliculus at one day postnatally. Note the well-developed crossed (medial) and uncrossed (lateral) efferent projection systems of the colliculus at this age. (From Stein et al 1982.)

cervical spinal cord to control movement of the head and limbs (see Huerta & Harting 1982 for a description of these projections in adult cats), and projections to the paralemniscal region involved in pinna movements (see Henkel & Edwards 1978, Henkel 1981). Some postnatal anatomical maturation was observed in some of these pathways of kittens between one and six days of age; however, they seemed much like the adult quite early in postnatal development.

At least some of the efferents are already functional in two-day-old animals and presumably earlier, so there is already a means by which the earliest sensory SC cells (see below) may influence behavior. Electrical stimulation of the SC of two-day-old animals produces movements of the eyes, pinnae, neck, vibrissae, and limbs (Stein et al 1980), although the thresholds for evoking these movements are higher than in the adult and the response reliability is lower. However, the evoked movements already show some topographical organization: stimulating the left SC produces movements to the right, and stimulation of the right SC produces movements to the left. The eye movements evoked from homotopic loci are approximate mirror images of one another, as in adult cats (Stein et al 1976a). But unlike the results from studies with adults, many stimulation loci within the neonatal SC were observed to be without effect. With eye movements, this failure appeared to be due to SC immaturity, because stimulation of the oculomotor nucleus always evoked eye movements (though thresholds here, too, were higher than in adults). Some of the maturational changes in eye movements may have been due to mechanical factors, since the size and orientation of the eyes change considerably during the first few weeks of life (Sherman 1972, Olson & Freeman 1978), and the contractile characteristics of the extraocular muscles also change (Lennerstrand & Hanson 1978a,b).

The presence of functional efferents in the SC of neonatal kittens has important implications for the topographical and functional organization of this structure. Sparks & Porter (1983) and Jay & Sparks (1982) have suggested that the SC is organized in motor rather than sensory coordinates. Since the development of a motor map may precede the sensory maps, perhaps it has an influence on early sensory organization.

The presence of a motor map, albeit an incomplete one, also has important implications for theories of visuomotor development. Hein et al (1979) have indicated that displacement of an image across the retina by eye movements is necessary for visuomotor development. These self-generated eye movements appear to be necessary to "make sense" of the visual input, and the demonstration that eye movements can be evoked from the SC before cells respond to natural visual stimuli is consistent with this hypothesis. It would be of great interest to determine which stimuli can evoke eye movements in neonatal kittens, for a stimulus must first induce eye movements if ever motor commands are later to be associated with visual cues, an association critical for normal visuomotor integration. Since the different sensory modalities represented in the SC are thought to have access to the same motor outputs that orient receptor organs, evoking eye movements via the early-appearing nonvisual sensory cells could set the stage for later visual-motor associations. This seems reasonable since both auditory and somatosensory stimuli are capable of driving SC cells before visual stimuli are (see below). On the other hand, it is possible that this process occurs on the basis of spontaneous eye movements, or via vestibular stimulation. In fact, vestibular cues have already been shown to be capable of producing eye movements in neonatal cats (Fish & Windle 1932).

Development of Somatosensory and Auditory Cells

In an early study of the development of neuronal properties of the SC, we (Stein et al 1973a,b) recorded from cells in kittens ranging in age from late fetal stages to two months. Unlike findings in the adult, no difference in neuronal activity marked the entrance of an electrode into the SC of kittens less than one week of age. In fact, there was little ongoing electrical activity at all. Instead, the superficial laminae appeared to be remarkably "silent" except for an occasional spontaneously active cell, and even in these rare examples, the activity was atypical in that discharges were at unusually long intervals and had long spike durations. Visual stimuli were ineffective in evoking discharges in any of these cells.

Deep lamina cells were also unusually inactive, and although more cells exhibited spontaneous activity here, many could not be driven by sensory stimuli. However, of those that were activated by sensory stimuli, nearly all were somatosensory cells, and the stimuli were capable of evoking discharges in some deep lamina cells even in fetal animals. The responses of these early somatosensory cells, though, consisted of comparatively few impulses, even to stimuli that are found to be optimal in adults. Furthermore, responses fatigued rapidly, and an interval of several seconds between stimuli was required for activation on successive stimulus presentations. Despite their sluggishness and tendency for response fatigue, the cells did not appear to have any obvious receptive field features that distinguished them from their adult counterparts. This may have been due to the comparative crudeness of the tests. For example, we now know that somatosensory cells in the SC are preferentially activated by high-velocity stimuli (Stein et al 1976b, Nagata & Kruger 1979) moved along a particular axis (Clemo & Stein 1982a). However, no tests of velocity or axis preferences were administered in the early study, nor was it possible to evaluate whether or not a topographic organization of the somatosensory representation was present at this time.

The presence of a somatosensory representation in the SC of newborn kittens gives rise to some intriguing speculation about its functional role at this time.

One possibility is that this representation is important in neonatal orientation and localization behavior. That a newborn kitten can orient to the mother and nipple is obvious, but whether the neonate uses tactile cues in this orientation behavior has been open to some controversy. Early investigations in rat indicated that olfactory, not tactile, cues are critical for locating the nipple and suckling, so that removing olfactory cues from the nipple interferes with nipple attachment in these animals (Blass et al 1977). Meanwhile, recent data in cat (Larson & Stein 1982) and rat (Kenyon et al 1982) illustrate the critical nature of tactile cues. Larson & Stein (1982) found that anesthetizing the vibrissal pads eliminates the newborn kitten's ability to locate the nipple, even when it is placed on the mother's dorsum. Yet when the kitten is placed several centi-meters from the mother, it locates her and crawls to her as efficiently as it did before. On the other hand, minimizing olfactory cues severely disrupts the kitten's ability to locate the mother from a distance, but when placed on her dorsum the kitten has no difficulty locating the nipple. In both cases there is no interference with the motor act of suckling. Apparently, the initiation of suckling is an integrated series of behaviors that consists of first locating the mother (olfactory cues) and then locating the nipple (tactile cues) and suckling. While it is not yet clear that the latter stages of this behavior sequence involve the circuitry of the SC, the presence of cells responsive to perioral tactile stimuli in neonatal kittens, coupled with the presence of some functional SC efferents in these animals, make this a reasonable possibility.

It appears that deep layer cells responsive to auditory stimuli develop later than the somatosensory cells. The earliest auditory cell in the SC was not recorded until five days of age (Stein et al 1973a), and their incidence increased gradually thereafter. Although the developmental sequence of auditory cells correlates with the chronology of sensory cues involved in orientation (Fox 1970, Villablanca & Olmstead 1979), we still have no information regarding the maturation of the interaural characteristics that would facilitate auditory localization.

Development of Visual Cells

Despite observations that electrical activity can be evoked in the visual cortex of kittens two to four days of age (Huttenlocher 1967, Rose & Lindsley 1968), single-unit visual activity in the SC does not begin until seven days of age (Stein et al 1973a,b). Thus, the developmental sequence of sensory cells in the SC is somatosensory first (prenatal), auditory second (five or more postnatal days), and then visual (one postnatal week), a chronology that parallels the maturation of the sensory cues used in orientation (Fox 1970, Norton 1974, Villablanca & Olmstead 1979). [A similar chronology is apparent in the cat cortex as evi-denced by electrocortical maturation (Marty & Scherrer 1964).] In the SC of neonatal rabbits (Spear et al 1972) the earliest visually responsive cells appear at approximately the same time as in cat, but unlike cat, many of these cells show mature receptive field properties within the second week. Still, full development in both rabbit (Fox et al 1978) and cat (Stein et al 1973a,b, Norton 1974, Stein & Gallagher 1981) requires a protracted postnatal developmental period.

The properties of the earliest visual cells in the kitten SC (7–9 days of age) are strikingly different from those observed in adults. Characteristically, these early cells

1. exhibit strong preferences for stationary flashed light;
2. often require long stimulus durations (e.g. 400 msec) and high intensities;
3. prefer moving stimuli at much lower velocities than are effective in adults;
4. are influenced by the contralateral eye only (monocular);
5. show an absence of directional selectivity;
6. have receptive fields that are not systematically subdivided into "on" and "off" regions;
7. show receptive fields larger than those in adults;
8. appear to have no suppressive flanks bordering the receptive field;
9. require optimal stimuli larger than those in adults;
10. display transient responses even to maintained stimuli;
11. show more pronounced response habituation than adults;
12. have very long latencies—always longer than 100 msec and sometimes exceeding 1 sec;
13. exhibit lower response reliability and fewer impulses/response than adults.

Apart from characteristics 6 and 10, the physiological profile of kitten SC cells is different from that of the adult cat. The first binocular neuron was recorded in a ten-day-old kitten, followed by a rapid rise in the proportion of binocular cells. Similarly, as moving targets became more effective in eliciting discharges, stationary flashed stimuli became less effective (Figure 3). Directional selectivity was first apparent toward the end of the second postnatal week and increased rapidly thereafter. Latencies to visual stimuli decreased markedly between three and eight weeks; this is probably a reflection of myelination in the optic pathways (Moore et al 1976). Other properties of adult SC cells, like suppressive flanks and greater responsivity, also develop during the six–eight-week postnatal maturational period.

During the first two postnatal months of life there is a concomitant change in visually guided behavior. Although demonstrating a causal relationship between specific visual capacities and specific neuronal response properties in a given structure has obvious inherent difficulties, there is correlational evidence for parallel development (Norton 1981).

INFLUENCES OF RETINAL DEVELOPMENT   Undoubtedly, some of the difficulties in activating visual cells in young animals result from the poor optical quality of the kitten eye (Thorn et al 1976, Bonds & Freeman 1978). However, it is unlikely that these factors contributed significantly to estimates of the maturation of complex receptive field properties because an admixture of cells with mature and immature properties has often been observed in the same neonatal animals and at the same retinal eccentricities (Stein et al 1973a,b, Daniels et al 1978).

On the other hand, many of the immature response properties of SC cells appear to reflect the immaturity of retinal ganglion cells. Rusoff & Dubin (1977) have shown that retinal ganglion cells in the peripheral retina of three-week-old kittens are very large, lack surround inhibition, and often respond inconsistently. More mature properties are exhibited in four-week-old animals, but the adult-like condition is not achieved until considerably later. The central retina appears to mature more rapidly than the peripheral retina (Donovan 1966, Johns et al 1979, Stone et al 1982), but even at three weeks of age central ganglion cells have long latencies and weak responses (Hamasaki & Flynn 1977).

The immaturity of ganglion cells in neonatal kittens is reflected in the lateral geniculate nucleus (lgn) as well as the SC; a particularly long maturational period has been noted for Y-cells in the lgn by Daniels et al (1978), and for particular properties of Y-cells in the medial intralaminar nucleus by Wilson et al (1982). However, these are not strictly a reflection of ganglion cell immaturity, for at certain ages, some Y-cells had more immature properties than would be expected on the basis of the maturity of their ganglion cell inputs (Daniels et al 1978). Furthermore, Hamasaki & Sutija (1979) found many Y-cells in the retina even in reasonably young kittens. The immaturity of Y-cells has special relevance for the SC, since it receives a Y-cell input directly from the retina, and indirectly via the lgn and visual cortex (Hoffmann 1973).

INFLUENCES OF CORTICOTECTAL DEVELOPMENT   This indirect Y-cell pathway is of particular importance for the appearance of complex receptive field properties in SC cells. When Stein & Magalhãs-Castro (1975) ablated visual cortex in neonates, the maturation of many complex properties in the ipsilateral SC was permanently arrested. Thus, later, when recordings were made from these animals when they were physically mature, the percentage of directionally selective, binocular, and movement-sensitive cells was markedly lower than in adults. In addition, surround inhibition never reached adult levels. (Similar observations have been made by others, e.g. Berman & Cynader 1976, Mize & Murphy 1976, Flandrin & Jeannerod 1977). In short, the properties of neurons in the colliculi of these animals resembled those of neonates. Similarly, ablation of visual cortex in adult cats renders SC cells unspecialized so that they, too, resemble those of neonates (Wickelgren & Sterling 1969a, Palmer et al 1972). The profound influence of visual cortex on SC cells is further illustrated by experiments in which visual cortex is reversibly deactivated by cooling. In these studies the cooling of visual cortex depressed or completely eliminated the responses of many SC cells to visual stimuli (Wickelgren & Sterling 1969a, Stein & Arigbede 1972b, Stein 1978b, Ogasawara et al 1983a).

This evidence for a functional corticotectal input as a prerequisite for the specialized properties of SC cells made it seem likely that much of the maturation of these properties was, in fact, a reflection of corticotectal development. In an attempt to trace the development of corticotectal pathways, Stein & Edwards (1979) made injections of radioactive leucine in the visual cortex of kittens 6 hours to 12 days of age. They found that a dense corticotectal projection was already present in the youngest animals studied, and noted no increase in this projection in older animals. Although some corticotectal projections have been demonstrated in prenatal animals by the use of degeneration methods (Anker 1977), the apparent completeness of the projection observed using autoradiographic methods in two-day-old neonates was surprising, given the total absence of visually active cells in the SC at this time. A further conceptual difficulty was the demonstration by Hubel & Wiesel (1963) of adult-like properties in many visual cortical cells in eight-day-old kittens, a time when visual SC cells are remarkably unspecialized. However, others have since shown that immature cells exist in visual cortex at this time as well (Pettigrew 1974, Blakemore & Van Sluyters 1975).

These difficulties have recently been resolved by using the method of reversible cortical deactivation (cooling) to evaluate the maturation of corticotectal influences in kittens from eight days to eight weeks of age (Stein & Gallagher 1981). In animals 8–13 days of age, the cooling of the visual cortex rarely influenced the firing rates of visual SC cells despite the presence of the anatomically well-developed corticotectal projection. However, at the end of the second week of life and thereafter, increasing proportions of SC cells were depressed by this procedure (Figure 3A). At the same time, there was a similar increase in the proportion of cells exhibiting binocularity and directional selectivity (Figure 3B). In fact, the presence of these two receptive field characteristics in any given SC cell was a reliable predictor that the cell would be affected by cortical cooling regardless of the age of the kitten in which it was recorded (Figure 3C). Therefore, a cell exhibiting these characteristics in a 13–14-day-old kitten stood essentially the same chance of being affected by cortical cooling as a cell with these charcteristics in an adult cat. The profound inhibition of these cells induced by cortical cooling indicates that they owe not

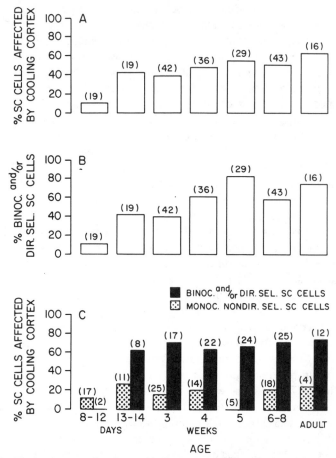

Figure 3   The development of corticotectal influence on superior colliculus cells. In each graph, the number in parenthesis above a bar represents the sample size in each case. In A, the percentage of cells affected by cortical cooling is plotted as a function of age. Note the dramatic increase in the percentage of cells affected by cortical cooling at 13–14 days of age and more gradual and irregular changes thereafter. In B, the percentage of binocular and directionally selective cells shows a similar developmental course. These changes in A and B are not independent. As shown in C, the presence of binocularity or directional selectivity is a reliable predictor that a given superior colliculus cell will be affected by cortical cooling regardless of age. Thus, at least 62% of the cells with these properties were depressed by cortical cooling at 13–14 days of age (when these properties were first evident) and this percentage changed very little during later maturation. (From Stein & Gallagher 1981.)

only their specialized receptive field characteristics to the cortex, but also their ability to respond vigorously (or at all) to any visual stimulus. This was true of both superficial and deep lamina cells.

These data indicate that the visual corticotectal system does not begin functioning until nearly two weeks of age. It is interesting to note that no evidence was obtained to indicate a gradual increase in the influence of cortex over individual SC cells. Rather, it seemed as if a "gate" suddenly opened allowing cortex to determine the receptive field characteristics and activity of a given SC cell in an "all-or-none" fashion. Because this point was reached at different times in different SC cells during the eight-week maturational period, it appeared as if the development of the corticotectal system as a whole was protracted.

It is important to determine where this critical postnatal change in the corticotectal system takes place to allow it to control visual SC cells. Presently, we can point to two obvious loci: the cortical cell itself and the corticotectal synapse. Perhaps the adult-like cells described in the cortex of the eight-day-old kitten by Hubel & Wiesel (1963) are not those specialized lamina V cells that project to the SC (Palmer & Rosenquist 1974, Gilbert & Kelly 1975). Many cortical cells are now known to exhibit a comparatively long developmental time course (Pettigrew 1974, Blakemore & Van Sluyters 1975). These observations are consistent with the long period of postnatal synaptogenesis described in visual cortex by Cragg (1975).

Until there is more information about the maturation of specific cortical cell types (e.g. corticotectal, corticothalamic, cortico-cortical), it is unclear whether the corticotectal cell itself or the corticotectal synapse is the more likely site of this critical change. These are not mutually exclusive possibilities, and both deserve close scrutiny in future developmental studies. Since the corticotectal projection depends on the input from lgn Y-cells, and this population develops slowly during the first two-to-three months of postnatal life (see INFLUENCES OF RETINAL DEVELOPMENT above), this might be the most profitable first site to explore. An abstract by Wolf & Albus (1981) is consistent with this speculation. They reported that lamina V (corticotectal) cells with complex receptive field properties developed later than other cortical cells and were first recorded at the end of the third week of postnatal life. However, Tsumoto & Suda (1982) concluded that lamina V cells develop mature properties earlier than cells in other laminae. Yet regardless of whether lamina V cells develop normal response properties first or last, mature corticotectal synapses may develop gradually and may not yet be functional in young kittens (Mize & Sterling 1977).

Susceptibility to Environmental Manipulation

The sensitivity of geniculocortical cells to early sensory restriction was demonstrated some years ago in an elegant series of studies performed by David Hubel and Torsten Wiesel (Hubel & Wiesel 1962, 1970, Wiesel & Hubel 1963, 1965). Using the techniques of monocular and binocular lid suture, these investigators found a critical developmental period (three weeks to three months of age) during which the inputs from the two eyes "compete" for control of cortical cells. Since that time, there has been a tremendous growth in the number of studies investigating the "plasticity" of geniculocortical visual cells, and a variety of different rearing conditions have been utilized. The results of these studies have recently been reviewed in detail by Movshon & Van Sluyters (1981) and Sherman & Spear (1982).

The normal maturation of visual SC cells is also dependent on an animal's early visual experience. However, as is described below, the effects of early visual manipulations on SC cells are, in large part, reflections of changes in the corticotectal (or Y-indirect) pathway.

MONOCULAR DEPRIVATION   Rearing kittens with one eye sewed closed seriously compromises the ability of that eye to activate SC cells. Although the exact proportion of cells influenced from this eye is subject to some controversy (see Wickelgren & Sterling 1969b, Hoffmann & Sherman 1974, Berman & Sterling 1976), it is generally agreed that these cells are abnormal. Specifically, they lack directional selectivity—a property dependent on a normal corti-cotectal system. Since the corticotectal pathway from the deprived eye has become nonfunctional (Hoffmann & Sherman 1974), those SC cells that can be activated by the deprived eye are being driven via direct retinotectal inputs.

It seems paradoxical to claim that despite the paucity of cells activated by the deprived eye, the retinotectal input from this eye develops normally. Yet the claim is based on the observation that removal of visual cortex in monocularly deprived animals results in an SC indistinguishable from one in an animal reared normally before decortication: Cells are driven from the contralateral eye (whether or not it was previously deprived), they lack directional selectiv-ity, and they respond most vigorously to a stationary flashed stimulus. One possible explanation is that during development the retinotectal inputs from the deprived eye and the corticotectal inputs from the normal eye "compete" for control of SC cells. Monocular deprivation gives the corticotectal projection from the normal eye some competitive advantage, and produces an anomalous corticotectal system. Thus, comparatively few SC cells can be driven by the retinotectal inputs of the deprived eye because a suppressive influence from cortex holds most of the SC cells receiving an input from this eye in check. Removing the cortex "releases" these cells so that they can now be driven by the direct retinotectal input (Berman & Sterling 1976, Sherman & Spear 1982).

In ongoing experiments in my own laboratory we have noted that suppres-sive corticotectal influences are present in some SC cells in normal animals as well. These tonic inhibitory influences are capable of completely suppressing responses to stimuli when cortex is functional. However, when cortex is temporarily deactivated by cooling, they respond vigorously to visual stimulation (Ogasawara et al 1983b).

The effects of monocular deprivation are restricted to the binocular segment of the SC (Hoffmann & Sherman 1974). This is consistent with behavioral studies in which these animals exhibit visually guided behavior only in the monocular segment of the deprived eye (Sherman 1973, 1974a). However, once visual cortex is removed, the visual field of the deprived eye is markedly expanded to include the entire ipsilateral visual field (Sherman 1974b).

BINOCULAR DEPRIVATION   Binocular lid suture and dark rearing appear to have similar influences on SC neurons. Both conditions reduce the number of SC cells that are visually responsive, directionally selective, and binocular (Sterling & Wickelgren 1970, Flandrin & Jeannerod 1975, 1977, Hoffmann & Sherman 1975), properties very much like those seen after removal of visual cortex in normally reared animals. The similarity is no coincidence: in both cases the functional integrity of the Y-indirect corticotectal pathway has been compromised (Hoffmann & Sherman 1975). Since the normal complement of Y-cells is present in the retina (Sherman & Stone 1973), and the pathway from cortex to SC is also functional (Hoffmann & Sherman 1975, Singer & Tretter 1976), the failure in this circuit is at the input to the cortical cells themselves. Unlike the effects of monocular deprivation, binocular deprivation appears to have some disruptive influence on the direct Y-cell input, and may even have similar effects in the monocular and binocular segments of the SC (Hoffmann & Sherman 1975).

Visually guided behaviors in these deprived animals appear to depend on the direct retinotectal input and are restricted to use of the ipsilateral visual field (subserved by the nasal hemiretinae) (Sherman 1973, 1974a, 1977). This agrees with the physiological finding that the corticotectal input has been deactivated by the rearing condition because removal of visual cortex in these animals does not change their visually guided capability. This observation does indicate that animals binocularly deprived and then decorticated differ from animals normally reared and then decorticated in one important aspect: in the latter preparation, cortical removal precludes visually guided behavior unless the inhibitory intercollicular pathway is removed. Apparently, binocular deprivation interferes not only with corticotectal development, but also with the development of intercollicular inhibition (Sherman 1977).

ALTERNATING MONOCULAR DEPRIVATION AND STRABISMUS   Raising animals with one eye closed and alternating the deprived eye each day makes binocular experience impossible. Thus, the normal symmetry of inputs from the two eyes is precluded. Similarly, rendering an eye strabismic (e.g. surgically disrupting the tension in different extraocular muscles to divert the visual axes from the two eyes) disrupts the symmetry of the signals received by the two eyes. However, in both cases, each eye has equal visual input. Although either rearing condition will render most cortical cells monocular, each eye drives many cells (see Movshon & Van Sluyters 1981 for a detailed discussion of the effects of strabismus on cortical cells).

Surprisingly, these changes in cortical physiology are not seen in SC cells. Instead, alternating monocular deprivation produces a small decrease in the effectiveness of the ipsilateral eye in the SC (Gordon & Presson 1977), and monocular strabismus decreases the effectiveness of the strabismic eye in both colliculi (Gordon & Gummow 1975). The reasons for this are not entirely clear, though Gordon (1975) has suggested that a cortical cell driven only by the ipsilateral eye and one driven by the contralateral eye may converge on the same SC cell. And since many cortical cells are driven by either eye (though they are monocular) after rearing animals with asymmetric visual inputs, these rearing conditions might not affect the normal binocularity of SC cells.

CONTROLLED STIMULUS MOVEMENT   The effects on SC cells of raising animals so that they see movement in only one direction are unclear. Although this rearing condition increases the proportion of cortical cells preferring the direction of movement viewed (Cynader et al 1975, Tretter et al 1975, Daw & Wyatt 1976), a variety of different results have been described for SC cells. For example, Vital-Durand & Jeannerod (1974) found that SC cells preferred directions of movement opposite those viewed during rearing, while Flandrin & Jeannerod (1975) found that cells that would normally have had preference opposite those seen were unresponsive, and Cynader et al (1975) found that SC cells did not change their directional preference as a consequence of these rearing conditions. At present, it is difficult to reconcile these observations.

Another technique for manipulating stimulus movement is to eliminate it. Cats raised in stroboscopic illumination have binocularly symmetrical visual experience, but do not experience visual movement. Visual cortical cells lose directional selectivity as a consequence of stroboscopic rearing (Cynader et al 1973, Olson & Pettigrew 1974, Cynader & Chernenko 1976). Apparently, corticotectal cells are among those affected, for Flandrin et al (1976) reported that the SC cells of animals reared stroboscopically are very similar to those raised with binocular deprivation: they lack directional selectivity and binocularity. Similarly, stroboscopic rearing reduces the incidence of directionally selective cells in hamster superior colliculus (Chalupa & Rhoades 1978).

CONCLUDING REMARKS

It is now obvious that the maturation of visual cells in cortex and in the SC are intimately related. Thus, the appearance of complex receptive field properties in SC cells during early maturation is evidence of the influence the cortex is now exercising over them. Consequently, the anomalies developed in cortical cells of animals subjected to restricted visual experiences are also reflected in the SC cells they control. The processes by which this control develops and is exerted are by no means clear, but given the complexity of the problems addressed, progress has already been impressive. However, progress on questions relating to nonvisual SC cells is somewhat less impressive, particularly with regard to development and corticotectal influences.

Some of the unresolved issues of immediate concern have been pointed out in the text, and there is little doubt that many of them will be resolved in the near future, since they require no major changes in experimental design or available techniques. On the other hand, many of the issues of SC maturation are fundamental questions concerning the general development of the nervous system, such as the mechanisms by which receptive fields and functional topographies are constructed, and how sensorimotor transduction takes place. Whether the answers are sought in the SC or in some other structure is immaterial if the underlying principles are the same. Yet, an admittedly biased observation is that the SC is particularly amenable to the exploration of such questions and will continue to be a useful model for studying central nervous system maturation.

ACKNOWLEDGMENTS

This chapter is dedicated to the memory of Stephen B. Edwards, an excellent scientist and a devoted friend. His absence is felt acutely by many. For several years we had planned to write a review such as this; ironically, it was not until the day before his untimely death that the opportunity finally arose. Yet aside from the reservations I have expressed about accepting a reparcellation of the SC at this time, I don't believe this chapter contains anything to which Steve would have objected.

The preparation of this review and some of the work presented in it has been supported by USPHS grant EY04119, NSF grant BNS-8209857, and a grant from the Jeffress Foundation.

Literature Cited

Abrahams, V. C., Rose, P. K. 1975. Projections of extraocular, neck muscle, and retinal afferents to superior colliculus in the cat: Their connections to cells of origin of tectospinal tract. *J. Neurophysiol.* 38:10–18

Adamük, E. 1872. Über angeborene und erworbene. Association von F. C. Donders, Albrecht V. Graefes. *Arch. Ophthalmol.* 18:153–64

Albano, J. E., Mishkin, M., Westbrook, L. E., Wurtz, R. H. 1982. Visuomotor deficits following ablation of monkey superior colliculus. *J. Neurophysiol.* 48:338–51

Allon, N., Wollberg, Z. 1978. Responses of cells in the superior colliculus of the squirrel monkey to auditory stimuli. *Brain Res.* 159:321–30

Altman, J., Malis, L. 1962. An electrophysiological study of the superior colliculus and visual cortex. *Exp. Neurol.* 5:233–49

Anderson, F. D., Berry, C. M. 1959. Degeneration studies of long ascending fiber systems in the cat brain stem. *J. Comp. Neurol.* 111:195–229

Anker, R. 1977. The prenatal development of some of the visual pathways in the cat. *J. Comp. Neurol.* 173:185–204

Apter, J. T. 1945. Projection of the retina on

superior colliculus of cats. *J. Neurophysiol.* 8:123–34

Baleydier, C., Mauguiere, F. 1978. Projections of the ascending somesthetic pathways to the cat superior colliculus visualized by the horseradish peroxidase technique. *Exp. Brain Res.* 31:43–50

Ballam, G. O. 1982. Bilateral and multimodal sensory interactions of single cells in the pigeon's midbrain. *Brain Res.* 245:27–34

Bastian, J. 1982. Vision and electroreception: Integration of sensory information in the optic tectum of the weakly electric fish *Apteronotus albifrons. J. Comp. Physiol.* 147:287–97

Bell, C., Sierra, G., Buendia, N., Segundo, J. P. 1964. Sensory properties of neurons in the mesencephalic reticular formation. *J. Neurophysiol.* 27:961–87

Berman, N., Cynader, M. 1972. Comparison of receptive-field organization of the superior colliculus in Siamese and normal cats. *J. Physiol. London* 224:363–89

Berman, N., Cynader, M. 1976. Early versus late visual cortex lesions: Effects on receptive fields in cat superior colliculus. *Exp. Brain Res.* 25:131–37

Berman, N., Sterling, P. 1976. Cortical suppression of the retino-collicular pathway in the monocularly deprived cat. *J. Physiol.* 255:262–73

Blakemore, C., Van Sluyters, R. C. 1975. Innate and environmental factors in the development of the kitten's visual cortex. *J. Physiol. London* 248:663–716

Blass, E. M., Teicher, M. H., Cramer, C. P., Bruno, J. P., Hall, W. G. 1977. Olfactory, thermal and tactile controls of suckling in preauditory and previsual rats. *J. Comp. Physiol. Psychol.* 91:1248–60

Blomqvist, A., Flink, R., Bowsher, D., Griph, S., Westman, J. 1978. Tectal and thalamic projections of dorsal column and lateral cervical nuclei: A quantitative study in the cat. *Brain Res.* 141:335–41

Bonds, A. B., Freeman, R. D. 1978. Development of optical quality in the kitten eye. *Vision Res.* 18:391–98

Casagrande, V. A., Harting, J. K., Hall, W. C., Diamond, I. T., Martin, G. F. 1972. Superior colliculus of the tree shrew: A structural and functional subdivision into superficial and deep layers. *Science* 177:444–47

Cavalcante, L. A., Rocha-Miranda, C. E. 1978. Postnatal development of retinogeniculate, retinopretectal and retinotectal projections in the opossum. *Brain Res.* 146:231–48

Chalupa, L. M., Rhoades, R. W. 1977. Responses of visual, somatosensory, and auditory neurons in the golden hamster's superior colliculus. *J. Physiol.* 270:595–626

Chalupa, L. M., Rhoades, R. W. 1978. Modification of visual response properties in the superior colliculus of the golden hamster following stroboscopic rearing. *J. Physiol.* 274:571–92

Chow, K. L., Mathers, L. H., Spear, P. D. 1973. Spreading of uncrossed retinal projection in superior colliculus of neonatally enucleated rabbits. *J. Comp. Neurol.* 151:307–22

Clemo, H. R., Stein, B. E. 1981. Regions of the somatic cortex that affect superior colliculus cells in the cat. *Soc. Neurosci. Abstr.* 7:758

Clemo, H. R., Stein, B. E. 1982a. Response of somatic neurons in the superior colliculus to moving tactile stimuli. *Soc. Neurosci. Abstr.* 8:1025

Clemo, H. R., Stein, B. E., 1982b. Somatosensory cortex: A "new" somatotopic representation. *Brain Res.* 235:162–68.

Clemo, H. R., Stein, B. E. 1983. The organization of a fourth somatosensory area of cortex (siv) in cat. *J. Neurophysiol.* In press.

Constantine-Paton, M., Ferrari-Eastman, P. 1981. Topographic and morphometric effects of bilateral embryonic eye removal on the optic tectum and nucleus isthmus of the leopard frog. *J. Comp. Neurol.* 196:645–61

Cotter, J. R. 1976. Visual and nonvisual units recorded from the optic tectum of *Gallus domesticus. Brain Behav. Evol.* 13:1–21

Coulter, J. D., Bowker, R. M., Wise, S. P., Murray, E. A., Castiglioni, A. J., Westlund, K. N. 1979. Cortical, tectal and medullary descending pathways to the cervical spinal cord. *Prog. Brain. Res.* 50:263–79

Cragg, B. G. 1975. The development of synapses in the visual system of the cat. *J. Comp. Neurol.* 160:147–66

Cunningham, T. J., Mohler, I. M., Giordano, D. L. 1981. Naturally occurring neuron death in the ganglion cell layer of the neonatal rat: Morphology and evidence for regional correspondence with neuron death in the superior colliculus. *Dev. Brain Res.* 2:203–15

Currie, J., Cowan, W. M. 1975. The development of the retino-tectal projection in *Rana pipiens. Dev. Biol.* 46:103–19

Cynader, M., Berman, N., Hein, A. 1973. Cats reared in stroboscopic illumination: Effects on receptive fields in visual cortex. *Proc. Natl. Acad. Sci. USA* 70:1353–54

Cynader, M., Berman, N., Hein, A. 1975. Cats raised in a one-directional world: Effects on receptive fields in visual cortex and superior colliculus. *Exp. Brain Res.* 22:267–80

Cynader, M., Chernenko, G. 1976. Abolition

of direction selectivity in the visual cortex of the cat. *Science* 93:504–05

Daniels, J. D., Pettigrew, J. D., Norman, J. L. 1978. Development of single-neuron responses in kitten's lateral geniculate nucleus. *J. Neurophysiol.* 41:1373–93

Daw, N. W., Wyatt, H. J. 1976. Kittens reared in a unidirectional environment: Evidence for a critical period. *J. Physiol.* 257:155–70

Diamond, I. T., Hall, W. C. 1969. Evolution of neocortex. *Science* 164:251–62

Diamond, I. T., Jones, E. G., Powell, T. P. S. 1969. The projection of the auditory cortex upon the diencephalon and brain stem in the cat. *Brain Res.* 15:305–40

Donovan, A. 1966. The postnatal development of the cat's retina. *Exp. Eye Res.* 5:249–54

Dräger, U. C. 1977. Abnormal neural developments in mammals. In *Function and Formation of Neural Systems*, ed. G. S. Stent, pp. 111–38. Berlin: Dahlem Konferenzen

Dräger, U. C., Hubel, D. 1975. Responses to visual stimulation and relationship between visual, auditory, and somatosensory inputs in mouse superior colliculus. *J. Neurophysiol.* 38:690–713

Dreher, B., Hoffmann, K.-P. 1973. Properties of excitatory and inhibitory regions in the receptive fields of single units in the cat's superior colliculus. *Exp. Brain Res.* 16:333–53

Edwards, S. B. 1980. The deep cell layers of the superior colliculus: Their reticular characteristics and structural organization. In *The Reticular Formation Revisited*, ed. A. Hobson, M. Brazier. New York: Raven

Edwards, S. B., Ginsburgh, C. L., Henkel, C. K., Stein, B. E. 1979. Sources of subcortical projections to the superior colliculus in the cat. *J. Comp. Neurol.* 184:309–30

Edwards, S. B., Henkel, C. K. 1978. Superior colliculus connections with the extraocular motor nuclei in the cat. *J. Comp. Neurol.* 179:451–68

Feldon, S., Feldon, P., Kruger, L. 1970. Topography of the retinal projection upon the superior colliculus of the cat. *Vision Res.* 10:135–43

Finlay, B. L., Schneps, S. E., Wilson, K. G., Schneider, G. E. 1978. Topography of visual and somatosensory projections to the superior colliculus of the golden hamster. *Brain Res.* 142:223–35

Finlay, B. L., Wilson, K. G., Schneider, G. E. 1979. Anomalous ipsilateral retinotectal projections in Syrian hamsters with early lesions: Topography and functional capacity. *J. Comp. Neurol.* 183:721–40

Fish, M. W., Windle, W. F. 1932. The effect of rotatory stimulation on the movements of the head and eyes in newborn and young kittens. *J. Comp. Neurol.* 54:103–7

Flandrin, J. M., Jeannerod, M. 1975. Superior colliculus: Environmental influences on the development of directional responses in the kitten. *Brain Res.* 89:348–52

Flandrin, J. M., Jeannerod, M. 1977. Lack of recovery in collicular neurons from the effects of early deprivation or neonatal cortical lesion in the kitten. *Brain Res.* 120:362–66

Flandrin, J. M., Kennedy, H., Amblard, B. 1976. Effects of stroboscopic rearing on the binocularity and directionality of cat superior colliculus neurons. *Brain Res.* 101:576–81

Fox, M. W. 1970. Reflex development and behavior organization. In *Developmental Neurobiology*, ed. W. A. Himwich, pp. 553–80. Springfield, Ill.: Thomas

Fox, P., Chow, K. L., Kelly, A. S. 1978. Effects of monocular lid closure on development of receptive-field characteristics of neurons in rabbit superior colliculus. *J. Neurophysiol.* 41:1359–72

Frost, D. O., Schneider, G. E. 1976. Normal and abnormal uncrossed projections in Syrian hamsters as demonstrated by Fink-Heimer and autoradiographic techniques. *Soc. Neurosci. Abstr.* 2:812

Frost, D. O., So, K. F., Schneider, G. E. 1979. Postnatal development of retinal projections in Syrian hamster: A study using autoradiographic and anterograde degeneration techniques. *Neuroscience* 4:1649–77

Gaither, N. S., Stein, B. E. 1979. Reptiles and mammals use similar sensory organization in the midbrain. *Science* 205:595–97

Garey, L. J., Jones, E. G., Powell, T. P. S. 1968. Interrelationships of striate and extrastriate cortex with the primary relay sites of the visual pathway. *J. Neurol. Neurosurg. Psychiatry* 31:135–57

Gilbert, C. D., Kelly, J. P. 1975. The projection of cells in different layers of the cat's visual cortex. *J. Comp. Neurol.* 163:81–106

Godement, P. 1979. Connectivity of retinal projections in uniocular mice. In *Developmental Neurobiology of Vision*, ed. R. D. Freeman. pp. 371–80. New York: Plenum

Goldberg, M. E., Robinson, D. L. 1978. Visual system: Superior colliculus. In *Handbook of Behavioral Neurobiology*, ed. R. B. Masterton, pp. 119–64. New York: Plenum

Goodale, M. A., Murison, R. C. C. 1975. The effects of lesions of the superior colliculus on locomotor orientations and the orienting reflex in the rat. *Brain Res.* 88:243–61

Gordon, B. G. 1973. Receptive fields in deep layers of cat superior colliculus. *J. Neurophysiol.* 36:157–78

Gordon, B. G. 1975. Superior colliculus: Structure, physiology and possible functions. *MTP Int. Rev. Sci.* 3:185–230

Gordon, B. G., Gummow, L. 1975. Effects of

extraocular muscle section on receptive fields in cat superior colliculus. *Vision Res.* 15:1011–19

Gordon, B. G., Presson, J. 1977. Effects of alternating occlusion on receptive fields in cat superior colliculus. *J. Neurophysiol.* 40:1406–14

Graham, J. 1977. An autoradiographic study of the efferent connections of the superior colliculus in the cat. *J. Comp. Neurol.* 173:629–54

Graybiel, A. M. 1975. Anatomical organization of retinotectal afferents in the cat: An autoradiographic study. *Brain Res.* 96:1–23

Hamasaki, D. I., Flynn, J. T. 1977. Physiological properties of retinal ganglion cells of 3-week-old kittens. *Vision Res.* 17:275–84

Hamasaki, D. I., Sutija, V. G. 1979. Development of X- and Y-cells in kittens. *Exp. Brain Res.* 35:9–23

Harris, L. R. 1980. The superior colliculus and movements of the head and eyes in cats. *J. Physiol.* 300:367–91

Harris, L. R., Blakemore, C., Donaghy, M. 1980. Integration of visual and auditory space in mammalian superior colliculus. *Nature* 288:56–59

Harris, W. A. 1982. The transplantation of eyes to genetically eyeless salamanders: Visual projections and somatosensory interactions. *J. Neurosci.* 2:339–53

Harting, J. K., Guillery, R. W. 1976. Organization of retinocollicular pathways in the cat. *J. Comp. Neurol.* 166:133–44

Harting, J. K., Hall, W. C., Diamond, I. T., Martin, G. F. 1973. Anterograde degeneration study of the superior colliculus in *Tupaia glis:* Evidence for a subdivision between superficial and deep layers. *J. Comp. Neurol.* 148:361–86

Hartline, P. H. 1983. The optic tectum of reptiles: Neurophysiological studies. In *Comparative Neurology of the Optic Tectum,* ed. H. Vanegas. New York: Plenum. In press

Harutiunian-Kozak, B., Dec, K., Wrobel, A. 1973. The organization of visual receptive fields of neurons in the cat colliculus superior. *Acta Neurobiol. Expl. Warsaw* 33:563–73

Hein, A., Vital-Durand, F., Salinger, W., Diamond, R. 1979. Eye movements initiate visual-motor development in the cat. *Science* 204:1321–22

Henkel, C. K. 1981. Afferent sources of a lateral midbrain tegmental zone associated with the pinnae in the cat as mapped by retrograde transport of horseradish peroxidase. *J. Comp. Neurol.* 203:213–26

Henkel, C. K., Edwards, S. B. 1978. The superior colliculus control of pinna movements in the cat: Possible anatomical connections. *J. Comp. Neurol.* 182:763–76

Hoffmann, K.-P. 1973. Conduction velocity in pathways from retina to superior colliculus in the cat: A correlation with receptive-field properties. *J. Neurophysiol.* 36:409–24

Hoffmann, K.-P., Sherman, S. M. 1974. Effects of early monocular deprivation on visual input to cat superior colliculus. *J. Neurophysiol.* 37:1276–86

Hoffmann, K.-P., Sherman, S. M. 1975. Effects of early binocular deprivation on visual input to cat superior colliculus. *J. Neurophysiol.* 38:1049–59

Holcombe, V., Hall, W. C. 1981. The laminar origin and distribution of the crossed tectoreticular pathways. *J. Neurosci.* 1:1103–12

Horn, G., Hill, R. M. 1966. Responsiveness to sensory stimulation of units in the superior colliculus and subjacent tectotegmental regions of the rabbit. *Exp. Neurol.* 14:199–223

Hubel, D. H., Wiesel, T. N. 1962. Receptive fields, binocular interaction and functional architecture in the cat's visual cortex. *J. Physiol. London* 160:106–54

Hubel, D. H., Wiesel, T. N. 1963. Receptive fields of cells in striate cortex of very young, visually inexperienced kittens. *J. Neurophysiol* 26:994–1002

Hubel, D. H., Wiesel, T. N. 1970. The period of susceptibility to the physiological effects of unilateral eye closure in kittens. *J. Physiol.* 206:419–36

Hubel, D. H., Wiesel, T. N., LeVay, S. 1977. Plasticity of ocular dominance columns in monkey striate cortex. *Philos. Trans. R. Soc. London Ser. B* 278:377–409

Huerta, M. F., Frankfurter, A. J., Harting, J. K. 1981. The trigeminocollicular projection in the cat: Patch-like endings within the intermediate gray. *Brain Res.* 211:1–13

Huerta, M. F., Harting, J. K. 1982. Tectal control of spinal cord activity: Neuroanatomical demonstration of pathways connecting the superior colliculus with the cervical spinal cord grey. *Prog. Brain Res.* 51:293–328

Huttenlocher, P. R. 1967. Development of cortical neuronal activity in the neonatal cat. *Exp. Neurol.* 17:247–62

Innocenti, G. M. 1981. Growth and reshaping of axons in the establishment of visual callosal connections. *Science* 212:824–27

Ivy, G. O., Killackey, H. P. 1981. The ontogeny of the distribution of callosal projection neurons in the rat parietal cortex. *J. Comp. Neurol.* 195:367–89

Jassik-Gerschenfeld, D. 1965. Somesthetic and visual responses of superior colliculus neurones. *Nature* 208:898–900

Jay, M. F., Sparks, D. L. 1982. Auditory and saccade-related activity in the superior colliculus of the monkey. *Soc. Neurosci. Abstr.* 8:951

Jeffery, G., Perry, V. H. 1981. Evidence for

ganglion cell death during development of the ipsilateral retinal projection in the rat. *Dev. Brain Res.* 2:176–80

Johns, P. R., Rusoff, A. C., Dubin, M. W. 1979. Postnatal neurogenesis in the kitten retina. *J. Comp. Neurol.* 187:545–56

Kanaseki, T., Sprague, J. M. 1974. Anatomical organization of pretectal nuclei and tectal laminae in the cat. *J. Comp. Neurol.* 158:319–38

Kawamura, K., Brodal, A., Hoddevik, G. 1974. The projection of the superior colliculus onto the reticular formation of the brain stem. An experimental anatomical study in the cat. *Exp. Brain Res.* 19:1–19

Kawamura, K., Hashikawa, T. 1978. Cell bodies of origin of reticular projections from the superior colliculus in the cat: An experimental study with the use of horseradish peroxidase as a tracer. *J. Comp. Neurol.* 182:1–16

Kawamura, K., Konno, T. 1979. Various types of corticotectal neurons of cats as demonstrated by means of retrograde axonal transport of horseradish peroxidase. *Exp. Brain Res.* 35:161–75

Kawamura, S., Sprague, J. M., Niimi, K. 1974. Corticofugal projections from the visual cortices to the thalamus, pretectum and superior colliculus in the cat. *J. Comp. Neurol.* 158:339–62

Kenyon, C. A. P., Keeble, S., Cronin, P. 1982. The role of perioral sensation in nipple attachment by weanling rat pups. *Dev. Psychobiol.* 15:409–21

Knudsen, E. I. 1982. Auditory and visual maps of space in the optic tectum of the owl. *J. Neurosci.* 2:1177–94

Kruger, L. 1970. The topography of the visual projection to the mesencephalon: A comparative survey. *Brain Behav. Evol.* 3:169–77

Kuwabara, T., Weidmann, T. A. 1974. Development of the prenatal rat retina. *Invest. Opthalmol.* 13:725–39

Land, D. W., Lund, R. D. 1979. Development of rat's uncrossed retinotectal pathway and its relation to plasticity studies. *Science* 205:698–700

Land, P. W., Polley, E. H., Kernis, M. M. 1976. Patterns of retinal projections to the lateral geniculate nucleus and superior colliculus of rats with induced unilateral congenital eye defects. *Brain Res.* 103:394–99

Lane, R. H., Kaas, J. H., Allman, J. M. 1974. Visuotopic organization of the superior colliculus in normal and Siamese cats. *Brain Res.* 70:413–30

Larson, M., Stein, B. E. 1982. The use of tactile and olfactory cues in neonatal orientation. *Soc. Neurosci. Abstr.* 8:751

Lennerstrand, C., Hanson, J. 1978a. The postnatal development of the inferior oblique

muscle of the cat. I. Isometric twitch and tetanic properties. *Acta Physiol. Scand.* 103:132–43

Lennerstrand, G., Hanson, J. 1978b. The postnatal development of the inferior oblique muscle of the cat. II. Effects of repetitive stimulation on isometric tension responses. *Acta Physiol Scand.* 103:144–53

Lent, R., Mendez-Otero, R. 1980. Plasticity of the ipsilateral retinotectal projection in early enucleated opossums: Changes in retinotopy and magnification factors. *Neurosci. Lett.* 18:37–43

LeVay, S., Stryker, M. D. 1979. The development of ocular dominance columns in the cat. In *Soc. Neurosci. Symp. Vol. 4: Aspects of Developmental Neurobiology,* ed. J. A. Ferrendelli. pp. 83–98. Bethesda, MD: Soc. Neurosci.

Lund, R. D. 1978. *Development and Plasticity of the Brain.* New York: Oxford

Lund, R. D., Lund, J. S. 1971. Modification of synaptic patterns in the superior colliculus of the rat during development and following deafferentation. *Vision Res. Suppl.* 11(3): 281–98

Mandl, G. 1974. The influence of visual pattern combinations on responses of movement sensitive cells in the cat's superior colliculus. *Brain Res.* 75:215–40

Marburg, O., Warner, J. F. 1947. The pathways of the tectum (anterior colliculus) of the midbrain in cats. *J. Nerv. Ment. Dis.* 106:415–46

Marchiafava, P. L., Pepeu, G. C. 1966. Electrophysiological study of tectal responses to optic nerve volley. *Arch. Ital. Biol.* 104:406–20

Martin, G. F. 1969. Efferent tectal pathways of the opossum *Didelphis virginiana. J. Comp. Neurol.* 135:209–24

Marty, R., Scherrer, J. 1964. Critères de maturation des systèmes afferents corticaux. *Prog. Brain Res.* 4:222–35

Mason, C. A. 1980. Postnatal development of retino-geniculate axon arbors in the kitten. *Soc. Neurosci. Abstr.* 6:669

Mason, R. 1979. Responsiveness of cells in the cat's superior colliculus to textured visual stimuli. *Exp. Brain Res.* 37:231–40

McHaffie, J. G., Ogasawara, K., Stein, B. E. 1983. Ascending nonvisual afferents to the superior colliculus in neonatal kittens. *Soc. Neurosci. Abstr.* 9: In press

McHaffie, J. G., Stein, B. E. 1982. Eye movements evoked by electrical stimulation in the superior colliculus of rats and hamsters. *Brain Res.* 247:243–53

McIlwain, J. T. 1972. Central vision: Visual cortex and superior colliculus. *Ann. Rev. Physiol.* 34:291–314

McIlwain, J. T. 1975. Visual receptive fields

and their images in superior colliculus of the cat. *J. Neurophysiol.* 38:219–30

McIlwain, J. T., Buser, P. 1968. Receptive fields of single cells in cat's superior colliculus. *Exp. Brain Res.* 5:314–25

McIlwain, J. T., Fields, H. L. 1971. Interactions of cortical and retinal projections on single neurons of the cat's superior colliculus. *J. Neurophysiol.* 34:763–72

Mehler, W., Feferman, M. E., Nauta, W. J. H. 1960. Ascending axon degeneration following antero-lateral cordotomy. An experimental study in the monkey. *Brain Res.* 83:718–52

Meredith, M. A., Stein, B. E. 1983. Interactions among converging sensory inputs in the superior colliculus. *Science* 221:389–91

Mize, R. R., Murphy, E. H. 1976. Alterations in receptive field properties of superior colliculus cells produced by visual cortex ablation in infant and adult cats. *J. Comp. Neurol.* 168:393–424

Mize, R. R., Sterling, P. 1977. Synaptic development in the superficial gray layer of the cat superior colliculus. *Anat. Rec.* 187:658

Moore, C. L., Kalil, R., Richards, W. 1976. Development of myelination in optic tract of the cat. *J. Comp. Neurol.* 165:125–36

Movshon, A. J., Van Sluyters, R. C. 1981. Visual neural development. *Ann. Rev. Psychol.* 132:477–522

Nagata, T., Kruger, L. 1979. Tactile neurons of the superior colliculus of the cat: Input and physiological properties. *Brain Res.* 174:19–37

Newman, E. A., Hartline, P. H. 1981. Integration of visual and infrared information in bimodal neurons of the rattlesnake optic tectum. *Science* 213:789–91

Norton, T. T. 1974. Receptive-field properties of superior colliculus cells and development of visual behavior in kittens. *J. Neurophysiol.* 37:674–90

Norton, T. T. 1981. Development of the visual system and visually guided behavior. In *Development of Perception,* ed. R. Aslin, J. Alberts, M. R. Petersen, Sec. 6, pp. 113–52. New York: Academic

Norton, T. T., Lindsley, D. 1971. Visual behaviour after bilateral superior colliculus lesions in kittens and cats. *Fed. Proc.* 30:615

Ogasawara, K., Kawamura, K. 1982. Cells of origin and terminations of the trigeminotectal projection in the cat as demonstrated with the horseradish peroxidase and autoradiographic methods. *Okajimas Folia Anat. Jpn.* 58:247–64

Ogasawara, K., McHaffie, J. G., Stein, B. E. 1983a. Differential influences of striate and extrastriate cortex on the response of visual cells in cat superior colliculus. *Invest. Ophthalmol. Suppl.* 24:224

Ogasawara, K., McHaffie, J. G., Stein, B. E. 1983b. Excitatory and inhibitory influences of striate and extrastriate cortex on superior colliculus cells. *Soc. Neurosci. Abstr.* 9: In press

O'Leary, D. D. M., Cowan, W. M. 1980. Observations on the effects of monocular and binocular eye removal on the development of the chick visual system. *Soc. Neurosci. Abstr.* 6:297

Olson, C. R., Freeman, R. D. 1978. Eye alignment in kittens. *J. Neurophysiol.* 41:848–59

Olson, C. R., Pettigrew, J. D. 1974. Single units in visual cortex of kittens reared in stroboscopic illumination. *Brain Res.* 70:189–204

Palmer, A. R., King, A. J. 1982. The representation of auditory space in the mammalian superior colliculus. *Nature* 299:248–49

Palmer, L. A., Rosenquist, A. C. 1974. Visual receptive fields of single striate cortical units projecting to the superior colliculus in the cat. *Brain Res.* 67:27–42

Palmer, L. A., Rosenquist, A. C., Sprague, J. M. 1972. Corticotectal systems in the cat: Their structure and function. In *Corticothalamic Projections and Sensorimotor Activities,* ed. T. Frigyesi, E. Rinvik, M. D. Yahr, pp. 491–523. New York: Raven

Paula-Barbosa, M. M., Sousa-Pinto, A. 1973. Auditory cortical projections to the superior colliculus in the cat. *Brain Res.* 50:47–61

Pettigrew, J. D. 1974. The effect of visual experience on the development of stimulus specificity by kitten cortical neurones. *J. Physiol. London* 237:49–74

Pinter, R. B., Harris, L. R. 1981. Temporal and spatial response characteristics of the cat superior colliculus. *Brain Res.* 207:73–94

Poirier, L. J., Bertrand, C. 1955. Experimental and anatomical investigation of the lateral spino-thalamic and spino-tectal tracts. *J. Comp. Neurol.* 102:745–58

Rakic, P. 1977. Prenatal development of the visual system in rhesus monkey. *Philos. Trans. R. Soc. London Ser. B* 278:245–60

Rhoades, R. W., Chalupa, L. M. 1980. Effects of neonatal enucleation on receptive-field properties of visual neurons in superior colliculus of the golden hamster. *J. Neurophysiol.* 43:595–611

Rhoades, R. W., DellaCroce, D. D., Meadows, I. 1981. Reorganization of somatosensory input to superior colliculus in neonatally enucleated hamsters: anatomical and electrophysiological experiments. *J. Neurophysiol.* 46:855–77

Rhoades, R. W., Jacquin, M., Mooney, R. 1982. Complex somatosensory receptive fields in the deep layers of the hamster's superior colliculus. *Soc. Neurosci. Abstr.* 8:1026

Rizzolatti, G., Camarda, R., Grupp, L. A., Pisa, M. 1974. Inhibitory effect of remote visual stimuli on visual responses of cat superior colliculus: spatial and temporal factors. *J. Neurophysiol.* 37:1262–75

Robinson, D. A. 1972. Eye movements evoked by collicular stimulation in the alert monkey. *Vision Res.* 12:1795–1808

Rose, G. H., Lindsley, D. B. 1968. Development of visually evoked potentials in kittens: Specific and nonspecific responses. *J. Neurophysiol.* 31:607–23

Roucoux, A., Crommelinck, M. 1976. Eye movements evoked by superior colliculus stimulation in the alert cat. *Brain Res.* 106:349–63

Rusoff, A. C., Dubin, M. W. 1977. Development of receptive-field properties of retinal ganglion cells in kittens. *J. Neurophysiol.* 40:1188–98

Sanderson, K. J., Pearson, L. J., Dixon, P. G. 1978. Altered retinal projections in brush-tailed possum *Trichosurus vulpecula* following removal of one eye. *J. Comp. Neurol.* 180:841–68

Schaefer, K.-P. 1970. Unit analysis and electrical stimulation in the optic tectum of rabbits and cats. *Brain Behav. Evol.* 3:222–40

Schiller, P. H. 1972. The role of the monkey superior colliculus in eye movement and vision. *Invest. ophthalmol.* 11:451–60

Schneider, G. E. 1969. Two visual systems: Brain mechanisms for localization and discrimination are dissociated by tectal and cortical lesions. *Science* 163:895–902

Schneider, G. E. 1975. Two visuomotor systems in the Syrian hamster. *Neurosci. Res. Prog. Bull.* 13:255–58

Sengelaub, D. R., Finlay, B. L. 1981. Early removal of one eye reduces normally occurring cell death in the remaining eye. *Science* 213:573–74

Shatz, C. J. 1983. The prenatal development of the cat's retinogeniculate pathway. *J. Neurosci.* 3:482–99

Sherman, S. M. 1972. Visual development in cats. *Invest. Ophthalmol.* 11:394–401

Sherman, S. M. 1973. Visual field defects in monocularly and binocularly deprived cats. *Brain Res.* 49:25–45

Sherman, S. M. 1974a. Permanence of visual perimetry deficits in monocularly and binocularly deprived cats. *Brain Res.* 73:491–501

Sherman, S. M. 1974b. Monocularly deprived cats: Improvement of the deprived eye's vision by visual decortication. *Science* 186:267–69

Sherman, S. M. 1977. The effect of cortical and tectal lesions on the visual fields of binocularly deprived cats. *J. Comp. Neurol.* 172:231–46

Sherman, S. M., Spear, P. D. 1982. Organization of visual pathways in normal and visually deprived cats. *Physiol. Rev.* 63:738–855

Sherman, S. M., Stone, J. 1973. Physiological normality of the retina in visually deprived cats. *Brain Res.* 60:224–30

Singer, W., Tretter, F. 1976. Receptive-field properties and neuronal connectivity in striate and parastriate cortex of contour-deprived cats. *J. Neurophysiol.* 39:613–30

Sireteanu, R., Maurer, D. 1982. The development of the kitten's visual field. *Vision Res.* 22:1105–11

So, K.-F., Schneider, G. E. 1978. Postnatal development of retinogeniculate projections in Syrian hamsters: An anterograde HRP study. *Soc. Neurosci. Abstr.* 4:127

Sparks, D. L., Pollack, J. G. 1977. The neural control of saccadic eye movements: The role of the superior colliculus. In *Eye Movements*, ed. B. A. Brooks, F. J. Bajandas, pp. 179–219. New York: Plenum

Sparks, D. L., Porter, J. D. 1983. The spatial localization of saccade targets. II. Activity of superior colliculus neurons preceding compensatory saccades. *J. Neurophysiol.* 49:64–74

Spear, D. D., Chow, K. L., Masland, R. H., Murphy, E. H. 1972. Ontogenesis of receptive-field characteristics of superior colliculus neurons in the rabbit. *Brain Res.* 45:67–86

Sprague, J. M. 1966. Visual, acoustic, and somesthetic deficits in the cat after cortical and midbrain lesions. In *The Thalamus*, ed. D. D. Purpura, M. Yahr. New York: Columbia Univ. Press

Sprague, J. M. 1972. The superior colliculus and pretectum in visual behavior. *Invest. Ophthalmol.* 11:473–82

Sprague, J. M. 1975. Mammalian tectum: Intrinsic organization, afferent inputs and integrative mechanisms. Anatomical substrate. *Neurosci. Res. Progr. Bull.* 13:204–13

Sprague, J. M., Marchiafava, P. L., Rizzolatti, G. 1968. Unit responses to visual stimuli in the superior colliculus of the unanesthetized mid-pontine cat. *Arch. Ital. Biol.* 106:169–93

Sprague, J. M., Meikle, T. H. Jr. 1965. The role of the superior colliculus in visually guided behavior. *Exp. Neurol.* 11:115–46

Stein, B. E. 1978a. Development and organization of multimodal representation in cat superior colliculus. *Fed. Proc.* 37:2240–45

Stein, B. E. 1978b. Nonequivalent visual, auditory and somatic corticotectal influences in cat. *J. Neurophysiol.* 41:55–64

Stein, B. E., Arigbede, M. O. 1972a. A parametric study of movement detection properties of neurons in the cat's superior colliculus. *Brain Res.* 45:437–54

Stein, B. E., Arigbede, M. O. 1972b. Unimodal and multimodal response properties of neurons in the cat's superior colliculus. *Exp. Neurol.* 36:179–96

Stein, B. E., Clamann, H. P. 1981. Control of pinna movements and sensorimotor register in cat superior colliculus. *Brain Behav. Evol.* 19:180–92

Stein, B. E., Clamann, H. P., Goldberg, S. J. 1980. Superior colliculus: Control of eye movements in neonatal kittens. *Science* 210:78–80

Stein, B. E., Dixon, J. 1978. Superior colliculus cells respond to noxious stimuli. *Brain Res.* 158:65–73

Stein, B. E., Dixon, J. 1979. Properties of superior colliculus neurons in the golden hamster. *J. Comp. Neurol.* 183:269–84

Stein, B. E., Edwards, S. B. 1979. Corticotectal and other corticofugal projections in neonatal cat. *Brain Res.* 161:399–409

Stein, B. E., Gaither, N. S. 1981. Sensory representation in reptilian optic tectum: Some comparisons with mammals. *J. Comp. Neurol.* 202:69–87

Stein, B. E., Gaither, N. S. 1983. Receptive-field properties in reptilian optic tectum: Some comparisons with mammals. *J. Neurophysiol.* 50:102–24

Stein, B. E., Gallagher, H. 1981. Maturation of cortical control over superior colliculus cells in cat. *Brain Res.* 223:429–35

Stein, B. E., Goldberg, S. J., Clamann, H. P. 1976a. The control of eye movements by the superior colliculus in the alert cat. *Brain Res.* 118:469–74

Stein, B. E., Gordon, B. G. 1981. Maturation of the superior colliculus. In *The Development of Perception: Psychobiological Perspectives,* Vol. 2, ed. R. N. Aslin, J. R. Alberts, M. R. Petersen. New York: Academic

Stein, B. E., Kruger, L. 1983. Somatosensory corticotectal and corticothalamic projections from anterior ectosylvian sulcus in newborn cats. *Soc. Neurosci. Abstr.* 9: In press

Stein, B. E., Labos, E., Kruger, L. 1973a. Sequence of changes in properties of neurons of superior colliculus of the kitten during maturation. *J. Neurophysiol.* 36:667–79

Stein, B. E., Labos, E., Kruger, L. 1973b. Determinants of response latency in neurons of superior colliculus in kittens. *J. Neurophysiol.* 36:680–89

Stein, B. E., Magalhães-Castro, B. 1975. Effects of neonatal cortical lesions upon the cat superior colliculus. *Brain Res.* 83:480–85

Stein, B. E., Magalhães-Castro, B., Kruger, L. 1975. Superior colliculus: Visuotopic-somatotopic overlap. *Science* 189:224–26

Stein, B. E., Magalhães-Castro, B., Kruger, L. 1976b. Relationship between visual and tactile representations in cat superior colliculus. *J. Neurophysiol.* 34:401–19

Stein, B. E., Spencer, R. F., Edwards, S. B. 1982. Efferent projections of the neonatal superior colliculus: Extraoculomotor-related brain stem structures. *Brain Res.* 239:17–28

Stein, B. E., Spencer, R. F., Edwards, S. B. 1983. Corticotectal and coricothalamic efferent projections of SIV somatosensory cortex in cat. *J. Neurophysiol.* In press

Sterling, P., Wickelgren, B. G. 1969. Visual receptive fields in the superior colliculus of the cat. *J. Neurophysiol.* 32:1–15

Sterling, P., Wickelgren, B. G. 1970. Function of the projection from the visual cortex to the superior colliculus. *Brain Behav. Evol.* 3:210–18

Stewart, W. A., King, R. B. 1963. Fiber projections from the nucleus caudatus of the spinal trigeminal nucleus. *J. Comp. Neurol.* 121:271–95

Stone, J., Rapaport, D. H., Williams, R. W., Chalupa, L. M. 1982. Uniformity of cell distribution in the ganglion cell layer of prenatal cat retina: Implications for mechanisms of retinal development. *Dev. Brain Res.* 2:231–42

Straschill, M., Hoffmann, K.-P. 1969. Functional aspects of localization in the cat's tectum opticum. *Brain Res.* 13:274–83

Straschill, M., Rieger, P. 1973. Eye movements evoked by focal stimulation of the cat's superior colliculus. *Brain Res.* 59:211–27

Straschill, M., Taghavy, H. 1967. Neuronale reactionen im Tectum Opticum der Katze aug bewegte und stationäre lichtreize. *Exp. Brain Res.* 3:353–67

Syka, J., Popelar, J., Bozkov, V. 1979. Responses of neurons in the superior colliculus of the cat to stationary and moving visual stimuli. *Vision Res.* 19:213–19

Syka, S., Radil-Weiss, T. 1971. Electrical stimulation of the tectum in freely moving cats. *Brain Res.* 28:567–72

Tamai, M. 1973. Inflows from the somatosensory cortex in the cat's superior colliculus. *Tohoku J. Exp. Med.* 109:7–11

Thorn, F., Gollender, M., Erickson, P. 1976. The development of the kitten's visual optics. *Vision Res.* 16:1145–50

Tortelly, A., Reinoso-Suarez, F., Llamas, A. 1980. Projections from non-visual cortical areas to the superior colliculus demonstrated by retrograde transport of HRP in cat. *Brain Res.* 188:543–49

Tretter, F., Cynader, M., Singer, W. 1975. Modification of direction selectivity of neurons in the visual cortex of kittens. *Brain Res.* 84:143–49

Tsumoto, T., Suda, K. 1982. Laminar differences in development of afferent innervation to striate cortex neurones in kittens. *Exp. Brain Res.* 45:433–46

Updyke, B. V. 1977. Topographic organization of the projections from cortical areas 17, 18, and 19 onto the thalamus, pretectum and superior colliculus in the cat. *J. Comp. Neurol.* 173:81–122

Villablanca, J. R., Olmstead, C. E. 1979. Neurological development of kittens. *Dev. Psychobiol.* 12:101–27

Vital-Durand, F., Jeannerod, M. 1974. Role of visual experience in the development of optokinetic response in kittens. *Exp. Brain Res.* 20:297–302

Weber, J. T., Martin, G. G., Behan, M., Huerta, M. F., Harting, J. K. 1979. The precise origin of the tectospinal pathway in three common laboratory animals: A study using the horseradish peroxidase method. *Neurosci. Lett.* 11:121–27

Wickelgren, B. G., Sterling, P. 1969a. Influence of visual cortex on receptive fields in the superior colliculus of the cat. *J. Neurophysiol.* 32:16–23

Wickelgren, B. G., Sterling, P. 1969b. Effect on the superior colliculus of cortical removal in visually deprived cats. *Nature* 224:1032–33

Wiesel, T. N., Hubel, D. H. 1963. Single-cell responses in striate cortex of kittens deprived of vision in one eye. *J. Neurophysiol.* 26:1003–17

Wiesel, T. N., Hubel, D. H. 1965. Comparison of the effects of unilateral and bilateral eye closure on cortical unit responses in kittens. *J. Neurophysiol.* 28:1029–40

Williams, R. W., Bastiani, M. J., Chalupa, L. M. 1983. Loss of axons in the cat optic nerve following fetal unilateral enucleation: An electron microscopic analysis. *J. Neurosci.* 3:133–44

Williams, R. W., Chalupa, L. M. 1982. Prenatal development of retinocollicular projections in the cat: An anterograde tracer transport study. *J. Neurosci.* 2:604–22

Wilson, J. R., Tessin, D. E., Sherman, M. 1982. Development of the electrophysiological properties of Y-cells in the kitten's medial interlaminar nucleus. *J. Neurosci.* 2:562–71

Wise, L. Z., Irvine, D. R. F. 1983. Auditory response properties of neurons in deep layers of cat superior colliculus. *J. Neurophysiol.* 49:674–85

Wolf, W., Albus, K. 1981. Postnatal development of receptive-field properties in the kitten's visual cortex. *Neurosci. Lett. Suppl.* 7:S202

Wurtz, R. H., Albano, J. E. 1980. Visual-motor function of the primate superior colliculus. *Ann. Rev. Neurosci.* 3:189–226

Ann. Rev. Neurosci. 1984. 7:127–47

THE NEURAL BASIS OF LANGUAGE

Antonio R. Damasio

Division of Behavioral Neurology, University of Iowa College of Medicine, Iowa City, Iowa 52242

Norman Geschwind

Department of Neurology, Harvard Medical School, and Harvard Neurological Unit, Beth Israel Hospital, Boston, Massachusetts 02215

INTRODUCTION

Until the middle of the 1960s, the understanding of the anatomical basis of language in humans had grown on the strength of a handful of methods. The oldest and admittedly the most fruitful of them was the neuropathological study of focal cerebral lesions in patients with aphasia. A closely related method was the behavioral study of patients who had undergone surgical interventions, such as the ablation of a cerebral lobe or even an entire hemisphere. Other approaches were provided by (*a*) the electrical stimulation of the human cerebral cortex, in awake patients, during surgical interventions, aimed at studying the effect of stimulation on language behaviors; (*b*) the use of established neuroradiological techniques, such as cerebral angiography and pneumoencephalography, for indirect localization of lesions related to language disturbance; and (*c*) the study of language behavior after transient hemispheric inactivation caused by a barbiturate injected into the carotid artery of one side (the Wada Test). Unquestionably, the advances brought about by the combination of those methods were remarkable. For instance, from about the middle of the nineteenth century and for a period of nearly 100 years, neurologists collected a large number of single case studies in which distinctive disturbances of the use of language were correlated with specifically located

127

0147-006X/84/0301-0127$02.00

lesions found at autopsy. The analyses, however, went beyond mere correlation since investigators used the growing knowledge of the connectional patterns of the brain to develop flow-diagrams for various activities within and between hemispheres. The neuropathological substrate of many syndromes of aphasia and of such associated disturbances as pure alexia and the apraxias were described during that period and the syndromes of callosal lesions were recognized and interpreted in terms of hemispheric disconnection. But the anatomical basis of several other language syndromes could not be defined on the basis of isolated and occasionally conflictual observations, and for decades the fruitful approach of anatomo-clinical correlation came to an impasse. One problem resided with the method of behavioral analysis, the quality and comprehensiveness of which varied from author to author and from decade to decade. In the postwar period it appeared that the problem might be solved with the introduction of more standard instruments of neuropsychological assessment. But by then the availability of autopsy material had declined so much that a significant number of observations capable of accounting for case-to-case variablity could not be attained. Yet another important problem was the prevailing antilocalizationist attitude, which succeeded in reducing the impact and even suppressing the facts and ideas of the early neurology of aphasia. Meanwhile, the study of patients following neurosurgical procedures continued to confirm and amplify many of the findings of the neuropathological studies, especially in relation to cerebral dominance for language. This was true for studies of callosally sectioned patients and also for the electrical stimulation of human cerebral cortex, which lent support to the concept of cerebral dominance, confirmed the existence of cortical areas specialized in language processing, and revealed the existence of additional cortical areas, e.g. supplementary motor area, involved in the process of language production. But the impasse was only broken in the past 15 years, during which the understanding of the anatomical basis of language has developed dramatically. The change can be traced to the following factors: (a) a gradual reversal in the attitude toward cerebral localization spearheaded by a reappraisal of the work of the classic aphasiologists and by the availability of a systematized theory on the anatomical basis of complex behavior (cf Geschwind 1965), (b) the development of relatively standard instruments of neuropsychological assessment, (c) the introduction of powerful new methods of radiological cerebral imaging, and (d) the study of cerebral asymmetries in man and in animals.

The panorama had begun to change by the late 1960s as a result of the introduction of radioisotope scanning, and it changed dramatically with the introduction, in 1973, of computerized tomography (CT). By the late 1970s the first studies of large series of brain-damaged patients with language disturbances began to be published. Current developments in other techniques of neuroimaging, such as tomographic regional cerebral blood flow (SPECT), positron emission tomography (PET), and nuclear magnetic resonance (NMR), together with more refined CT technology, hold promise for the continuation of anatomical studies in humans with neurological disease.

Also by the late 1960s, the first statistically confirmed findings of cerebral asymmetries in humans led to the discovery that specific cerebral areas, clearly related to language, were often larger in the dominant hemisphere. In recent years the presence of asymmetries has been repeatedly confirmed, for a variety of language-related cerebral structures, using diverse methods, and cerebral asymmetries have now been studied in animals as well, paving the way for animal models of cerebral dominance. This review concentrates on the advances brought about by the introduction of the new radiological methods and by the systematic study of cerebral asymmetry.

RADIOISOTOPE CORRELATES OF THE APHASIAS

Radioisotope brain scanning is an unsatisfactory technique for cerebral localization. Nonetheless, in the days prior to the advent of CT, radioisotope scan allowed relatively reliable localization of lesions in relation to the major quadrants of a hemisphere and also permitted the first indirect anatomical studies of large series of patients evaluated with standardized behavioral methods. Two studies using this method provided important information.

An association between the different linguistic characteristics of aphasic speech and the localization of the lesions had been suggested by several of the early writers on aphasia. Benson (1967) clarified that relationship in a study of 100 aphasic patients who were evaluated with radioisotope scans and with a rating scale for ten features of aphasic speech. The features were rate of speech, prosody, pronunciation, phrase length, effort, pauses, press of speech, perseveration, word use, paraphasia. The pattern of these features in the different aphasic syndromes was such that they formed two distinct clusters in the majority of patients. The patients belonging to one of these clusters had almost without exception lesions located anteriorly to the central sulcus. The patients belonging to the other cluster had lesions located posterior to the central sulcus. The speech of the anterior group was characterized by a low rate of verbal output, dysprosody, dysarthria, effort—in short, nonfluent speech. The other group with posterior lesions had normal or higher than normal verbal output, normal prosody and articulation, effortless and syntactically well-organized speech—in short, fluent speech, which was also abnormal because of incorrect choice of content words. Thus, this study helped to confirm the belief suggested by the descriptions of Wernicke himself, that linguistically different aphasic syndromes are strongly associated with remarkably different lesions within the dominant hemisphere.

The study by Kertesz et al (1977) attempted to identify the radioisotope correlates of the principal aphasia syndromes. In spite of a certain amount of overlapping of lesions due to the technical limitations of the method, the prevalent loci of Broca's aphasia, Wernicke's aphasia, and conduction aphasia (to be described below) were clearly distinguishable. Less consistent correlations were obtained for the transcortical aphasias.

COMPUTERIZED TOMOGRAPHY CORRELATES OF THE APHASIAS

The era of computerized tomography (CT) began in 1973 and the first major studies on the aphasias began to appear in 1977. The technique is noninvasive and the results highly consistent. In vascular lesions, the best type of specimen for anatomical correlation, the images become remarkably stable within three to four weeks after onset and permit a detailed morphological study. Although CT cannot provide many of the fine macroscopic details of a postmortem examination and none of the microscopic ones, it permits inferences about anatomical structures that were, until now, only obtainable with autopsy. Thus, our concepts of the more frequent lesions associated with the aphasias have been sharpened. In addition, the anatomical bases of some previously unrecognized syndromes of aphasia have been described as well as alternative anatomical loci for some classical syndromes. Currently available information permits a fairly comprehensive mapping of different left hemisphere structures involved in the processes of speech and language. We discuss the progress made in the past four years in relation to each specific entity.

Wernicke's Aphasia

In Wernicke's aphasia the salient distinguishing features are fluently aphasic speech of the type described above, poor repetition, and a significant auditory comprehension deficit. According to CT data, most patients with Wernicke's aphasia resulting from cerebrovascular disease have lesions that involve Wernicke's area (Brodmann's area 22, lying in the posterior aspect of the left superior temporal gyrus) to a greater or lesser extent. The only exception to this rule, in this as well as any other aphasia syndrome, are right-handed patients with inverted dominance ("crossed aphasia") and a minority of left-handed or ambidextrous patients. The primary left auditory cortex (areas 41, 42) is not involved necessarily (see Figure 1). The extent of the lesion is variable and parallels the presence or absence of associated linguistic disturbances (alexia, agraphia) and nonlinguistic disturbances (constructional apraxia, acalculia), as well as the severity of the auditory comprehension and speech defects. Not uncommonly the lesion extends into the parietal lobe, especially into the angular gyrus, Brodmann's area 39, and into the postero-inferior temporal lobe, the second and third temporal gyri in which areas 21, 20 and 37 are

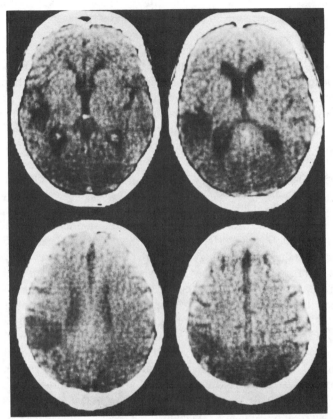

Figure 1   CT of a patient with Wernicke's aphasia. The photos show a low density image in the left temporo-parietal region, involving Wernicke's area. The lesion was caused by a nonhemorrhagic infarction.

located (Damasio 1981). In patients with vascular lesions, CT suggests that the lesions of Wernicke's aphasia are the result of infarction or hemorrhage in the territory of posterior temporal branches of the middle cerebral artery; the angular gyrus artery may be involved as well. Occasionally, the lesions may be exclusively subcortical and may act by disrupting input and output of an otherwise intact auditory association cortex. The remarkable consistency of the anatomical findings in patients with the clinical syndrome of Wernicke's aphasia cannot be overemphasized. It lends support to the concept of Wernicke's area as a crucial processing center for language.

Broca's Aphasia

In Broca's aphasia the salient features are nonfluent aphasic speech with poor repetition but with considerable preservation of auditory comprehension. Controversy regarding the anatomical correlates of Broca's aphasia has repeatedly been found in the literature on the anatomical bases of the aphasias. It formed the basis for the confrontation between Pierre Marie and Jules Dejerine in their renowned discussions at the Société de Medicine. It has been fueled by occasional descriptions of cases of Broca's aphasia in which it was reported that no lesion in Broca's area could be found, and by cases of lesions in Broca's area in patients who failed to show the typical clinical picture of Broca's aphasia. Curiously, this type of controversy has been much less common in relation to Wernicke's aphasia and even the issues regarding conduction aphasia or the transcortical aphasias have been of a less conflicting nature. CT has now helped clarify the issue.

CT was crucial in finally revealing the limits of the lesion in Broca's original case, Leborgne, one of the bones of contention in the debate between Marie and Dejerine. The brain of that patient had never been cut but had been preserved as a museum specimen. Broca's description of the lesion had been made from inspection of the lateral surface of its left hemisphere. He emphasized the damage to the third left frontal gyrus, an area which he proposed as the seat of the faculty of articulate language. Marie disagreed, contending that the lesion most probably involved not only the third frontal gyrus but also the insula and basal ganglia, as well as Wernicke's area itself. In turn, Dejerine was convinced that the lesion spared both Wernicke's area and the basal ganglia region. Recently CT cuts of the Leborgne brain were obtained, in the horizontal level of Dejerine's classical cerebral sections. The results show that the truth lies in between. The lesion did spare Wernicke's area, as Broca and Dejerine claimed. But the involvement of the basal ganglia and insular cortex was massive (Castaigne et al 1980).

CT scan findings were also crucial in the study of Mohr and associates, who reviewed autopsy records of the Massachusetts General Hospital (Mohr 1976). Mohr concluded that infarction confined to the inferior frontal gyrus causes a brief period of mutism, which is later replaced by effortful speech, but that it does not cause a significant linguistic defect. On the other hand, he noted that a more severe syndrome characterized by persistent mutism, verbal stereotypes, and agrammatism is associated with a considerably larger infarct involving most of the frontal operculum and insula. Findings from four other studies (those of Naeser & Hayward 1978, Kertesz et al 1979, Damasio et al 1979, and Damasio 1981) support to a great extent Mohr's contention. Lesions confined to the midfrontal convolution and to the nearby motor area rarely produce the typical picture of persistent Broca's aphasia. More extensive lesions involving adjoining regions of the premotor cortex and of the orbital aspects of the frontal operculum commonly produce typical Broca's aphasia. Extensive lesions of the frontal operculum extending into the insula and into the basal ganglia are associated with severe language and motor deficits of the type described by

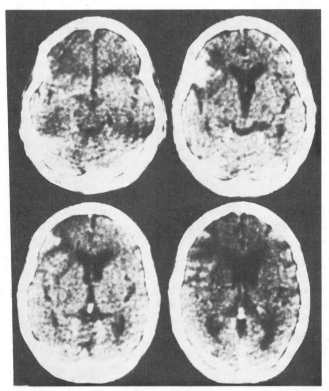

Figure 2   CT of a patient with Broca's aphasia. The photos show a high density image in the lower aspect of the left frontal operculum. The lesion was a nonhemorrhagic infarction and the high density quality was the result of "gray matter enhancement." Such a phenomenon is noted during the period of natural repair of a lesion when iodine contrast is injected intravenously prior to scanning.

Broca in his original case. On occasion, such lesions cause at first what amounts to a global aphasia, which may in time turn into a Broca-like aphasia (see Figure 2 for an example). At any rate, current CT evidence confirms the notion that Broca's aphasia is related to damage of the anterior language zone, and that it is not, as Marie and others would have it, the result of a combination of lesions in Wernicke's area and in the motor system. Past controversy clearly stems from the fact that the anatomo-clinical correlation in this syndrome is not as consistent as in Wernicke's aphasia, in itself an interesting issue quite worthy of further inquiry.

Conduction Aphasia

Conduction aphasia resembles Wernicke's aphasia in the presence of mostly fluent abnormal speech and poor repetition, but relatively well-preserved auditory comprehension. The controversy surrounding conduction aphasia had to do with the distinctiveness of the entity that had been all but ignored in some classification systems. The entity was revived by Konorski (1961) and Geschwind (1965), who noted its importance for the understanding of the physiopathology of the aphasias. But although the syndrome became accepted, the matter of whether or not it had an equally distinctive anatomical correlate rested on the study by Benson and associates (1973) in which the results of three postmortem cases were analyzed. More recently, CT scan findings from several studies, covering a large number of cases, have proved that conduction aphasia is commonly associated with an independent lesion located between the area frequently damaged in Wernicke's aphasia and the area in the frontal operculum commonly associated with Broca's aphasia and its variants. In some cases the lesion sits exclusively in the lower parietal region (the supramarginal gyrus also known as Brodmann's field 40), involving the underlying white matter extensively. In other cases the lesion involves a part of the auditory and insular cortices and the white matter underlying both, sparing the supramarginal gyrus. In such cases the lesion spares most of the auditory association cortex (Wernicke's area) while consistently involving the primary auditory cortex. In some instances the lesion encompasses all three loci; the supramarginal, auditory, and insular cortices (see Figure 3). Rarely, damage to the insular cortex and underlying white matter alone has been associated with this syndrome. Such findings presented in studies by Naeser & Hayward 1978, Kertesz et al 1979, and Damasio & Damasio 1980 led to the inescapable conclusion that anatomically as well as behaviorally, conduction aphasia is an intermediate syndrome between Wernicke's and Broca's aphasias. The strategic placement of the lesion, with its complete or partial sparing of both Wernicke's and Broca's areas, results in remarkably preserved auditory comprehension and speech production. On the other hand, the lesion compromises structures capable of transferring auditory information into the motor system, a physiological step required for the act of repeating a verbal sentence verbatim.

Global Aphasia

Patients with global aphasia have marked nonfluent speech, poor repetition, and poor comprehension. They have traditionally been thought to have extensive lesions involving, in contiguous fashion, all of the perisylvian territory of the left hemisphere. CT has largely confirmed that this is indeed the most common basis for that most devastating of the aphasias in which comprehension and expression of language, as well as repetition, are severely disturbed. In vascular cases, the lesion encompasses all the territories supplied by cortical branches of the middle cerebral artery and often involves the lenticulo-striate territory too, damaging the basal ganglia and capsule. But CT has shown also that slightly different patterns can cause global aphasia. For instance, two

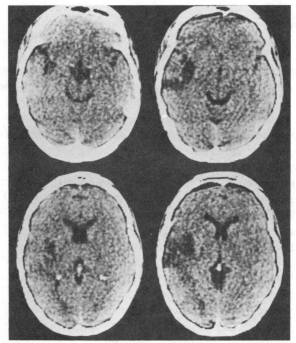

Figure 3   CT of a patient with conduction aphasia. The photos show a lesion involving the left primary auditory cortex and extending anteriorly into the insular region. It was caused by a nonhemorrhagic infarction.

noncontiguous lesions, one placed in Broca's area and another in Wernicke's area, can cause global aphasia with or without involvement of the underlying white matter and basal ganglia (Damasio 1981). Furthermore, the patients with basal ganglia component tend to have a more severe speech defect.

Transcortical Motor Aphasia and Mutism

Transcortical aphasia is a curious term used to designate an aphasic disorder in which repetition is preserved. Patients with transcortical motor aphasia (TMA) have good repetition and good comprehension despite poor language output. The pathological basis of the transcortical aphasias had not been fully elucidated prior to the age of the CT scan. The original cases of transcortical aphasia had been identified clinically and related to a hypothesized anatomical basis long before the first postmortem cases became available. Traditionally, it had been accepted that lesions located anteriorly and superiorly to Broca's area could cause transcortical motor aphasia. Radioisotope scan data (Rubens 1975, Kertesz et al 1977) supported this association and extended the anatomical locus to include not only lateral frontal lesions but also mesial frontal lesions, the common anatomical denominator being that they spared Broca's area and were located in the left hemisphere. CT has changed this panorama remark- . ably. Most cases of true transcortical motor aphasia appear with lesions of the lateral frontal lobe that spare Broca's area totally or in part. In vascular cases damage is clearly the result of infarction in the prefrontal or vascular border zone territories (i.e. the border between the anterior and middle cerebral distributions). In a significant number of cases the lesions are exclusively subcortical, located deep in the frontal lobe, anteriorly and superiorly to the left frontal horn. Such lesions result from infarctions in the deep aspect of the border zone between the anterior and middle cerebral vascular territories. But the patients with lesions located in the mesial aspect of the frontal lobe, which involve the supplementary motor area, the cingulate gyrus, or both, do not have transcortical motor aphasia and present, rather, with a syndrome of mutism and akinesia. Although those patients share some characteristics with TMA such as little spontaneous speech, comparatively good comprehension, and preserved repetition, their general behavior is entirely different in that they show no intent to communicate by word or gesture and have a marked disturbance of affect and emotion. In other words, their drive to communicate, not just to speak, is disturbed, while there is no conclusive evidence that their language processing is altered. In keeping with the fact that their lesions compromise the mesial limbic system cortices and not the lateral regions of neocortex, their disorder is motor and affective and not primarily linguistic.

Transcortical Sensory Aphasia

In transcortical sensory aphasia (TSA), repetition is preserved despite poor comprehension. CT has not only shown that the most prominent lesion in Wernicke's aphasia is distinct from that in conduction aphasia but also that it is distinct from that of transcortical sensory aphasia. In TSA the lesion spares most of the auditory association cortex but involves either the nearby parietal or the nearby inferior temporal cortices, or both. Not uncommonly it involves visual association cortex too, in the lateral aspect of the adjoining occipital lobe and in the white matter underneath (Damasio 1981). The lesions associated with transcortical sensory aphasia are generally the result of border zone infarcts, i.e. infarcts in the territory supplied by terminal branches of both the middle and posterior cerebral arteries.

CT clearly confirmed the notion that most cases of transcortical aphasia (motor or sensory) are related to lesions outside the immediate perisylvian circle. The reader is referred elsewhere for an appraisal of the physiopathologic significance of these "isolation of speech area" syndromes (Geschwind et al 1968). As for the clinical significance, it is obvious: The finding of transcortical aphasia is diagnostic of a specific cerebral location and, in vascular cases, is suggestive of a specific type of disease process.

Atypical Aphasias

Systematic observation of CTs revealed that some syndromes of aphasia that could not be classified according to the Boston Diagnostic Aphasia Examination categories were associated with lesions that did not involve the cerebral cortex and that, instead, compromised structures of the basal ganglia and thalamus of the dominant hemisphere. Damasio et al (1982) and Naeser et al (1982) have noted that the head of the caudate, the anterior aspect of the putamen, and the anterior limb of the capsule are often involved in those aphasias. Some patients have fluent speech, others have nonfluent speech, but most have disturbed auditory comprehension and, invariably, have disturbances of articulation and right hemiparesis. It is too early to know the mechanisms that lead to these language disturbances and, in fact, whether those mechanisms are at all comparable to those which produce the cortical aphasias. It is also too early to decide whether or not more than one individual syndrome exists, or to comment on the relative severity of these aphasias in comparison with the more typical ones related to lesions involving cortical loci. Our current impression is that these are, in general, milder aphasias. Previously, Hier et al (1977) and Alexander & LoVerme (1980) had found atypical aphasias in association with hemorrhages of the basal ganglia in patients with hypertension. Those cases share some features with the ones described above but had, in addition, major disturbances of attention, which precluded a clear analysis of the linguistic deficits. Furthermore, those large hemorrhagic lesions, by virtue of their space-occupying nature, did not permit the same kind of anatomical correlation attempted in the studies by Damasio and by Naeser. Many of these reservations apply to the cases of atypical aphasias described with hemorrhages in the thalamus. A detailed anatomical localization is difficult on the basis of CT data. The clinical characteristics of those cases are highly variable, although they are generally suggestive of transcortical sensory aphasia. [Mohr et al (1975) and Cappa & Vignolo (1979) have described those clinical characteristics in detail.] A similar picture has been found recently in patients with nonhemorrhagic infarctions of the thalamus (Graff-Radford et al 1983).

The Alexias

Some of the finest discoveries of classic aphasiology concern the anatomical underpinnings of reading disorders, the fundamentals of which were proposed by Dejerine based on the autopsy findings in two patients. One had lost the ability to read and write (acquired alexia-with-agraphia) (Dejerine 1891). The other had also lost the ability to read but his ability to write remained intact (acquired alexia-without-agraphia, also known as "pure alexia") (Dejerine 1892). The reader is referred elsewhere for information on the physiological and clinical significance of these discoveries (Geschwind 1965). CT has confirmed beyond a shadow of a doubt that the two syndromes have markedly different pathological correlates. Alexia-with-agraphia, in its rare and pure form, is associated with lesions of the vast extent of association cortex that lies in the left parietal lobe, behind the auditory association cortex. Most commonly those lesions involve the angular gyrus or area 39 (quite often alexia and agraphia are associated with Wernicke's aphasia in patients in whom the underlying lesion extends from Wernicke's area to the angular gyrus). The syndrome of alexia-without-agraphia (pure alexia) is associated with a left occipital lobe lesion or lesions that compromise the flow of visual information from both the left visual cortex and the right visual cortex, in the direction of the left language cortex (information from the latter courses in the splenium of the corpus callosum). CT has been of help in demonstrating that more than one pattern of damage can accomplish such a complex pattern of disturbance (Damasio & Damasio 1983). For instance, patients can have an isolated lesion of the splenium of the corpus callosum interrupting leftward visual information outflow, combined with another lesion located more posteriorly in the occipital lobe and capable of disconnecting the left visual cortex from the left temporo-parietal language cortices. More commonly, patients have a continuous lesion that severs connections from both visual systems and thus cuts off both visual cortices from the left temporo-parietal region. It is interesting to note that, as might be predicted from the physiological mechanism, the lesion need not lie within the splenium of the corpus callosum itself, provided that the callosal fibers are indeed damaged in their intrahemispheric course after passing through the splenium. CT has also been crucial in clarifying the relation between pure alexia and a variety of possible accompanying manifestations, such as color "agnosia" (a condition in which the patient is unable to name colors shown to him or to identify colors given their names, but is capable of matching colors), achromatopsia (a condition in which the patient is unable to perceive color and reports the perception of shades of gray while he maintains normal form vision), and the visual agnosias. About half of all patients with pure alexia also have color "agnosia." A small percentage have achromatopsia but not color agnosia, and the remainder have alexia in the absence of either. The association between pure alexia and visual agnosia is uncommon. Its presence indicates that in addition to the left hemisphere lesion, a supplemental lesion of the right hemisphere has occurred in a fairly comparable location of the nondominant hemisphere's visual system (Damasio & Damasio 1982).

Concluding Remarks

At a time in which autopsy studies have become much less common, the availability of accurate, in vivo, neuroimaging techniques is most welcome. When used proficiently, with an awareness of the many methodological limitations, CT has provided precious information concerning the neural basis of language. In addition to confirming cerebral localizations proposed on the basis of a handful of single case studies, CT has permitted experimental behavioral studies in relation to the lesions detected. Furthermore, it has made possible the discovery of several new, previously unsuspected, anatomo-clinical entities.

But, perhaps more importantly, by repeatedly revealing the consistency of brain/behavior correlations in aphasia, obtained in different research centers and with different technical equipment, CT has contributed to the acceptance and consolidation of a neuroanatomical way of conceptualizing behavior against which there was widespread resistance as late as a decade ago.

The introduction of PET scanning, tomographic regional cerebral blood flow (SPECT), nuclear magnetic ressonance (NMR), all currently in their preliminary stages, will reinforce and amplify the results obtained with CT, by adding a dynamic component to the images now available.

ANATOMY OF HUMAN CEREBRAL DOMINANCE

As late as the mid 1960s the standard view regarding cerebral dominance for language stated that it had no anatomical correlates, that it did not exist in other species, and that its evolution in humans could not be studied. Such a position (exemplified by Von Bonin 1962) implied the neglect or active rejection of the older anatomical studies on dominance (Pfeiffer 1936). But the discoveries of the past 15 years have proven that each of these standard views was false and have opened entirely new avenues of study. We present some data on human asymmetries as well as a brief discussion of findings in other species.

Geschwind & Levitsky (1968) showed not only that anatomical asymmetry was present in the human brain but also that it was readily visible even to the naked eye. They studied the planum temporale, i.e. the area lying between the posterior margin of the first transverse gyrus of Heschl and the posterior end of the sylvian fossa, in the brains of 100 normal individuals. This area was larger on the left in 65%, larger on the right in 12%, and approximately equal in 23%—results that were highly significant statistically, and have since been replicated repeatedly. The area at issue is the portion of Wernicke's area contained within the sylvian fissure, on the upper surface of the temporal lobe. Most of Brodmann's cytoarchitectural field/22 is contained in the planum temporale (see Figure 4).

This asymmetry is present in the fetus and newborn infant (Wada et al 1975, Witelson & Pallie 1973) and appears by the thirty-first week of gestation (Chi, et al 1977). The right side develops about a week earlier than the left. Galaburda et al (1978) showed that this gross asymmetry corresponded to asymmetry of the cytoarchitectonic area Tpt (in the nomenclature of Sanides). The left planum temporale is on the average one-third larger than the right planum, and is often many times as large. The magnitude of the microscopic asymmetry is even more striking. In the first brain studied, the left Tpt was

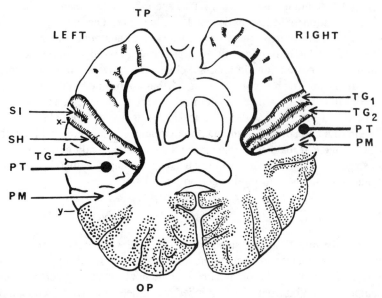

Figure 4 Upper surfaces of human temporal lobes exposed by a cut in the plane of the Sylvian fissure. Anatomical landmarks and typical left-right differences are shown. The posterior margin (PM) of the planum temporale (PT) slopes backward more sharply on the left than on the right, so that end y of the left Sylvian fissure lies posterior to the corresponding point on the right. The anterior margin of the planum formed by the sulcus of Heschl (SH) slopes forward more sharply on the left. In this brain there is a single transverse gyrus of Heschl (TG) on the left, but two on the right (TG$_1$, TG$_2$). TP= Temporal pole; OP= occipital pole; SI= sulcus intermedius. Reproduced from Geschwind & Levitsky (1968).

more than 700% larger than the corresponding area on the right. Even when the right side is large, which Wada (1981) finds, interestingly, to be more frequent in females, it rarely attains this magnitude (see Figure 5).

Another human asymmetry described by LeMay & Culebras (1972) involves the sylvian fissures. The middle cerebral artery travels in the plane of this fissure and can be visualized by arteriography. In most people the right sylvian fissure turns up more sharply at its end than the left. This finding was of importance since it showed a marked asymmetry that could be visualized in the living human, thus creating the possibility of correlating asymmetries with functional characteristics, such as handedness. In right handers, they found this pattern in 67%, with the reverse pattern in 8%, and fissures at equal levels in 25%. In left handers the corresponding figures were 20%, 10%, and 70%.

Utilizing CT, LeMay (see Galaburda et al 1978) showed that in most individuals there is a wider occipital region on the left and a wider frontal region on the right, and she claimed that the distributions differed with handedness. There seems to be agreement that the left posterior region is generally wider but

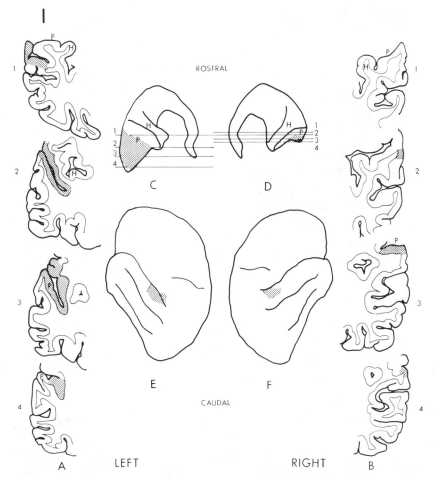

Figure 5   Coronal sections of a human brain are shown in A and B. The superior temporal planes are shown in C and D with planum temporale (P) and Heschl's gyrus (H). Lateral surfaces of brain show area Tpt (*shaded*) in E and F. *Arrowheads* point to Sylvian fissures. Note large planum asymmetry (C and D). Note buried temporal cortex on coronal sections (*arrows*). Reproduced from Galaburda et al (1978).

not that the distribution differs in left handers (Chui & Damasio 1980). The reasons for the discrepancies are not entirely clear. In one study, one of the "normal" groups contained several patients with early brain disease, a group that might well have an anomalous pattern of asymmetry. There may also be variations in technique, e. g. different angles of the cut through the brain might give variable results. Another possibility now under investigation is that some computer programs may foreshorten the brain and thus obscure certain asymmetries. Further study is obviously needed in this area (see Figure 6).

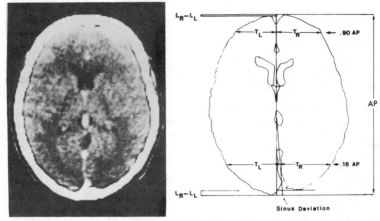

Figure 6   CT of a normal male with left occipital petalia and diagram of reference points used to measure petalia and other brain asymmetries with CT. AP= antero-posterior diameter. 0.16 AP and 0.90 AP= levels at 16% and 90% of AP length (measured from most posterior point) used to determine transverse diameter. TL and TR= left and right transverse diameters of cerebrum (occipital TL and TR are determined at 0.16 AP; frontal TL and TR at 0.90). LL and LR = length of left and right hemisphere, used to determine petalia.

Another of LeMay's findings replicated in the study by Chui & Damasio (1980) was that of predominant left occipital "petalia." Petalia is an anthropological term used to designate an impression in the inner table of the skull, produced by the bulging of the adjacent cerebral lobe.

Evolution of Human Asymmetry

In many individuals, some major fissures and sulci may leave an imprint in the skull. Aware of this fact, LeMay & Culebras (1972) studied the endocranial casts of human fossils. It is apparent that in the cast of the Neanderthal man of LaChapelle-aux-Saints (dating back 30,000 to 50,000 years) the imprint of the right sylvian fissure is higher than that of the left, i.e. the same asymmetry seen in modern humans. It was also found in Peking man, probably dating back about 300,000 years. Because most of the cortical regions related to language processing lie near the sylvian fissure, these findings suggest an early evolution of language.

The significance of these data is increased by studies in nonhuman primates. LeMay & Geschwind (1978) found that there were patterns of asymmetry in the great apes identical to those of modern humans but that brains of rhesus monkeys showed no comparable asymmetry. Yeni-Komshian & Benson (1976) reported that in the chimpanzee the left planum temporale is typically larger than the right, as in humans. These results, showing for the first time a definite anatomical asymmetry in another mammal, raise an important question: Do the great apes possess language abilities, or do their perisylvian regions contain structures destined for some still unspecified function that is a forerunner of language?

Functional Asymmetry in Other Species

Nottebohm (1970) was responsible for overturning the standard belief that functional asymmetry existed in no other species. He has shown that in several species of singing birds, although not in all, cutting the left hypoglossal nerve will almost entirely abolish song, while cutting the right hypoglossal nerve will cause only minor alterations of song production. If the left nerve is cut in early life, compensation takes place and singing will occur in adult life. Since bird song (only present in males) is related to sex, the nucleus HVc, which contains testosterone receptors, is larger in males. It is of interest that there is no significant side asymmetry in this nucleus; this suggests that some forms of cerebral dominance might be based on biochemical differences. The hypoglossal nucleus is, however, significantly larger on the left side, which perhaps is important in determining asymmetrical function within the central nervous system.

Many examples of functional or anatomical asymmetry have been found in other species. We believe that both anatomical and functional asymmetry extend well back in the phylogenetic scale and that they are standard features of the animal world.

The most extensive experiments on functional asymmetry in a nonhuman mammalian species are those of Denenberg (1981) and his colleagues. He found that in the laboratory rat, lesions of the right or left hemispheres can have different effects. These studies provide evidence for right hemisphere advantage for spatial and emotional functions, a finding of special interest since the human right hemisphere has similar advantages for precisely the same functions. Studies on functional asymmetry in other species are summarized by Denenberg (1981), to which we refer the reader.

Anatomical asymmetry is not confined to primates. Diamond (1981) has shown greater cortical thickness in parts of the right posterior hemisphere in the rat and A. M. Galaburda (unpublished observations) has found larger cytoarchitectonic areas on the right side in the rat. Wada & Kolb have also found right-sided asymmetry in other species.

There are thus both anatomical and functional findings in other mammals that suggest the existence of right hemisphere dominance. Interestingly, in humans, right hemisphere dominance appears to be more consistent than language dominance. Thus, some adults develop crossed aphasias, i.e. aphasia with right hemisphere lesions. But it is not common to find visuospatial defects following left posterior lesions that are as severe as those seen after right posterior lesions. (The typical right hemisphere functions thus reverse sides less often than language.) We suspect that right hemisphere dominance appeared earlier in evolution than left hemisphere dominance, which only developed to its highest levels at a later period.

Anatomical asymmetry has been known for years in submammalian species. For instance, the pineal eye of lizards is asymmetrically placed. The right habenular nucleus of the lamprey is much larger than the left (Kappers et al 1936) and there is asymmetry of the same nuclei in the frog (Kemali & Braitenberg 1969). Such subcortical asymmetries may well be the forerunners of the cortical asymmetries.

Chemical Asymmetries in the Nervous System

After studying epileptic patients undergoing surgery, Serafetinides (1965) suggested that LSD acted primarily on the right temporal lobe, thus raising the issue of neurochemical asymmetry. The work of Glick and his co-workers (1977) has clearly documented chemical asymmetries. They found that uni- lateral injection of dopaminergic drugs will cause the rat to circle to one side consistently. The dopamine content of the caudate nucleus that is opposite the direction of circling is higher than the one on the other side. Circling tendencies differ in males and females, a fact of interest since human females tend to be somewhat more right-handed than males. Animals that are strongly consistent circlers learn mazes better than those who are not, presumably because they have an inborn marker for right and left. The predominance of chemical asymmetry on one side is correlated with a consistent anatomical pattern of other chemical asymmetries. Neurochemical asymmetry in the human thala- mus has been found by Oke et al (1978), who have described higher nore- pinephrine content in the left pulvinar and right VPL. Eidelberg & Galaburda (1982) have now reported anatomical asymmetries in the human thalamus.

Asymmetries in Human Disease

Among children with dyslexia, stuttering, autism, and other developmental disorders, all of which are predominantly seen in males, there is some evidence for an increased proportion of left-handedness or ambidexterity. Galaburda & Kemper (1979) studied the asymmetries of cellular architecture in the brain of a severe childhood dyslexic with a strong family history of dyslexia. They discovered many microscopic anomalies confined to the left hemisphere that involved, in particular, area Tpt (part of Wernicke's area). Groups of nerve cells were found in abnormal locations and normal cortical structure was disrupted. There was distinct polymicrogyria, i.e. islands of cortex in the white matter. These disturbances clearly resulted from disordered migration of neurons during development. Such left hemisphere anomalies could well have been the cause of the patient's left-handedness. A careful review of the literature suggests that similar findings had been present in other cases, although they were not specifically commented on as the cause of the learning disability. A second case now being examined by Galaburda & Kemper (1979) shows similar anomalies.

It is interesting to note that Galaburda & Kemper found that the abnormal left hemisphere was larger, not smaller, than the right hemisphere. This could have been the result of slowed neuronal migration in that hemisphere, and is comparable, in milder degree, to what is found in megalencephaly in which striking neuronal migration deficits lead to a very enlarged hemisphere. If that is the case, the study of CT asymmetries may be useful in the investigation of variants in normal populations, but may not be helpful in cases of developmental disturbance.

Another type of disease association with cerebral dominance has recently been reported. Geschwind & Behan (1982) have shown that immune diseases were two and one-half times as common in 500 strong left-handers as in 900 strong right-handers, and have advanced an experimentally testable theory regarding the neurobiological basis of the development of dominance and of its pathology.

Concluding Remarks

For most biological traits it would be ideal to know the relevant anatomy, genetics, fetal and childhood development, evolution, physiology, chemistry, immunology, associations with other traits, and its forms in other species. This type of information has been notably lacking for the language systems even though language ranks among the most distinctive human traits. The recent neurological discoveries in cerebral dominance have opened up areas previously considered inaccessible.

Human biology and medicine always proceeds more rapidly when an animal model is available, and this is now the case for cerebral dominance. Many people would still assert that we will learn little from other species that do not have language. We believe, however, that animal models will lead to experiments that will cast significant light on the evolution of language and other human functions, as well as to new approaches to the study of disordered dominance, either developmental or acquired.

Literature Cited

Alexander, M. P., Lo Verme, S. R. 1980. Aphasia after left hemispheric intracerebral hemorrhage. *Neurology* 30:1193–1202

Benson, D. F. 1967. Fluency in Aphasia: Correlation with radioactive scan localization. *Cortex* 3:373–94

Benson, D. F., Sheremata, W. A., Buchard, R., Segarra, J., Price, D., Geschwind, N. 1973. Conduction aphasia. *Arch. Neurol.* 28:339–46

Cappa, S. F., Vignolo, L. A. 1979. "Transcortical" features of aphasia following left thalamic hemorrhage. *Cortex* 15:121–30

Castaigne, P., Lhermitte, F., Signoret, J. L., Abelanet, R. 1980. Description et étude scannographique du cerveau de Leborgne (la découverte de Broca). *Rev. Neurol.* 136(10):563–83

Chi, J. G., Dooling, E. C., Gilles, F. H. 1977. Gyral development of the human brain. *Ann. Neurol.* 1:86–93

Chui, H. C., Damasio, A. R. 1980. Human cerebral asymmetries evaluated by computerized tomography. *J. Neurol. Neurosurg. Psychiatry* 43:873–78

Damasio, A., Damasio, H. 1982. Cerebral

localization of complex visual manifestations: Clinical and physiologic significance. *Neurology* 32(4):96

Damasio, A., Damasio, H. 1983. The anatomical basis of pure alexia and color "agnosia." *Neurology.* In press

Damasio, A., Damasio, H., Rizzo, M., Varney, N., Gersh, F. 1982. Aphasia with nonhemorrhagic lesions in the basal ganglia and internal capsule. *Arch. Neurol.* 39:15–20

Damasio, H. 1981. Cerebral localization of the aphasias. In *Acquired Aphasia*, ed M. Taylor Sarno. New York: Academic

Damasio, H., Damasio, A., Hamsher, K., Varney, N. 1979. CT scan correlates of aphasia and allied disorders. *Neurology* 29:572

Damasio, H., Damasio, A. 1980. The anatomical basis of conduction aphasia. *Brain* 103:337–50

Dejerine, J. 1981. Sur un cas de cecité verbale avec agraphie, suivi d' autopsie. *Mem. Soc. Biol.* 3:197–201

Dejerine, J. 1982. Contribution a l'étude anatomo-pathologique et clinique des differentes varietés de cecité verbale. *Mem. Soc. Biol.* 4:61–90

Denenberg, V. H. 1981. Hemispheric laterality in animals and the effects of early experience. *Behav. Brain Sci.* 4:1–49

Diamond, M. C., Dowling, G. A., Johnson, R. E. 1981. Morphologic asymmetry in male and female rats. *Exp. Neurol.* 71:261–68

Eidelberg, D., Galaburda, A. M. 1982. Symmetry and asymmetry in the human posterior thalamus. Part I: Cytoarchitectonic analysis in normal persons. *Arch. Neurol.* 39:325–32

Galaburda, A. M., Kemper, T. 1979. Cytoarchitectonic abnormalities in developmental dyslexia: A case study. *Ann. Neurol.* 6:94

Galaburda, A. M., Sanides, F., Geschwind, N. 1978. Human brain: Cytoarchitectonic left-right asymmetries in the temporal speech region. *Arch. Neurol.* 35:812–17

Geschwind, N. 1965. Disconnexion syndromes in animals and man. *Brain* 88:237–94, 585–644

Geschwind, N., Behan, P. 1982. Left-handedness: Association with immune disease, migraine, and developmental learning disorder. *Proc. Natl. Acad. Sci USA* 79:5097–5100

Geschwind, N., Levitsky, W. 1968. Human brain: Left-right asymmetries in temporal speech region. *Science* 161:186–87

Geschwind, N., Quadfasel, F., Segarra, J. 1968. Isolation of the speech area. *Neuropsychologia* 6:327–40

Glick, S. D., Jerussi, T. P., Simmerberg, B. 1977. Behavioral and neuropharmacological correlates of nigrostriatal asymmetry in rats.

In *Lateralization in the Nervous System,* ed. S. Harnad, R. W. Doty, L. Goldstein, J. Jaynes, G. Krauthamer, pp. 213–49. New York: Academic

Graff-Radford, N., Eslinger, P., Yamada, T., Damasio, A. 1984. Nonhemorrhagic infarction of the thalamus: Behavioral, anatomical and physiological correlates. *Neurology.* In press

Hier, D. B., Davis, K. R., Richardson, E. P., et al 1977. Hypertensive putaminal hemorrhage. *Dev. Neurol. Behav. Pediatr.* 1:54–57

Kappers, C. U. A., Huber, G. C., Crosby, E. C. 1936. *The Comparative Anatomy of the Nervous System of Vertebrates Including Man,* Vol. 1. New York: Macmillan

Kemali, M., Braitenberg, V. 1969. *Atlas of the Frog Brain.* Berlin/New York: Springer

Kertesz, A., Lesk, D., McCabe, P. 1977. Isotope location of infarcts in aphasia. *Arch. Neurol.* 34:590–601

Kertesz, A., Harlock, W., Coates, R. 1979. Computer tomographic localization, lesion size, and prognosis in aphasia and nonverbal impairment. *Brain Language* 8:34–50

Konorski, J., Kozniewska, H., Stepien, L. 1961. Analysis of symptoms and cerebral localization of audio-verbal aphasia. *Proc. 7th Int. Congr. Neurol.* 2:234–36

LeMay, M., Culebras, A. 1972. Human brain—morphological differences in the hemispheres demonstrable by carotid angiography. *N. Engl. J. Med.* 287:168–70

LeMay, M., Geschwind, N. 1978. Asymmetries of the human cerebral hemispheres. In *Language Acquisition and Language Breakdown,* ed. A. Caramazza, E. B. Zurif, pp. 311–28. Baltimore: Johns Hopkins Univ. Press

Mohr, J. P. 1976. Broca's area and Broca's aphasia. In *Studies in Neurolinguistics,* ed. H. Whitaker. New York: Academic

Mohr, J. P., Watters, W. C., Duncan, G. W. 1975. Thalamic hemorrhage and aphasia. *Brain Language* 2:3–17

Naeser, M. A., Hayward, R. W. 1978. Lesion localization in aphasia with cranial computed tomography and the Boston diagnostic aphasia exam. *Neurology* 28(6):545–51

Naeser, M. A., Alexander, M. P., Helm-Estabrooks, N., Levine, H. L., Laughlin, S. A., Geschwind, N. 1982. Aphasia with predominantly subcortical lesion sites: Description of three capsular/putaminal aphasia syndromes. *Arch. Neurol.* 39:2–14

Nottebohm, F. 1970. Ontogeny of bird song. *Science* 167:950–56

Oke, A., Keller, R., Mefford, I., Adams, R. N. 1978. Lateralization of norepinephrine in human thalamus. *Science* 200:1411–13

Pfeifer, R. A. 1936. Pathologie der Horstrahlung und der corticalen Horsphare. *Hand. Neurol.* 6:533–626

Rubens, A. B. 1975. Aphasia with infarction in the territory of the anterior cerebral artery. *Cortex* 11:239–50

Serafetinides, A. E. 1965. The significance of the temporal lobes and of hemispheric dominance in the production of LSD-25 symtomatology in man. *Neuropsychologia* 3:69–79

Von Bonin, G. 1962. In *Interhemispheric Relations and Cerebral Dominance*, ed. V. Mountcastle, p. 6. Baltimore: Johns Hopkins Univ. Press

Wada, J. A. 1981. Presentation at Epilepsy Int. Congr., Kyoto, Japan

Wada, J. A., Clarke, R., Hamm, A. 1975. Cerebral hemispheric asymmetry in humans. *Arch. Neurol.* 32:239–46

Witelson, S. F. Pallie, W. 1973. Left hemisphere specialization for language in the newborn: Neuroanatomical evidence of asymmetry. *Brain* 96:641–46

Yeni-Komshian, G. H., Benson, D. A. 1976. Anatomical study of cerebral asymmetry in the temporal lobe of humans, chimpanzees, and rhesus monkey. *Science* 192:387–89

Ann. Rev. Neurosci. 1984. 7:149–70

NEW NEURONAL GROWTH FACTORS

Darwin K. Berg

Department of Biology, University of California at San Diego, La Jolla, California 92093

Introduction

The complexity and remarkable specificity found in the nervous system challenge the investigator to identify molecular mechanisms guiding neuronal development. Cell-cell interactions mediated by diffusible substances have frequently been suggested to play an important role. Studies on naturally occurring cell death in the nervous system provided early evidence for such interactions by demonstrating that neuronal survival beyond a critical period in development requires interaction between the neurons and the postsynaptic target tissue (for a review see Oppenheim 1981). The discovery of Nerve Growth Factor (NGF) raised the possibility that neuronal development, including target-dependent neuronal survival, might be mediated by diffusible molecules. NGF can be isolated as a 140,000 dalton protein complex (7S NGF) of heterologous subunits or as a 26,500 dalton component (beta-NGF) containing two identical polypeptide chains. Experiments in vivo indicate that NGF can reverse naturally occurring as well as experimentally induced cell death in sympathetic and sensory neurons; moreover, anti-NGF antibodies in vivo can block the development of the sympathetic nervous system (for reviews see Thoenen & Barde 1980, Hamburger et al 1981, Levi-Montalcini 1982). That NGF might also influence the choice of targets contacted by neurons is suggested by studies in cell culture showing that NGF can change the direction of neuritic growth (Gundersen & Barrett 1979) and can help shape the neuritic tree maintained by the neuron (Campenot 1982).

Naturally occurring cell death has been documented for almost every population of vertebrate neurons studied (Oppenheim 1981), and every neuron is faced with the problem of finding appropriate target cells. Only sensory and sympathetic neurons, however, have been convincingly shown to depend on

149

0147-006X/84/0301-0149$02.00

NGF in normal development. Accordingly, after the discovery of NGF nearly four decades ago, it seemed likely that further research would soon identify new factors that served other neuronal populations. Until recently, however, the success with NGF remained unique. In retrospect, this can be partially attributed to the fortuitous discovery of an extremely rich (though still unexplained) source of NGF in the male mouse submaxilliary gland that aided greatly in the purification and characterization of the factor.

The search for new factors that regulate neuronal growth and development has mostly employed in vitro assays similar to those used for NGF. In the last few years numerous studies with tissue extracts and cell-conditioned media have revealed activities that differ from NGF although they also promote neuronal survival and development in culture. The activities do not appear to stimulate neuronal cell division, unlike most growth factors operative on non-neuronal cells, perhaps because the neurons used for the assays have been post-mitotic. This work has been reviewed in several recent articles (Varon & Adler 1981, Berg 1982, Barde et al 1983). A new factor that stimulates survival of sympathetic neurons in culture has now been purified to homogeneity from pig brain (Barde et al 1982). This achievement supports the notion that a family of growth factors exists that helps orchestrate neuronal development, and raises hopes that additional factors will soon be purified.

This review examines recent progress in identifying new neuronal growth factors that affect vertebrate neurons. Emphasis is given to components that have been at least partially purified. The material is organized into four sections based on the types of assays used. The first, titled "Neuronal Survival Factors," surveys components that promote neuronal survival in dissociated cell culture. The second section, "Neurite Extension Factors," deals with components that act through the culture medium to stimulate neurite extension. The third, "Substrate Adhesion/Neurite Promoting Factors," reviews neurite-promoting components that act by adsorbing to the culture substratum. The final section, "Developmental Factors," describes components that influence development and possibly differentiation of neurons in culture without having obvious effects on growth per se. Dividing the material according to assay in this manner is convenient but somewhat arbitrary. Like NGF, a component may have more than one kind of stimulatory activity. For example, some of the "survival" and "neurite extension" activities may represent the same component. Further purification and characterization of the factors can be expected to result in some convergence among the numerous stimulatory activities currently being described.

Not included in the present review are factors that influence non-neuronal cells in the nervous sytem such as a glial growth factor purified from bovine pituitary (Brockes et al 1980; for a review see Varon & Manthorpe 1982). Also omitted are components of neural origin that act on muscle or other tissues outside the nervous system, and components such as N-CAM which are thought to mediate cell-cell adhesion in the nervous system (for review see Edelman 1983).

Neuronal Survival Factors

CILIARY GANGLION NEURONS   The chick ciliary ganglion has been a useful preparation for identifying new putative neuronal growth factors, partly because survival of the neurons appears not to be responsive to NGF. Helfand et al (1976) first showed that chick ciliary ganglion neurons can survive and extend neurites in dissociated cell culture when supplied with culture medium conditioned by heart cells. Though half of the neurons normally die in vivo during early development (Landmesser & Pilar 1974), all of the neurons in the ganglion prior to the period of death can survive and develop in dissociated cell culture when grown with muscle cells (Nishi & Berg 1977) or with culture medium supplemented tissue extracts (Nishi & Berg 1979, Tuttle et al, 1980). The fact that the neurons form synapses in culture (Nishi & Berg 1977, Margiotta & Berg 1982) and develop high levels of choline acetyltransferase activity (Nishi & Berg 1979, Tuttle et al 1980) suggest that the neurons continue to develop in culture beyond the state characteristic of neurons in vivo during the period of cell death. These results provide encouraging evidence that components identified in the culture system enhancing neuronal survival and growth may be important for development in vivo as well.

Varon and colleagues have used the ciliary ganglion cell culture system to devise a short-term, rapid assay for neuronal survival factors. The dissociated neurons are put in culture with the test fractions, and the number of surviving neurons are determined after a 24-hr period. Extracts prepared from embryonic eye tissue, which contains all of the normal synaptic targets for the neurons, were found to have the highest levels of survival factor activity in the assay (Adler et al 1979). The amount of activity in eye tissue is itself developmentally regulated and increases markedly during the period in which the neurons become dependent on the target tissue for survival (Landa et al 1980). Moreover, studies with explants of ciliary ganglia in culture suggest that the time course of neuronal dependence on the survival activity parallels the onset of cell death in vivo. Thus neurons in five-day embryonic ganglia do not require the survival factor for maintenance in explant culture for a three-day period, whereas neurons in eight-day ganglia do (Adler & Varon 1982). These properties of the survival activity are consistent with those predicted for a target-derived component mediating survival of the neurons in vivo. Subsequent studies have shown that high levels of survival activity for ciliary ganglion neurons can also be found in extracts of chick heart, another parasympathetic target tissue, when tested later in development (Hill et al 1981), and in CNS wounds in developing and young adult rats (Nieto-Sampedro et al 1982).

Bovine cardiac muscle also contain components that stimulate survival of chick ciliary ganglion neurons in the 24-hr assay (Bonyhady et al 1980). Gel filtration of cardiac extract under different conditions reveals a complex pattern of interactions for the active components (Bonyhady et al 1982). In low ionic strength buffers the activity migrates as a large molecular weight component (greater than 40,000 daltons), while in high ionic strength buffers or in the presence of EDTA the activity migrates primarily as a component of about 20,000 daltons. Recoveries of activity are poor in high ionic strength buffers until one recombines large and small molecular weight fractions, suggesting that the 20,000 dalton component requires a larger component for full expression of activity. The active component also appears capable of aggregating with smaller material, but aggregation in this case does not lead to the same enhancement of activity. The ciliary survival activity from embryonic chick eye tissue may undergo equally complex interactions. Gel filtration of the eye tissue extract in low ionic strength buffers produces a spread of activity in large molecular weight fractions (35,000 daltons and greater) with low recoveries of activity (Manthrope et al 1980). The activity can also be recovered as a component of 20,000 daltons (Varon et al 1983). These results suggest that more than one component in an extract may influence neuronal survival. Fractionation of the extract in this case can lead to losses of activity as the components are separated. The results also illustrate some of the difficulties associated with determining physical properties of impure factors. Interaction with other material in the extracts can radically alter the apparent properties of the active component.

Isoelectric focusing has been used to compare the properties of survival factors in extracts prepared from chick embryo eye tissue and chick and bovine cardiac tissue. Values between pH 4.5 and 5.5 were found for the isoelectric points of the active components from chick tissue (Hill et al 1981, Manthorpe et al 1982a), and 6.2 was reported for the isoelectric point of the component associated with the bovine activity (Bonyhady et al 1980). From the size estimates and isoelectric points, it seems that similar components have been identified in these cases.

A long-term growth assay with ciliary ganglion neurons has revealed a stimulatory factor that may be related to the survival factors described above (Nishi & Berg 1981). Neurons were maintained in dissociated cell culture for a one to two week period using basal conditions chosen to permit survival of all the neurons present originally in the ganglion in the absence of tissue extracts and conditioned media. Extracts prepared from embryonic chick eye tissue were then tested for stimulatory activities that increased growth and development of the neurons beyond the low levels occurring in the basal conditions alone. Two kinds of stimulatory activities were found. One activity stimulates development of choline acetyltransferase (CAT) activity in the cultures without influencing neuronal growth, and is discussed below ("Developmental Factors"). The other stimulates neuronal growth in the cultures while having no effect on levels of CAT activity. This activity, termed *growth-promoting activity* or GPA, appears to be associated with a component of about 20,000 daltons, as indicated by gel filtration of the crude extract. Nearly all of the stimulatory activity can be recovered from the gel filtration column, and titration experiments indicate that GPA can stimulate neuronal growth to the same maximal levels obtained with unfractionated eye extract. Preliminary experiments suggest that the partially purified ciliary survival factor provided by Varon and colleagues from embryonic chick eye tissue has properties similar to those of GPA in the growth assay (J. Fujii and D. Berg, unpublished results), and that a GPA fraction has effects similar to those of the survival factor in the survival assay (M. Manthorpe and S. Varon, unpublished results).

PARAVERTEBRAL SENSORY NEURONS   Perhaps the most impressive achievement in the search for new neuronal growth factors has been the purification of a factor from pig brain that supports the survival of paravertebral sensory neurons in culture (Barde et al 1982). Thoenen and collaborators first demonstrated that a component in glioma-conditioned medium (GCM) can promote the survival of chick dorsal root ganglion (DRG) neurons in dissociated cell culture (Barde et al 1978). The component is not inactivated by anti-NGF antibodies and appears to act on neurons different from those affected by NGF. With the culture conditions used, NGF supports the survival of only a fraction of the neurons present in the ganglion, and the fraction changes with ganglion age. The fraction of neurons supported by glioma-conditioned medium also varies with culture age but in a roughly complementary fashion to that observed with NGF. For example, few if any neurons from 16-day embryonic ganglia survive with NGF, whereas nearly 75% do so with glioma-conditioned medium. Synergistic effects also occur. For eight-day ganglia the combination of glioma-conditioned medium and NGF causes more neurons to survive (80%) than predicted from experiments with glioma-conditioned medium and NGF alone (10% + 25%) (Barde et al 1980). These results demonstrate not only that the GCM factor is different from NGF but also that dorsal root ganglion neurons respond to more than one survival factor and at some stages may require two or more factors.

The finding that rat brain contains an activity similar to that described for glioma-conditioned medium raised the possibility that mammalian brain might be a convenient source for purification of the factor (Lindsay & Tarbit 1979, Barde et al 1980). Starting with pig brain and employing a number of conventional purification steps, including the perhaps not so conventional preparative two-dimensional gel electrophoresis first in urea and then in SDS buffers, Barde et al (1982) were able to purify the active component to homogeneity.

They obtained 1 μg of purified factor from 1.5 kg of brain tissue with a purification of about 1.4×10^6 fold. The component has a molecular weight of 12,300 daltons as determined by gel electrophoresis in SDS under nonreducing conditions or by gel filtration in the absence of SDS. It has an isoelectric point equal to or greater than 10.1, and is active at a concentration of 0.4 ngm/ml. In several respects the component is similar to NGF, e.g. effective concentration, isoelectric point, and molecular weight (though beta-NGF contains two such chains). The component is clearly different from NGF, however, as indicated by its failure to cross-react with anti-NGF antibodies known to inactivate mammalian NGFs and by its specificity. It promotes the survival of populations of dorsal root ganglion neurons not supported by NGF, and does not support the survival of sympathetic neurons in culture (Barde et al 1982).

Given its size and isoelectric point, the sensory neuron survival factor also appears to be different from the factors that support ciliary ganglion neurons. This is a tentative conclusion, however, since the physical properties of the ciliary ganglion factors have only been determined with partially purified material and may not represent the properties of the isolated component. It should be noted that partially purified fractions of ciliary ganglion survival factor from eye tissue have also been reported to promote the survival of sensory and sympathetic neurons in culture (Manthorpe et al 1982c). The survival activity for sensory and sympathetic neurons was found in this case to co-migrate with the ciliary survival activity during step-wise elution from a DEAE-cellulose column and during isoelectric focusing, and displayed a similar sensitivity to heat, pH, and trypsin. Further purification of these activities will be necessary to determine whether they are associated with the same component and to examine the relationship between them and the sensory neuron survival factor purified from pig brain. The finding that some dorsal root ganglion neurons can be maintained in culture as long as one week with only insulin or somatemedin multiplication stimulating activity (MSA) in 2% fetal calf serum suggests that sensory neurons may be responsive to many components, and that the cell culture assay is only the first step in evaluating the in vivo role of such components (Bothwell 1982).

SYMPATHETIC NEURONS   An even more complex pattern emerges for sympathetic neurons, for which at least three survival factors can be distinguished (Edgar et al 1981). Survival of sympathetic neurons in response to NGF and glioma-conditioned medium in cell culture is qualitatively similar to that observed for dorsal root ganglion neurons. NGF supports the survival of few neurons from eight-day or 18-day ganglia, for example, but permits 37% of the neurons from 12-day ganglia to survive. Glioma-conditioned medium supports very few neurons from 8 to 12-day ganglia but permits survival of more than 50% of the neurons from 18-day ganglia. The effects of NGF and glioma-conditioned medium are roughly additive when simultaneously supplied to neurons from 12-day embryos. A factor present in heart cell-conditioned medium (HCM) produces a pattern of survival different from that obtained with either NGF or glioma-conditioned medium. HCM has its maximal effect on 14-day ganglia, permitting 57% of the neurons to survive in cell culture, and supports fewer neurons from younger and older ganglia. The effects of heart-cell-conditioned medium are additive with those of NGF on neurons from 12-day ganglia but are more than additive with those of GCM on the same population of neurons. Neither the GCM nor the heart-cell-conditioned medium factors are inactivated by anti-NGF antibodies. It is not yet clear whether the different factors serve distinct populations of sympathetic neurons or whether individual neurons express changing requirements for the factors as a function of developmental age. The fact that the neurotransmitter enzymes associated with the surviving neurons differ according to the presence of NGF or heart-cell-conditioned medium suggests the possibility that different populations of neurons are selected by each of the survival factors. Alternatively, the factors may induce different developmental destinies in the surviving neurons (see "Developmental Factors").

CNS NEURONS   Diffusible factors present in tissue extracts and conditioned media can enhance survival of spinal cord neurons in culture. Using a back-filling technique to prelabel chick motoneurons in vivo with horseradish peroxidase (HRP), Bennett et al (1980) were able to identify the motoneurons in culture after a two-day period and demonstrate that survival of the cells in this case was increased by culture medium conditioned by skeletal muscle cells but not by medium conditioned by kidney or smooth muscle cells. Slack & Pockett (1982) used the same labeling method and culture system to demonstrate that motoneuron survival during the two-day test period could be supported by extracts prepared from the endplate regions of denervated soleus muscle but not by extracts from non-endplate regions of innervated muscle. Schnaar & Schaffner (1981) fractionated spinal cord cells by buoyant density and showed that cells with the highest specific activities of choline acetyltransferase (CAT), putative motoneurons, required components supplied in muscle-conditioned media for survival in cell culture. Though the active components have yet to be characterized, these studies do suggest a pattern of specificity consistent with the hypothesis that target-derived components influence survival and development of motoneurons.

In apparent conflict with these findings are results of others reporting that muscle-conditioned medium does not enhance survival of spinal cord neurons. Giller and colleagues showed that although mouse muscle cells and muscle-conditioned medium increase the levels of CAT activity that develop in cultures of mouse spinal cord cells, the numbers of cells with neuronal morphologies remain comparable to those in control cultures (Giller et al 1973, 1977; see "Developmental Factors"). Henderson et al (1981) found that muscle-conditioned medium promotes neurite extension of chick spinal cord neurons in culture (see "Neurite Extension Factors") but does not change neuronal survival over a 66-hr test period. No clear resolution emerges from the results, though differences in technique may be responsible (Tanaka et al 1982). These include differences in culture substrata used for the assay (e.g. collagen vs uncoated plastic), differences in serum content in the culture medium (e.g. horse serum, fetal calf serum, or no serum), and differences in the age of embryo used to obtain spinal cord neurons. Of these differences, embryo age seems most interesting since it might reflect a changing dependence of motoneurons on muscle components. That is, motoneurons might not require muscle factors in culture if taken from embryos prior to the period of naturally occurring cell death in vivo. Because motoneurons were not specifically labeled in the studies by Henderson et al and Giller et al, however, it can also be argued that variations in the numbers of motoneurons might have gone undetected within the larger population of spinal cord neurons present.

Recent studies on retinal ganglion neurons suggest a regulation of neuronal survival that parallels results described for chick ciliary ganglion neurons. Bennett and co-workers have used the HRP back-filling technique to label retinal ganglion neurons of chick and rat in situ (Nurcombe & Bennett 1981, McCaffery et al 1982). Histochemistry on transverse sections of embryonic chick retina confirmed that the HRP label was confined to the ganglion cell lamina. After back-filling the retinal ganglion cells, the retinal tissues were dissociated and transferred to cell culture, where the number of HRP-labeled neurons were compared with the number of retinal ganglion neurons predicted from axon counts on the optic nerve. A good correlation was found, indicating that most of the neurons present at the time of dissociation survived the treatment and attached to the culture substratum. The number of retinal ganglion neurons, as indicated both by axon counts in vivo and by initial cell counts in culture, reaches a maximum by day ten in the chick embryo and declines thereafter to about half the maximum number through naturally occurring cell death in vivo. In retinal cultures from ten-day embryos, all of the labeled neurons die over a four-day period when grown alone. Addition of optic tectum fragments results in rescue of nearly all the neurons for the entire four-day period, and culture medium conditioned by the tectal fragments produces a partial rescue. Fragments of cerebellum have no effect. These results indicate that the optic tectum produces a component that can support the survival of retinal ganglion neurons in dissociated cell culture, and that the component rescues neurons that would have died in vivo. Similar though less striking results were obtained with rat retinal ganglion neurons. In this case repeated inoculation of the cultures with diencephalon and mesencephalon fragments (which include the normal synaptic targets of the retinal ganglion cells) produced a partial rescue of the neurons, while repeated inoculation with cerebellum fragments caused a significant though smaller rescue. Again the findings are consistent with the hypothesis that naturally occurring cell death in the retina reflects competition among retinal ganglion neurons for a target-derived component available in limited amounts.

Other neurons within the CNS also appear to respond to diffusible growth factors. Survival of hippocampal cells in dissociated cell culture is increased by culture medium conditioned by astroglial cells (Banker 1980, Müller & Seifert 1982).

Neurite Extension Factors

Stimulation of neuritic outgrowth from ganglionic neurons in culture has been a traditional assay for NGF and was instrumental in its purification and characterization. Similar assays have revealed new factors in a variety of tissues that stimulate neurite production by ganglionic neurons. The components appear to be different from NGF since they are not blocked by anti-NGF antibodies and in some cases act on cell populations that either are unresponsive to or are influenced differently by NGF. The neurite extension assays have also revealed a second class of components that stimulate neurite growth by adsorbing to the culture substratum; these are discussed in a separate section (see "Substrate Adhesion/Neurite Promoting Factors").

PERIPHERAL GANGLION NEURONS   New factors that promote neuritic outgrowth from chick ganglionic neurons have been identified as activities in extracts from a number of chick and rat tissues, including heart, brain, hippocampus, gizzard, sciatic nerve, and even chick yolk sac at early stages (Ebendal et al 1978, Ebendal 1979, McLennan & Hendry 1980, Ebendal 1981, Hill et al 1981, Miki et al 1981, Riopelle et al 1981, Riopelle & Cameron 1981, Collins & Dawson 1982, Crutcher & Collins 1982, Ebendal et al 1982, Richardson & Ebendal 1982). Collins & Dawson (1982) identified two types of components in heart cell-conditioned medium that stimulate production of neurites by chick ciliary ganglion neurons in cell culture. One of the components had previously been shown to act by adsorbing to the polyornithine-coated culture substratum (Collins 1978a,b, 1980; see "Substrate Adhesion/Neurite Promoting Factors"). The second active component does not adsorb to the substratum and has no detectable effect on neurite extension by itself. In conjunction with the substrate adhesion factor, however, it causes a rapid acceleration of neurite extension and a change in growth cone morphology. The second component is nondialyzable and heat- and trypsin-sensitive.

Ebendal and co-workers have described an activity in extracts of embryonic chick tissue that stimulates neurite production by cilary, sympathetic, and dorsal root ganglion explants in culture. Gel filtration of cardiac tissue extract in low ionic strength buffer results in most of the recovered activity eluting in the range of 40,000 dalton material, while a small amount elutes with much larger material (Ebendal et al 1979). The activity from chick tissue is reported to be associated with a component having an isoelectric point of five (Ebendal et al 1982). These properties are similar to those reported for the ciliary ganglion survival factor from chick eye tissue when examined at low ionic strength (Manthorpe et al 1980, 1982a).

It has been reported in several instances that fractionation of tissue extracts yields a co-distribution of stimulating activities for neurite extension and neuronal survival (Collins & Dawson 1982, Ebendal et al 1982, Hill et al 1981). Although it seems reasonable that the two kinds of activity are associated with the same component, given the precedent of NGF with its multiple effects on target neurons, further purification of the neurite extension and neuronal survival activities will be necessary in each case to determine their relationship.

CNS NEURONS   Culture media conditioned by a number of cell types have been reported to stimulate neurite production by spinal cord neurons both in explant and in dissociated cell culture (Dribin & Barrett 1980, Obata & Tanaka 1980, Henderson et al 1981, Pollack et al 1981, Schnaar & Schaffner 1981). No species specificity was found between rat, mouse, and chick tissues, though rat- and mouse-conditioned media were more effective at stimulating neurite extension from all spinal cord explants examined than was chick-conditioned medium (Dribin 1982). Gel filtration of rat fibroblast-conditioned medium revealed two active components on rat spinal cord explants (Dribin & Barrett 1982a). One component with a size of about 300,000 daltons stimulates neurite extension by adsorbing to the culture substratum (see "Substrate Adhesion/Neurite Promoting Factors"). The other component with a size of about 50,000 daltons acts through the culture medium and cannot be preadsorbed to the substratum. Combining the two produces a synergistic effect: More neuritic outgrowth occurs when both components are supplied to the spinal cord explant than predicted by the sum of the two responses obtained with each alone. The extent of neuritic outgrowth obtained with both components together is comparable to that induced by the unfractionated conditioned medium. Both components are trypsin- and chymotrypsin-sensitive and resistant to neuraminidase treatment. (Dribin & Barrett 1980, 1982b).

Studies with chick spinal cord neurons in dissociated cell culture have indicated a similar pattern (Henderson et al 1981). In this case gel filtration of chick muscle-conditioned medium revealed a broad spread of neurite-promoting activity with peaks corresponding to components of approximately 40,000, 500,000, and greater than 10^6 daltons. Again, the activity was found to be trypsin-sensitive. It is not yet known whether any of the components acts by adsorbing to the culture substratum as is true of the large molecular weight component in rat fibroblast-conditioned medium. In contrast to the studies with rat tissues (Dribin & Barrett 1980, 1982a,b), however, medium conditioned by heart and lung were found to be ineffective. Only muscle, liver, and, to some extent, skin were effective at conditioning media to stimulate neurite production. Also, the chick muscle-conditioned medium did not enhance neuronal survival in the dissociated spinal cord cultures, in contrast to results obtained by others (Bennett et al 1980, Slack & Pockett 1982). Possible reasons for this are discussed above (see "Neuronal Survival Factors").

Chick cerebral neurons in dissociated cell culture can be induced to extend neurites by components present in bovine brain extracts (Kligman 1982a) and in chick heart-conditioned medium (Kligman 1982b). The bovine activity is trypsin-sensitive but has the unusual property of being resistant to heating at 100°C for five minutes. The active component was partially purified by taking advantage of its heat-resistance to denature and remove most of the other protein in the extract. The material was then purified further by ion exchange chromatography and gel filtration to yield material 90% homogeneous by sodium dodecyl sulfate (NaDodSO$_4$) polyacrylamide gel electrophoresis and active at 40–100 ngm/ml. Under nonreducing conditions, the major component in the purified material had a molecular weight of about 75,000 daltons on the gels, and under reducing conditions, 37,000 daltons. Minor components in the range of 15,000 to 30,000 daltons were also revealed on the gels under the reducing conditions. It will be important to determine whether the activity is associated with the major component. Despite its affect on neurite extension, the purified material does not enhance survival of cerebral neurons in culture. Other material in the initial bovine brain extract does increase cerebral neuron survival, but the survival activity is lost during purification. The heat-resistance of the neurite extension component is at present unique. No test of heat sensitivity was reported for the neurite-promoting factor in heart cell-conditioned medium that acts on cerebral neurons, though it was shown to be trypsin-sensitive. Heart-conditioned medium does contain temperature-sensitive components that stimulate neurite extension from other neuronal populations, as described above.

Mammalian brain also contains components that stimulate neurite production by retinal tissue in culture. Extracts prepared from pig brain cause a significant increase in the amount of neuritic outgrowth from rat retinal explants at all times examined over a two-week period in culture (Turner et al 1983). Unlike goldfish retinal explants, in which neurite outgrowth may be induced by NGF (Turner et al 1981), rat retinal explants do not respond to NGF, and anti-NGF antibodies do not block neurite production by the explants in response to pig brain extract (Turner et al 1983). The relationship of the stimulatory component to the sensory neuron survival factor recently isolated from the same tissue (Barde et al 1982) and described above (see "Neuronal Survival Factors") will be important to determine.

Chick retinal explants also respond to components from brain tissue. In this case, extracts prepared from chick optic lobe and forebrain hemispheres were found to increase significantly neuritic outgrowth from the retinal explants over a four-day test period (Carrie & Ebendal 1983). Histology of the cultured explants confirmed that the neurites originated from the ganglion cell layer. The activity in optic lobe extract is associated with macromolecular material and cannot be replaced by NGF in the assay. Little effect was seen with extracts prepared from skeletal muscle, heart, and yolk sac. Apparently retinal ganglion cells are unable to respond to components in these latter tissues, previously shown to induce neurite extension from peripheral ganglionic neurons, as described above. This again must be regarded as a tentative conclusion, however, until the active components are purified, since complex extracts may contain inhibitory components that mask the action of stimulatory factors (Manthorpe et al 1982b).

Substrate Adhesion/Neurite-Promoting Factors

Neurite extension assays have revealed a class of stimulatory components that differ from NGF both in their physical properties and mode of action. The components appear to be factors of high molecular weight that stimulate neuritic outgrowth by adsorbing to the culture substratum, and are most effective with polyornithine or polylysine substrata. Neurite-promoting factors of this type can be distinguished by showing that they are removed from the medium when incubated with polyornithine-or polylysine-coated dishes. A "depleted" medium produced in this manner no longer retains the capacity to promote neurite extension when assayed, for example, on neurons in untreated polyornithine dishes. Treated dishes, on the other hand, are especially effective in supporting neurite extension even in conjunction with a "depleted" medium.

Using the chick ciliary ganglion culture system, Collins (1978a,b, 1980) first demonstrated that a culture medium conditioned by heart cells contains a neurite-promoting factor of this type. The component adsorbs to a polyornithine substratum and allows nearly all the neurons to extend at least one neurite in the first hour. Varon and colleagues termed the component *polyornithine-attachable neurite-promoting factor* (or PNPF) and demonstrated that a wide range of cell types can produce PNPF-like material (Varon et al 1979, Adler & Varon 1980, Adler et al 1981). PNPF can stimulate neurite production from a number of neuronal populations of peripheral origin, including parasympathetic, sympathetic, and sensory ganglia, but appears to be ineffective on most neurons tested from the CNS with the exception of embryonic retinal neurons and a small fraction of spinal cord neurons (Adler et al 1981, Adler

1982). Media conditioned by several cell types have been shown to contain components that can adsorb to a polylysine substratum and induce neurite extension from spinal cord explants in culture (Dribin & Barrett 1982a, Nurcombe & Bennett 1982). PNPF does not appear to be required for neurite extension by neurons grown on a collagen substratum.

The molecular properties of the partially purified stimulatory component have been examined in several cases. Lander et al (1982) have shown that bovine corneal endothelial cells release a component into the culture medium that enhances neurite production by adsorbing to the substratum, and the stimulatory activity appears to be associated with a heparan sulfate proteoglycan. The active component is trypsin-sensitive, stable at 60°C and inactivated at 80°C. If the component has the properties of a globular protein, its behavior during gel filtration would suggest a size just under 4×10^6 daltons. Equilibrium sedimentation of the conditioned medium in CsCl density gradients indicates a density for the active component between that of pure proteins and polysaccharides. When a conditioned medium is prepared from cells grown in (^{35}S)sulfate and (^3H)leucine, equilibrium density sedimentation demonstrates that sulfate- and leucine-labeled material indicative of sulfated proteoglycan is found in the same region of the gradient as the neurite-promoting activity. Moreover, activity recovered from the CsCl gradient can be destroyed by a crude heparinase preparation but not by chondroitinase ABC. Passage of the heparinase preparation over a heparan sulfate-Sepharose column removes the material responsible for destroying the neurite-promoting activity. These results are consistent with the suggestion that the activity is associated with a heparan sulfate proteoglycan. Neither heparan sulfate nor any other purified glycosaminoglycan, however, is able to substitute in the assay for the neurite-promoting activity. Accordingly, heparan sulfate may be necessary for the activity, but it is not sufficient. The identification of heparan sulfate as part of the active complex must be considered provisional until the component is purified further. The present evidence depends heavily on the specificity of the heparinase preparation used.

The neurite-promoting components found in culture media conditioned by mouse heart cells (Coughlin et al 1981) and by a rat Schwannoma cell line (Manthorpe et al 1981) that act via the substratum have also been shown to be trypsin-sensitive, stable at 60°C, and inactivated by boiling. The rat Schwannoma component apparently contains sugar moieties since it has been found to bind to the lectins concanavalin A and wheat germ agglutinin (Adler et al 1982). Gel filtration of the partially purified component from mouse heart cells would indicate a size of about 5×10^6 daltons for the component were it to have the shape of a globular protein (Coughlin & Kessler 1982). Agarose/acrylamide gel electrophoresis of the material shows that the activity migrates behind the major bands of protein and glycosaminoglycan-like material. Though dyes known to stain glycosaminaglycans did not stain the active region of the gel, lack of staining may reflect the amount of material present rather than its composition.

The role of these large molecular weight neurite-promoting factors in vivo remains to be determined. It is possible that the components normally promote cellular adhesion, and that the neurite production induced by the factors in culture results from an increased adhesion of the neuron to the substratum. Neurite production in culture is known to depend on the adhesiveness of the substratum (Letourneau 1975a,b, 1979). In this context it would be interesting to know the relationship between these neurite-promoting components and factors termed "adherons" that are released by muscle cells in culture (Schubert & LaCorbiere 1980, 1982). Adherons are large molecular weight complexes containing glycosaminoglycans and protein, and they influence adhesion of cells to the culture substratum. Adherons from smooth muscle cell lines, for example, enhance the adhesion of the pheochromocytoma cell line PC12 to the substratum, whereas adherons from skeletal muscle preparations inhibit it.

A more interesting possibility is that in vivo the neurite-promoting components may serve to guide neurite extension, contributing to the pattern and density of tissue innervation as has previously been suggested (Collins & Garrett 1980, Adler and Varon 1981, Lander et al 1982). Studies on chick ciliary ganglion neurons have shown that the disposition of adsorbed material on the substratum can guide the advancing neurites in culture (Collins & Garret 1980, Adler & Varon 1981). Antiserum raised against the partially purified component from mouse heart-conditioned medium blocks NGF-induced neurite extension in cultures of mouse superior cervical ganglia, but does not prevent NGF from inducing increases in the levels of tyrosine hydroxlase activity in the ganglia (Coughlin & Kessler 1982). Moreover, monoclonal antibodies that recognize a related neurite-promoting factor produced by rat cells (again a proteoglycan complex) cross-react with a component on the surface of rat superior cervical ganglion neurons in vivo (Greif & Reichardt 1982).

Factors in the conditioned medium that act by adsorbing to the substratum may serve other developmental roles as well. It has been found that heart cell-conditioned medium contains a component that adsorbs to the polyornithine substratum and changes the survival requirements of chick sympathetic neurons. Thus NGF normally supports only a fraction of the neurons present in the ganglion at some developmental stages, and additional factors are required to obtain survival of all the neurons, as described above (see "Neuronal Survival Factors"). In the presence of the adsorbed factor, however, NGF is able to support survival of all the neurons (Edgar & Thoenen 1982). The adsorbed material does not support neuronal survival by itself. It is not known whether the active component is the same as that which adsorbs to the substratum and stimulates neurite extension. Again it will be important to purify the activities to determine their relationships, specificities, and modes of action.

Developmental Factors

Studies on neurotransmitter synthesis by neurons in culture have led to the discovery of factors that specifically control the type and amount of transmitter made by the neurons. Since neuronal survival and growth seems not to be altered by these components, they are often thought of as "instructive" or "developmental" factors that influence neuronal differentiation and guide neuronal development in culture. The components may serve similar functions in vivo, acting as environmental "cues" provided by other cell types—including the postsynaptic target tissue, for example—to shape the developmental fate of the presynaptic neurons.

The best characterized factor of this type is a component that acts on rat superior cervical ganglion (SCG) neurons in culture to induce the normally noradrenergic neurons to express cholinergic function. These results are well-established and discussed at length in excellent reviews (Patterson 1978, Le Douarin et al 1981), so they are summarized only briefly here. A number of cell types, including heart and ganglionic nonneuronal cells, make and release the cholinergic factor into the culture medium (Patterson & Chun 1974, 1977). When incubated with rat superior cervical ganglion neurons, the component can induce a 1000-fold increase in the amount of ACh made by the neurons (Patterson & Chun 1974) and permits them to form functional cholinergic synapses on cells with ACh receptors such as skeletal myotubes, heart cells, and other superior cervical ganglion neurons (O'Lague et al 1974, Nurse & O'Lague 1975, Ko et al 1976). The neurons make reduced levels of catecholamines under these conditions, but some cells continue to express a "dual" function over a number of weeks in the sense that the same neuron forms both cholinergic and noradrenergic synapses on target cells (Furshpan et al 1976). These conditions also alter the types of surface proteins on the neurons (Braun et al 1981), and the proteins released into the medium (Sweadner 1981). Neuronal activity prevents the component from inducing a cholinergic phenotype (Walicke et al 1977), apparently by promoting Ca^{2+} entry (Walicke & Patterson 1981). Production of the component by nonneuronal cells is in turn hormonally regulated (Fukada 1980). A 1500-fold purification of the component using ammonium sulfate precipitation followed by ion exchange chromatography and gel filtration yields material active at 1 μg/ml with a size of 40,000–45,000 daltons by gel filtration (Weber 1981). Starting with serum-free conditioned medium, Fukada has recently purified the component 10,000-fold to obtain material active at about 50 ngm/ml, though it still appears to be heterogenous (Fukada 1983; K. Fukada and P. Patterson, personal communication).

Culture medium conditioned by chick heart cells has been found to influence the choice of neurotransmitters synthesized in cultures of chick sympathetic neurons as well, but in this instance it has been suggested that the component acts by permitting the survival of a specific subpopulation of the ganglionic neurons expressing cholinergic properties (Edgar et al 1981). Sympathetic neurons grown with heart-cell conditioned medium develop significantly higher levels of CAT activity and lower levels of tyrosine hydroxylase activity than do neurons grown with NGF. In the presence of both NGF and heart-cell-conditioned medium the levels of CAT and tyrosine hydroxylase activity are each roughly additive with those obtained with NGF and heart-cell-conditioned medium alone. Neurons grown in NGF for 48 hr do not express increased levels of CAT activity when subsequently exposed to heart-cell-conditioned medium. The levels of CAT activity induced by heart-cell-conditioned medium alone are similar to those found in sympathetic ganglion chains in vivo at a time when little preganglionic development has occurred, giving rise to the speculation that the early levels of endogenous CAT activity can be attributed to ganglionic neurons rather than preganglionic terminals in vivo (Edgar et al 1981). Although these results are consistent with heart-cell-conditioned medium selecting a "cholinergic" population of neurons for survival, additional studies will be necessary, as the authors suggest, to exclude the possibility that heart-cell-conditioned medium actually induces cholinergic development in the cultures.

A component has been identified in extracts of embryonic chick eye that stimulates cholinergic development of ciliary ganglion neurons in cell culture (Nishi & Berg 1981). Neuronal survival is not a consideration in this case since all of the neurons in the ganglion can be accounted for in culture and survive throughout the test period even in the absence of the factor. The component induces a two- to four-fold increase in the levels of choline acetyltransferase (CAT) activity expressed by the neurons over a two-week period, but has no effect on the growth rate of the neurons. Although the relative increase is much less dramatic than the one induced by the rat cholinergic factor with superior cervical ganglion neurons, it should be noted that the ciliary ganglion neurons already produce significant levels of CAT activity in the absence of the factor. Gel filtration of the eye extract indicates that the stimulatory activity, termed *CAT-stimulating activity* or CSA, migrates as a component of 40,000–45,000 daltons and elutes from DEAE columns at an ionic strength similar to that found for the rat cholinergic factor (Weber 1981). It will be of interest to determine whether the component shares other properties with the rat cholinergic factor. A rabbit antiserum raised against a crude fraction of CSA specifically blocks the increase in CAT activity induced by CSA in culture without influencing neuronal survival, neuronal growth rate, or the basal levels of CAT activity obtained in the absence of CSA (Stollberg & Berg 1982). A partially purified fraction of CSA active at 50–100 ngm/ml still contains a number of components visible after NaDodSO$_4$ gel electrophoresis.

A possible in vivo role for developmental factors of this type is suggested by an elegant series of transplantation experiments on avian neural crest tissue by Le Douarin and colleagues. Though different regions of the neural crest produce sympathetic and parasympathetic neurons normally, transplanting a piece of the crest to a new location can alter the choice of neurotransmitters expressed by the progeny cells so that, for example, cholinergic neurons now arise from tissue that would have produced adrenergic neurons. These results indicate that the local environment can play an important role in directing neuronal development (for review see Le Douarin et al 1981). The cholinergic factors described here may represent mechanisms by which nonneuronal cells in the local environment can influence neuronal development.

An example of conditioned medium factors that influence neurotransmitter synthesis in CNS neurons comes from studies on mouse spinal cord cells in dissociated cell culture. A medium conditioned by skeletal muscle cells as well as a number of other cell types dramatically increases the levels of CAT activity that develop in the cultures (Giller et al 1973, 1977, Godfrey et al 1980). The effect appears to be specific in the sense that levels of glutamic acid decarboxylase, another neurotransmitter enzyme present in the cultures, are not increased by the conditioned medium. It is argued that the increase in CAT levels represents an induction of the enzyme in existing cells rather than an increased survival of cholinergic neurons, since no significant difference was seen in the number of surviving neurons under the two conditions. This conclusion should be regarded as tentative, however, since cholinergic neurons were not specifically labeled and could not be readily distinguished from other neurons in the cultures.

Summary

The interest in identifying new neuronal growth factors derives largely from the expectation that the local environment and synaptic target tissue help determine neuronal survival and development. Diffusable growth factors represent an attractive mechanism for mediating cell-cell interactions of this type. The purification of a new growth factor from pig brain that supports survival of sensory neurons in culture lends further support to existing evidence for this point of view.

Individual neurons may normally be guided in development by a complex interaction of many factors. For example, some sensory neurons cannot be rescued in culture by NGF unless the culture substratum is modified by a component present in heart-conditioned medium, although that component by itself does not permit neuronal survival. Similarly, some sensory neurons early in development cannot survive in cell culture with either NGF or the new sensory neuron survival factor from pig brain, but do survive in culture when supplied with both. Separate factors appear to regulate neuronal growth and neurotransmitter synthesis by autonomic neurons, but interaction of the two types of factors can produce synergistic effects in some cases. The existence of a family of regulatory factors with overlapping influences and multiple inter-dependencies would provide machinery both for creating a number of possible phenotypes for a neuron and for fine tuning neuronal development within the phenotype.

A wide array of activities that stimulate various aspects of neuronal growth and development have been described in tissue extracts and cell-conditioned media. It seems likely that in some cases further purification will demonstrate that a single component is responsible for diverse activities, as is true for NGF, which supports both neuronal survival and neurite extension. It is also probable that in some instances in which partially purified material has been found to act on several types of neurons, further purification will reveal a mixture of active components, each with a unique specificity. In fact, one may question the value of carrying out extensive assays on impure components to test their specificity and distribution, since the results obtained are often critically dependent on obscure features of the assay and may even be compromised by toxic or inhibitory components in the test sample. The collective experience to date offers a clear message: Unambiguous identification and characterization of a component requires that the component be purified.

An equally important point is that the effect of a component must be determined in vivo before its role in development can be considered estab-lished. Two approaches appear promising in this regard. The first strategy alters the concentration of the component in vivo and then analyzes the consequent effects on neuronal development. Such alterations can be made either by injecting antibodies that block the activity of the component or by directly introducing into the animal high concentrations of the purified mate-rial. This method proved very successful with NGF. A second approach is now possible once the component has been purified or antibodies are obtained against it. The gene for the component can be cloned and used to probe tissues during development as a means of determining when and where the gene is expressed and how its expression is regulated. These developments and the immediate prospects promise considerable excitement in the field of neuronal growth factors for the foreseeable future.

ACKNOWLEDGMENTS

I wish to thank numerous colleagues for providing copies of their manuscripts in press, and Dr. Nicholas C. Spitzer, Dr. William A. Harris, and Dr. Susan Kirkpatrick for editorial comments. Grant support was provided by the Nation-al Institutes of Health (NS 12601), the Muscular Dystrophy Association, and the American Heart Association with funds contributed in part by the California Heart Association.

Literature Cited

Adler, R. 1982. Regulation of neurite growth in purified retina neuronal cultures: Effects of PNPF, a substratum-bound, neurite-promoting factor. *J. Neurosci. Res.* 8:165–77

Adler, R., Landa, K. B., Manthorpe, M., Varon, S. 1979. Cholinergic neuronotrophic factors: Intraocular distribution of trophic activity for ciliary neurons. *Science* 204:1434–36

Adler, R., Manthorpe, M., Skaper, S. D., Varon, S. 1981. Polyornithine-attached neurite-promoting factors (PNPFs). Culture sources and responsive neurons. *Brain Res.* 206:129–44

Adler, R., Manthorpe, M., Varon, S. 1982. Lectin reactivity of PNPF, a polyornithine-binding neurite-promoting factor. *Dev. Brain Res.* 6:69–75

Adler, R., Varon, S. 1980. Cholinergic neuronotrophic factors: V. Segregation of survival- and neurite-promoting activities in heart-conditioned media. *Brain Res.* 188:437–48

Adler, R., Varon, S. 1981. Neuritic guidance by polyornithine-attached materials of ganglionic origin. *Dev. Biol.* 81:1–11

Adler, R., Varon, S. 1982. Neuronal survival in intact ciliary ganglia in vivo and in vitro: Ciliary neuronotrophic factor as a target surrogate. *Dev. Biol.* 92:470–75

Banker, G. A. 1980. Trophic interactions between astroglial cells and hippocampal neurons in culture. *Science* 209:809–10

Barde, Y.-A., Edgar, D., Thoenen, H. 1980. Sensory neurons in culture: Changing requirements for survival factors during embryonic development. *Proc. Natl. Acad. Sci. USA* 77:1199–1203

Barde, Y.-A., Edgar, D., Thoenen, H. 1982. Purification of a new neurotrophic factor from mammalian brain. *EMBO J.* 1:549–53

Barde, Y.-A., Edgar, D., Thoenen, H. 1983. New neurotrophic factors. *Ann. Rev. Physiol.* 45:601–12

Barde, Y.-A., Lindsay, R. M., Monard, D., Thoenen, H. 1978. New factor released by cultured glioma cells supporting survival and growth of sensory neurons. *Nature* 274:818

Bennett, M. R., Lai, K., Nurcombe, V. 1980. Identification of embryonic motoneurons in vitro: Their survival is dependent on skeletal muscle. *Brain Res.* 190:537–42

Berg, D. K. 1982. Cell death in neuronal development: Regulation by trophic factors. In *Neuronal Development*, ed. N. C. Spitzer, pp. 297–331. New York: Plenum

Bonyhady, R. E., Henry, I. A., Hill, C. E. 1982. Reversible dissociation of a bovine cardiac factor that supports survival of avian ciliary ganglionic neurons. *J. Neurosci. Res.* 7:11–21

Bonyhady, R. E., Hendry, I. A., Hill, C. E., McLennan, I. S. 1980. Characterization of a cardiac muscle factor required for the survival of cultured parasympathetic neurons. *Neurosci. Lett.* 18:197–201

Bothwell, M. 1982. Insulin and somatemedin MSA promote nerve growth factor-independent neurite formation by cultured chick dorsal root ganglionic sensory neurons. *J. Neurosci. Res.* 8:225–31

Braun, S. J., Sweadner, K. J., Patterson, P. H. 1981. Neuronal cell surfaces: Distinctive glycoproteins of cultured adrenergic and cholinergic sympathetic neurons. *J. Neurosci.* 1:1397–1406

Brockes, J. P., Lemke, G. E., Balzer, D. R. Jr. 1980. Purification and preliminary characterization of a glial growth factor from the bovine pituitary. *J. Biol. Chem.* 255:8374–77

Campenot, R. B. 1982. Development of sympathetic neurons in compartmentalized cultures. II. Local control of neurite survival by nerve growth factor. *Dev. Biol.* 93:13–21

Carrie, N. G., Ebendal, T. 1983. Organotypic cultures of neural retina: Neurite outgrowth stimulated by brain extracts. *Dev. Brain Res.* 6:219–29

Collins, F. 1978a. Induction of neurite outgrowth by a conditioned-medium factor bound to the culture substratum. *Proc. Natl. Acad. Sci. USA* 75:5210–13

Collins, F. 1978b. Axon initiation by ciliary neurons in culture. *Dev. Biol.* 65:50–57

Collins, F. 1980. Neurite outgrowth induced by the substrate associated material from non-neuronal cells. *Dev. Biol.* 79:247–52

Collins, F., Dawson, A. 1982. Conditioned medium increases the rate of neurite elongation: Separation of this activity from the substratum-bound inducer of neurite outgrowth. *J. Neurosci.* 2:1005–10

Collins, F., Garrett, J. E. Jr. 1980. Elongating nerve fibers are guided by a pathway of material released from embryonic nonneuronal cells. *Proc. Natl. Acad. Sci. USA* 77:6226–28

Coughlin, M. D., Bloom, E. M., Black, I. B. 1981. Characterization of a neuronal growth factor from mouse heart-cell-conditioned medium. *Dev. Biol.* 82:56–68

Coughlin, M. D., Kessler, J. A. 1982. Antiserum to a new neuronal growth factor: Effects on neurite outgrowth. *J. Neurosci. Res.* 8:289–302

Crutcher, K. A., Collins, F. 1982. In vitro evidence for two distinct hippocampal growth factors: Basis of neuronal plasticity? *Sci.* 217:67–68

Dribin, L. B. 1982. On the species and substrate specificity of conditioned medium enhancement of neuritic outgrowth from spinal cord explants. *Dev. Brain Res.* 3:300–4

Dribin, L. B., Barrett, J. N. 1980. Conditioned medium enhances neuritic outgrowth from rat spinal cord explants. *Dev. Biol.* 74:184–95

Dribin, L. B., Barrett, J. N. 1982a. Two components of conditioned medium increase neuritic outgrowth from rat spinal cord explants. *J. Neurosci. Res.* 8:271–80

Dribin, L. B., Barrett, J. N. 1982b. Characterization of neuritic outgrowth-promoting activity of conditioned medium on spinal cord explants. *Dev. Brain Res.* 4:435–41

Ebendal, T. 1979. Stage-dependent stimulation of neurite outgrowth exerted by nerve growth factor and chick heart in cultured embryonic ganglia. *Dev. Biol.* 72:276–90

Ebendal, T. 1981. Control of neurite extension by embryonic heart explants. *J. Embryol. Exp. Morphol.* 61:289–301

Ebendal, T., Belew, M., Jacobson, C.-O., Porath, J. 1979. Neurite outgrowth elicited by embryonic chick heart: Partial purification of the active factor. *Neurosci. Lett.* 14:91–95

Ebendal, T., Hedlund, K.-O., Norrgren, G. 1982. Nerve growth factors in chick tissues. *J. Neurosci. Res.* 8:153–64

Ebendal, T., Jordell-Kylberg, A., Söderström, S. 1978. Stimulation by tissue explants on nerve fibre outgrowth in culture. *Zoon* 6:235–43

Ebelman, G. M. 1983. Cell adhesion molecules. *Science* 219:450–57

Edgar, D., Barde, Y.-A., Thoenen, H. 1981. Subpopulations of cultured chick sympathetic neurons differ in their requirements for survival factors. *Nature* 289:294–95

Edgar, D., Thoenen, H. 1982. Modulation of NGF-induced survival of chick sympathetic neurons by contact with a conditioned medium factor bound to the culture substrate. *Dev. Brain Res.* 5:89–92

Fukada, K. 1980. Hormonal control of neurotransmitter choice in sympathetic neuron culture. *Nature* 287:553–55

Fukada, K. 1983. *Intl. Soc. Neurochem. Abstr.* 41:589 (Suppl.)

Furshpan, E. J., MacLeish, P. R., O'Lague, P. H., Potter, D. D. 1976. Chemical transmission between rat sympathetic neurons and cardiac myocytes developing in microcultures: Evidence for cholinergic, adrenergic, and dual-function neurons. *Proc. Natl. Acad. Sci. USA* 73:4225–29

Giller, E. L. Jr., Neale, J. H., Bullock, P. N., Schrier, B. K., Nelson, P. G. 1977. Choline acetyltransferase activity of spinal cord cell cultures increased by co-culture with muscle and by muscle-conditioned medium. *J. Cell Biol.* 74:16–29

Giller, E. L. Jr., Schrier, B. K., Shainberg, A., Fisk, H. R., Nelson, P. G. 1973. Choline acetyltransferase activity is increased in combined cultures of spinal cord and muscle cells from mice. *Science* 182:588–89

Godfrey, E. W., Schrier, B. K., Nelson, P. B. 1980. Source and target cell specificities of a conditioned medium factor that increases choline acetyltransferase activity in cultured spinal cord cells. *Dev. Biol.* 77:403–18

Greif, K. F., Reichardt, L. F. 1982. Appearance and distribution of neuronal cell surface and synaptic vesicle antigens in the developing rat superior cervical ganglion. *J. Neurosci.* 2:843–52

Gundersen, R. W., Barrett, J. N. 1979. Neuronal chemotaxis: Chick dorsal-root axons turn toward high concentrations of nerve growth factor. *Science* 206:1079–80

Hamburger, V., Brunso-Bechtold, J. K., Yip, J. W. 1981. Neuronal death in the spinal ganglia of the chick embryo and its reduction by nerve growth factor. *J. Neurosci.* 1:60–71

Helfand, S. L., Smith, G. A., Wessells, N. K. 1976. Survival and development in culture of dissociated parasympathetic neurons from ciliary ganglia. *Dev. Biol.* 50:541–47

Henderson, C. E., Huchet, M., Changeux, J.-P. 1981. Neurite outgrowth from embryonic chicken spinal neurons is promoted by media conditioned by muscle cells. *Proc. Natl. Acad. Sci. USA* 78:2625–29

Hill, C. E., Hendry, I. A., Bonyhady, R. E. 1981. Avian parasympathetic neurotrophic factors: Age-related increases and lack of regional specificity. *Dev. Biol.* 85:258–61

Kligman, D. 1982a. Isolation of a protein from bovine brain which promotes neurite extension from chick embryo cerebral cortex neurons in defined medium. *Brain Res.* 250:93–100

Kligman, D. 1982b. Neurite outgrowth from cerebral cortical neurons is promoted by medium conditioned over heart cells. *J. Neurosci. Res.* 8:281–87

Ko, C.-P., Burton, H., Johnson, M. I., Bunge, R. P. 1976. Synaptic transmission between rat superior cervical ganglion neurons in dissociated cell cultures. *Brain Res.* 117:461–85

Landa, K. B., Adler, R., Manthorpe, M., Varon, S. 1980. Cholinergic neuronotrophic factors. III. Developmental increase of trophic activity for chick embryo ciliary ganglion neurons in their intraocular target tissues. *Dev. Biol.* 74:401–8

Lander, A. D., Fujii, D. K., Gospodarowicz, D., Reichardt, L. F. 1982. Characterization of a factor that promotes neurite outgrowth: Evidence linking activity to a heparan sulfate proteoglycan. *J. Cell Biol.* 94:574–85

Landmesser, L., Pilar, G. 1974. Synaptic transmission and cell death during normal ganglionic development. *J. Physiol.* 241:737–49

Le Douarin, N. M., Smith, J., Le Lièvre, C. S. 1981. From the neural crest to the ganglia of the peripheral nervous system. *Ann. Rev. Physiol.* 43:653–71

Letourneau, P. C. 1975a. Possible roles for cell-to-substratum adhesion in neuronal morphogenesis. *Dev. Biol.* 44:77–91

Letourneau, P. C. 1975b. Cell-to-substratum adhesion and guidance of axonal elongation. *Dev. Biol.* 44:92–101

Letourneau, P. C. 1979. Cell-substratum adhesion of neurite growth cones, and its role in neurite elongation. *Exp. Cell Res.* 124:127–38

Levi-Montalcini, R. 1982. Developmental neurobiology and the natural history of nerve growth factor. *Ann. Rev. Neurosci.* 5:341–62

Lindsay, R. M., Tarbit, J. 1979. Developmentally regulated induction of neurite outgrowth from immature chick sensory neurons (DRG) by homogenates of avian or mammalian heart, liver, and brain. *Neurosci. Lett.* 12:195–200

Manthorpe, M., Barbin, G., Varon, S. 1982a. Isoelectric focusing of the chick eye ciliary neuronotrophic factor. *J. Neurosci. Res.* 8:233–39

Manthorpe, M., Longo, F. M., Varon, S. 1982b. Comparative features of spinal neuronotrophic factors in fluids collected in vitro and in vivo. *J. Neurosci. Res.* 8:241–50

Manthorpe, M., Skaper, S., Adler, R., Landa, K., Varon, S. 1980. Cholinergic neuronotrophic factors: Fractionation properties of an extract from selected chick embryonic eye tissues. *J. Neurochem.* 34:69–75

Manthorpe, M., Skaper, S. D., Barbin, G., Varon, S. 1982c. Cholinergic neuronotrophic factors. Concurrent activities on certain nerve growth factor-responsive neurons. *J. Neurochem.* 38:415–21

Manthorpe, M., Varon, S., Adler, R. 1981. Neurite-promoting factor in conditioned medium from RN22 Schwannoma cultures: Bioassay, fractionation, and properties. *J. Neurochem.* 37:759–67

Margiotta, J. F., Berg, D. K. 1982. Functional synapses are established between ciliary ganglion neurons in dissociated cell culture. *Nature* 296:152–54

McCaffery, C. A., Bennett, M. R., Dreher, B. 1982. The survival of neonatal rat retinal ganglion cells in vitro is enhanced in the presence of appropriate parts of the brain. *Exp. Brain Res.* 48:377–86

McLennan, I. S., Hendry, I. A. 1980. Influence of cardiac extracts on cultured ciliary ganglia. *Dev. Neurosci.* 3:1–10

Miki, N., Hayashi, Y., Higashida, H. 1981. Characterization of chick gizzard extract that promotes neurite outgrowth in cultured ciliary neurons. *J. Neurochem.* 37:627–33

Müller, H. W., Seifert, W. 1982. A neurotrophic factor (NTF) released from primary glial cultures supports survival and fiber outgrowth of cultured hippocampal neurons. *J. Neurosci. Res.* 8:195–204

Nieto-Sampedro, M., Lewis, E. R., Cotman, C. W., Manthorpe, M., Skaper, S. D., Barbin, G., Longo, F. M., Varon, S. 1982. Brain injury causes a time-dependent increase in neuronotrophic activity at the lesion site. *Science* 217:860–61

Nishi, R., Berg, D. K. 1977. Dissociated ciliary ganglion neurons in vitro: Survival and synapse formation. *Proc. Natl. Acad. Sci. USA* 74:5171–75

Nishi, R., Berg, D. K. 1979. Survival and development of ciliary ganglion neurons grown alone in cell culture. *Nature* 277:232–34

Nishi, R., Berg, D. K. 1981. Two components from eye tissue that differentially stimulate the growth and development of ciliary ganglion neurons in cell culture. *J. Neurosci.* 1:505–13

Nurcombe, V., Bennett, M. R. 1981. Embryonic chick retinal ganglion cells identified "in vitro." *Exp. Brain Res.* 44:249–58

Nurcombe, V., Bennett, M. R. 1982. Evidence for neuron-survival and neurite-promoting factors from skeletal muscle: Their effects on embryonic spinal cord. *Neurosci. Lett.* 34:89–93

Nurse, C. A., O'Lague, P. H. 1975. Formation of cholinergic synapses between sympathetic neurons and skeletal myotubes in cell culture. *Proc. Natl. Acad. Sci. USA* 72:1955–59

O'Lague, P. H., Obata, K., Claude, P., Furshpan, E. J., Potter, D. D. 1974. Evidence for cholinergic synapses between dissociated rat sympathetic neurons in cell culture. *Proc. Natl. Acad. Sci. USA* 71:3602–6

Obata, K., Tanaka, H. 1980. Conditioned medium promotes neurite growth from both central and peripheral neurons. *Neurosci. Lett.* 16:27–33

Oppenheim, R. W. 1981. Neuronal cell death and some related regressive phenomena during neurogenesis: A selective historical review and progress report. In *Studies in Developmental Neurobiology: Essays in Honor of Viktor Hamburger*, ed. W. M. Cowan, pp. 74–133. New York: Oxford Univ. Press

Patterson, P. H. 1978. Environmental determination of autonomic neurotransmitter functions. *Ann. Rev. Neurosci.* 1:1–17

Patterson, P. H., Chun, L. L. Y. 1974. The influence of nonneuronal cells on catecholamine and acetylcholine synthesis and accumulation in cultures of dissociated sympathetic neurons. *Proc. Natl. Acad. Sci. USA* 71:3607–10

Patterson, P. H., Chun, L. L. Y. 1977. The induction of acetylcholine synthesis in primary cultures of dissociated rat sympathetic neurons. I. Effects of conditioned medium. *Dev. Biol.* 56:263–80

Pollack, E. D., Muhlach, W. L., Liebig, V. 1981. Neurotropic influence of mesenchymal limb target tissue on spinal cord neurite growth in vitro. *J. Comp. Neurol.* 200:393–405

Richardson, P. M., Ebendal, T. 1982. Nerve growth activities in rat peripheral nerve. *Brain Res.* 246:57–64

Riopelle, R. J., Boegman, R. J., Cameron, D. A. 1981. Peripheral nerve contains heterogeneous growth factors that support sensory neurons in vitro. *Neurosci. Lett.* 25:311–16

Riopelle, R. J., Cameron, D. A. 1981. Neurite growth promoting factors of embryonic chick—ontogeny, regional distribution, and characteristics. *J. Neurobiol.* 12:175–86

Schnaar, R. L., Schaffner, A. E. 1981. Separation of cell types from embryonic chicken and rat spinal cord: Characterization of motoneuron-enriched fractions. *J. Neurosci.* 1:204–17

Schubert, D., LaCorbiere, M. 1980. Role of a 16S glycoprotein complex in cellular adhesion. *Proc. Natl. Acad. Sci. USA* 77:4137–41

Schubert, D., LaCorbiere, M. 1982. The specificity of extracellular glycoprotein complexes in mediating cellular adhesion. *J. Neurosci.* 2:82–89

Slack, J. R., Pockett, S. 1982. Motor neurotrophic factor in denervated adult skeletal muscle. *Brain Res.* 247:138–40

Stollberg, J., Berg, D. K. 1982. Antiserum against a factor that stimulates cholinergic development of chick ciliary ganglion neurons. *Soc. Neurosci. Abstr.* 8:398

Sweadner, K. J. 1981. Environmentally regulated expression of soluble extracellular proteins of sympathetic neurons. *J. Biol. Chem.* 256:4063–70

Tanaka, H., Sakai, M., Obata, K. 1982. Effects of serum, tissue extract, conditioned medium, and culture substrata on neurite appearance from spinal cord explants of chick embryo. *Dev. Brain Res.* 4:303–12

Thoenen, H., Barde, Y.-A. 1980. Physiology of nerve growth factor. *Physiol. Rev.* 60:1284–1335

Turner, J. E., Barde, Y.-A., Schwab, M. E., Thoenen, H. 1983. Extract from brain stimulates neurite outgrowth from fetal rat retinal explants. *Dev. Brain Res.* 6:77–83

Turner, J. E., Delaney, R. K., Johnson, J. E. 1981. Retinal ganglion cell response to axotomy and nerve growth factor antiserum treatment in the regenerating visual system of the goldfish *(Carassius auratus):* an in vivo and in vitro analysis. *Brain Res.* 204:283–94

Tuttle, J. B., Suszkiw, J. B., Ard, M. 1980. Long-term survival and development of dissociated parasympathetic neurons in culture. *Brain Res.* 183:161–80

Varon, S., Adler, R. 1981. Trophic and specifying factors directed to neuronal cells. *Adv. Cell. Neurobiol.* 2:115–63

Varon, S., Manthorpe, M. 1982. Schwann cells: An in vitro perspective. *Adv. Cell. Neurobiol.* 3:35–95

Varon, S., Manthorpe, M., Adler, R. 1979. Cholinergic neuronotrophic factors: I. Survival, neurite outgrowth and choline acetyltransferase activity in monolayer cultures from chick embryo ciliary ganglia. *Brain Res.* 173:29–45

Varon, S., Manthorpe, M., Longo, F. M., Williams, L. R. 1983. Growth factors in regeneration of neural tissues. In *Nerve, Organ, and Tissue Regeneration: Research Perspectives*, ed. F. J. Seil, pp. 127–55. New York: Academic

Walicke, P. A., Campenot, R. B, Patterson, P. H. 1977. Determination of transmitter function by neuronal activity. *Proc. Natl. Acad. Sci. USA* 74:5767–71

Walicke, P. A., Patterson, P. H. 1981. On the role of Ca^{2+} in the transmitter choice made by cultured sympathetic neurons. *J. Neurosci.* 1:343–50

Weber, M. J. 1981. A diffusible factor responsible for the determination of cholinergic functions in cultured sympathetic neurons. *J Biol. Chem.* 256:3447–53

Ann. Rev. Neurosci. 1984. 7:171–88

CELL DEATH IN INVERTEBRATE NERVOUS SYSTEMS

James W. Truman

Department of Zoology, University of Washington, Seattle, Washington 98195

Cell death is a common feature of developing nervous systems. Questions as to the factors that determine which cells survive and which die have concerned vertebrate neurobiologists for over 40 years (Hamburger & Oppenheim 1982). Despite this interest in neuron death in vertebrates, relatively little attention had been given to this phenomenon in invertebrates except to indicate that it exists. This state of affairs has changed dramatically over the past five years as a number of invertebrate preparations emerged as favorable ones in which to study cell death. The features of the invertebrate CNS that made it useful for this type of study included some of the same features that facilitated studies in other areas; namely, the relatively small number of cells and the presence of unique neurons that could be reliably located, identified, and studied in successive individuals. In addition, detailed work on the development of a number of species have specified cellular origins and the times of synaptic contact with great precision (see Anderson et al 1980 for a review). Consequently, these animals can be used to ask questions about cell death in a cellular context that is not yet possible with vertebrate systems.

Functional Roles of Cell Death in the CNS

Neuronal death occurs in many contexts during the normal life of an animal. Different systems have been used to categorize the types of normal death that occur in the CNS (Glucksmann 1951, Hamburger & Oppenheim 1982, Horvitz et al 1982, Saunders 1966), but I have chosen to reduce these to three categories. Cell death occurs to regulate the size of neuronal populations, thereby allowing a close matching of cell numbers between two regions of the CNS or between the CNS and periphery. Neuron degeneration is occasionally

171

0147-006X/84/0301-0171$02.00

the necessary outcome of producing cells of a particular phenotype. Also, it provides a mechanism to rid the CNS of cells whose functions are no longer needed.

CELL DEATH FOR REGULATION OF THE SIZE OF NEURONAL POPULATIONS   During development of most vertebrates, cell death provides an important mechanism for matching cell numbers in various regions of the nervous system. One of the most extensively studied examples involves the motor centers of the developing chick spinal cord (see Hamburger & Oppenheim 1982 for a review). These centers initially overproduce motoneurons, but then competition between these cells for target muscles subsequently determines which neurons survive and which die. Although the nature of the competition is still in question, it is clear that the result is an adjustment of motoneuron numbers to accommodate the number of target cells that are present in the periphery. The use of cell death to adjust cell populations occurs not only between the CNS and the periphery but also at numerous levels within the CNS (see Oppenheim 1981 for review).

Motoneuron-muscle interactions   The importance of appropriate targets for the survival of developing motoneurons has received relatively scant attention in invertebrates. This question was directly addressed by Chiarodo (1963), who extirpated leg discs from blowflies in their last larval stage and then examined the effect of this removal on the number of cells in the thoracic ganglia after metamorphosis of the fly. Removal of one disc resulted in a 29% reduction in the number of small cells in the cortex of the ganglia—a loss that was assumed to reflect the death of the motoneurons expected to innervate the leg. However, the number of large cells was unaffected by removal of the disc and it is now known that motoneurons would be within the latter group. In addition, the small cells that were counted included both neurons (presumably interneurons) and glia, so it is not clear whether any neurons were actually lost after this manipulation.

The question of muscle-motoneuron interaction was recently examined in embryonic locusts by Whitington et al (1982). They were concerned with the fates of two neurons, the fast extensor tibia (Feti) and the dorsal unpaired median extensor tibia (Dumeti), both of which innervate the extensor tibia muscle in the third thoracic (T3) leg. Both neurons come from uniquely identifiable neuroblasts in ganglion T3 according to a stereotyped lineage. The same neurons are also generated in abdominal segments (where there are no legs), but the abdominal homologues die when their axons reach the edge of the CNS (Goodman & Bate 1981). Consequently, it was tempting to suggest that the presence or absence of appropriate targets was a cue for survival of the thoracic versus the abdominal homologues. This hypothesis was directly tested by Whitington et al (1982) by surgically removing the embryonic limb buds from T3 before the time that the motoneuron axons reached the periphery. Both Feti and Dumeti nevertheless survived and showed essentially normal physiological and morphological characteristics despite the lack of the extensor tibia muscles. The only abnormality observed was a disturbance in the peripheral branching pattern of the axon of some cells. These data suggest that the segment specific survival of these two neurons is due to factors other than the simple presence or absence of their appropriate target. These factors are presently unknown.

It is important that the studies on the locust and blowfly not be overinterpreted at this time. Both reports indicate that motoneurons can survive without their normal targets, but in neither case was it determined whether the cells found inappropriate targets to innervate. Consequently, the question as to whether contact with a postsynaptic target is essential for the survival of invertebrate motoneurons is still open.

Regulation of cell numbers in optic ganglia   The optic ganglia of many invertebrates present a clear case of the regulation of cell number through cell death. The first indication that the normal development of the insect brain required the presence of the eyes was provided by Kopec (1922) for the moth *Lymantria*. He noted that when eye imaginal discs were extirpated prior to differentiation of the adult, the resulting moths showed a reduction of cells in the "external layers" of the brain. Similar observations have been made after partial or complete eye removal in a number of other insects (Alverdes 1924, Anderson 1978, Hinke 1961, Maxwell & Hildebrand 1981, Meinertzhagen 1973, Mouze 1974) and in crustaceans (Macagno 1979). Analysis of various *Drosophila* mutants that have reduced eyes also showed a reduction in the number of cells in the outer layers of the optic ganglia (Johannsen 1924, Krafka 1924, Meyerowitz & Kankel 1978, Power 1943, Richards & Furrow 1925). In molluscs the suppression of eye development in the squid by injections of lithium chloride was associated with reduction of the optic ganglia (Ranzi 1928).

The positive relationship between the presence of all or part of the eye and the number of cells in the optic ganglia could occur through either of two mechanisms. The presence of photoreceptor axons might be required for the generation of optic lobe neurons from stem cells. Alternatively, the production of cells might be normal, but subsequent survival of the neurons could be dependent on successful innervation by photoreceptor axons. A number of studies on insects and crustaceans indicate that most of the reduction in cell number is explained by the latter hypothesis.

The visual system of most arthropods is a highly ordered structure. The retina and the lamina (the outer layer of the optic lobe) are each composed of repeating units, ommatidia and optic cartridges, respectively. Each ommatidium has eight photorecepter cells and each optic cartridge typically comprises five laminar neurons. The pattern of connectivity between photorecepter axons and the cells that make up the optic cartridges is highly stereotyped within a given species but may vary somewhat from group to group. This highly ordered arrangement of photoreceptors and first-order cells requires interactions during the growth of the arrays.

Naturally occurring cell death during the development of the optic lobes was first noted in the metamorphosing brain of the Monarch butterfly (Nordlander & Edwards 1968, 1969). In this insect, columns of laminar neurons are derived from neuroblasts in the outer optic anlagen. Three to six days after their birth a few of the cells begin to die as their sister cells are beginning to extend processes. Presumably, the cells that die are those which were not successfully contacted by photoreceptor axons.

An experimental test of this hypothesis was carried out in locusts, which have incomplete metamorphosis (Anderson 1978). Unlike the Monarch butterfly, in the locust a compound eye is present throughout larval life. With each new larval stage the size of the compound eye increases through addition of ommatidia to the anterior margin of the eye. The underlying lamina also grows by addition of new neurons to its anterior border. Some normal neuron death is seen in the lamina during this process. Removal of the area that produces new ommatidia has no effect on the generation of laminar neurons. These new cells do not then differentiate, but die. Similar results have been reported in dragonflies (Mouze 1974). Moreover, implantation of supernumerary ommatidia in the latter insect reduces the amount of normally occurring cell deaths (Mouze 1978).

The most detailed analysis of the relationship between photoreceptor axons and the survival of laminar neurons has been carried out on an isogenic strain of *Daphnia magna* (Macagno 1979). This small crustacean has a fused median compound eye comprising 22 ommatidia. The eight photoreceptor axons from each ommatidium project to their own discrete optic cartridge in the lamina, where they synapse on the five cells that make up the cartridge. In total, the lamina has 110 neurons, which then send axons to the next layer, the medulla, which has about 326 cells.

The importance of cellular interactions for the survival of laminar neurons was examined by deleting ommatidia in the embryo by means of an ultraviolet microbeam and later reconstructing the region from EM serial sections (Macagno 1979). The most extensive series of lesions were made at 28–29 hr after oviposition, a time before the photoreceptors sent axons into the lamina. The number of laminar neurons subsequently found in the adult was proportional to the number of photoreceptor axons that reached the lamina. The relationship was linear with a slope of five eighths, reflecting the relationship of eight photoreceptor axons to the five laminar cells in an optic cartridge. There were slight deviations from a strict five-eighths ratio because photoreceptor bundles down to five axons were competent to maintain an entire cartridge.

In embryos in which the photoreceptor cells were destroyed, the laminar neurons were nevertheless produced, but they failed to differentiate processes and started degeneration about 48 hr postoviposition (Macagno 1979). Survival of these neurons appears to be dependant on the formation of synaptic contacts with the photoreceptor axons. Interestingly, in the adult, after connections have been made and the laminar neurons have differentiated, lesions of ommatidia did not result in the degeneration of the laminar cells. Thus, contacts are necessary for the survival and differentiation of the immature cells, but once this differentiation is complete, this trophic input is no longer important. It should be noted that it is not clear whether the laminar cells in the adult might also die after long-term, chronic deafferentiation.

The survival of neurons in the second optic ganglion, the medulla, is to some extent dependent on synapses from laminar cells (Macagno 1979). Complete degeneration of the lamina is accompanied by the loss of many, although not all, medullary neurons. The number of surviving medullary cells is not linearly proportional to the number of surviving laminar neurons and the latter can be reduced to approximately half of their normal number before a substantial effect is seen. The difference between this relationship and that of the photoreceptors to the lamina undoubtedly arises from the fact that a given laminar cell receives input from one and only one ommatidium, whereas a medullary neuron receives input from laminar cells from a number of optic cartridges. Also, the medulla neurons may receive some centrifugal inputs from the brain.

Outside of the eye-optic ganglion system, there are no other examples of cell death being used to regulate cell numbers in developing invertebrate nervous sytems. Removal of other major sensory structures, such as the antennae of moths (Sanes et al 1977) or the cerci of crickets (Murphey et al 1975), seems to have no effect on the survival of the appropriate first-order interneurons. The potential importance of competitive interactions between interneurons within the CNS is completely unexplored.

CELL DEATH REQUIRED FOR THE GENERATION OF CELLS HAVING SPECIFIC CHARACTERISTICS   Cell lineage studies in both nematodes and locusts suggest that the production of certain cell types may demand the generation of other cells that are not otherwise needed and subsequently die. This is nicely illustrated by the development of the ventral nervous system of the nematode *Caenorhabdites elegans* (Sulston 1976, Sulston & Horvitz 1977). This region of the CNS is derived from 12 precursor cells (P1–P12). Each precursor (Pn) initially divides to form an anterior neuroblast (Pn.a) and a posterior hypodermal cell (Pn.p). Each neuroblast then undergoes an identical series of

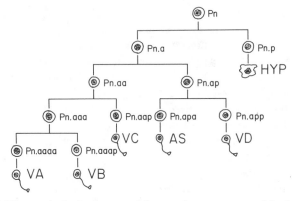

Figure 1 Cell lineages in the development of the ventral nervous system of *C. elegans*. Twelve precursor cells, P1 to P12 (here designated as Pn), undergo the indicated pattern of cell division. *Left branches* represent anterior (a) daughters and *right branches* are posterior (p) daughters. Progeny of the Pn.a neuroblast differentiate to become motoneurons of five distinct classes (VA, VB, VC, AS, and VD). The posterior product of the first division of the precursor cell (Pn.p) becomes a hypodermal cell (HYP). In the P1-P2 and P9-P12 lineages, the Pn.aap cell dies instead of becoming a VC motoneuron. In the P11 and P12 lineages the Pn.aaap cells die rather than differentiate into VB motoneurons. Modified from Horvitz et al (1982).

divisions to produce five daughter cells. The characteristics of each daughter derived from such a sequence is a strict function of its lineage (see Figure 1). The Pn.aap cells derived from precursor cells P3–P8 differentiate into VC motoneurons that innervate the vulva, which is located in midworm. By contrast, in the more anterior and posterior regions of the worm, the Pn.aap cells undergo programmed death. These cells are born, only then to die, because each is the product of a cell division that produces the sister cell Pn.aaa, which then divides to generate the VA and VB motoneurons needed in these regions. Thus it appears that the ventral nervous system is formed by producing 12 identical arrays of cells and then killing those that are inappropriate for a given region (White et al 1976).

A similar situation appears to exist in the development of the segmental nervous system of insects (Goodman & Bate 1981). In locusts each thoracic and abdominal ganglion comes from an identical array of 61 neuroblasts (Bate 1976) and seven midline precursor cells (Bate & Grunewald 1981). After embryonic development is complete, each thoracic ganglion contains about 3000 neurons, whereas the abdominal ganglia have about 500 cells each. The difference in cell number between the thoracic and abdominal ganglia reflects primarily differences in the extent of cell death (Goodman & Bate 1981). For example, the unpaired midline neuroblast in the third thoracic ganglion produces about 100 progeny, whereas that in the first abdominal ganglion generates about 90. Then half of those progeny die, while essentially all of the thoracic cells survive. The first four progeny of the midline neuroblast (including Dumeti discussed above) are among the abdominal cells that die. Presumably these cells must be generated in order to derive cells with the appropriate characteristics later in the lineage.

CELL DEATH TO RID THE CNS OF OBSOLETE NEURONS   As an animal develops and matures, neurons that were originally needed to perform a certain function may no longer be required and subsequently die. One such example is the Rohon-Beard cells found in the spinal cord of embryonic fish and amphibians. These neurons are thought to be primary sensory neurons, and they die at a particular stage of embyronic development as their function is apparently taken over by dorsal root ganglion cells (Hughes 1957, Lamborghini 1981). Neurons that have a transient function during embryonic development and then degenerate have also been seen in locusts (Bate et al 1981). As described above, seven midline precursor (MP) cells contribute to the formation of a segmental ganglion. In the case of one of these cells, MP3, its two progeny are among the first neurons to grow axons, which aid in laying out the basic pattern of connectives and commissures that is evident in the insect CNS (Bate & Grunewald 1981). Later in embryonic development, after they have fulfilled their presumed guidance role, many of the MP3 progeny die (Bate et al 1981). It should be noted that death is not the only option for a neuron whose function is no longer required. The MP3 progeny in segments posterior to A3 die, but those in the thorax grow new axons, lose their initial axon, and assume a new function in the animal (Goodman et al 1981). This reorganization of "obsolete" neurons for new functions is also seen during postembryonic development of *C. elegans* (White et al 1978) and during metamorphosis of the moth *Manduca sexta* (Truman & Reiss 1976).

A more massive removal of obsolete cells is seen in animals that undergo metamorphosis. The involution and death of the Mauthner cell is seen in some amphibians accompanying the loss of the lateral line system in the transition from an aquatic tadpole to a terrestrial frog (Zottoli 1978). In the ascidians, a lower chordate group, the transformation of the free-swimming tadpole stage into the sessile, filter-feeding adult is accompanied by the death of almost all of the larval nervous system. The small adult nervous system then arises from the larval "neurohypophysis" (Elwyn 1937). The larval neurons appear to undergo a programmed death and their remains are then consumed by phagocytes (Torrence 1983).

A detailed description of cell death during metamorphosis has been provided for the segmental abdominal ganglia of the moth, *Manduca sexta* (Truman 1983, Truman & Schwartz 1982a,b). The metamorphosis of holometabolous insects such as *Manduca* occurs in two stages, with first the production of a pupal stage and then the adult. The larval abdomen is supplied with a complex musculature; each segment bears sets of thick, longitudinal, intersegmental muscles and numerous groups of smaller external muscles. During the transition from larva to pupa, the larval external muscles die and are replaced by new external muscles, which subsequently differentiate during adult development (Finlayson 1956). The intersegmental muscles remain intact and are responsible for generating abdominal movements through the metamorphic transition. They are used by the moth to shed the old pupal cuticle (a process termed *ecdysis* or *eclosion*) and then they die (Finlayson 1956, Lockshin & Williams 1965a), leaving the new external muscles as the sole muscles responsible for adult behavior.

These changes in musculature during metamorphosis are accompanied by two major periods of motoneuron death. The first period occurs shortly after pupal ecdysis. At this time some of the larval motoneurons that had innervated the now degenerate external muscles also die (J. C. Weeks and J. W. Truman, unpublished). Interestingly, other larval motoneurons continue to live even though they have no peripheral target. Their axon terminals become associated with myoblasts, which subsequently differentiate into the adult muscles. Why one group of targetless motoneurons live and the other dies is unknown.

The second period of cell death occurs at the end of metamorphosis after adult ecdysis. The intersegmental muscles die at this time as do about 50% of the interneurons and motoneurons in the abdominal CNS (Truman 1983). Interneuron death begins about 2 hr before adult ecdysis. Peak numbers of dying cells are evident at 16 hr after ecdysis, and by 30 hr degeneration is substantially completed. Death among the motoneurons begins at about 8 hr and continues through the next 30. By following motoneurons that could be repeatedly identified in sectioned ganglia, it was shown that during adult development particular cells invariably become fated either to live or to die. Interestingly, among those that would die the time of death was a cell-specific characteristic. For example, the cell 11 always started degeneration at 8 hr after ecdysis, whereas cell 12 never began until 30–36 hr. Indeed, the degeneration of abdominal motoneurons followed a strict spatial and temporal program that specified which individuals would die and when the death of each would begin. The pattern of interneuron death suggests that a similar program may exist for these cells, but this has not yet been confirmed by following the fate of identified individual interneurons.

The motoneurons that die early are those that participate in the ecdysis behaviors. The later dying cells are involved in later behaviors, such as the postecdysial expansion of the new wings of the moth to their appropriate dimensions. This rough correlation between the order of cell death and the order of occurrence of the behaviors in which the various neurons participate does not indicate causation, however, since the cells die in their appropriate sequence even when the behaviors are prevented (J. W. Truman and L. M.

Schwartz, unpublished). The order of cell death appears to be a fail-safe mechanism to insure that the death of a particular neuron does not occur until the completion of the behavior with which it is involved. This last point is important since adult ecdysis in *Manduca* normally occurs underground and the insect has to delay wing expansion for a period of time until it digs to the surface. The performance of sustained digging behavior can selectively delay by a few hours the death of the motoneurons involved in this behavior (Truman 1983).

Signals for Cell Death

Numerous types of cell interactions are possible in a structure as complex as the nervous system. Information can be conveyed through direct contact, by chemicals that are released into local regions, or by factors that are broadcast throughout the organism. Spatial interactions are further complicated because in developing systems timing is of critical importance. Thus, cell survival may depend on communication only during a specific critical period, after which the cell is committed to one fate or another. The question of what kind of signals are important for cell survival is only beginning to be addressed in invertebrate nervous systems. Thus far, two types of signals are known to be important: trophic influences and hormones.

TROPHIC INFLUENCES   In vertebrate nervous systems trophic influences can either be *anterograde* or *retrograde* (Cowan 1970). In the former, the pre-synaptic cell is required for the growth or survival of the postsynaptic cell, whereas in the latter situation the roles are reversed. There are not any cases presently described of retrograde trophic influences in invertebrates. By analogy to the relationships seen in vertebrates, one might expect that the survival of embryonic motoneurons might be dependent on the presence of an appropriate target muscle. As described above, however, embryonic locust motoneurons survive and differentiate in the absence of their normal target muscles (Whitington et al., 1982). The absence of retrograde influences is also apparent in the differentiation of various invertebrate sensory structures. For example, the compound eyes of arthropods (eg. Kopec 1922, Eichenbaum & Goldsmith 1968) as well as the antennae of moths (Sanes et al 1976) can differentiate and produce sensory neurons that send out axons even after their normal targets in the CNS have been removed.

In contrast to retrograde influences, the presence of anterograde trophic actions are well established. One of the best known examples is the requirement for motoneuron innervation for the survival and differentiation of skeletal muscle. In insects this was first shown by Kopec (1923) in his studies of metamorphosis in the gypsy moth, *Lymantria,* and later was extensively exploited by Nuesch and his students in studying the development of the flight musculature in the moth *Antheraea polyphemus* (Nuesch 1968). A reversed role for trophic influences was postulated by Lockshin & Williams (1965b,c) in their study of the regulation of the death of the intersegmental muscles after adult ecdysis in silkmoths. Based on a series of drug and stimulation experiments, they suggested that muscle death might be triggered through the motoneurons withdrawing their trophic influence. More recent experiments, however, indicate that a circulating peptide hormone is the primary trigger for muscle death (Schwartz & Truman 1982), although chronic stimulation of the muscle through its motoneuron seems to provide some measure of protection to the muscle.

The best example of anterograde trophic influences to maintain central neurons is seen in the development of the visual system as described above. In these systems the trophic influence appears to be most important at the time that the postsynaptic cell is receiving synapses and undergoing differentiation [eg. *Daphnia* (Macagno 1979)].

The importance of trophic input for fully differentiated invertebrate neurons in mature nervous systems appears to be minimal. In *Daphnia* the destruction of the photoreceptor cells has little effect on the laminar neurons when done in the adult (Macagno 1979). In *Manduca* electrophysiological studies of motoneurons that were in the early stages of cell death showed that known synapses were still in place and functional at this time (Truman & Levine 1980). Death was not proceeded by the withdrawal of all of the synapses from the doomed cells.

ENDOCRINE TRIGGERS   Fully differentiated neurons that subsequentially undergo a natural death often do so in response to hormonal signals. For example, thyroid hormone has been implicated in the death of the Mauthner cells in amphibians (Kimmel & Model 1978, Pesetsky 1966). In *Manduca*, hormones play a major role in regulating the cell death that occurs at metamorphosis. As detailed above, both major periods of postembryonic cell death in this species occur after molts—the pupal and adult, respectively. Each molt results in the production of a new exoskeleton and is induced by the ecdysteroid, 20-hydroxyecdysone (20-HE). Steroid titers peak about 30–40% of the way through each molt and then gradually decline through the remainder until the termination at ecdysis. These declining steroid levels at the end of the adult molt trigger the subsequent death of the intersegmental muscles as well as the abdominal interneurons and motoneurons (Truman & Schwartz 1982a,b). This conclusion was initially suggested by the finding in *Manduca* that ligation of the abdomen from the rest of the body late in adult development resulted in precocious death of both neurons and muscles. This manipulation separates the abdominal tissues from the site of ecdysteroid production in the thorax and consequently the abdomen is exposed to a precipitous decline in steroid levels.

The precocious degeneration brought about by ligation could be prevented by injection or infusion with 20-HE or by implantation of prothoracic glands (the glands that secrete ecdysteroids). Moreover, injection or infusion of ecdysteroids into intact animals delayed or prevented the normal onset of degeneration.

A number of features of this endocrine-regulated degeneration are worthy of note. That the intersegmental muscles begin to die before the death of their motorneurons was consistent with a model of retrograde trophic influences. This possibility was excluded by experiments that examined the critical periods for rescuing the muscles and neurons by ecdysteroid treatments. For both tissues, 20-HE given before a critical time late in development prevented death, whereas after this time treatments were ineffective. Interestingly, the critical period for the intersegmental muscles occurs earlier than that for the motoneurons that innervate them. Consequently, it was possible to find a time when ecdysteroid treatment blocked motoneuron death, whereas the muscles died on schedule (Truman & Schwartz 1982a). Thus, neuron death is not a direct result of the loss of their target muscles.

A second point relates to variation in the critical periods for individual neurons within the CNS (J. W. Truman and L. M. Schwartz, unpublished). A study of the response of identified neurons to ecdysteroid treatments at various times showed that each cell had its own characteristic critical period. Interestingly, the temporal order of critical periods for the various motoneurons is later reflected in the order of death of these cells. How these critical periods relate to the biochemical changes that bring about cell death are unknown.

A third point relates to the requirement that 20-HE must be present in order for these cells to survive in the adult. As long as the steroid is present they will continue to live; when it is withdrawn, death ensues. This steroid requirement is interesting since the same cells at earlier stages in the animal's life do not need the hormone, as is strikingly illustrated in pupae that enter a period of developmental arrest termed *diapause*. The diapause condition can last for months and it is maintained by the prolonged absence of circulating ecdysteroids. Despite the lack of steroids, the motoneurons nevertheless survive without ill effect. The subsequent appearance of ecdysteroids, which terminates diapause and initiates adult development, apparently alters these cells so that they will henceforth require the steroid in order to live. In the case of the death of the Mauthner cell in amphibia, Pesetsky (1966) suggested that the nuerons become sensitized by the appearance of thyroxine and then degenerate once the hormone levels decrease.

The giant silkmoths also show intersegmental muscle death (Finlayson 1956, Lockshin & Williams 1965a), but, unlike *Manduca,* this is not accompanied by motoneuron degeneration (J. W. Truman, unpublished). These species also differ from *Manduca* in the endocrine cues that bring about the death (Schwartz & Truman 1982). Muscle breakdown is triggered by a peptide hormone, *eclosion hormone,* that is released from the CNS at the end of adult development (Truman 1980). Ecdysteroids are also involved in this system since the decline in the steroid titer at the end of development renders the muscles competent to respond to the peptide (Schwartz & Truman 1982). This relationship was experimentally exploited by treating silkmoths with ecdysteroids at a critical time late in development. The treatment did not interfere with ecdysis or with the release of eclosion hormone, but it rendered the muscles incapable of showing the rapid degenerative response to the peptide. Because the peptide is not released again during the life of the moth and the muscles were not in the state to respond to the "death signal" when it was presented, the muscles then survived into adult life.

Cell Death and Sexual Differentiation

An area presently of interest in vertebrate nervous systems is the role of cell death in establishing sexual differences in the CNS (eg. Jordan et al 1982). Relatively little information is now available concerning sexual differences in invertebrate nervous systems, let alone how cell death may contribute to these differences. One exception is the nematode *C. elegans.* This worm has two sexes: male and hermaphrodite. Only the hermaphrodite retains two hermaphrodite-specific neurons, which are involved in egg-laying (Sulston & Horvitz 1977). By contrat, the male possesses four putative chemosensitive neurons, which are apparently used for chemotaxis toward the hermaphrodite (Ward et al 1975, Sulston & Horvitz 1977). Cells lineally equivalent to chemosensory cells are generated in the hermaphrodite but then die. Similarly, the male generates two cells having the same lineage as the hermaphrodite-specific neurons, but these degenerate (Horvitz et al 1982). That the cells that die would indeed have differentiated into the type of cell seen in the other sex was confirmed in the mutant *ced-3,* which blocks programmed cell death (Horvitz et al 1982). Thus, sexual differences in the nervous system of *C. elegans* seems attributable entirely to differences in cell death between the sexes.

Sex-specific cell death is also important in *Manduca* for shaping the differences in the motoneurons of the terminal ganglion that innervates the genitalia (J. Giebultowicz and J. W. Truman, unpublished). Larvae have identical arrays of motoneurons in the terminal ganglia of both sexes. After pupal ecdysis, changes begin to occur as cells start to die according to a sex-dependent pattern. This sex-specific degeneration is unique with respect to the time of its occurrence. All of the larval motoneurons that die after pupal ecdysis in both sexes do so within two days after ecdysis so that death is completed before adult differentiation begins. By contrast, the sex-specific death, especially in the male, is postponed until adult differentiation is well underway (J. Giebultowicz and J. W. Truman, unpublished). The significance of a unique time for sex-specific degeneration is not known. Possibly, it may mean that these cells need information from the differentiating periphery in order to make the decision to live or to die.

Ultrastructural Studies of Neuronal Death

In *Daphnia*, cells in the lamina that do not receive contacts from photoreceptor axons do not grow processes and eventually degenerate (Macagno 1979). The earliest stages of death in these cells have not been described, but the neurons eventually end up as aggregations of membrane-bound, deeply staining, cellular debris.

The programmed degeneration of a particular identified neuron was followed in *C. elegans* by Robertson & Thomson (1982). During the development of the ventral nerve cord, cell P11.aaa divides to produce two daughter cells: P11.aaaa, which differentiates into a motoneuron, and P11.aaap, which dies. At the time of its birth, the latter cell is already surrounded by the process from a neighboring hypodermal cell. Unlike its sister, P11.aaap fails to extend an axon. The cytoplasm then condenses and the nuclear envelope dilates, accompanied by aggregation of the chromatin. Pieces of dying cell are split off into membrane-bound fragments that are surrounded by the arms of hypodermal cells. Eventually cell organelles appear in autophagic vacuoles and the remains of the cell are phagocytized by the surrounding hypodermis. Although the hypodermal cell is associated with the doomed cell from the moment of its birth, the former is probably not a "killer" cell. As described below, the mutants *ced-1* and *ced - 2* show normal cell death without hypodermal cell engulfment (E. Hedgecock, J. White, and J. Sulston, personal communication, as cited in Horvitz et al 1982). Consequently, engulfment appears to be a response to a dying cell rather than the cause of that death.

The death of cells in *Manduca* differs from that described for *Daphnia* and *C. elegans* in that *Manduca* cells are differentiated, functioning neurons at the time that they die. Three pairs of motoneurons in the D-IV group show similar time courses of degeneration and were chosen for ultrastructural study (Stocker et al 1978). The muscles innervated by these neurons accumulate lysosomes in preparation for degeneration (Lockshin & Beaulaton 1974), but similar accumulations are not seen in the motoneurons. Indeed, the ultrastructural appearance of these cells is identical to that of motoneurons that persist through the life of the adult. The first ultrastructural changes occur in the D-IV cell at 12 hr after ecdysis when the endoplasmic reticulum appears to release its ribosomes and breaks up into vesicles. Changes in other cytoplasmic organelles, such as the rounding of the mitochondria, are seen shortly thereafter. Unlike the neurons described from *C. elegans,* the nucleus of the moth cells do not show chromatin condensation until after the nuclear membrane ruptures at about 24 hr. By 36 hr numerous organelles are found in autophagic vacuoles and the surrounding glia begin to phagocytize fragments of the neurons. Thus, the first involvement of the glia in the cell death occurs quite late in the degeneration process. The last vestiges of the cell eventually appear as a tightly wrapped ball of membranes.

Biochemical Aspects of Neuron Death

There have been no studies of the early biochemical events that occur during neuron degeneration in invertebrates. By contrast, biochemical aspects of the programmed death of the intersegmental muscles in moths have received considerable study and provide insight into some of the processes that may be involved in cell death (eg. Lockshin 1981). As described above, the death of these muscles in the silkmoth, *Antheraea polyphemus,* is triggered by the peptide, eclosion hormone. The earliest known action of the peptide is to cause a 20–30-fold increase in cyclic guanosine 3'5'-monophosphate (cyclic GMP) in the muscles (Schwartz 1982, Truman & Schwartz 1980). Interestingly, cyclic GMP (but not cyclic adenosine 3',5'-monophosphate) and drugs that increase endogenous levels of cyclic GMP such as isobutylmethylxanthine or sodium nitroprusside mimic the action of the hormone in inducing death of the muscle. Thus, cyclic GMP is thought to be involved in the mediation of the action of the peptide.

The changes in cyclic GMP levels appear to be followed by nuclear events (Lockshin 1969). RNA synthesis inhibitors such as actinomycin D prevent cell death if given within 2–3 hr of eclosion hormone release, and protein synthesis inhibitors are effective for 6–7 hr. Indeed, the process of death appears to be an active cellular response that involves the production of new RNAs and proteins. The nature of these gene products are unknown. Possibly they could be analogous to the product of the *ced-3* gene described below.

Genetic Aspects of Cell Death in Invertebrates

Mutants are available that influence neuron death in both *C. elegans* and *Drosophila melanogaster.* In the nematode a number of mutants effect the death of specific cells. For example, the mutant *n709* causes the death of P3.aap - P8.aap cells, which normally differentiate into the six VC motoneurons (Horvitz et al 1982). Similarly, the mutation *e1611* causes the death of six mechanoreceptor cells (Chalfie & Sulston 1981). The most interesting mutants, however, are the *ced* (cell death) mutants.

The *ced-3* mutation blocks the death of all cells that would undergo programmed cell death (Horvitz et al 1982). Immature neurons that normally die, survive and undergo differentiation according to the instructions imposed by their lineage. The result is a worm with a supernumerary number of cells in its CNS; but this appears to have little impact on the behavior of the animal. Mutants *ced-1* and *ced-2* allow cell death to occur but not the engulfment of the remains by the hypodermis (E. Hedgecock, unpublished as cited in Horvitz et al 1982). As indicated above, this phenotype strongly suggests that degeneration is an endogenous program of the cell rather than an event imposed by the hypodermal cell.

The mutants available in *Drosophila* are ones that cause excessive neuron death rather than preserving cells that would normally die. The mutants thus far isolated affect primarily the visual pathway. A number of eye structure mutants are mentioned above. Those that reduce the size of the eye, also result in the excessive death of laminar neurons, a situation reflecting the dependence of the developing optic lobe on the presence of photoreceptor axons (Meyerowitz & Kankel 1978). Flies with the mutation optimotor blind *(omb)* have a specific lesion in that they lack the giant visual interneurons that extend from the lobula plate (Heisenberg et al 1978). In this case it is not certain whether the cells have actually died or have simply failed to grow to their normal size. A clear cell death mutation is small optic lobes *(sol)* (Fischbach & Heisenberg 1981). Mutant flies have a normal lamina, but the medulla has only 20,000 cells as compared with the wild-type complement of 40,000. This reduction in cell number is due to excessive cell death during metamorphosis rather than to the failure to produce normal numbers of cells. Genetic mosaic flies were generated, some of which showed a wild-type eye associated with a mutant optic lobe and *vice versa*. Thus, the phenotype of the optic lobe was not due to influences from the eye. The control over the death of the medulla medullary neurons is either intrinsic to the medulla or dependent on deeper brain structures.

There are no mutants in *Manduca* that affect the normal programmed death seen in its nervous sytem. Interestingly, though, in silkmoths such as *Antheraea polyphemus* and *Hyalophora cecropia* the abdominal motoneurons survive through the life of the adult even though their target muscles die. The reason for this interspecific variation is unknown.

Conclusions

Most of our information on cell death in invertebrate nervous systems is presently based on the arthropod visual system, early development of locusts and nematodes, and metamorphosis in moths. Obviously, it is premature to draw generalizations. The examples in this review illustrate the precision by which cell death can be predicted and subsequently followed in these preparations. The signals that trigger the death vary (hormones and trophic interactions are two examples), but the actual degeneration of the neurons appears to be an intrinsic response of the respective cells to these signals. The breakdown process is an orderly process which, as illustrated by the *ced-3* mutant in nematodes, is apparently overseen by specific genes that must be expressed in order for the phenomenon to occur. One of the exciting research areas in the future will be to determine the nature of these gene products that bring about cellular suicide. It should be cautioned, however, that all cell deaths, even in the same organism, do not necessarily occur according to the same mechanism. For example, the *ced-3* mutation prevents cell death in a number of tissues in the nematode, yet there are some cells that still die even in mutant individuals (Horvitz et al 1982).

The cell's decision to die should be thought of as a differentiative option just as is the production of an axon or the synthesis of a certain type of transmitter. As one might expect from a differentiative response, this pathway is not necessarily available to all cell types or even to a given cell at all times in its life. Thus, one finds that cells that die usually do so at specific times, for example at their birth, when they have sent an axon to the periphery, or after they are fully differentiated, functional neurons. It will be important to understand the mechanisms that open up the option of degeneration to these cells, in addition to knowing the factors that eventually trigger the degeneration itself.

ACKNOWLEDGMENTS

I thank Professor L. M. Riddiford for a critical reading of the manuscript. Unpublished work was supported by grants from NSF, NIH, and the McKnight Foundation.

Literature Cited

Alverdes, F. 1924. Die Wirkung experimenteller Eingriffe Inbesondere der Blendung, auf den histologischen Bau des Insektgehirns. *Z. Morphol. Oekol. Tiere* 2:189–216

Anderson, H. 1978. Postembryonic development of the visual system of the locust *Schistocera gregaria*. I. Patterns of growth and developmental interactions in the retina and optic lobe. *J. Embryol. Exp. Morphol.* 45:55–83

Anderson, H., Edwards, J. S., Palka, J. 1980. Developmental neurobiology of invertebrates. *Ann. Rev. Neurosci.* 3:97–139

Bate, C. M. 1976. Embryogenesis of an insect nervous system. I. A map of the thoracic and abdominal neuroblasts in *Locusta migratoria*. *J. Embryol. Exp.Morphol.* 35:107–123

Bate, C. M., Grunewald, E. B. 1981. Embryogenesis of an insect nervous sytem. II. A second class of neuronal precursor cells and the origin of the intersegmental connectives. *J. Embryol. Exp. Morphol.* 61:317–30

Bate, M., Goodman, C. S., Spitzer, N. C. 1981. Embryonic development of identified neurons: Segment-specific differences in the H cell homologues. *J. Neurosci.* 1:103–6

Chalfie, M., Sulston, J. 1981. Developmental genetics of the mechanosensory neurons of *Caenorhabditis elegans*. *Dev. Biol.* 82:358–70

Chiarodo, A. J. 1963. The effects of mesothoracic leg extirpation on the postembryonic development of the nervous system of the blowfly. *Sarcophaga bullata*. *J. Exp. Zool.* 153:263–77

Cowan, W. M. 1970. Anterograde and retrograde transneuronal degeneration in the central and peripheral nervous system. In *Contemporary Research and Methods in Neuroanatomy*, ed. W. J. Nauta, S. O. E. Ebbesson, pp. 217–51. New York: Academic

Eichenbaum, D. M., Goldsmith, T. H. 1968. Properties of intact photoreceptor cells lacking synapses. *J. Exp. Zool.* 169:15–32

Elwyn, A. 1937. Some stages in the development of the neural complex in *Ecteinascidia turbinata*. *Bull Neurol. Inst. NY* 6:163–77

Finlayson, L. H. 1956. Normal and induced degeneration of abdominal muscles during metamorphosis in the Lepidoptera. *Q. J. Microsc. Sci.* 97:215–34

Fischbach, K. F., Heisenberg, M. 1981. Structural brain mutant of *Drosophila melanogaster* with reduced cell numbers in the medulla cortex and with normal optomotor yaw response. *Proc. Natl. Acad. Sci. USA* 78:1105–9

Glucksmann, A. 1951. Cell deaths in normal vertebrate ontogeny. *Biol. Rev.* 26:59–86

Goodman, C. S., Bate, M. 1981. Neuronal development in the grasshopper. *Trends Neurosci.* 4:163–69

Goodman, C. S., Bate, M., Spitzer, N. C. 1981. Embryonic development of identified neurons: Origin and transformation of the H cell. *J. Neurosci.* 1:94–102

Hamburger, V., Oppenheim, R. W. 1982. Naturally occurring neuronal death in vertebrates. *Neurosci. Comment.* 1:39–55

Heisenberg, M., Wonneberger, R., Wolf, R. 1978. Optamotor-blind[H31] a *Drosophila* mutant of the lobula plate giant neurons. *J. Comp. Physiol.* 124:287–96

Hinke, W. 1961. Das relative postembryonale Wachstum der Hirnteite von *Culex pipiens, Drosophila melanogaster* und *Drosophila— Mutaten. Z. Morphol. Oekol. Tiere* 50:81–118

Horvitz, H. R., Ellis, H. M., Sternberg, P. W. 1982. Programmed cell death in nematode development. *Neurosci. Comment.* 1:56–65

Hughes, A. F. 1957. The development of the primary sensory system in *Xenopus laevis* (Davdin) *J. Anat.* 91:323–38

Johannsen, O. A. 1924. Eye structure in normal and "eyeless" *Drosophila. Biol. Bull.* 48:243–58

Jordan, C. L., Breedlove, S. M., Arnold, A. P. 1982. Neurogenesis of motoneurons in the sexually dimorphic spinal nucleus of the bulbocavernosus in rats. *Soc. Neurosci. Abstr.* 8:198

Kimmel, C. B., Model, P. 1978. Developmental studies of the Mauthner cell. In *Neurobiology of the Mauthner Cell,* ed. D. S. Faber, H. Korn, pp. 183–220. New York: Raven

Kopec, S. 1922. Mutual relationship in the development of the brain and eyes of Lepidoptera. *J. Exp. Zool.* 36:459–67

Kopec, S. 1923. The influence of the nervous system on the development and regeneration of muscles and integument in an insect. *J. Exp. Zool.* 37:15–25

Krafka, J. Jr. 1924. Development of the compound eye of *Drosophila melanogaster* and its Bar-eyed mutant. *Biol. Bull.* 47:143–48

Lamborghini, J. E. 1981. Kinetics of Rohon-Beard neuron disappearance in *Xenopus laevis. Soc. Neurosci. Abstr.* 7:291

Lockshin, R. A. 1969. Programmed cell death. Activation of lysis by a mechansim involving the synthesis of protein. *J. Insect Physiol.* 15:1505–16

Lockshin, R. A. 1981. Cell death in metamorphosis. In *Cell Death in Biology and Pathology,* ed. I. D. Bowen, R. A. Lockshin, pp. 79–121. London: Chapman & Hall

Lockshin, R. A., Beaulaton, J. 1974. Programmed cell death. Cytochemical evidence for lysosomes during normal breakdown of the intersegmental muscles. *J. Ultrastruc. Res.* 46:43–62

Lockshin, R. A., Williams, C. M. 1965a. Programmed cell death. I. Cytology of degeneration in the intersegmental muscles of the Pernyi silkmoth. *J. Insect Physiol.* 11:123–33

Lockshin, R. A., Williams, C. M. 1965b. Programmed cell death. III. Neural control of the breakdown of the intersegmental mus-

cles of silkmoths. *J. Insect Physiol.* 11:601–10

Lockshin, R. A., Williams, C. M. 1965c. Programmed cell death. IV. The influence of drugs on the breakdown of the intersegmental muscles of silkmoths. *J. Insect Physiol.* 11:803–9

Macagno, E. R. 1979. Cellular interactions and pattern formation in the development of the visual system of *Daphnia magna* (Crustacea, Branchiopoda). *Dev. Biol.* 73:206–38

Maxwell, G. D., Hildebrand, J. G. 1981. Anatomical and neurochemical consequences of deafferentiation in the development of the visual system of the moth *Manduca sexta. J. Comp. Neurol.* 195:667–80

Meinertzhagen, I. A. 1973. Development of the compound eye and optic lobe of insects. In *Developmental Neurobiology of Arthropods,* ed. D. Young, pp. 51–104. Cambridge: Cambridge Univ. Press

Meyerowitz, E. M., Kankel, D. R. 1978. A genetic analysis of visual system development in *Drosophila melanogaster. Dev. Biol.* 62:112–42

Mouze, M. 1974. Interactions de l'oeil et du lobe optique au cours de la croissance postembryonnaire des insects odonates. *J. Embryol. Exp. Morphol.* 31:377–407

Mouze, M. 1978. Role des fibres postretiniennes dans la croissance du lobe optique de la larve d' *Aeschna cyanea* Müll (Insecte Odonate) *Wilhelm Roux Arch.* 184:325–50

Murphey, R. K., Mendenhall, B., Palka, J., Edwards, J. S. 1975. Deafferentiation slows the growth of specific dendrites of identified giant interneurons. *J. Comp. Neurol.* 159:407–18

Norlander, R. H., Edwards, J. S. 1968. Morphological cell death in the postembryonic development of the insect optic lobes. *Nature* 218:780–81

Nordlander, R. H., Edwards, J. S. 1969. Postembryonic brain development in the monarch butterfly *Danaus plexippus plexippus,* L. I. Cellular events during brain morphogenesis. *Wilhelm Roux Arch. Dev. Biol.* 162:197–217

Nuesch, H. 1968. The role of the nervous system in insect morphogenesis and regeneration. *Ann. Rev. Entomol.* 13:27–44

Oppenheim, R. W. 1981. Neuronal cell death and some related regressive phenomena during neurogenesis: A selective historical review and progress report. In *Studies in Developmental Neurobiology: Essays in Honor of Viktor Hamburger,* ed. W. M. Cowan, pp. 74–133. New York: Oxford Univ. Press

Pesetsky, I. 1966. The role of the thyroid in the development of the Mauthner's neuron. A karyometric study in thyroidectomized larvae. *Z. Zellforsch. Mikrosk. Anat.* 75:138–45

Power, M. E. 1943. The effect of reduction in numbers of ommatidia upon the brain of *Drosophila melanogaster. J. Exp. Zool.* 94:33–72

Ranzi, S. 1928. Correlazioni tra organi di senso e centri nervosi in via di svillupo (Ricerche di morphologia sperimentale nei Cefalopodi). *Wilhelm Roux Arch. Dev. Biol.* 114:364–70

Richards, M. H., Furrow, E. Y. 1925. The eye and optic tract in normal and "eyeless" *Drosophila. Biol. Bull. Marine Biol. Lab. Woods. Hole* 48:243–58

Robertson, A. M., Thomson, J. N. 1982. Morphology of programmed cell death in the ventral nerve cord of *Caenorhabditis elegans* larvae. *J. Embryol. Exp. Morphol.* 67:89–100

Sanes, J. R., Hildebrand, J. G., Prescott, D. J. 1976. Differentiation of insect sensory neurons in the absence of their normal synaptic targets. *Dev. Biol.* 52:121–27

Sanes, J. R., Prescott, D. J., Hildebrand, J. G. 1977. Cholinergic neurochemical development of normal and deafferented antennal lobes during metamorphosis of the moth *Manduca sexta. Brain Res.* 119:389–402

Saunders, J. W. 1966. Death in embryonic systems. *Science* 154:604–12

Schwartz, L. M. 1982. *The endocrine coordination of developmentally programmed muscle degeneration in Lepidoptera.* PhD thesis. Univ. Washington, Seattle

Schwartz, L. M., Truman, J. W. 1982. Peptide and steroid regulation of muscle degeneration in an insect. *Science* 215:1420–21

Stocker, R. F., Edwards, J. S., Truman, J. W. 1978. Fine structure of degenerating moth abdominal motor neurons after eclosion. *Cell Tissue Res.* 191:317–31

Sulston, J. E. 1976. Post-embryonic development in the ventral cord of *Caenorhabdites elegans. Philos. Trans. R. Soc. London Ser. B* 275:287–97

Sulston, J. E., Horvitz, H. R. 1977. Postembryonic cell lineages of the nematode *Caenorhabditis elegans. Dev. Biol.* 56:110–56

Torrence, S. A. 1983. *Ascidian larval nervous system: Anatomy, ultrastructure and metamorphosis.* PhD thesis. Univ. Washington, Seattle

Truman, J. W. 1980. Eclosion hormone: Its role in coordinating ecdysial events in insects. In *Insect Biology in the Future,* ed. M. Locke, D. S. Smith, pp. 385–401. New York: Academic

Truman, J. W. 1983. Programmed cell death in the nervous system of an adult insect. *J. Comp. Neurol.* 216:445–52

Truman, J. W., Levine, R. B. 1980. Programmed cell death in the nervous system of an insect: Histological and physiological aspects. *Soc. Neurosci. Abstr.* 6:668

Truman, J. W., Reiss, S. R. 1976. Dendritic reorganization of an identified motorneuron during metamorphosis of the tobacco hornworm moth. *Science* 192:447–79

Truman, J. W., Schwartz, L. M. 1980. Peptide hormone regulation of programmed death of neurons and muscle in an insect. In *Peptides: Integrators of Cell and Tissue Function,* ed. F. E. Bloom, pp. 55–67. New York: Raven

Truman, J. W., Schwartz, L. M. 1982a. Insect systems for the study of programmed neuronal death. *Neurosci. Comment.* 1:66–72

Truman, J. W., Schwartz, L. M. 1982b. Programmed death in the nervous system of a moth. *Trends Neurosci.* 5:270–73

Ward, S., Thomson, N., White, J., Brenner, S. 1975. Electron microscopical reconstruction of the anterior sensory anatomy of the nematode, *Caenorhabditis elegans. J. Comp. Neurol.* 160:313–38

White, J. G., Albertson, D. G., Anness, M.A.R. 1978. Connectivity changes in a class of motoneurone during the development of a nematode. *Nature* 271:764–66

White, J., Southgate, E., Thomson, N., Brenner, S. 1976. The structure of the ventral nerve cord of *Caenorhabditis elegans. Philos Trans. R. Soc. London. Ser. B* 275:327–48

Whitington, P. M., Bate, M., Seifert, E., Ridge, K., Goodman, C. S. 1982. Survival and differentiation of identified embryonic neurons in the absence of their target muscles. *Science* 215:973–75

Zottoli, S. J. 1978. Comparative morphology of the Mauthner cell in fish and amphibians. In *Neurobiology of the Mauthner Cells,* ed. D. Faber, H. Korn, pp. 13–45. New York: Raven

Ann. Rev. Neurosci. 1984. 7:189–222

PROTEOLYSIS IN NEUROPEPTIDE PROCESSING AND OTHER NEURAL FUNCTIONS

Y. Peng Loh

Laboratory of Neurochemistry and Neuroimmunology, National Institute of Child Health and Human Development, National Institutes of Health, Bethesda, Maryland 20205

Michael J. Brownstein

Laboratory of Cell Biology, National Institute of Mental Health, National Institutes of Health, Bethesda, MD 20205

Harold Gainer

Laboratory of Neurochemistry and Neuroimmunology, National Institute of Child Health and Human Development, National Institutes of Health, Bethesda, Maryland 20205

INTRODUCTION

Proteolysis is commonly associated with degradation, inactivation, and turn-over of peptides and proteins in cells (Goldberg & St. John 1976, Lajtha & Dunlop 1978, Hershko & Ciechanover 1981). However, it also plays a more subtle role in the "functional modification" of protein substrates, through limited proteolysis (Holzer & Heinrich 1980, Reich et al 1975). In recent years it has become increasingly apparent that the presence of biologically active peptides and proteins in cells and tissues is a function not only of the biosynthetic mechanism per se but also of a variety of postsynthetic (i.e. post-translational) processing mechanisms. Several of these processes (e.g. gly-cosylation, methylation, and phosphorylation), because of their obvious in-

189

0147-006X/84/0301-0189$02.00

volvement in membrane phenomena and electrogenesis, have already captured the attention of neurobiologists. Other processes such as amidation, ADP-ribosylation, *N*-acylation, and limited proteolysis still remain relatively ignored by neuroscientists (for reviews about these processes see Freedman & Hawkins 1980, Wold 1981).

In this review we focus primarily on proteolytic processing (limited proteolysis) for several reasons. First, proteolysis (i.e. hydrolysis of peptide bonds) is considered the most general of all covalent modifications occurring in vivo (Wold 1981). Since all proteins are synthesized de novo using the universal initiation codon for methionine (AUG), but few proteins are found in vivo containing this amino acid at their amino-terminus, then at least proteolysis of this peptide bond must be occurring widely. Second, proteolysis differs from other post-translational modification mechanisms in that it is essentially irreversible, and hence represents a definite "commitment" to a cell biological function in molecular terms.

The term "limited proteolysis," which means selective peptide bond cleavage (as opposed to nonselective degradation), was first introduced by Linderstrom-Lang & Ottesen (1949). This process has since been shown to play a major role in many physiological events (e.g. in blood coagulation, the complement system, prohormone conversion, viral assembly, digestion, development, etc). It is beyond the scope of this review to discuss most of these subjects, and excellent reviews and books in this area are available (e.g. Reich et al 1975, Holzer & Heinrich 1980, Freedman & Hawkins 1980).

This review begins with a brief discussion on the classification of proteases and the nomenclature in this field.[1] The purpose of this first section is to introduce to the neophyte some of the conventions and general sources in this complex field. Our choice of specific subjects reflects both our cell biological approach and our specific interests in this field as it relates to neurobiology. Hence, we concentrate on proteolytic mechanisms in the biosynthesis and processing of neuropeptides. In addition, certain relevant nonproteolytic

[1]List of abbreviations: ACTH, adrenocorticotropin; APMSA, aminomercaptosuccinic acid; AVP, arginine vasopressin; βNE, β-neoendorphin; B₂ISA, benzylsuccinic acid; CANP, calcium dependent neutral protease; CCK, cholecystokinin; CLIP, corticotropin-like intermediate lobe peptide; CPB, carboxypeptidase B; DFP, diisopropyl fluorophosphate; EDTA, ethylenedinitrilo tetra-acetic acid; EGTA, ethyleneglycol-bis-(β-amino-ethyl ether) tetra-acetic acid; GP, glycopeptide; GPSA, guanidinopropylsuccinic acid; LE, [leu]enkephalin; LHRH, luteinising hormone releasing hormone; LPH, lipotropin; ME, [met]enkephalin; ME-RF, [met]enkephalin arg-phe; ME-RGL, [met]enkephalin arg-gly-leu; MGTA, 2-mercaptomethyl-3-guanidinoethyl thiopropanoic acid; MSH, melanotropic stimulating hormone; NpII, Neurophysin II; PCE, pro-protein converting enzyme; PCMB, *para*-chloromercuribenzoate; PCMPSA, *p*-chloromercuriphenylsulfonic acid; PTH, parathyroid hormone; RER, rough endoplasmic reticulum; SBTI, soybean trypsin inhibitor; TLCK, *N*-α-*p*-tosyl-L-lysine chloromethyl ketone; TPCK, L-1-tosylamide-2-phenylethyl chloromethyl ketone; TRH, thyrotropin releasing hormone.

mechanisms in neuropeptide formation are also discussed. Another area of emphasis is with reference to the calcium-activated proteases, which appear to be a vigorous and emerging field in neurobiology. It is obvious that protein degradation and turnover issues, and inactivation mechanisms to terminate the biological actions of neuropeptides, are of considerable importance in neurobiology. However, constraints of space allow us to provide only limited commentary in these areas. Finally, the reader should be warned at the outset that the field of proteolytic mechanisms in neural systems is still in its infancy. Hence, there may be a paucity of generalizations, and perhaps an excess of preliminary data and speculations. However, we write this review with the hope that it will serve as a catalyst for what promises to be an exciting and important area in the neurosciences.

CLASSIFICATION OF PROTEASES

The International Union of Biochemistry's Standing Committee on Enzymes is responsible for the standardization of protease classification. As of its last report (International Union of Biochemistry 1978), proteases can be placed into two general classes: (*a*) endopeptidases, which cleave peptide bonds within polypeptide chains, and (*b*) exopeptidases, which remove amino acids, one or a few at a time, from the amino or carboxy termini of the polypeptides. In addition, the individual proteases are grouped into subclasses based on catalytic mechanism (e.g. serine or cysteine proteinase), and accepted names are followed by "EC" numbers, which are code numbers made up of four parts followed by periods. The first number indicates the main division (e.g. hydrolases = 3), the second the subclass (e.g. 4 = peptide hydrolases), the third a sub-subclass (e.g. 21 = serine proteases), and the fourth the individual enzyme. Illustration of this point is made in Table 1, which lists some endopeptidases by classification and characteristics. Because this field is in constant flux, proteases are often reclassified or unclassified in the primary literature (for more details see IUB 1978, Barrett & McDonald 1980, Barrett 1980, Lorand 1981). Uncertainty with respect to the characteristics or purity of an individual enzyme results at least in the absence of a fourth code number (see Table 1).

During investigations on new proteolytic activities (or studies of proteolysis in new tissues), it is helpful to be bear in mind these standardized sources of information. Not only is there a standard nomenclature to attend to, but this literature also supplies valuable information for the characterization of the new enzyme activity. For example, proteases can be characterized by their specific inhibitors, pH range of activity, cofactors and, of course, the specificities of their cleavage sites (see Table 1). The ultimate aim of such enzymology is to purify the enzyme (e.g. protease) to homogeneity, in order to characterize it fully. Valuable information about an enzyme activity is often obtained before reaching this stage, however, and knowing the specific inhibitors of the activity can often facilitate its purification by affinity chromatography. The data in Table 1 is by no means complete. Many other endopeptidases and exopeptidases that are not shown here can be found listed in the standardized sources (e.g. IUB 1978, Lorand 1981, Barrett & McDonald 1980). These sources (particularly the *Methods in Enzymology* series, e.g. Lorand 1981) also provide many methodological hints, conceptual clues, and caveats for the investigator interested in proteases to ponder.

In the subsequent pages of this review, many of the biologically significant proteolytic activities we discuss are not as well characterized as those shown in Table 1. Hence, we resort to terms such as "trypsin-like," "cathepsin-B-like," or "carboxypeptidase-B-like" to describe proteolytic activities that have similar specificities of cleavages to the standard enzyme but that may differ in one or another property (e.g. inhibitor spectrum, pH optimum, molecular weight, etc). Often proteolytic activities (enzymes) are found that are highly specific in their cleavage sites (e.g. prohormone-converting enzymes, see below) and have no known counterparts. This situation is to be expected, since there are undoubtedly many unknown families of proteolytic enzymes yet to be discovered. The reader should be alerted to the fact that even for the "classified" enzymes (e.g. in Table 1), there are ambiguities. For example, pH optima of purified enzymes can vary depending on substrate. In the case of cathepsin B, the pH optimum is 6.0 for small synthetic substrates, but is 3.5 when collagen is used as the substrate. Cathepsin H has a pH optimum of 6.8 for small synthetic substrates, but a pH optimum of 5.0 when degrading azocasein. In addition, reaction specificity is often dependent on conditions, e.g. cathepsin B usually acts as an endopeptidase (see Table 1) but can also exhibit peptidyldipeptidase activity (at the C-terminus); and cathepsin H can exhibit either endopeptidase or aminopeptidase activity depending on specific conditions. Given these ambiguities, we have tried in this review to discuss the enzymes in the context of their natural microenvironments and their activities on endogenous (non-synthetic) substrates wherever possible.

INTRACELLULAR ORGANIZATION OF NEUROPEPTIDE PROCESSING

Peptidergic neurons are nerve cells specialized in the production and secretion of neuropeptides. Various studies have shown that such neuropeptides are synthesized from large precursors by a series of sequential proteolytic cleavages. These proteolytic events are highly organized within the cell and are programmed to occur within specific intracellular compartments, where the appropriate enzymes are localized and activated. The initial neuropeptide gene product (the largest form of the neuropeptide) synthesized on the rough endo-

Table 1 Classification of some endopeptidases[a]

Proteinase class	Protease (EC number)	Specificity of cleavage	pH optimum	Cofactors or activators	Inhibitors	Molecular weight (Kd)	Distribution
Serine proteinase (active serine, inhibited by DFP)	Chymotrypsin (EC 3.4.21.1)	Tyr-X, Tryp-X Phe-X, Leu-X	7.5		DFP, TPCK	26	Zymogen granules in pancreas as chymotrypsinogen, activated in intestine
	Trypsin (EC 3.4.21.4)	Arg-X, Lys-X	8.0	Ca^{2+}	DFP, TLCK soybean trypsin inhibitor, leupeptin	23.5	Zymogen granules in pancreas as trypsinogen
	Thrombin (EC 3.4.21.5)	Arg-Gly in fibrinogen	6.4 (to clot) 9.0 (for proteolysis)	Ca^{2+} phospholipid	DFP, TLCK	38	Plasma as prothrombin, converted by factor X_a and V_a
	Plasmin (EC 3.4.21.7)	Lys-X, Arg-X	7.0–7.4		DFP, TLCK leupeptin, aprotinin	85	Plasma as plasminogen (93 Kd)
	Plasma kallikrein (EC 3.4.21.8)	Lys-Arg, Arg-Ser Kininogen→ bradykinin	7.0–9.0		DFP, α_2-macroglobulin, aprotinin	90	Plasma as prokallikrein
	Acrosin (EC 3.4.21.10)	Arg-X, Lys-X	8.0		DFP, TLCK	38	Acrosome of mammalian spermatozoan
	Cathepsin G (EC 3.4.21.20)	Broad specificity Leu, Tyr, Phe, Met, Tryp, Glu, Asn-X	7.5		DFP, chymostatin	30	Azurophil granules in human neutrophils
	Plasminogen activator (E.C. 3.4.21.-)	Lys-X plasminogen→ plasmin	7.0–7.4		DFP	50	Plasma as proenzyme
	Urokinase (EC 3.4.21.31)	Arg-Val, Lys-X Arg-X	7.8		DFP	33, 47	Kidney cells (acts as renin plasminogen activator)

Table 1 Classification of some endopeptidases[a]

Proteinase class	Protease (EC number)	Specificity of cleavage	pH optimum	Cofactors or activators	Inhibitors	Molecular weight (Kd)	Distribution
Cysteine (thiol) proteinases (active cysteine, inhibited by thiol blocking agents)	Cathepsin B (EC 3.4.22.1)	Arg-X, Lys-X	3.5–6.0	Cysteine EDTA	Thiol blocking agents (TBA) leupeptin, TLCK	25	Intralysosomal
	Cathepsin H (EC 3.4.22.-)	Arg-X, Lys-X and can act as amino-peptidase	5.6–6.8	Glutathione EDTA	TBA, leupeptin TLCK	28	Intralysosomal
	Cathepsin L (EC 3.4.22.15)	Arg-X	5.0–5.6	Cysteine EDTA	TBA, leupeptin TLCK	24	Intralysosomal
	Cathepsin N (EC 3.4.22.-)	Like Cathepsin B, but with higher specificity for collagen	3.5	Cysteine	TBA, leupeptin TLCK	20	Intralysosomal
	Cathepsin S (EC 3.4.22.-)	Arg-X	3.5	Cysteine	TBA, leupeptin	25	Intralysosomal (specifically in lymph nodes and spleen)
Calcium dependent Neutral protease (EC 3.4.22.-)		Degrades very specific proteins	7.5	Cysteine Ca^{2+}	TBA, EGTA, EDTA, leupeptin	80–100	Cytosolic
ATP dependent (EC 3.4.22.-) protease		Degrades hemoglobin	7.5–9.5	5 mM ATP, cysteine, Triton X-100	TBA	550	Cytosolic

Class	Enzyme	Specificity	pH optima	Inhibitor	Metal	MW (×10³)	Source
Aspartic (acid or carboxyl) proteinases (acidic pH optima)	Pepsin A (EC 3.4.23.1)	Phe, Met, Leu, or Trp-X	1.8–2.2	Pepstatin		34.5	Gastric mucosal cells
	Renin (EC 3.4.23.-) or (EC 3.4.99.19)	Very specific for angiotensinogen Leu-Leu	3.5, 6.0 (depending on substrate)	Pepstatin		36.4–42	Kidney: nonlysosomal organelles, and in plasma as pro-enzyme
	Cathepsin D (EC 3.4.23.5)	Similar to Renin (called isorenin, pseudorenin)	2.8–5.0	Pepstatin		42	Intralysosmal
	Cathepsin E (EC 3.4.23.-)	?	2.5	Pepstatin		50, 100	Rabbit bone marrow, blood cells
Metallo-proteinases (activated by metals)	Vertebrate collagenase (EC 3.4.24.7)	Specific for triple helix of collagen Gly-Leu, Gly-Ileu	7.5	EDTA $\alpha_2$macroglobulin B_1-anticollagenase	Ca^{2+}	35, 65	Secreted by fibroblasts as proenzyme
	Microbial metallo-enzymes (EC 3.4.24.4)	X-Leu, X-Phe	7.0	EDTA	Zn^{2+}		Secreted by microorganisms

[a]Adapted from Barrett & McDonald (1980) and Lorand (1981).

plasmic reticulum (RER) is the pre-pro-protein, e.g. pre-pro-vasopressin, pre-pro-enkephalin, pre-pro-opiocortin, pre-pro-dynorphin (Nakanishi et al 1979, Drouin & Goodman 1980, Policastro et al 1981, Noda et al 1982, Comb et al 1982, Gubler et al 1982, Kakidani et al 1982, Sabol et al 1983). These pre-pro-proteins consist of a "pre" or "signal" sequence at the N-terminus which is composed of between 15–30 amino acids rich in hydrophobic residues, similar to pre-sequences found in newly synthesized secretory protein precursors such as pre-pro-albumin (Strauss et al 1977), pre-pro-immunoglobulin (Burnstein & Schechter 1978), pre-pro-mellitin (Sachanek et al 1978), and pre-pro-insulin (Villa-Komaroff et al 1978). The "pre" sequence participates in directing the pre-pro-protein through the RER membrane into the cisternae, thus segregating this class of proteins from other non-secretory proteins. The "pre" sequence is immediately cleaved off the protein (i.e. co-translational processing) prior to completion of synthesis of the protein (Blobel & Dobberstein 1975, Lingappa et al 1978, Harwood 1980, Kreil 1981), giving rise to the pro-protein. Cleavage of the "pre" sequence is carried out by a metallo-endopeptidase associated with the inner membrane of the RER cisternae (Jackson & Blobel 1977, Zwizinski & Wickner 1980). Biochemical, electron microscopic, and autoradiographic evidence (Patzett et al 1980, Potts et al 1980, Van Heldon 1980) suggest that the pro-protein is then translocated from the RER cisterna into the Golgi apparatus, where it is packaged into secretory granules (Palade 1975). Depending on the system, some pro-proteins undergo their first proteolytic cleavage in the Golgi apparatus, and the final steps of proteolytic processing occur after packaging into secretory granules. Other pro-proteins, however, are primarily cleaved after packaging into secretory granules.

Biosynthesis and axonal transport studies on the neuropeptides, vasopressin and oxytocin, in the rat hypothalamo-neurohypophysial system have implicated the secretory granule as the site of processing of pro-vasopressin and pro-oxytocin. Kinetic studies showed that newly synthesized pro-vasopressin and pro-oxytocin were progressively cleaved, with time, to vasopressin, oxytocin, and their related neurophysins within the axons of these neurons located in the median eminence (Gainer et al 1977). Thus, these pro-proteins appear to be processed during axonal transport within secretory granules.

Other examples of the cellular organization of pro-protein processing have been derived from nonneuronal systems. It is commonly stated that processing of pro-insulin begins at the Golgi level and continues within the secretory granules (Steiner et al 1974, Rodriguez-Boulan et al 1978, Okada et al 1979). Processing of pro-opiocortin in the toad and mouse intermediate lobe of the pituitary has been shown to occur within the secretory granule (Loh & Gainer 1979, Van Heldon 1980, Loh & Gritsch 1981), whereas in dissociated cells from rat intermediate lobe (Glembotski 1981) and mouse anterior pituitary tumor cells (Gumbiner & Kelly 1981), the first cleavage step may occur in the Golgi. In the case of pro-albumin, proteolytic processing also occurs after packaging into vesicles (Quinn & Judah 1978, Judah & Quinn 1978). The available experimental evidence from neuronal and non-neuronal systems, therefore, indicates that the secretory granule (or vesicle) is a major site for proteolytic processing of pro-proteins. In those cases, in which proteolysis of the pro-proteins may begin at the Golgi, the final proteolytic processing steps have been shown to occur intragranularly (Glembotski 1981, Gumbiner & Kelly 1981). Thus it would be expected that the subsequent modifications of the peptides prior to secretion (e.g. amidation, acetylation) would also occur within granules. Indeed Stoeckel et al (1983) have immunocytochemical evidence showing that acetylation of α-MSH occurs intragranularly in rabbit pituitary intermediate lobe cells.

The mechanism underlying the packaging of the pro-proteins and the processing enzymes in the same subcellular processing compartment are poorly understood at the present time. One possibility is that the enzyme and precursors are cosegregated and copackaged. This would suggest that there may be "signals" on the precursors and enzymes that would bind to certain Golgi membrane regions where subsequent formation of the secretory granule occurs. Consistent with this idea is the finding that newly synthesized pro-opiocortin in toad and mouse intermediate lobe is associated with the secretory granule membrane (Y. P. Loh, unpublished data), whereas the cleaved products are not. Another idea is the fusion model suggested by the experimental work of Judah & Quinn (1978). These authors found that processing of endogenously labeled pro-albumin is Ca^{2+} dependent. Reagents that promote membrane fusion greatly increased the rate of processing and those that block fusion (e.g. colchicine) inhibited processing. Hence, they proposed that intragranular conversion of pro-albumin requires the fusion of vesicles containing the processing enzymes with the granules containing the pro-albumin. A third possible mechanism is the fusion of vesicles containing the processing enzymes with the Golgi, at the budded off area, just prior to the complete formation of the secretory granules. All three possibilities provide working hypotheses for future studies on mechanisms.

NEUROPEPTIDE PRO-PROTEIN SEQUENCES

Recent advances in molecular cloning techniques have greatly accelerated the task of identifying and determining the primary structures of neuropeptide pro-proteins (Nakanishi et al 1979, Gubler et al 1982, Land et al 1982, Kakidani et al 1982). Given these primary structures (e.g. in Figure 1), it is possible to deduce patterns of amino acid sequences that are characteristic of pro-protein processing. Virtually all of the pro-proteins have pairs of basic

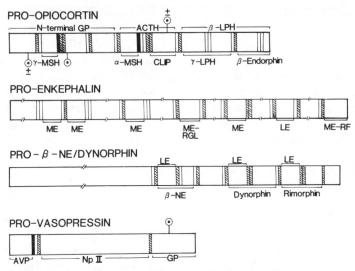

Figure 1 **Diagramatic representation of the** structure of several neuropeptide pro-proteins, showing the paired basic amino acid residues (lysine □, arginine ▨), the glycine residues (■) located at the *N*-termini of the pairs of basic amino acids, and glycosylation sites (⊙). In pro-opiocortin, the ○ sites indicated with ± are not always glycosylated in different forms of this molecule (Crine et al 1980, Phillips et al 1981). Bovine proopiocortin is not glycosylated at the ACTH site (Bennett et al 1982). The lengths of the peptides and spacer regions have not been drawn to scale. Regions shortened are indicated //. GP = glycopeptide, MSH = melanotropic stimulating hormone, ACTH = adrenocorticotropin 1–39, LPH = liptropin, CLIP = corticotropin-like intermediate lobe peptide, ME = [met]enkephalin, ME-RGL = [met]enkephalin arg-gly-leu. ME-RF = [met]enkephalin arg-phe, β-NE = β-neoendorphin, LE = [leu]enkephalin. AVP = arginine vasopressin, NpII = Neurophysin II. The positions of the paired basic aminio acids and glycine were taken from the reported sequences for bovine pro-opiocortin (pro-opiomelanocortin) (Nakanishi et al 1979), bovine pro-enkephalin (Gubler et al 1982), porcine pro-β-neo-endorphin/ dynorphin (Kakidani et al 1982), and bovine pro-vasopressin (Land et al 1982).

amino acids that flank the neuropeptide sequences that are to be cleaved. The most common pair of basic residues is the Lys-Arg sequence, although Arg-Lys, Arg-Arg, and Lys-Lys can also be found (Figure 1). This structural pattern suggests that the pro-protein converting enzymes exhibit a specificity for pairs of basic amino acid residues. Some exceptions to the rule are illustrated in Figure 1 for pro-vasopressin, which contains a single Arg at the cleavage site between neurophysin and the C-terminal glycopeptide, and pro-β-neo-endorphin/dynorphin, which contains a single Arg at the C-terminal cleavage site of rimorphin. Another structural feature of the pro-proteins is that amidation of the C-terminal amino acids in peptides appears to be signaled by a glycine residue on the N-terminal side of the basic amino acid pair, e.g. in vasopressin by Gly-Lys-Arg, α-MSH by Gly-Lys-Lys-Arg-Arg (see Figure 1), and calcitonin by Gly-Lys-Lys-Arg (Jacobs et al 1981).

Several other post-translational modifications of the pro-proteins also occur in situ. Some of these (e.g. acetylation of the N-terminal, and amidation of the C-terminal) occur after proteolytic cleavage of the peptide from the pro-protein. Others, such as glycosylation, formation of S-S bridges, and possibly phosphorylation, occur on the intact pro-protein before cleavage. The enzymes for these events appear to be located in the RER and Golgi (Czichi & Lennarz 1977, Phelps 1980, England 1980). Glycosylation has been found to be asparagine-linked, occurring characteristically on Asn-X-Ser(Thr) sequences as in pro-opiocortin (Eipper & Mains 1976, 1977, Loh & Gainer 1979, Bennett et al 1982) and pro-vasopressin (Land et al 1982). Phosphorylation was found to occur at the serine residue in pro-opiocortin (Mains & Eipper 1982, Bennett et al 1982). Neuropeptide pro-proteins may contain more than one biologically significant product (Figure 1). Pro-opiocortin may be converted into MSH, ACTH, β-LPH, and β-endorphin (Eipper & Mains 1980). Pro-enkephalin contains six Met-enkephalin sequences as well as one Leu-enkephalin (Gubler et al 1982, Comb et al 1982), while pro-β-neo-endorphin/dynorphin contains three Leu-enkephalins, one distributed in each of β-neo-endorphin, dynorphin 1–17, and rimorphin sequences (Kangawa et al 1981, Goldstein et al 1981, Tachibana et al 1982, Kilpatrick et al 1982, Cox 1982, and Kakidani et al 1982). Thus pro-proteins can be "common precursors" or multivalent. It is apparent for pro-opiocortin, pro-enkephalin, and pro-β-neo-endorphin/dynorphin that the nature of the processing enzymes found in situ will be important factors determining the types of peptides secreted by any neuron that synthesizes these pro-proteins.

Variability in the processed products formed from a common precursor, in different neurons, may also arise from alterations at the genetic level. These include the deletion of certain segments of nucleotides in the mRNA, resulting in the elimination of a pair of basic amino acid residues. An example of deletion of pairs of basic residues has been found in the bovine and rat pro-opiocortin sequences (Nakanishi et al 1979, Drouin & Goodman 1980). Recent studies of Herbert et al (1982) have also demonstrated that the length of the pro-opiocortin mRNA in amygdala is shorter than in hypothalamus and pituitary by about 25–50 nucleotides. Changing of one nucleotide in the codon coding for Asn could result in the replacement of that amino acid by an Asp, resulting in the elimination of a glycosylation site. An example is found in the ACTH sequence in the pro-opiocortin molecule of bovine and rat (Nakanishi et al 1979, Drouin & Goodman 1980). The Asp (residue 29) in the bovine ACTH sequence is replaced by an Asn in rat, giving rise to nonglycosylated and glycosylated forms of ACTH and CLIP in these two species, respectively (Bennett et al 1981, 1982).

Finally, although the primary sequence of the pro-protein may greatly dictate the processing pattern, the conformation of the molecule may also play a role.

Geisow & Smyth (1980) have made extensive theoretical calculations based on the primary structure of the pro-proteins and proposed that certain pairs of basic amino acid residues are not cleaved on some pro-proteins because they are found in the α-helix structure of the protein, where the regular pattern of hydrogen bonds can stablize the polypeptide against attack [e.g. The Arg-Arg in proglucagon (Blundell et al 1976) and the Arg-Lys in relaxin (Bedarkar et al 1977, Isaacs et al 1978)]. the basic residues that are cleaved are probably located in superficial and aperiodic regions of the globular pro-proteins. Extra lengths of peptides in the pro-proteins with no known biological function may play a role in the folding of the pro-protein so as to provide the appropriate conformation for specific cleavage.

ENZYMOLOGY OF POST-TRANSLATIONAL PROCESSING OF NEUROPEPTIDE PRO-PROTEINS

As discussed in the above section, the structure of the neuropeptide pro-proteins suggests that two types of proteolytic activity are necessary for the conversion of the pro-protein in the final products. The first is an enzyme that would recognize the paired basic residues flanking the peptide sequences to be cleaved (Figure 1), and the second is a carboxypeptidase-B-like enzyme that would remove the basic residues from the C-terminus of the cleaved peptide. Often, the cleaved peptides undergo further modifications, usually at the N or C termini. Examples of C-terminal modifications are amidation found in α-MSH, thyrotropin-releasing factor (TRH), luteinzing hormone releasing hormone LHRH, substance P, and cholecytokinin (CCK) (Eberle & Schwyzer 1975, Leeman et al 1977, Burgus et al 1969, Potts et al 1971. Anastasi et al 1968, Baba et al 1971), and proteolytic cleavages that eliminate C-terminal amino acids from the peptide to form truncated products, [e.g. β-endorphin 1–27, β-endorphin 1–26, CLIP 18–38 (ACTH 18–38) (Smyth et al 1980, Bennett et al 1982)]. The most commonly found N-terminal modifications are acetylation [e.g. β-endorphin and α-MSH (Eberle & Schwyzer 1975, Smyth et al 1980)] and cyclization of the glutamic acid to form pyroglutamate [e.g. TRH, gastrin, bombesin (Burgus et al 1969, Rehfeld et al 1979, Endean et al 1975)]. Other modifications include sulfation of tyrosine in CCK (Anastasi et al 1968) and phosphorylation of ACTH (Bennett et al 1981, Mains & Eipper 1982). Current knowledge of the nature of the enzymes involved in these post-translational processing events is summarized in the following sections.

Converting Enzymes Specific for Paired Basic Residues

Several enzymatic activities that are specific for paired basic amino acid residues of pro-proteins have been found in various neuronal and nonneuronal tissues. These include pro-protein enzyme converting activities in pancreatic islet cells and the parathyroid. The converting enzyme activities for pro-insulin, pro-glucagon, and pro-somatostatin found within the pancreatic islet cell granules of anglerfish (Fletcher et al 1980, 1981, Noe 1981) and rat (Docherty et al 1982) cleave at pairs of basic amino acids in the respective prohormones to yield insulin, glucagon, and somatostatin. These enzymes appear to be acid, thiol proteases distinct from cathepsin B (based on inhibitor studies), but similar to a pro-opiocortin coverting enzyme activity reported in secretory granules from rat neurointermediate lobes (Loh & Gainer 1982a; also see below).

Two groups have reported the presence of a parathyroid hormone (PTH) converting activity in bovine parathyroid that cleaves at paired basic residues to convert pro-PTH to PTH (McGregor et al 1976, 1978, Habener et al 1977). Unlike the pro-opiocortin and pro-insulin converting enzymes, this enzyme activity has a pH optimum of 7 to 9, is inhibited by benzamidine and choloro-quine, and is Ca^{2+} sensitive. The inhibitor studies suggest that the enzyme is not pancreatic trypsin.

Although albumin is not a neuropeptide or peptide hormone, it is perhaps worth mentioning that a pro-albumin converting activity that has all the characteristics of lysosomal cathepsin B has been found in the large granules of rat liver (Quinn & Judah 1978). The enzyme is specific for the Arg-Arg residues of pro-albumin; has a pH optimum of 6.0 and is inhibited by leupeptin and TLCK, but not pepstatin.

In the rat and bovine pituitary neural lobe secretory granules, an enzyme activity that cleaves specifically at paired basic amino acid residues has been reported (Loh & Chang 1982, Chang et al 1982). In these studies, toad pro-opiocortin was used as a model substrate. The major products formed were a 21K (K = 1000) molecular weight ACTH/β-LPH peptide and a 16K N-terminal glycopeptide (see Figure 1), indicating a cleavage at the pair of Lys-Arg residues between the N terminal glycopeptide and ACTH. This converting activity has a pH optimum of 5.0 and is a thiol protease that is distinct from lysosomal cathepsin B, since it is not inhibited by several cathepsin B inhibitors and is unable to cleave typical cathepsin B substrates, including small peptides with paired basic residues. This latter observation suggests that in addition to paired basic residues, the substrate must have a conformation that matches the enzyme active site for specific cleavage to occur. The converting activity appears to be present in a membrane-associated and soluble form. Analysis of the molecular weight of the soluble enzyme activity from bovine neural lobe granules suggests that it is a protein of about 20K–60K daltons in size (Chang et al 1982). In vivo, this enzyme probably serves to cleave the vasopressin and oxytocin pro-proteins.

A similar pro-protein converting enzyme (PCE) activity has been found in rat intermediate lobe secretory granules (Loh & Chang 1982; T.-L. Chang and

Y. P. Loh, in preparation). This enzyme activity, which was found both in a soluble and a membrane-associated form within the granules, catalyzed the cleavage of toad pro-opiocortin to 21K ACTH (N-terminal glycopeptide + ACTH sequence), 13K glycosylated ACTH, β-LPH, α-MSH, β-endorphin-like peptides, and the 16K N-terminal glycopeptide (see Figure 1). These in vitro generated products are similar to those synthesized by the toad intermediate lobe in situ (Loh et al 1982), and their structure indicates cleavages at the Lys-Arg residues of pro-opiocortin (Figure 1). When [^{3}H]-arginine-labeled products generated by the rat intermediate-lobe-converting activity were subsequently treated with carboxypeptidase B, free [^{3}H] arginine was liberated, suggesting that the enzyme cleaved at the peptide bond just after the arginine of the Lys-Arg pairs (T. -L. Chang and Y. P. Loh, in preparation). Specific cleavage of the toad pro-opiocortin by rat intermediate lobe PCE activity does not depend on the glycosylation state of the prohormone (Loh & Gainer 1982b). Analysis of the nature of the intermediate lobe PCE activity shows that it is an acid thiol protease, distinct from lysosomal cathepsin B, with a pH optimum of 5.0, similar to the neural lobe coverting enzyme. More extensive characterization of rat intermediate lobe soluble and membrane-associated PCE activity shows that the nature of the two activities are very similar (as determined by their protease inhibitor profiles), but they favor cleavages of different Lys-Arg residues (T. -L. Chang and Y. P. Loh, in preparation) in the pro-opiocortin molecule. The membrane-associated activity favors the first cleavage of pro-opiocortin at the Lys-Arg residues between ACTH and β-LPH (see Figure 1), while the soluble activity favors cleavage of the basic residues between the 16K N-terminal glycopeptide and ACTH. Since the first cleavage of pro-opiocortin in the pars intermedia in situ occurs between ACTH and β-LPH (Eipper & Mains 1980, Loh et al 1982), it is likely that the membrane-associated activity may be responsible for the initial cleavage of pro-opiocortin in the intact granule. The membrane activity is readily liberated by 1 M NaCl and therefore is not an integral membrane enzyme. Solubilization of the membrane activity resulted in an alteration in the preference of the pair of Lys-Arg residues to be cleaved to that characteristic of the soluble enzyme; this indicates that the association of the converting enzyme with the granule membrane may influence the enzyme-substrate interaction in a manner that favors cleavage of the Lys-Arg pair of pro-opiocortin, between ACTH and β-LPH.

Studies have also revealed the presence of a paired basic amino acid specific, acid-thiol protease in rat anterior lobe secretory granules (Chang & Loh 1983). This converting enzyme activity cleaved toad pro-opiocortin primarily to 13K glycosylated ACTH and β-LPH. The inhibitor profile of the anterior lobe converting activity is very similar to the intermediate lobe except for its sensitivity to PCMB, a thiol protease inhibitor. Although pro-opiocortin converting activity in the intermediate lobe was inhibited by the thiol inhibitors, PCMB and dithiodipyridine, the anterior lobe activity was inhibited only by the latter inhibitor.

Another enzyme activity that is specific for pairs of basic amino acids, but has properties unlike trypsin, has been found in porcine pituitary granules (Bradbury et al 1976, Smyth et al 1977). However, this porcine converting enzyme activity is quite different in its pH optimum (pH 8.8) from the rat intermediate neural and anterior lobe granule enzymes described by Loh and her co-workers. The assay for this enzyme was carried out using a synthetic hexapeptide, and the relationship of this activity to the pro-opiocortin conversion described above remains to be determined.

Several laboratories have reported enzyme activities in bovine adrenal medulla chromaffin granules that cleave, at the paired basic residues of larger forms of enkephalin, including BAM_{12} and peptide E, to produce Met-enkephalin (Evangelista et al 1982, Mizuno et al 1982, Troy & Musacchio 1982, Lindberg et al 1982). Three of the groups (Evangelista et al 1982, Mizuno et al 1982, Troy & Musacchio 1982) have characterized the chromaffin granule converting enzyme activity to be a thiol protease with a pH optimum of 5–6. The enzyme activity reported by Mizuno et al (1982) has a molecular weight of about 220K and appears to make two simultaneous cleavages at both sides of consecutive basic residues. This is in contrast to the enzyme activity studied by Evangelista et al (1982), which cleaves on the carboxy side of the paired basic amino acids. However, Mizuno et al may be observing the action of more than one enzyme. The enkephalin generating enzyme reported by Lindberg et al (1982) is different again, having a pH optimum of 7.5–8.0 and is a serine protease. All these enkephalin converting enzymes remain to be evaluated for their ability to generate enkephalin from endogenous pro-enkephalin (Figure 1).

Docherty & Steiner (1982) have proposed several criteria for the identification of pro-protein converting proteases. Among these are the following:

1. whether the enzyme activity correctly cleaves the pro-protein to all known products;
2. whether the enzyme activity is localized in an appropriate organelle, i.e. where conversion occurs in vitro;
3. whether the pH optimum and stability of the enzyme activity is consistent with the internal pH of the organelle in which the activity is localized.

Several studies have considered these criteria. The prohormone converting activities found in the rat and anglerfish pancreatic islet and rat pituitary intermediate and anterior lobe secretory granules fulfill all these criteria. The action of these enzymes is very specific, and appears to cleave only at paired basic residues to yield the proper cleavage products from the respective prohormones. The activities are found in secretory granules, consistent with the observations implicating the granules as the site for pro-hormone processing in these tissues in vivo. The acidic pH optima and ranges (between pH 5–6) of these enzyme activities correlate with the acidic intragranular pHs (between pH 5–6) found in virtually all secretory granules that have been studied (Scarpa & Johnson 1976, Pollard et al 1979, Russell & Hotz 1981, Carty et al 1982). The converting activity found in the neural lobe secretory granules and chromaffin granules (Troy & Mussachio 1982, Evangelista et al 1982, Mizuno 1982) clearly fulfills criteria (2) and (3). Although it has been shown that the enzyme activity in neural lobe granules does specifically cleave at the paired basic amino acid residues of pro-opiocortin, experiments have not yet been reported showing that the pro-proteins for vasopressin and oxytocin (the relevant substrates) are correctly processed by this enzyme activity. Similarly, the adrenal medulla chromaffin granule enkephalin converting enzyme activity has to be shown to cleave pro-enkephalin correctly to all the enkephalin peptides found in the adrenal medulla in situ. Current studies indicate that many of the paired basic residue-specific enzymes have properties quite unlike trypsin. Hence they should be referred to as pro-protein converting enzymes rather than as "trypsin-like" enzymes, a term frequently used for these paired basic residue-specific converting activities.

Carboxypeptidase-B-like Converting Enzymes

Early studies from Steiner's group (Kemmler et al 1973, Steiner et al 1975) have provided evidence for the existence of a carboxypeptidase B-like (CPB-like) activity in the secretory granules of rat pancreatic islet cells that is involved in pro-insulin processing. The activity cleaved the basic residues from the C terminus of the B chain of insulin and the C-peptide (split from pro-insulin) at pH 7.0. EDTA and ortho-phenanthroline inhibited the activity and Co^{2+} stimulated the activity, whereas DFP, an inhibitor of serine proteases, and thiol reagents were without effect. The inhibitor profile suggested that the activity was due to a metallo enzyme, similar to pancreatic carboxypeptidase B (Marinkovic et al 1977).

More recently, the presence of a specific carboxypeptidase in bovine adrenal chromaffin granules, pituitary, and brain that converts enkephalin precursors ([Met]-enkephalin and [Leu]-enkephalin hexapeptides containing carboxyl terminal arginine or lysine residues [Met]-enkephalin Arg-Arg) to enkephalin have been reported (Fricker & Snyder 1982, Fricker et al 1982). The regional variation of the enzyme activity in the brain correlates with the distribution of enkephalin. The enzyme has a pH optimum of 5.5–6.0 and is activiated by Co^{2+} and is significantly inhibited by EDTA, 1–10 phenanthroline, and the thiol inhibitors $HgCl_2$, p-chloromercuriphenylsulfonate and iodoacetamide. The enzyme has been purified to apparent homogeneity and characterized as a glycoprotein of about 50,000 daltons in molecular weight. In the chromaffin granules, the activity is in the soluble and membrane fractions. V. Y. H. Hook and L. E. Eiden (in preparation) have also found a similar carboxypeptidase B-like activity in highly purified bovine adrenal medulla chromaffin granules that converts [Met]-enkephalin-Arg to [Met]-enkephalin. This activity has a very similar inhibitor profile and pH optimum to those reported by Fricker & Synder (1982) and is also activated by Co^{2+}. It is likely that both groups are assaying the same activity.

CPB-like activity has also been detected in purified secretory granules from rat neural, intermediate, and anterior lobes of the pituitary using [^{125}I] enkephalin-Arg as a model substrate (V. Y. H. Hook & Y. P. Loh, in preparation). In addition to cleaving the Arg and Lys from [Met]-enkephalin-Arg and [Met]-enkephalin-Lys respectively, the intermediate and anterior lobe secretory granule lysate also cleaved $ACTH_{1-17}$ (which has a Lys-Lys-Arg sequence at the C terminal) to $ACTH_{1-16}$, $ACTH_{1-15}$, and $ACTH_{1-14}$, sequentially. Neural lobe granule lysate cleaved Arginine-vasopressin (AVP)-Gly-Lys-Arg to AVP-Gly-Lys and AVP-Gly. Both these substrates ($ACTH_{1-17}$ and AVP-Gly-Lys-Arg) are peptides cleaved from the respective pro-proteins by the paired basic amino acid residue specific enzymes. The CPB-like activity in all three lobes of the pituitary has a pH optimum of 5.5–6., and is stimulated by Co^{2+} and inhibited by "by product" analogues of arginine and lysine (GPSA, MGTA, B$_2$ISA, AMPSA, EDTA, phenanthroline (metalloenzyme inhibitors), and the thiol inhibitors $HgCl_2^{2+}$ and PCMPSA. The inhibitor profiles of the pituitary granule CPB activities are very similar to the adrenal medulla chromaffin granule activity reported by Fricker & Snyder (1982), and different from rat lysosomal CPB activity. The CPB activity in the granules of all three lobes of the pituitary are present in the soluble and membrane fractions as in chromaffin granules.

The CPB-like activities in the pituitary and adrenal medulla secretory granules have many properties similar to pancreatic CPB but differ in the pH optimum and inhibition by thiol protease inhibitors. The pituitary secretory granule CPB-like activities are also different from lysosomal CPB, which is not stimulated by Co^{2+}. Thus the CPB-like enzyme involved in neuropeptide pro-protein processing may be a subset of a family of intracellular CPB activities.

Specific Proteases Acting at Nonbasic Residues

A number of proteases that cleave peptide bonds not involving basic amino acid residues are reviewed briefly below. These enzymes cleave larger neuropeptides to smaller ones. Some of them may be classified as synthetic or converting enzymes, while others may perhaps have an inactivating role since they are often not very specific enzymes. The subcellular localization of many of these enzymes are also unknown.

Two enzymes that cleave CCK_{33} to either CCK_{12} or CCK_8 at peptide bonds between Arg-Isoleu and Arg-Asp of CCK_{33}, respectively, have been found in brain. These enzymes are distinct from trypsin in size and lack inhibition by soybean trypsin inhibitor (SBTI). They will also cleave synthetic dipeptides such as Arg-Val and Arg-Leu and are therefore not CCK_{33} specific (Malesci et al 1980, Ryder et al 1980).

An enkephalin generating enzyme has been partially purified from rat brain (Knight et al 1982). This enzyme is a thiol endopeptidase with chymotryptic-like specificity for peptide bonds following hydrophobic residues Met and Leu. The enzyme is able to cleave a model substrate Leu-Try-Met-Arg-Phe-Ala and endogenous enkephalin precursors from striatum. It has been proposed that such an enzyme may be involved in the cleavage of pro-enkephalin or its tryptic fragments to yield enkephalin.

A number of other proteinases that may be relevant in pro-protein conversion have been found in pituitary and brain. One of these enzymes is a neutral proteinase reported by Graf & Kenessey (1981) to be present in porcine anterior pituitary granules. The enzyme cleaves at the Arg of the synthetic substrate Z-Lys-Pro-Arg-Nap and the Arg^{60}-Tyr^{61} bond of β-LPH_{1-91} to release β-LPH_{1-60}. Acid-thiol proteinases in pituitary and brain that are able to cleave Bz-Arg-Nap have been reported by Marks et al (1981). However, these enzymes also degrade a number of other protein substrates such as histones, glucagon, the lipotropins, and myelin basic protein. In addition to these proteinases, Cathepsin D-like enzymes have been found in anterior pituitary and brain. These are able to cleave the Leu^{77}-Phe^{78} bond in β-endorphin to form γ-endorphin (Graf & Kenessey 1981). In the neural lobe of the rat pituitary, North et al (1977) have reported the presence of an acidic (pH optimum 4.5), chymotryptic-like enzyme in neurosecretory granules that converts rat neurophysin II to neurophysin III and neurophysin I to neurophysin I'.

A proteolytic enzyme(s)-substrate system that is normally considered as a blood system and cardiovascular regulator, but has recently been associated with the central nervous system, is the renin-angiotensin system. Angiotensin II, which is the effector substance in the periphery to produce increased blood pressure in response to hypotension, is generated by the sequential action of two enzymes. Renin liberated from the kidney cleaves plasma angiotensinogen (Arg-Arg-Val-Tyr-Ile-His-Pro-Phe-His-Leu-Leu-Val-Tyr-Ser-R) at the Leu-Leu bond to give angiotensin I. Angiotensin converting enzyme (E.C.3.4.15.1) converts angiotensin I (a decapeptide) to angiotensin II (an octapeptide) in blood, by splitting a Phe-His bond, resulting in the removal of a

His-Leu sequence at the C terminal. Angiotensin converting enzyme may be a relatively nonspecific dipeptidylcarboxypeptidase that removes COOH-terminal dipeptides from a variety of substrates, some as small as tripeptides. Recently, angiotensin converting enzyme has been shown to convert met-enkephalin Arg-Phe to [Met]-enkephalin (Marks et al 1982).

Virtually all the components of the renin-angiotension system have been reported as being present in the central nervous system (Yang & Neff 1972, Igic et al 1977, Peach 1977, Ganten et al 1978, Ganten & Speck 1978, Hirose et al 1980, Dzau et al 1980, Hermann et al 1982). However, it is still not clear whether these components are identical in molecular character to those in the periphery. Indeed, one group has suggested that brain angiotensinogen differs at least in its carbohydrate moiety from the circulating molecule (Ito et al 1980). Recently it has been shown that cultured neurons can synthesize an angiotensin II-like molecule (Fishman et al 1981, Raizada et al 1983), thus increasing the credibility of this system as being endogenous to the brain. Furthermore, there is evidence for the biological action of angiotensin II in the brain (Reid et al 1977, Phillips 1980, Van Houten et al 1980), although the possibility that peripheral angiotensin II may be involved in some of these actions (particularly around circumventricular organs) cannot be discounted (Van Houten et al 1980).

Other Post-Translational Modification Enzymes

N-ACETYLTRANSFERASES   Much of the knowledge on peptide N-acetyltransferases have come from studies on the N terminal acetylation of α-MSH and β-endorphin in the intermediate lobe of the pituitary. Glembotski (1982a, b) reported the presence of an acetylating enzyme in bovine intermediate lobe secretory granules that acetylated desacetyl α-MSH (ACTH$_{1-13NH_2}$), α-MSH, and β-endorphin to form α-MSH, α-N-o-diacetyl α-MSH, and acetylated β-endorphin, respectively. Acetylation of desacetyl α-MSH and α-MSH occurred at serine residues, and β-endorphin at a tyrosine residue. The enzyme is soluble and has a pH optimum of 7.0. Kinetic studies on the crude enzyme preparation showed similar K_m values for ACTH$_{1-13NH_2}$ and β-endorphin$_{1-31}$ but a much higher K_m for α-MSH. Competition studies suggest that the same N-acetylation enzyme acetylates all three peptides.

Studies on the rat have also revealed a similar α-MSH/β-endorphin (opio-melanotropin) acetyltransferase activity in neurointermediate lobe homogenate, a crude neurointermediate lobe secretory granule preparation, and in the hypothalamus (Chappell et al 1982, O' Donohue 1983). The activity was shown to be distinct from a general acetyltransferase found in the cytosol of pituitary cells and in other tissues (Chappell et al 1982, Woodford & Dixon 1979, Pease & Dixon 1981). Further analysis using highly purified secretory granules from rat intermediate lobe revealed two α-N-acetyltransferase activities in these organelles (Y. P. Loh, M. Chappell, and T. L. O'Donahue, in preparation). One has a pH optimum of 7.0–7.4 and acetylates both $ACTH_{1-13H_2}$ and β-endorphin$_{1-31}$. The other, which has a pH optimum of 6.0, acetylates $ACTH_{1-13NH_2}$ to form a α-MSH. However, β-endorphin is a very poor substrate for this acidic enzyme activity. Both activities are found in the soluble fraction of the granule lysate. The neutral opiomelanotropin-β-N-acetyltransferase activity found in rat intermediate lobe secretory granules appears to be very similar to the activity reported in bovine intermediate lobe secretory granules (Glembotski 1982b) and they are probably the same enzyme. Whether the acidic activity in rat intermediate lobe secretory granules is in fact due to a different enzyme remains to be determined. Since the internal pH of the granule is acid (pH 5–6), and given the current knowledge of α-N-acetyltransferase activities localized in the rat intermediate lobe secretory granules, it would be expected that acetylation of $ACTH_{1-13NH_2}$ would be more rapid than β-endorphin in this tissue. Indeed, kinetic studies show that in rat intermediate lobe cells, α-MSH sized moledulces were acetylated twice as rapidly α-endorphin sized molecules (Glembotski 1982a).

Neurointermediate lobe extracts has been tested for its ability to acetylate serotonin and choline. Both these compounds were not substrates for the neurointermediate lobe acetyltransferase(s), suggesting that α-N-acetylation of serotonin in the pineal gland and the acetylation of choline are due to a different enzyme (T. L. O'Donohue, personal communication). Thus, these peptide α-N-acetyltransferase(s) may represent another subclass of a family of transferases in the nervous system.

AMIDATION ENZYMES   The recent studies of Bradbury et al (1982) have provided evidence for a specific enzyme that amidates peptides with a C-terminal glycine. This enzyme activity assayed using D-Tyr-Val-Gly was found in porcine pituitary granules and was characterized as a neutral enzyme with a molecular weight of ~60,000. Glycine was found to be a mandatory amino acid in the C-terminal position of the peptide to be amidated, although the valine in the tripeptide substrate can be replaced by any neutral amino acid. From their data, they proposed that amidation occurs by the removal of hydrogen from the C-terminal glycine, followed by spontaneous hydrolysis of the resulting amino linkage. By such a mechanism, the lysine residue, which is consistently the next amino acid following the glycine in all the peptides that are eventually amidated at the C-terminal, does not appear to participate in the reaction. Further analysis is necessary to determine whether this amidation enzyme activity can amidate naturally occurring substrates such as $ACTH_{1-14}$ and AVP-Glycine found in the pituitary.

OTHER INTRACELLULAR PROTEOLYTIC MECHANISMS

Lysososomal and Nonlysosomal Degradation of Proteins

Protein turnover is a general property of all living cells (Poole 1971, Rechigal 1971, Lajtha & Dunlop 1981), and degradation occurs by both lysosomal (Barrett & Heath 1977, Holzer & Heinrich 1980, Bainton 1981, Pfeiffer 1981) and nonlysosomal mechanisms (Goldberg & St. John 1976, Hershko & Ciechanover 1982). Although it is generally accepted that "deprivation-enhanced" degradation of proteins in cells occurs entirely in lysosomes, the situation for the basal metabolic state is less clear. However, even in the latter case it is believed that lysosomal autophagy (Pfeiffer 1981) plays a major role in intracellular protein breakdown. Lysosomal-based proteolysis studies have a long history (see Bainton 1981); much about this mechanism is known, and antibodies to specific mammalian proteases have been used in immunolocalization studies (Poole 1977, 1981). However, little work has been done on lysosomal enzymes specifically in neural tissues, although it is commonly observed that exogenous proteins (e.g. horseradish peroxidase) taken up by nerve cells are usually routed to lysosome in the cell bodies (Holtzman & Mercurio 1980).

It is generally believed that degradation of most long-lived intracellular proteins occurs within lysosomes, whereas short-lived and abnormal proteins are degraded by nonlysosomal proteases. Recent findings of ATP-dependent proteolytic systems in reticulocytes and *E. coli* provide a number of candidate proteases for nonlysosomal mechanisms (see Hershko & Ciechanover 1982 for general discussions of this subject and the fascinating ubiquitin pathway of protein degradation in reticulocytes). Although there is ample protein turnover in the nervous sytem (Lajtha & Dunlop 1981), very few studies have been directed at the cell biological organization of this turnover.

Calcium-Dependent Neutral Proteases

Calcium-dependent neutral proteases (also referred to as calcium-activated proteases, calcium-activated neutral proteases, and calpain) are thiol-proteases, with pH optima around neutrality, which are found in the cytosol of a wide variety of tissues and cells (Ishiura 1981, Murachi et al 1981a, b). Their activation in situ often leads to the proteolysis of only one or few specific intracellular proteins, with selective physiological consequences depending on cell-type. Calcium-dependent neutral proteases have been implicated in specific intracellular enzyme activation (Huston & Krebs 1968, Nishizuki et al 1979, Hamon & Bourgoin 1979), myoblast differentiations (Kaur & Sanwal 1981), myofibrillar protein turnover (Dayton et al 1976a, b, 1981, Ishiura et al

1978, Kameyama & Etlinger 1979), regulation of steroid receptors (Puca et al 1977, Vedeckis et al 1980), peptide receptors (Cassell & Glasser 1982), and glutamate receptors (Baudry & Lynch 1980, Vargas et al 1980, Baudry et al 1981), and modification of cytoskeletal proteins (Anderson et al 1977, Phillips & Jakabova 1977, Triplett et al 1972, Wallach et al 1978, Gilbert et al 1975, Lucas et al 1979, Traub & Nelson 1981, Nelson & Traub 1982, Collier & Wang 1982).

The presence of calcium-dependent neutral proteases (CANP) in nervous tissue was first reported by Guroff (1946). The first evidence, however, that CANP was of neuronal origin and had a specific endogenous protein substrate came from the work of Gilbert and his colleagues on axoplasm isolated from the giant axons of the fan worm, *Myxicola,* and the squid, *Loligo* (Gilbert 1975, Gilbert et al 1975, Anderton et al 1976). This and subsequent work on both invertebrates and vertebrate axons showed that the CANP was a neutral, thiol protease specifically activated by Ca^{2+}, which selectively degraded neurofilament proteins (Gilbert et al 1975, Schlaepfer & Micko 1978, Schlaepfer & Freeman 1980, Pant et al 1979, Pant & Gainer 1980, Schlaepfer et al 1981, Eagles et al 1981, Tashiro & Ishizaki 1982, Malik et al 1981).

It is known that there are two types of CANPs extractable from various tissues, which have different elution postions in DEAE-cellulose chromatography, one of which has a high sensitivity (<40 μM) and the other a low sensitivity (>500 μM) to Ca^{2+} (Mellgren 1980, Dayton et al 1981. Kubota et al 1981, DeMartino 1981, Suzuki et al 1981, Murachi et al 1981a,b, Mellgren et al 1982, Hathaway et al 1982). It appears that the form of CANP found in axons requires >500 μM of Ca^{2+} (Pant & Gainer 1980, Pant et al 1982), whereas the form of CANP found acting on postsynaptic glutamate receptors (Baudry & Lynch 1980, Baudry et al 1981) requires lower levels of Ca^{2+} (<40 μM). Murachi et al (1981b) have shown that of 11 tissues studied that had CANP activity, nervous tissue (brain) was the least abundant in the high Ca^{2+} sensitivity form (referred to by those workers as Calpain I). Although some evidence is available for Calpain I in brain (see Baudry & Lynch 1980, Baudry et al 1981), it is unlikely that this is the type of CANP associated with the neurofilaments in axons. In contrast, Calpain II (the low Ca^{2+} sensitivity form of CANP) was quite abundant in brain tissue (Murachi et al 1981b), and would appear to be the type of CANP found in axoplasm (Pant & Gainer 1980, Pant et al 1982).

Given these two forms of CANP, the question can be raised as to whether the low sensitivity form of CANP (i.e. Calpain II) serves as a proenzyme form in the cell for Calpain I. Since physiological levels of intracellular Ca^{2+} are rarely above 10^{-6}–10^{-7}M, one would expect that Calpain II is effectively inactive under normal circumstances. One experimental procedure has been found to convert Calpain II to Calpain I in vitro (Dayton et al 1981, Kubota et al 1981,

Suzuki et al 1981, Hathaway et al 1982, Mellgren et al 1982). This turns out to be an accidental consequence of the affinity purification of Calpain II on a casein CH sepharose 4B column in the presence of 24 mM of Ca^{2+} (Suzuki et al 1981, Kubota et al 1981). The affinity purified CANP I is slightly smaller (i.e. 76 Kd) than the unpurified CANP II (i.e. 80 Kd), and the authors suggest that this change in Ca^{2+} sensitivity of the CANP was caused by a self-induced limited proteolysis (autolysis) of the CANP on the column to the CANP I (or Calpain I) form of the enzyme. It is well known that CANPs when activated by Ca^{2+} in vitro undergo autolysis to inactive enzymes (Puca et al 1977, Gilbert et al 1975, Truglia & Stracher 1981, Murachi et al 1981a,b, Kubota et al 1981, Suzuki et al 1981, Hathaway et al 1982, Pant et al 1982), and that CANP I is more rapidly inactivated than CANP II. It is still unclear why this autolysis is "limited" on the affinity column, but not in the in vitro case even in the presence of excess casein substrate. Nevertheless, these data indicate that CANP II can be converted to CANP I by limited proteolysis (i.e. limited autolysis). The problem is to find how this is regulated within the cell. In a recent review, Ishiura (1981) points out that although endogenous inhibitors of CANP have been found (Murachi et al 1981a, Tashiro & Ishizaki 1982), no endogenous factor has yet been found that could transform CANP II to CANP I. (Some that have been tried without success are AMP, ADP, ATP, trifluoroperazine, calmodulin, and tissue extracts; see also Pant et al 1982.). The only candidate at present is "unphysiological" levels of Ca^{2+} (see Pant et al 1982 for further discussion of the issue).

There are three hypotheses with respect to the functions of CANP in neurons. It is clear that although axons contain large amounts of CANP, it is in the CANP II form and is not activated in the axon under normal circumstances. This makes biological sense in the axon, where the neurofilaments are stable structural components, and are degraded only after injury to the nerve when intracellular calcium concentrations approach extracellular levels (Schlaepfer 1974, Schlaepfer & Micko 1978). Hence, one hypothesis is that CANP is involved in Wallerian degeneration, specifically in the destruction of neurofilaments preceding nerve degeneration (after injury) (Schlaepfer 1974, Schlaepfer & Micko 1978). Another hypothesis suggested by Lasek & Hoffman (1976) was that the turnover of neurofilaments in nerve terminals may involve the CANP, and that one mechanism regulating the growth of axons might be via the action of CANP in the developing nerve terminals (Lasek & Black 1977). In this regard it should be remembered that axonal CAP directed at neurofilament degradation requires an unphysiological level (>500 μM) of calcium for activation (see above). Even in the synaptic terminal, where during activity the Ca^{2+} influx is relatively high, the internal calcium concentration rarely exceeds 10 μM (Llinas & Heuser 1977), and hence the mechanism by which activation of CANP would occur in the nerve terminal is still unclear. However, recent studies on the distribution of CANP and endogenous protein substrates in the nerve cell showed that there was no endogenous substrate or CANP in the nerve terminal, whereas the immediately contiguous area (i.e. the axon) contained both components abundantly (Pant et al 1982). These data were, therefore, interpreted as support for the idea of a programmed turnover of neurofilaments as they enter the nerve terminals, with a subsequent inactivation of the CAP (shown to occur with an appropriate time course for axonal CANP, see Pant et al 1982). The third hypothesis involves the action of CANP I on glutamate receptors in the hippocampus. Baudry & Lynch (1979) first found that Ca^{2+} appeared to regulate glutamate receptor efficacy, and Vargas et al (1980) showed that a cysteine proteinase was involved in the Ca^{2+} induced increased binding of glutamate to hippocampal synaptic membranes. Subsequent work (Baudry & Lynch 1980, Baudry et al 1981) using the protease inhibitor, leupeptin, indicated that the protease involved was CANP I. The fundamental problem with CANP in general is to determine how, in neuronal and non-neuronal cells, its activity is regulated in situ.

EXTRACELLULAR PEPTIDE HYDROLASES

Owing to limitation of space, we present only a brief commentary about this very important subject in this review. Two functions of extracellular peptide hydrolases that are undoubtedly of fundamental significance in the nervous system are (*a*) the termination of neuropeptide synaptic transmitter action by enzymatic inactivation and (*b*) the modification of secreted structural (e.g. collagen, fibronectin) and informational (e.g. basal lamina matrix) polypeptides by the extracellular environment. These enzymatic events can influence synaptic efficacy and connectivity. In the space below, we provide only a few examples and leave these issues to other reviewers for a more comprehensive exposition.

Inactivation of Neuropeptides

Most peptide neurotransmitters are believed to be enzymatically inactivated in the synaptic cleft. Peptide hydrolysis, however, does not always simply inactivate a peptide, instead it may change its biological activity. For example, the action of pyroglutamyl aminopeptidase on TRH (i.e. pyroglu-His-ProNH$_2$) produces His-prodiketopiperazine, which inactivates the TRH property (i.e. secretion of TSH) of this peptide, but also produces an inhibitor of prolactin secretion (Bauer et al 1978). The idea that peptide action is terminated enzymatically in the synaptic cleft, although appealing, has still not been critically tested. (The alternate views tht the reuptake of the peptide or internalization of the receptor-ligand complex may play such a role have even less compelling evidence in their favor.) The credence of the enzymatic inactivation hypothesis has primarily been based on studies showing the breakdown of peptides by intact tissue or crude enzyme preparations. Little work has been done to show that the particular enzymes being studied are relevant to the terminations of peptide actions in situ. Some criteria that must be fulfilled for such a demonstration are the following:

1. the enzyme candidate should be active on the peptide at physiological pH (i.e. extracellular pH, around 7.4);
2. the enzymes should be membrane (or basal lamina) associated or bound;
3. the enzymes should be located at least within 30 nm from the peptide receptor site on the postsynaptic cell;
4. the enzyme should inactivate the peptide;
5. specific inhibitors of the enzyme should result in the prolongation or enhancement of the peptide's action;
6. the degradative products of the peptide found in vivo around the active site should be the same or related to the degradative products found in vitro.

Several candidates for inactivating enzymes have surfaced in the literature. Many hydrolases have been found in tissues, but very few have been critically examined with respect to the above criteria. Some hydrolases that appear to have sufficient selectivity to recommend them for futher analysis have been reported (Rupnow et al 1979, Orlowski et al 1979, Hersh & McKelvy 1979, Blumberg et al 1980, Lee et al 1981). Because of its neurobiological importance, it is anticipated that this will be a particularly active area of research in the future.

Plasminogen-Activator-Plasmin System

Plasminogen is a component of the fibrinolytic (blood clot lysing) system in blood, and is the precursor of the serine protease, plasmin (see Table 1). Plasminogen is converted to plasmin by the action of another tryptic-like serine protease(s) in plasma (Table 1) called plasminogen activator. Current evidence indicates that there are a variety of plasminogen activators in many tissues other than blood. Furthermore, plasminogen-like precursors may be similarly widely distributed (see Lorand 1981 for the current state of the art).

This system (or a related one) may be of great interest to neurobiologists since it has recently been found in a variety of tissues associated with tissue remodeling and cell migration. It has been implicated in rupture of the follicle during mammalian ovulation (Strickland & Beers 1976), blastocyst implantation (Strickland et al 1976), macrophage activation (Unkeless 1974), and in tumor invasiveness (Ossowski et al 1979). These findings in nonneuronal tissue have led a number of neurobiologists to examine whether related proteases are involved in growth and development in nervous tissue. The results of these studies (Kalderon 1979, Krystock & Seeds 1981a,b, Moonen et al 1982) suggest that the plasminogen-activator-plasma system may facilitate the migration of neurons and Schwann cells, the outgrowth of axons, and the wrapping of neurites by supporting elements. All these studies were performed on *in vitro* cultured cells, and comparable investigations *in vivo* should be done. The data in this field is still preliminary. However, previous studies that have suggested extracellular protease involvement in synaptic remodeling (O'Brien et al 1978) may benefit from an examination of this particular system.

CONCLUDING REMARKS

In this review we have tried to introduce to the reader some of the current directions in neurobiology- and neuroscience-related fields with respect to peptide hydrolases. We have focused on the "creative" character of limited proteolysis as opposed to the degradation function traditionally assigned to tissue proteases. Although the latter functions are certainly important in the nervous system (e.g. for peptide neurotransmitter inactivation, and various extracellular functions such as "clearing space" for axonal regeneration, etc), the roles of proteolytic enzymes to fashion specific neuropeptides from protein precursors and to regulate cell growth and shape and receptor sensitivity are of equal importance. A related research area that we did not address was that of endogenous protease inhibitors (Umezawa & Aoyagi 1977, Laskowski & Kato 1980, Lorand 1981, Aoyagi & Umezawa 1981). Various cells, tissues, and humours contain endogenous inhibitors of specific proteases, and these molecules may play a very important role in the regulation (expression) of specific protease activity in various circumstances.

Many questions remain open for investigation. Among these are those that relate to the selectivity of intracellular proteolysis. Several mechanisms could account for such selectivity: (*a*) a stringent conformational matching between some proteases and their substrates, (*b*) the routing of specific proteases into specific subcellular locations, so that a protease with a broad specificity would only encounter and act upon a colocalized substrate, and (*c*) regulation of enzyme activity via an activator (cofactor), or endogenous inhibitor, or the microenvironment (e.g. pH). As we pointed out above in the discussion of precursor processing to neuropeptides, where these issues are clearly encountered, the final peptide products to emerge from the precursor are a function of the specificities of cleavages of the converting enzymes, which are copackaged with the precursor in the secretory granule, and the pH of the microenvironment. Because the processing of a precursor in the granule occurs in a specific sequence, e.g. endopeptidase cleavage, followed by carboxypeptidase-B-like activity, and finally by amidation, it will be important to determine the molecular organization within the secretory granule that allows for such an efficient cascade. Finally, it is already apparent that proteases play important roles in a number of biological functions in the nervous system. However, neurobiologists have only begun to explore this area, and future studies will undoubtedly uncover new significant functions of proteases in the nervous system.

Literature Cited

Anastasi, A., Erspamer, V., Cei, J. M. 1968. Isolation and amino acid sequence of Caerulein, the active decapeptide of the skin of *hyla caerulea. Arch. Biochem. Biophys.* 125:57–68

Anderson, D. R., Davis, J. L., Carraway, K. L. 1977. Calcium-promoted changes of the human erthrocyte membrane. Involvement of spectrin, trans-glutaminase, and membrane bound protease. *J. Biol. Chem.* 252: 6617–23

Anderton, B. H., Bell, C. W., Newby, B. J., Gilbert, D. S., 1976. Neurofilaments. *Biochem. Soc. Trans., 53rd Meet., London* 4:544–47

Aoyagi, T., Umezawa, H. 1981. The relationship between enzyme inhibitors and function of mammalian cells. *Acta Biol. Med. Ger.* 40:1523–29

Baba, Y., Matsuo, H., Schally, A. V. 1971. Structure of the Porcine LH - and FSH-releasing hormone, II. Conformation of the proposed structure by conventional sequential analysis. *Biochem. Biophys. Res. Commun.* 44:459–63

Bainton, D. F. 1981. The discovery of lysosomes. *J. Cell Biol.* 91(Pt. 2):66a-76a

Barrett, A. J. 1980. Introduction: The classification of proteinases. In *Protein Breakdown in Health and Disease, Ciba Found. Symp.* 75:1–13

Barrett, A. J., M. F. 1977. Lysosomal Enzymes. In *Lysosomes: A Laboratory Handbook,* ed. J. T. Dingle, pp. 19–145. Amsterdam: Elsevier-North Holland. 2nd ed.

Barrett, A. J., MacDonald, J. K., eds. 1980. *Mammalian Proteases,* Vol. 1. New York: Academic. 416 pp.

Baudry, M., Bundman, M. C., Smith, E. R., Lynch, G. S. 1981. Micromolar calcium stimulates proteolysis and glutamate binding in rat brain synaptic membranes. *Science* 212:937–38

Baudry, M., Lynch, G. 1979. Regulation of glutamate receptors by cations. *Nature* 282:748–50

Baudry, M., Lynch, G. S. 1980. Regulation of hippocampal glutamate receptors: Evidence for the involvement of a calcium activated protease. *Proc. Natl. Acad. Sci. USA* 77:2298–2302

Bauer, K., Graff, K. J., Faivre-Bauman, A., Beier, S., Tixier-Vidal, A., Kleinkauf, H. 1978. Inhibition of prolactin secretion by histidyl-proline-diketopiperazine. *Nature* 274:174–75

Bedarkar, S., Turnell, W. G., Blundell, T. L., Schwabe, C. 1977. Relaxin has conformational homology with insulin. *Nature* 270:449–51

Bennett, H. P. J., Browne, C. A., Solomon, S. 1981. Biosynthesis of phosphorylated forms of corticotropin-related peptides. *Proc. Natl. Acad. Sci. USA* 78:4713–17

Bennett, H. P. J., Browne, C. A., Solomon, S. 1982. Characterization of eight forms of corticotropin-like intermediary lobe peptide from rat intermediary pituitary. *J. Biol. Chem.* 257:10096–10102

Blobel, G., Dobberstein, B. 1975. Transfer of proteins across membranes. I. Presence of proteolytically processed and unprocessed nascent immuno-globulin light chains on membrane-bound ribosomes of murine myeloma. *J. Cell. Biol.* 67:835–51

Blumberg, S., Teichberg, V. I., Charli, J. L., Hersh, L. B., McKelvy, J. F. 1980. Cleavage of substance P by an N-terminal tetrapeptide and C-terminal heptapeptide by a post-proline cleaving enzyme from bovine brain. *Brain Res.* 92:477–85

Blundell, T. L., Dockerill, S., Sasaki, K., Tickle, I. J., Wood, S. P. 1976. The relation of structure to storage and receptor binding of glucagon. *Metabolism* 25:1331–41

Bradbury, A. F., Finnie, M. D. A., Smyth, D. G. 1982. Mechanisms of C-terminal amide formation by pituitary enzymes. *Nature* 298:686–88

Bradbury, A. F., Smyth, D. G., Snell, C. R. 1976. Prohormones of α-melanotropin (α-Melanocyte-stimulating hormone, α-MSH) and corticotropin (adrenocorticotrophic hormone, ACTH): Structure and activation. In *Polypeptide Hormones: Molecular and Cellular Aspects, Ciba Found. Symp.* 4:61–75

Burgus, R., Dunn, T. F., Desidero, D. M., Guillemin, R. 1969. Structure Moleculaire du factur hypothalamique hyphophysiotrope TRF d'origine ovine: Mise en de la sequence

PCA-His-Pro-NH$_2$. *C. R. Acad. Sci. Paris* 269:1870–73

Burstein, Y., Schechter, I. 1978. Primary structures of N-terminal extra peptide segments linked to the variable and constant regions of immunoglobulin light chain precursors: Implications on the organization and controlled expression of immunoglobulin genes. *Biochemistry* 17:2392–400

Carty, S. E., Johnson, R. G., Scarpa, A. 1982. Electrochemical proton gradient in dense granules isolated from anterior pituitary. *J. Biol. Chem.* 257:7269–73

Cassell, D., Glasser, L. 1982. Proteolytic cleavage of epidermal growth factor receptor. A Ca^{2+} - dependent, sulfhydryl-sensitive proteolytic system in A 431 cells. *J. Biol. Chem.* 257:9845–48

Chang, T. -L., Gainer, H., Russell, J. T., Loh, Y. P. 1982. Pro-opiocortin converting enzyme activity in bovine neurosecretory granules. *Endocrinology* 11:1607–14

Chang, T. -L., Loh, Y. P. 1983. Characterization of pro-opiocortin converting activity in rat anterior pituitary secretory granules. *Endocrinology* 112:1832–38

Chappell, M., Loh, Y. P., O'Donohue, T. L. 1982. Evidence for an opiomelanotropin acetyltransferase in the rat pituitary neurointermediate lobe. *Peptides* 3:405–10

Collier, N. C., Wang, K. 1982. Purification and properties of human platelet P235. A high molecular weight protein substrate of endogenous calcium-activated protease(s). *J. Biol. Chem.* 257:6937–43

Comb, M., Seeburg, P. H., Adema, J., Eiden, L., Herbert, E. 1982. Primary structure of the human Met- and Leu-enkephalin precursor and its mRNA. *Nature* 295:663–66

Cox, B. M. 1982. Endogenous opiod peptides: A guide to structures and terminology. *Life Sci.* 31:1645–58

Crine, P., Seidah, N. G., Heanotte, L., Chretien, M. 1980. Two glycoprotein fragments related to the NH$_2$ terminal part of the adrenocorticotropin-β-lipotropin precursor are the end products of the maturation process in the rat pars intermedia. *Can. J. Biochem.* 58:1318–22

Czichi, U., Lennarz, W. J. 1977. Localization of the enzyme system for glycosylation of proteins via the lipid-linked pathway in rough endoplasmic reticulum. *J. Biol. Chem.* 252:7901–4

Dayton, W. R., Goll, D. E., Zeece, M. G., Robson, R. M., Reville, W. J. 1976a. A Ca^{2+} activated protease possibly involved in myofibrillar protein turnover. Purification from porcine muscle. *Biochemistry* 15:2150–58

Dayton, W. R., Reville, W. J., Goll, D. E.,

Storms, M. G. 1976b. Ca^{2+} activated protease possibly involved in myofibrillar protein turnover. Partial characterization of purified enzyme. *Biochemistry* 15:2159–67

Dayton, W. R., Schollmeyer, J. V., Lepley, R. A., Cortex, L. R. 1981. A calcium-activated protease possibly involved in myofibrillar protein turnover: Isolation of a low calcium requiring form of the protease. *Biochim. Biophys. Acta* 659:48–61

DeMartino, G. N. 1981. Calcium-dependent proteolytic activity in the rat liver: Indentification of two proteases with different calcium requirements *Arch. Biochem. Biophys.* 211:253–57

Docherty, K., Carroll, R. J., Steiner, D. F. 1982. Conversion of pro-insulin to insulin: Involvement of a 31,500 molecular weight thiol protease *Proc. Natl. Acad. Sci. USA* 79:4613–17

Docherty, K., Steiner, D. F. 1982. Posttranslational proteolysis in polypeptide hormone biosynthesis. *Ann. Rev. Physiol.* 44:625–38

Drouin, J., Goodman, H. M. 1980. Most of the coding region of rat ACTH/β-LPH precursor gene lacks intervening sequences. *Nature* 288:610–13

Dzau, V. J., Brenner, A., Emmett, N., Haber, E. 1980. Identification of renin and renin like enzymes in rat brain by a renin specific antibody. *Clin. Sci.* 59(Suppl. 6):455–87

Eagles, P. A. M., Gilbert, D. S., Maggs, A. 1981. The location of phosphorylation sites and calcium dependent proteolytic cleavage sites on the major neurofilament polypeptides from Myxicola infundibulum. *Biochem. J.* 199:101–11

Eberle, A., Schwyzer, R. 1975. Hormone receptor interactions. Demonstration of two message sequences (active sites) in α–melanotropin. *Hel Chim. Acta* 58:1528–2535

Eipper, B. A., Mains, R. E. 1977. Peptide analysis of a glycoprotein form of adrenocorticotropin hormone. *J. Biol. Chem.* 252:8821–32

Eipper, B. A., Mains, R. E. 1980. Structure and biosynthesis of pro-adrenocorticotropin/endorphin and related peptides. *Endocr. Rev.* 1:1–27

Eipper, B. A., Mains, R. E. 1982. Phosphorylation of pro-adrenocorticotropin endorphin-derived peptides. *J. Biol. Chem.* 257:4907–15

Eipper, B. A., Mains, R. E., Guenzi, R. E. 1976. High molecular weight forms of adrenocorticotropic hormone are glycoproteins. *J. Biol. Chem.* 251:4121–26

Endean, R., Erspamer, V., Falconieri-Erspamer, G., Improta, G., Melchiorri, F.,

Negri, L., Sapranzi, N. 1975. Parallel bioassay of bombesin and litorin, a bombesin-like peptide from the skin of *litoria aurea*. *Br. J. Pharmcol.* 55:213–19

England, P. J. 1980. Regulation of phosphorylation and dephosphorylation. In *The Enzymology of Post-translational Modifications of Proteins* ed. R. B. Freedman, H. C. Hawkins, pp. 1–423. London: Academic

Evangelista, R., Ray, P., Lewis, R. V. 1982. A "trypsin-like" enzyme in adrenal chromaffin granules: A proenkephalin processing enzyme. *Biochem Biophys. Res. Commun.* 106:895–902

Fishman, M. C., Zimmerman, E. A., Slater, E. E. 1981. Renin and angiotensin: The complete system within the neuroblastoma glioma cell. *Science* 214:921–23

Fletcher, D. J., Noe, B. D., Bauer, G. E., Quigley, J. P. 1980. Characterization of the conversion of a somatostatin precursor to somatostatin by islet secretory granules. *Diabetes* 29:593–99

Fletcher, D. J., Quigley, J. P., Bauer, G. E., Noe, B. D. 1981. Characterization of proinsulin and proglucagon-converting activities in isolated islet secretory granules. *J. Cell Biol.* 90:312–22

Freedman, R. B., Hawkins, H. C., eds. 1980. *The Enzymology of Post-translational Modification of Proteins*, Vol. 1. London: Academic. 456 pp.

Fricker, L. D., Supattapone, S., Snyder, S. H. 1982. Enkephalin convertase: A specific enkephalin synthesizing carboxypeptidase in adrenal chromaffin granules, brain, and pituitary gland. *Life Sci.* 31:1841–44

Fricker, L. D., Snyder, S. H. 1982. Enkephalin convertase: Purification and characterization of a specific enkephalin-synthesizing carboxypeptidase localized to adrenal chromaffin granules. *Proc. Natl. Acad. Sci. USA* 79:3886–90

Gainer, H., Sarne, Y., Brownstein, M. J. 1977. Biosynthesis and axonal transport of rat neurhypophyseal proteins and peptides. *J. Cell Biol.* 73:366–81

Ganten, D., Fuxe, K., Phillips, M. J., Mann, J. F. E., Ganten, U. 1978. The brain isorenin-angiotensin system: Biochemistry, localization and possible role in drinking and blood pressure regulation. *Front. Neuroendocrinol.* 5:61–69

Ganten, D., Speck, G. 1978. The brain renin-angiotensin system: A model for the synthesis of peptide in the brain. *Biochem. Pharmacol.* 27:2379–89

Geisow, M. J., Smyth, D. G. 1980. Proteolysis of prohormones and pro-proteins. See Freedman & Hawkins 1980, pp. 259–87

Gilbert, D. S. 1975. Axoplasm chemical compositon in *Myxicola* and solubility properties

of its structural proteins. *J. Physiol.* 253:303–19

Gilbert, D. S., Newby, B. J., Anderton, B. H. 1975. Neurofilament disguise, destruction and discipline. *Nature* 256:586–89

Glembotski, C. C. 1981. Subcellular fractionation studies on the post-translational processing of pro-adrenocorticotropic hormone/endorphin in rat intermediate pituitary. *J. Biol. Chem.* 256:7433–39

Glembotski, C. C. 1982a. Acetylation of α-melanotropin and β-endorphin in the rat intermediate pituitary. *J. Biol. Chem.* 257:10493–10500

Glembotski, C. C. 1982b. Characterization of the peptide acetyltransferase activity in bovine and rat intermediate pituitaries responsible for the acetylation of β-endorphin and α-melanotropin. *J. Biol. Chem.* 257:10501–9

Goldberg, A. L., St. John, A. C. 1976. Intracellular protein degradation in mammalian and bacterial cells, Pt. 2. *Ann. Rev. Biochem* 45:747–803

Goldstein, A., Fischli, W., Lowney, L. I., Hunkapiller, M., Hood, L. 1981. Porcine pituitary dynorphin: Complete amino acid sequence of the biologically active heptadecapeptide. *Proc. Natl. Acad. Sci. USA* 78:7219–23

Graf, L., Kenessey, A. 1981. Characterization of proteinases involved in the generation opioid peptides from β-lipotropin. *Horm. Proteins Peptides* 10:35–63

Gubler, U., Seeburg, P., Hoffman, B. J., Gage, L. P., Udenfriend, S. 1982. Molecular cloning establishes pro-enkephalin as precursor of enkephalin-containing peptides. *Nature* 295:206–8

Gumbiner, B., Kelley, R. B. 1981. Secretory granules of an anterior pituitary cell line, atT-20 contain only mature forms of corticotropin and β-lipotropin. *Proc. Natl. Acad. Sci. USA* 78:318–22

Guroff, G. 1964. A neutral, calcium-activated protease from the soluble fraction of the rat brain. *J. Biol. Chem.* 239:149–55

Habener, J. F., Chang, H. T., Potts, J. T. Jr. 1977. Enzymic processing of parathyroid hormone by cell free extracts of the parathyroid glands. *Biochemistry* 16:3910–17

Hamon, M., Bourgoin, S. 1979. Characterization of the Ca^{2+} induced proteolytic activation of tryptophan hydroxylase from the rat brain stem. *J. Neurochem.* 32:1837–44

Harwood, R. 1980. Protein transfer across membranes: The role of signal sequences and signal peptidase activity. See Freedman & Hawkins 1980, pp. 3–52

Hathaway, D. R., Werth, D. K., Haeberle, J. R. 1982. Limited autolysis reduces the Ca^{2+} requirements of a smooth muscle Ca^{2+}-

activated protease. *J. Biol. Chem.* 257:-9072–77

Herbert, E., Birnberg, N., Lissitsky, J-C., Civelli, O., Uhler, M. 1982. Regulation of expression of pro-opiomelanocrotin and related genes in various tissues of the rat and mouse. In *Pituitary Hormones and Related Peptides*, ed. M. Motta, M. Zanisi, F. Piva. pp. 30–42. London/New York: Academic

Hermann, K., Ganten, D., Bayer, C., Unger, T., Lang, R. E., Rascher, W. 1982. Definite evidence for the presence of (Il e⁵)-Angiotension I and (Ile⁵)-Angiotension II in the brain of rats. In *The Renin-Angiotension System in the Brain*, ed. M. Printz, M. I. Phillips, B. A. Scholkens, pp. 192–207. Heidelberg; Springer

Hersh, L. B., McKelvy, J. F. 1979. Enzymes involved in the degradation of thyrotropin releasing hormone (TRH) and luteinizing hormone (LHRH) in bovine brain. *Brain Res.* 168:553–64

Hershko, A., Ciechanover, A. 1982. Mechanisms of intracellular protein breakdown. *Ann. Rev. Biochem.* 51:335–64

Hirose, S., Yokosawa, H., Inagam, I., Workman, R. J. 1980. Renin and prorenin in hog brain: Ubiquitous distribution and high concentration in the pituitary and pineal. *Brain Res.* 191:489–99

Holzer, H., Heinrich, P. C. 1980. Control of proteolysis. *Ann. Rev. Biochem.* 49:63–91

Holtzman, E., Mercurio, A. M. Membrane circulation in neurons and photoreceptors: Some unresolved issues. *Int. Rev. Cytol.* 67:1–67

Huston, R. B., Krebs, E. G. 1968. Activation of skeletal muscle phosphorylase kinase by Ca²⁺. II. Identification of the kinase activating factor as a proteolytic enzyme. *Biochemistry* 7:2116–21

Igic, R. P., Robinson, C. J. G., Erdos, E. G. 1977. Angiotension I converting enzyme activity in the choroid plexus and in the retina. In *Central Actions of Angiotensin and Related Hormones*, ed. J. P. Buckley, C. M. Ferrario, pp. 23–27. New York: Pergammon

International Union of Biochemistry. 1978. *Enzyme Nomenclature.* New York: Academic. 443 pp.

Isaacs, N., James, R., Niall, H., Bryant-Greenwood, G., Dodson, G., Evans, A., North, A.C.T. 1978. Relaxin and its structural relationship to insulin. *Nature* 271:278–80

Ishiura, S. 1981. Calcium dependent proteolysis in living cells. *Life Sci.* 29:1079–88

Ishiura, S., Murofushi, H., Suzuki, K., Imahori, K. 1978. Studies of a calcium-activated protease from chicken skeletal muscle. I. Purification and characterization. *J. Biochem.* 84:225–30

Ito, T., Eggenmor, P., Barretta, J. D., Katz, D., Metter, J., Sambhi, M. P. 1980. Studies on angiotensinogen of plasma and cerebrospinal fluid in normal and hypertensive human subjects. *Hypertension* 2:432–36

Jackson, R., Blobel, G. 1977. Post-translational cleavage of presecretory proteins with an extract of rough microsomes from dog pancreas containing signal peptidase activity. *Proc. Natl. Acad. Sci. USA* 74:5598–602

Jacobs, J. W., Goldman, R. H., Chin, W. W., Dee, P. C., Habener, J. F., Bell, N. H., Potts, J. T. Jr. 1981. Calcitonin messenger RNA encodes multiple polypeptides in a single precursor. *Science* 213:457–59

Judah, J. D., Quinn, P. S. 1978. Calcium ion-dependent vesicle fusion in the conversion of proalbumin to albumin. *Nature* 271:384–85

Kakidani, H., Furutani, Y., Takahashi, H., Noda, M., Morimoto, Y., Hirose, T., Asai, M., Inayama, S., Nakanishi, S., Numa, S. 1982. Cloning and sequence analysis of cDNA for porcine β-neo-endorphin/dynorphin precursor. *Nature* 298:245–49

Kalderon, N. 1979. Migration of Schwann cells and wrapping of neurites in vitro: a function of protease activity (plasmin) in the growth medium. *Proc. Natl. Acad. Sci. USA* 76:5992–96

Kameyama, T., Etlinger, J. D. 1979. Calcium-dependent regulation of protein synthesis and degradation of muscle. *Nature* 279:344–46

Kangawa, K., Minamino, N., Chino, N., Sakakibara, S., Matsuo, H. 1981. The complete amino acid sequence of α-neo-endorphin. *Biophys. Res. Commun.* 99:871–78

Kaur, H., Sanwal, B. D. 1981. Regulation of activity of a calcium-activated neutral protease during differentiation of skeletal myoblasts. *Can. J. Biochem.* 59:743–48

Kemmler, W., Steiner, D. F., Borg, J. 1973. Studies on the conversion of pro-insulin to insulin. *J. Biol. Chem.* 248:4544–51

Kilpatrick, D. L., Wahlstrom, A., Lahm, H. W., Blacher, R., Ezra, E., Fleminger, G., Udenfriend, S. 1982. Characterization of Rimorphin, a new [leu] enkephalin-containing peptide from bovine posterior pituitary glands. *Life Sci.* 31:1849–52

Knight, M., Plotkin, C., Tamminga, C. 1982. Brain endopeptidase generates enkephalin from striatal precursors. *Peptides* 3:461–68

Kreil, G. 1981. Transfer of proteins across membranes. *Ann. Rev. Biochem.* 50:317–48

Krystock, A., Seeds, N. W. 1981a. Plasminogen activator secretion by granule neurons in cultures of developing cerebellum. *Proc. Natl. Acad. Sci. USA* 78:7810–14

Krystock, A., Seeds, N. W. 1981b. Plasminogen activator release at the neural growth cone. *Science* 213:1532–34

Kubota, S., Suzuki, K., Imahori, K. 1981. A new method for the preparation of a calcium-activated neutral protease highly sensitive to calcium ions. *Biochem. Biophys. Res. Commun.* 100:1189–94

Lajtha, A., Dunlop, D. 1981. Turnover of protein in the nervous sytem. *Life Sci.* 29:755–67

Land, H., Schutz, G., Schmale, H., Richter, D. 1982. Nucleotide sequence of cloned cDNA encoding bovine arginine vasopressin II precursor. *Nature* 295:299–303

Lasek, R. J., Black, M. M. 1977. How do axons stop growing? Some clues from the metabolism of the proteins in the slow component of axonal transport. In *Mechanisms, Regulations, and Special Functions of Protein Synthesis in the Brain*, ed. E. Roberts, pp. 161–69. New York: Elsevier-North Holland

Lasek, R. J., Hoffman, P. M. 1976. The neuronal cytoskeleton, axonal transport and axonal growth. In *Cell Motility*, ed. R. Goldman, pp. 1021–49. New York: Cold Spring Harbor Lab.

Laskowski, M., Jr., Kato, I. 1980. Protein inhibitors of proteinases. *Ann. Rev. Biochem.* 49:593–626

Lee, C. M., Sandberg, B. E. B., Hanley, M. R., Iverson, L. 1981. Purification and characterization of a membrane bound substance-P-degrading enzymes from human brain. *Eur. J. Biochem.* 114:315–27

Leeman, S. E., Mroz, E. A., Carraway, R. E. 1977. Substance P and neurotensin. In *Peptides in Neurobiology*, ed. H. Gainer, pp. 99, 144. New York/London: Academic

Lindberg, I., Yang, H. -Y. T., Costa, E. 1982. An enkephalin generating enzyme in bovine adrenal medulla. *Biochem. Biophys. Res. Commun.* 106:186–93

Linderstrom-Lang, K. U., Ottesen, M. 1949. Formation of plakalbumin from ovalbumin. *C. R. Trav. Lab. Carlsborg* 26:403–12

Lingappa, V. A., Katz, F. N., Lodish, H. F., Blobel, G. 1978. A signal sequence for the insertion of a transmembrane glycoprotein. *J. Biol. Chem.* 253:8667–70

Llinas, R. R., Heuser, J. E. 1977. Depolarization-release coupling systems in neurons. *Neurosci. Res. Prog. Bull.* 15:557–687

Loh, Y. P., Gainer, H. 1979. The role of the carbohydrate in the stabilization, processing, and packaging of the glycosylated adrenocorticotropin/endorphin common precursor in toad pituitaries. *Endocrinology* 105:474–87

Loh, Y. P., Gritsch, H. A. 1981. Evidence for intragranular processing of proopiocortin in the mouse pituitary intermediate lobe. *Eur. J. Cell Biol.* 26:177–83

Loh, Y. P., Chang, T. -L. 1982. Proopiocortin converting activity in rat intermediate and neural lobe secretory granules. *FEBS Lett.* 137:57–62

Loh, Y. P., Gainer, H. 1982a. Characterization of pro-opiocortin converting activity in purified secretory granules from rat pituitary neurointermediate lobe, *Proc. Natl. Acad. Sci. USA* 79:108–12

Loh, Y. P., Gainer, H. 1982b. Processing of normal and nonglycosylated forms of toad pro-opiocortin by rat intermediate (pituitary) lobe pro-opiocortin converting enzyme activity. *Life Sci.* 31:3043–50

Loh, Y. P., Gritsch, H. A., Chang, T. -L. 1982. Pro-opiocortin processing in the pituitary: A model for neuropeptide biosynthesis. *Peptides* 3:397–404

Lorand, L., ed. 1981. *Meth. Enzymol.* Vol. 80, P. C

Lucas, R. C., Rosenberg, S., Lawrence, J., Shafig, S., Stracher, S. 1979. The isolation of a cytoskeleton and contractile apparatus from human platelets. *Cytoskeleton Struct. Protides Biol. Fluids* 26:465–70

Malesci, A., Straus, E., Yalow, R. S. 1980. Cholecystokinin-converting enzymes in brain. *Proc. Natl. Acad. Sci. USA* 77:597–99

Malik, M. N., Meyers, L. A., Igbal, K., Sheikh, A. M., Scotto, L., Wisniewiski, H. M. 1981. Calcium activated proteolysis of fibrous protein in central nervous system. *Life Sci.* 29:795–802

Marinkovic, D., Marinkovic, J. N., Erdos, E. G., Robinson, C. V. G. 1977. Purification of carboxypeptidase B from human pancreas. *J. Biochem.* 163:253–60

Marks, N., Benuck, M., Berg, M. J. 1982. Metabolism of heptapeptide opioid by rat brain and cardiac tissue. *Life Sci.* 31:1845–48

Marks, N., Suhar, A., Benuck, M. 1981. Peptide processing in the central nervous system. In *Neurosecretion and Brain Peptides*, pp. 49–52. New York: Raven

McGregor, R. R., Chin, L. L. H., Cohn, D. V. 1976. Conversion of proparathroid hormone to parathyroid hormone by a particular enzyme of the parathyroid gland. *J. Biol. Chem.* 251:6711–16

McGregor, R. R., Hamilton, J. W., Cohn, D. V. 1978. The mode of conversion of proparathyroid to parahromone by a particulate converting enzymic activity of the parathyroid gland. *J. Biol. Chem.* 253:2012–17

Mellgren, R. L. 1980. Canine cardiac calcium-dependent proteases: Resolution of two forms with different requirements for calcium. *FEBS Lett.* 109:129–33

Mellgren, R. L., Repetti, A., Muck, T. C., Easly, J. 1982. Rabbit skeletal muscle calcium-dependent protease requiring millimolar Ca^{2+}. Purification, subunit structure and Ca^{2+} dependent autolysis. *J. Biol. Chem.* 257:7203–9

Mizuno, K., Miyata, A., Kangawa, K., Matsuo, H. 1982. A unique proenkephalin-converting enzyme purified from adrenal chromaffin granules. *Biochem. Biophys. Res. Commun.* 108:1235–42

Moonen, G., Grau-Wagemans, M. P., Sclak, I. 1982. Plasminogen activator-plasmin system and neuronal migration. *Nature* 298:753–55

Murachi, T., Hatanaka, M., Yasumoto, Y., Nakayama, N., Tamaka, K. 1981b. A quantitative distribution study on calpain and calpastatin in rat tissue and cells. *Biochem. Int.* 2:651–56

Murachi, T., Tanaka, K., Hatanaka, M., Murakami, T. 1981a. Intracellular Ca^{2+}-dependent protease (calpain) and its high molecular weight endogenous inhibitor (calpastatin). *Adv. Enzyme Regul.* 19:407–24

Nakanishi, S., Inone, A., Kita, T., Nakamura, M., Chang, A. C. Y., Cohen, S. N., Numa, S. 1979. Nucleotide sequence of cloned cDNA for bovine cortiocotropin-β-lipotropin precursor. *Nature* 278:423–27

Nelson, W. J., Traub, P. 1982. Purification and further characterization of the Ca^{2+}-activated proteinase specific for the intermediate filament proteins vimentrin and desmin. *J. Biol. Chem.* 257:5544–53

Nishizuki, Y., Takai, Y., Hashimoto, E., Kishimoto, A., Kuroda, Y., Sakai, K., Yamamura, H. 1979. Regulation and functional compartment of three multifunctional protein kinase systems. *Mol. Cell Biochem.* 23:153–65

Noda, M. Teranishi, Y., Takahashi, H., Toyosato, M., Notake, M., Nakanishi, S., Numa, S. 1982. Isolation and structural organization of the human preproenkephalin gene. *Nature* 297:431–34

Noe, B. D. 1981. Inhibition of islet prohormone to hormone conversion by incorporation of arginine and lysine analogs. *J. Biol. Chem.* 256:4940–46

North, W. G., Valtin, H., Morris, J. F., LaRochelle, F. T. 1977. Evidence for metabolic conversions of rat neurophysins within neurosecretory granules of the hypothalamo-neurohypophysial system. *Endocrinology* 101:110–18

O'Brien, R. A. D., Osteberg, A. J. C., Vrbova, G. 1978. Observations on the elmination of polyneuronal innervation in developing mammalian skeletal muscles. *J. Physiol* 282:571–82

O'Donohue, T. L. 1983. Identification of endorphin acetyltransferase in rat brain and pituitary gland. *J. Biol. Chem.* 258:2163–67

Okada, Y., Sabatini, D. D., Kreibich, G. 1979. Sites of synthesis of rat liver cytochrome B_5 and NADPH-cytochrome P-450 reductase. *J. Cell Biol.* 83:437a

Orlowski, M., Wilk, E., Pearce, S., Wilk, S. 1979. Purification and properties of a prolyl endopeptidase from rabbit brain. *Neurochemistry* 33:461–69

Ossowski, L., Beigel, D., Reich, E. 1979. Mammary plasminogen activator: Correlation with involution, hormonal modulation and comparison between normal and neoplastic tissue. *Cell* 16:929–40

Palade, G. 1975. Intracellular aspects of the process of protein secretion. *Science* 189:347–58

Pant, H. C., Gainer, H. 1980. Properties of calcium-activated protease in squid axoplasm which selectively degrades neurofilament proteins. *J. Neurobiol.* 11:1–12

Pant, H. C., Gallant, P. E., Gould, R., Gainer, H. 1982. Distribution of calcium-activated protease activity and endogenous substrates in the squid nervous system. *J. Neurosci.* 2:1578–87

Pant, H. C., Terakawa, S., Gainer, H. 1979. A calcium activated protease in squid axoplasm. *J. Neurochem.* 32:99–102

Patzelt, C., Tager, H. S., Carroll, R. J., Steiner, D. F. 1980. Identification of prosomatostatin in pancreatic islets. *Proc. Natl. Acad. Sci. USA* 77:2410–14

Peach, M. J. 1977. Renin-angiotension system: Biochemistry and mechanisms of action. *Physiol. Rev.* 57:313–70

Pease, K. A., Dixon, J. E. 1981. Characterization of N-α-acetylation of corticotropin fragments by a rat pituitary enzyme. *Arch. Biochem. Biophys.* 212:177–85

Pfeiffer, U. 1981. Morphological aspects of intracellular protein degradation: Autophagy. *Acta Biol. Med. Ger.* 40:1619–24

Phelps, C. F. 1980. Glycosylation. See Freedman & Hawkins 1980, p. 105

Phillips, D. R., Jakabova, M. 1977. Ca^{2+} dependent protease in human platelets. *J. Biol. Chem.* 252:5602–5

Phillips, M. A., Budarf, M. L., Herbert, E., 1981. Glycosylation events in the processing and secretion of pro-ACTH-endorphin in mouse pituitary tumor cells. *Biochemistry* 20:1666–75

Phillips, M. I. 1980. Biological effects of angiotension II in the brain. In *Enzymatic Release of Vasoactive Peptides*, ed. F. Gross, H. G. Vogel, pp. 335–63. New York: Raven

Policastro, P., Phillips, M., Oates, E., Herbert, E., Roberts, J. L., Seidah, N., Chretien, M. 1981. Evidence for a signal sequence at the N-terminus of the common precursor to adrenocorticothropin and β-lipotropin in mouse pituitary cells. *Eur. J. Biochem.* 116:255–59

Pollard, H. B., Pazoles, C. J., Creutz, C. E., Zinder, O. 1979. The chromaffin granule and possible mechanisms of exocytosis. *Int. Rev. Cytol.* 58:159–98

Poole, A. R. 1977. Antibodies to enzymes and their uses, with particular reference to lysosomal enzymes. In *Lysosomes: A Laboratory Handbook*, ed. J. T. Dingle, pp. 245–312. Amsterdam: Elsevier-North Holland. 2nd ed.

Poole, A. R. 1981. Immunological studies of tissue proteinases. *Subcell. Biochem.* 8:311–46

Poole, B. 1971. Synthesis and degradation of proteins in relation to cellular structure. In *Enzyme Synthesis and Degradation in Mammalian Systems*, ed, M. Reichigal, pp. 375–402. Basel: Karger. 447 pp.

Potts, J. T. Jr., Keutman, H. T., Niall, H. D., Treager, G. W., 1971. The chemistry of parathyroid hormone and the calcitonins. *Vitam. Horm.* 29:41–93

Potts, J. T. Jr., Kronenberg, H. M., Habener, J. F., Rich, A. 1980. Biosynthesis of parathyroid Hormone. *Ann. Acad. Sci.* 343:38–55

Puca, G. A., Nola, E., Sica, V., Breciani, F. 1977. Estrogen binding of proteins of calf uterus. Molecular and functional characterization of the receptor transforming factor: A Ca^{2+}-activated protease. *J. Biol. Chem.* 252:1358–70

Quinn, P. S., Judah, J. D. 1978. Calcium-dependent Golgi-vesicle fusion and cathespin B in the conversion of proalbumin into albumin in rat liver. *Biochem. J.* 172:301–9

Raizada, M. K., Phillips, M. I., Gerndt, J. S. 1983. Primary cultures from fetal rat brain incorporate [³H-]-isoleucine and ³H-valine into immunoprecipitable Angiotensin II. *Neuroendocrinolgy* 36:64

Rechigal, M. 1971. Intracellular protein turnover and the roles of synthesis and degradation in regulation of enzyme levels. In *Enzyme Synthesis and Degradation in Mammalian Systems*, ed. M. Rechigal, pp. 236–310. Basel: Karger. 447 pp.

Rehfeld, J. F., Goltermann, N., Larsson, L. I., Emson, P. M., Lee, C. M. 1979. Gastrin and cholecystokinin in central and peripheral neurons. *Fed. Proc.* 38:2325–29

Reich, E., Rifkin, D. B., Shaw, E. 1975. *Proteases and Biological Control*, Vol. 2. New York: Cold Spring Harbor Lab. 1021 pp.

Reid, I. A., Day, R. P. 1977. Interaction and properties of some components of the renin-angiotensin system. In *Central Actions of Angiotension and Related Hormones*, ed. J. P. Buckley, C. M. Ferrario, pp. 267–83, New York, Pergammon

Rodriguez Boulan, E., Sabattini, D. D., Pereyra, B., Dreibich, G. 1978. Spatial orientation of glycoproteins in membranes of rat liver rough microsomes. II. Transmembrane disposition and characterizations of glycoproteins. *J. Cell Biol.* 78:894–909

Rupnow, J. H., Taylor, W. L., Dixon, J. E. 1979. Purification and characterization of a thyrotropin-releasing hormone deamidase from rat brain. *Biochemistry* 18:1206–12

Russell, J. T., Hotz, R. 1981. Measurement of Δph and membrane potential in isolated neurosecretory vesicles from bovine neurohypophyses. *J. Biol. Chem.* 256:5950–53

Ryder, S. W., Straus, E., Yalow, R. S. 1980. Further characterization of brain cholecystokinin-converting enzyme. *Proc. Natl. Acad. Sci. USA* 77:3669–71

Sabol, S. L., Liang, C. -M., Dandekar, S., Kranzler, L. S. 1983. *In vitro* biosynthesis and processing of immunologically identified methionine-enkephalin precursor protein. *J. Biol. Chem.* 258:2697–2704

Sachanek, G., Kreil, G., Hermodson, M. A. 1978. Amino acid sequence of honeybee pre-promelittin synthesized *in vitro*. *Proc. Natl. Acad. Sci. USA* 75:701–4

Scarpa, A., Johnson, R. G. 1976. Internal pH of isolated chromaffin vesicles. *J. Biol. Chem.* 251:2189–91

Schlaepfer, W. W. 1974. Calcium-induced degeneration of axoplasm in isolated segment of rat peripheral nerve. *Brain Res.* 69:203–15

Schlaepfer, W. W., Freeman, L. A. 1980. Calcium-dependent degradation of mammalian neurofilaments by soluble tissue factor(s) from rat spinal cord. *Neuroscience* 5:2305–14

Schlaepfer, W. W., Micko, S. 1978. Chemical and structural changes of neurofilaments in transected rat sciatic nerve. *J. Cell. Biol.* 7 8:369–78

Schlaepfer, W. W., Zimmerman, U. J. P., Micko, S. 1981. Neurofilament proteolysis in rat peripheral nerve. Homologous with calcium-activated proteolysis of other tissues. *Cell Calcium* 2:235–50

Smyth, D. G., Austen, B. M., Bradbury, A. F., Geisow, M. J., Snell, C. R., 1977. Biogenesis and metabolism of opiate active peptides. In *Centrally Acting Peptides* ed. J. Hughes, pp. 231–39. London: Macmillan

Smyth, D. G., Zakarian, S., Deakin, L. F. W., Massey, D. E. 1980. β-Endorphin-related peptides in the pituatry gland: Isolation, identification and distribution. In *Intermediate Lobe of the Pituitary, Ciba Symp.* 81:79–76

Steiner, D. F., Kemmler, W., Tager, H. S., Peterson, J. D. 1974. Proteolytic processing in the biosynthesis of insulin and other proteins. *Fed. Proc.* 33:2105–15

Steiner, D. F., Kemmler, W., Tager, H. S., Rubenstein, H. H., Lernmark, A., Zühlke H. 1975. Proteolytic mechanisms in the biosynthesis of polypeptide hormones. In *Proteases and Biological Control,* ed. E. Reich, D. B. Rifkin, E. Shaw. New York: Cold Spring Harbor Lab. 531 pp.

Stoeckel, M. E., Schimchowitsch, S., Garand, J. C., Schmitt, G., Vandry, H., Porte, A. 1983. Immunocytochemical evidence of intragranular acetylation of α-MSH in the melatotropic cells of the rabbit. *Cell Tissue Res.* 230:511–15

Strauss, A. W. Bennett, C. D., Donohue, A. M., Rodkey, J. A., Alberts, A. W. 1977. Rat liver pre-proalbumin: Complete aminoacid sequence of the pre-piece. *J. Biol. Chem.* 252:6846–55

Strickland, S., Beers, W. H. 1976. Studies on the role of plasminogen activator in ovulation. *J. Biol. Chem.* 251:5694–702

Strickland, S., Reich, E., Sharman, M. I. 1976b. Plasminogen activator in early embyrogenesis: Enzyme production by tropoblast and parietal endoderm. *Cell* 9:231–40

Suzuki, K., Tsuji, S., Kubota, S., Kimura, Y., Imahori, K. 1981. Limited autolysis of Ca^{2+} activated neutral protease (CANP) changes its sensitivity to Ca^{2+} ions. *J. Biochem.* 90:275–78

Tachibana, S., Araki, K., Ohya, S., Yoshida, Y. 1982. Isolation and structure of dynorphin, an opoid peptide, from porcine duodenum. *Nature* 295:339–40

Tashiro, T., Ishizaki, Y. 1982. A calcium-dependent protease selectively degrading the 160,000 Mr component of neurofilaments is associated with the cytoskeletal prearation of the spinal cord and has an endogenous inhibitory factor. *FEBS Lett.* 141:41–44

Traub, P., Nelson, W. J. 1981. Occurrence in various mammalian cells and tissues of the Ca^{2+} activated protease specific for the intermediate-sized filament proteins vimentin and desmin. *Eur. J. Cell. Biol.* 26:61–67

Triplett, R. B., Wingate, J. M., Carraway, K. L. 1972. Calcium effects on erythrocyte membrane proteins. *Biochem. Biophys. Res. Commun.* 49:1014–20

Troy, C. M., Musacchio, J. M. 1982. Processing of enkephalin precursors by chromafin granule enzymes. *Life. Sci.* 31:1717–20

Truglia, J. A., Stracher, A. 1981. Purification and characterization of a calcium dependent sulfhydryl protease from human platelets. *Biochem. Biophys. Res. Commun.* 100:814–22

Umezawa, H., Aoyagi, T. 1977. Activities of proteinase inhibitors of microbial origin. In *Proteinases in Mammalian Cells and Tissues,* ed. A. J. Barrett, pp. 637–722. New York: Elsevier-North Holland

Unkeless, J. C., Gordon, S., Reich, E. 1974. Secretion of plasminogen activator by stimulated macrophages. *J. Exp. Med.* 139:834–50

Van Heldon, H. P. M. 1980. *Aspects of biosynthesis and release of peptides by the pars intermedia of the pituitary gland in Xenopus laevis.* PhD thesis. Catholic Univ., Nijmegen, The Netherlands. 44 pp.

Van Houten, M., Schiffrin, E. L., Mann, J. F. E., Posner, B. I., Boucher, R. 1980. Radioautographic localization of specific binding sites for blood-borne angiotensin II in the rat brain. *Brain Res.* 186:480–85

Vargas, V., Greenbaum, L., Costa, E. 1980. Participation of cysteine proteinase in the high affinity Ca^{2+}-dependent binding of glutamate to hippocampal synaptic membranes. *Neuropharmacology* 19:791–94

Vedeckis, W. V., Freeman, M. R., Schrader, W. T., O'Malley, B. W. 1980. Progesterone-binding components of check oviduct: partical purification and characterization of a calcium-activated protease which hydrolyses the progesterone receptor. *Biochemistry* 19:335–43

Villa-Komaroff, L,, Efstratida, A., Broome, S., Lomedico, P., Tizard., R., Naber, S. P., Chick, W. L., Gilbert, W. 1978. A bacterial clone synthesizing proinsulin. *Proc. Natl. Acad. Sci. USA* 75:3727–31

Wallach, D. P., Davies, J. A., Pastan, I. 1978. Purification of mammalian filamin. *J. Biol. Chem.* 253:3328–35

Wold, F. 1981. *In vivo* chemical modification of proteins (post-translational modification). *Ann. Rev. Biochem.* 50:783–814

Woodford, T. A., Dixon, J. E. 1979. The acetylation of corticotropin and fragments of corticotropin by a rat *N*-acetyltransferase. *J. Biol. Chem.* 254:4993–99

Yang, H. -Y. T., Neff, N. H. 1972. Distribution and properties of angiotensin converting enzyme of rat brain. *J. Neurochem.* 19:2443–50

Zwizinski, C., Wickner, W. 1980. Purification and characterization of leader (signal) peptidase from *escherichia coli. J. Biol. Chem.* 255:7973–77

Ann. Rev. Neurosci. 1984. 7:223–55
Copyright © 1984 by Annual Reviews Inc. All rights reserved

ENDOGENOUS OPIOIDS:
Biology and Function

Huda Akil, Stanley J. Watson, Elizabeth Young, Michael E. Lewis, Henry Khachaturian, and J. Michael Walker

Mental Health Research Institute, University of Michigan, Ann Arbor, Michigan 48109

INTRODUCTION

While most would agree that the mid-seventies were vintage years for endorphin research, 1982 is certain to be "a very good year." Less obvious to the public eye, it is nevertheless a turning point in endorphin research, because it is the year in which all the brain opioids "found a home." Prior to that point we could count more than eight different endogenous ligands containing the opioid core Tyr-Gly-Gly-Phe-Met (or Leu). However, their relationships were unclear, and their anatomical distribution in the central nervous system (CNS) even less so. It was therefore impossible for the functionally minded neuroscientist to design and execute experiments taking this heterogeneity into consideration. Even pharmacological studies with these peptides were difficult to interpret in a physiological framework. In 1982, thanks mainly to the recombinant DNA techniques, we learned that all these peptides belong to three genetically distinct peptide families, as we describe below. We now know a great deal more about their anatomy, and we are beginning to clarify their biosynthetic pathways. Suddenly, we can think in terms of circuits rather than in "humors" or in black boxes. In our study of function, we can no longer ignore the multiple systems, any more than the student of monoamines can ignore differences between dopaminergic and noradrenergic systems.

Less clear, but equally critical, is the issue of multiple opioid receptors. Unquestionably, the heterogeneity exists. What remains to be established is whether each of the three families of opioids has its own receptor, or whether a given family can interact with more than one subtype, and each receptor subtype with more than one family. More critical to physiology is whether

223

these unique combinations result in different biological events and are involved in different functions.

This review focuses on these issues. It is certainly not an exhaustive review of the past year's studies on endorphins and behavior. Rather, we update the reader on the biology of the multiple systems, their biosynthesis, post-translational processing, anatomy, and multiple receptors. We then proceed to raise some issues of relevance to the physiology and function of endorphins, choosing our examples from two areas in which the endogenous opioids have clear functions: (a) stress and analgesia, (b) cardiovascular control. We point out the most recent strategies used in these fields, show the complexities introduced by multiple opioids and multiple receptors, and discuss future directions for research in a biological framework.

ENDOGENOUS LIGANDS

Three Opioid Peptide Gene Families

In the last five years several laboratories have elucidated the biosynthetic origin of all known opioid peptides at the level of protein, messenger RNA, or even the gene. These data grew out of the rather impressive list of opioid peptides (Table 1) extracted and sequenced from nervous tissue. As can be seen from Table 1, the known opioid peptides come from one of three precursors (Figure 1): the beta-endorphin/ACTH precursor (also known as Proopiomelanocortin or POMC), the enkephalin precursor (known as proenkephalin or proenkephalin A), and the dynorphin/neo-endorphin precursor (also known as prodynorphin or proenkephalin B).

A schematic drawing of the *beta-endorphin/ACTH precursor* is shown in Figure 1 A. At its carboxy terminus is the 31 amino acid peptide beta-endorphin (β-END) and its 91 amino acid precursor beta-lipotropin (β-LPH), which contains β-MSH in some species; the midregion of this precursor contains ACTH-(1-39), which can also be cleaved into α-MSH and CLIP (Corticotropin-Like Intermediate Lobe Peptide) or ACTH-(18–39) (Bradbury et al 1976, Chretien et al 1976, Eipper & Mains 1978, Guillemin et al 1976, Li & Chung 1976, Mains et al 1977, Roberts & Herbert 1977a). At the amino terminus of this precursor is a recently discovered repeat of the active ACTH/MSH core, known as γ-MSH, and finally, the extreme amino terminus is another highly conserved (and therefore biologically active?) peptide fragment. Perhaps the simplest summary of this precursor would be that it contains one opioid peptide (β-END) and potentially three MSH-like peptides (α-MSH from ACTH, β-MSH, γ-MSH) (cf Drouin & Goodman 1980, Kita et al 1979, Nakanishi et al 1979, Nakanishi et al 1981, Roberts & Herbert 1977a,b, Roberts et al 1979, Tsukada et al 1981, Whitfeld 1982).

Table 1 Opioid peptide family

β-END/ACTH (porcine)	Proenkephalin (human)	Prodynorphin (porcine)
β-Endorphin:	[Met]enkephalin:	α-Neo-endorphin:
Try-Gly-Gly-Phe-Met- Thr-Ser-Glu-Lys-Ser- Gln-Thr-Pro-Leu Val-Thr-Leu-Phe-Lys- Asn-Ala-Ile-Val-Lys- Asn-Ala-His-Lys Lys-Gly-Gln	Tyr-Gly-Gly-Phe-Met	Try-Gly-Gly-Phe-Leu- Arg-Lys-Try-Pro-Lys
	[Leu]enkephalin:	β-Neo-endorphin:
	Try-Gly-Gly-Phe-Leu	Try-Gly-Gly-Phe-Leu- Arg-Lys-Tyr-Pro
	[Met]enkephalin-8:	Dynorphin A-(1–8):
	Try-Gly-Gly-Phe-Met- Arg-Gly-Leu	Try-Gly-Gly-Phe-Leu- Arg-Arg-Ile
	[Met]enkephalin-Arg[6]-Phe[7]:	Dynorphin A-(1–17):
	Try-Gly-Gly-Phe-Met- Arg-Phe	Try-Gly-Gly-Phe-Leu- Arg-Arg-Ile-Arg-Pro- Lys-Leu-Trp-Asp- Asn-Gln
	Peptide E:	Dynorphin B-(1–13):
	Try-Gly-Gly-Phe-Met- Arg-Arg-Val-Gly-Arg- Pro-Glu-Trp-Trp-Met- Asp-Tyr-Gln-Lys-Arg- Tyr-Gly-Gly-Phe-Leu	Tyr-Gly-Gly-Phe-Leu- Arg-Arg-Gln-Phe-Lys- Val-Val-Thr

Proenkephalin (Figure 1 B) also contains the coding for several active peptides (Comb et al 1982, Gubler et al 1981, Hughes et al 1975, Kimura et al 1980, Mizuno et al 1980, Noda et al 1982a,b). In contrast to the β-END/ACTH precursor, all of the known active peptides from proenkephalin are opioid in nature. Proenkephalin contains within its structure seven peptides with the [Met]- or [Leu]enkephalin active core. Four of the seven peptides produced are simply [Met]enkephalin; two are carboxyl extended [Met]enkephalin-Arg[6]-Phe[7] and -Arg[6]-Gly[7]-Leu[8]. Finally, one copy of [Leu]enkephalin is produced.

Prodynorphin (the neo-endorphin/dynorphin precursor) is a somewhat simpler precursor than proenkephalin (Fischli et al 1982, Goldstein et al 1979, Goldstein et al 1981, Kakidani et al 1982, Kangawa et al 1981). Prodynorphin (Figure 1 C) produces three main [Leu]enkephalin-containing peptides: α/β-neo-endorphin, dynorphin A, and dynorphin B. The final processing and forms

A: POMC

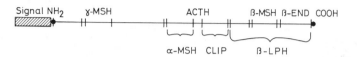

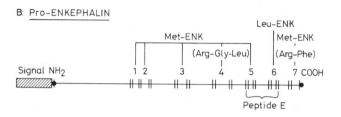

B: Pro-ENKEPHALIN

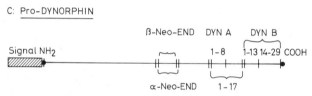

C: Pro-DYNORPHIN

Figure 1   Schematic representation of the protein precursor structures of the three opioid peptide families. The β-END/ACTH precursor was primarily derived with protein biochemistry, but it was confirmed by DNA cloning methods. Both proenkephalin and prodynorphin precursor structures were very largely determined in DNA cloning studies. The *double vertical lines* represent di-basic amino acid clevage sites.

of the neo-endorphin/dynorphin family are an area of active investigation (see below).

Anatomy of the Three Opioid Systems

Peptides from the β-*END/ACTH precursor* can be found in four major loci in nervous tissue. The major site of production of these peptides is the pituitary (Bloom et al 1977, Moon et al 1973, Pelletier et al 1977). A small percentage of anterior lobe cells produce ACTH (hence the name corticotroph), β-LPH, and β-END. In most species, excluding man, the pituitary contains an intermediate lobe. All the cells of this lobe produce β-END/ACTH related peptides, the main end points being β-END and α-MSH.

In brain there are also two cell groups that produce β-END/ACTH peptides (Bloch et al 1978, Bloom et al 1978, Jacobowitz & O'Donohue 1978, Krieger et al 1977, Nilaver et al 1979, Pelletier 1980, Watson & Akil 1979, 1980,

Watson et al 1977b, 1978a,b). The main cell group is found in the region of the arcuate nucleus of the medial basal hypothalamus. Its fibers project very widely to include many areas of the limbic system and brain stem (see Figure 2). The second group of brain cells is found in the nucleus of the solitary tract and the nucleus commissuralis (Schwartzberg & Nakane 1981). This is a modest group of cells and is not well described in terms of its projections.

The second major opioid neuronal pathway *(proenkephalin)* is very widespread and also includes endocrine and CNS distributions (Elde et al 1976, Hökfelt et al 1977a,b, Khachaturian et al 1982a, 1983, Pickel et al 1980, Sar et al 1978, Uhl et al 1978, 1979, Watson et al 1977a, 1978a, 1982b). Peripherally, proenkephalin-produced peptides are found in the adrenal medulla (with catecholamines), the gastrointestinal tract, and several other structures. In brain, proenkephalin-produced peptides are very widely spread at every level of the neuraxis, including cells in the cortex all the way to cells in the spinal cord (Figure 2). Fiber projections have not been clearly mapped, but seem to include both local and long tract systems and are associated with many functions (pain, respiration, endocrine actions, motor actions, and general limbic activity).

Finally, the most recently described precursor *(pro-neo-endorphin/dynorphin)* is found in gut, posterior pituitary, and brain (Khachaturian et al 1982b, Maysinger et al 1982, Vincent et al 1982, Watson et al 1981, 1982a,b, 1983, Weber et al 1982b). It has been found in several hypothalamic cell groups, including the vasopressin-producing cells of the magnocellular neurosecretory nuclei. It too is found in many widely scattered cell groups in the brain stem. Some pathways are known (supraoptic nucleus to posterior pituitary), whereas others are just being studied (Figure 2).

Post-translational Modifications, Processing, and Multiple Forms

The final products produced by and stored within a given neuron depend not only on the genetic code for the precursor, but also on the program that directs enzymes to process the precursor in certain ways. The action of cutting specific peptides out of the precursor protein, and modifying these products by acetylation, amidation, phosphorylation, methylation, glycosylation, or further cleavage, is part of the biological program of a given cell. These post-translation events (i.e. events following the translation of messenger RNA into a protein precursor) are capable of determining the exact mix of peptides in a given neuron. They appear to vary from one tissue to the next, in spite of the existence of a common gene for the precursor. Since these changes result in peptides of widely differing potencies, pharmacological profiles, and receptor selectivities, they are critical in determining function and may constitute a critical step in the regulation and homeostasis of a given opioid system in a particular region

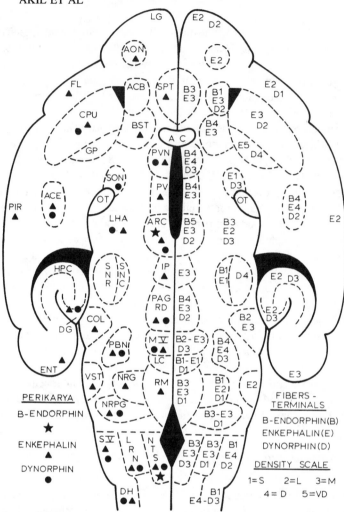

Figure 2 Schematic horizontal view of the rat brain showing the comparative distribution of all three opioid peptide neuronal systems (β-END/ACTH, enkephalin, and dynorphin). Perikarya are shown on the *left* and fibers and terminals on the *right*. Fiber-terminal densities (but not perikaryal densities) are indicated in a scale: sparse (S), light (L), medium (M), dense (D), and very dense (VD).

AC, anterior commissure; ACB, n. accumbens; AON, anterior olfactory n.; ARC, arcuate n.; BST, bed n. stria terminalis; COL, colliculi; CPU, caudate-putamen; DG, dentate gyrus; DH, dorsal horn spinal cord; ENT, entorhinal cortex; FL, frontal lobe; GP, globus pallidus; HPC, hippocampus; IP, interpeduncular n.; LC, locus coeruleus; LG, lamina glomerulosa olfactory bulb; LHA, lateral hypothalamic area; LRN, lateral reticular n.; MV, mesencephalic n. trigeminal; NRG, n. reticularis gigantocellularis; NRPG, n. reticularis paragigantocellularis; NTS, n. tractus solitarius; OT, optic tract; PAG, periaqueductal gray; PBN, parabrachial n.; PIR, piriform cortex; PV, periventricular n. thalamus; PVN, paraventricular n.; RD, raphe dorsalis; RM, raphe magnus; SV, spinal n. trigeminal; SNC, substantia nigra pars compacta; SNR, substantia nigra pars reticulata; SON, supraoptic n.; SPT, septum; VST, vestibular complex.

of the CNS. We therefore describe them briefly, although our knowledge is very limited in the CNS.

POMC   Processing and post-translation events are best understood in the POMC system, particularly in the pituitary (Zakarian & Smyth 1982, Eipper & Mains 1980, Akil et al 1981a). It is now abundantly clear that the POMC cells of the anterior lobe and those of the intermediate lobe of the pituitary process the precursor quite differently. In general, the intermediate lobe produces smaller products, which have undergone more post-translational modifications, than does the anterior lobe. Thus, one can detect in the anterior lobe a small amount of unprocessed precursor (mol wt = 31,000 daltons). Most of the β-END-related immunoreactivity is in fact the intermediate peptide, β-LPH, and the β-END-sized material is primarily made up of unmodified β-END-(1–31). In contrast, the intermediate lobe stores little or no POMC or β-LPH. Further, the β-END-sized material is further modified in two ways: it is alpha-N-acetylated on the tyrosine, or it is cleaved at the carboxyl-terminus to remove the last four or five residues, or both. In fact, the main product in the intermediate lobe is not β-END-(1–31), but rather N-acetyl-β-END-(1–27), a peptide totally devoid of activity at the opioid receptor (Akil et al 1981c).

Similar differences between the intermediate and anterior lobes are seen in the processing of ACTH. Whereas ACTH-(1–39) is the major product of the anterior lobe, the peptide is further processed in the intermediate lobe to produce the highly modified α-MSH [N-acetyl ACTH-(1–13)-NH$_2$] and CLIP or ACTH-(18–39). In the same vein, the third domain of the POMC molecule, the N-terminus region or so-called 16 K fragment, is processed to yield peptides smaller in the intermediate lobe than in the anterior lobe (Ling et al 1979).

The processing of POMC in the brain is a matter of some controversy. It was established early on that the brain may resemble the intermediate lobe in that it shows a preponderance of smaller products over large ones, and exhibits the existence of α-MSH-like immunoreactivity, a product typical of the intermediate lobe (Watson & Akil 1980). However, more recent work suggests that N-acetylation, as the predominant modification in the intermediate lobe, does not take place in the brain (Evans et al 1982). These results are at variance with those of other groups showing the existence of "true α-MSH" as well as des-acetyl α-MSH [i.e. ACTH-(1–13)-NH$_2$] in the brain (O'Donohue et al 1979, Loh et al 1980), as well as the existence of acetylated β-END (Zakarian & Smyth 1982). Our own work suggests that indeed both contentions are, at least partially, right. N-acetylation is a great deal less predominant in the brain than it is in the intermediate lobe. Nevertheless, in certain regions of the brain, especially in terminal areas such as the midbrain and colliculi, N-acetylated forms of β-END can be detected, and the opioid active β-END-(1–31) constitutes only 50–60% of the total activity. Furthermore, cleavage at the C-terminus is evident and may be a more active way of altering β-END in brain than is acetylation. It is therefore evident that the brain has its own pattern of post-translational events, and that these may indeed vary from region to region (Akil 1982).

While such differences may appear hopelessly confusing, they may provide the neuroscientist with handles on the study of the dynamics of POMC in the CNS. Measuring total content of peptides has proven a rather disappointing tool for investigating their function. We have relearned the lesson previously learned by the monoamine investigator—that dynamic measures of neuronal activity are necessary for investigating their function. It is conceivable that the modification of β-END-(1–31) into β-END-(1–27), which is ten-fold less active as an opioid, may serve a physiological function (Akil et al 1981c). Furthermore, such a modification appears to require time, and may not take place if β-END-(1–31) is freshly synthesized. Thus, a measurement of the ratio of these two forms may yield an index of activity in the system and may correlate with various functional states of the animal.

PROENKEPHALIN   Although the enkephalins were the first to be discovered, little is known about the details of their post-translational modifications in the CNS. We know that the precursor contains four distinct forms ([Leu]enkephalin, [Met]enkephalin and two extended [Met]enkephalins), but we do not understand the orderly events that lead to their cleavage from the precursor, even within the adrenal gland. Neither do we know whether peptides F and E give rise to forms of intermediate sizes with the enkephalin core. We also do not know the fate of the nonopioid peptides, which bridge between enkephalins (e.g. mid-portion of peptides F and E), or of the peptides at the N-terminus of the precursor. We (Alessi et al 1982) have put forward some preliminary evidence suggesting that a portion of peptide F (between two [Met]enkephalins) appears to be processed into a smaller fragment and has an interesting regional distribution in brain. Clearly, this whole question deserves further attention. Its delay possibly stems from the controversy as to whether the adrenal enkephalin precursor is exactly identical to what is in brain. Our work (Watson et al 1982b) and that of Rossier and his collaborators (Liston et al 1983) clearly suggests that the products of the enkephalin precursor found in the adrenal gland are conserved enough in brain to exhibit substantial immunological similarities.

PRO-DYNORPHIN   Here again, the processing of the precursor into its end products has not been delineated. Furthermore, the production of any free peptides from the nonopioid region of the precursor has not been studied. What is clear is that there are two possible neo-endorphins, alpha and beta, differing by a lysine (Kangawa et al 1981), and that there are several sizes of dynorphin

A, including 6000 daltons (Seizinger et al 1981b), dynorphin A(1–17) (Gold-stein et al 1981), dynorphin A(1–8) (Seizinger et al 1981a), and other peptides of intermediate sizes (Suda et al 1982). Recently, Weber and co-workers (1982a) have shown that dynorphin A(1–8) is a major product in brain regions and that the ratio of dynorphin A(1–17) to dynorphin A(1–8) can vary from one area to the next. There is no evidence to date as to how dynorphin B is processed (Fischli et al 1982, Kilpatrick et al 1982), and no evidence regarding the possible acetylation of the pro-dynorphin products in various tissues, although Seizinger et al (1982) have shown that dynorphin in the magnocellular system does not appear to be N-acetylated.

RECEPTORS

Multiple Opioid Receptors

The existence of multiple forms of opiate receptors was first postulated by Martin et al (1976), based on addiction, cross tolerance, and abstinence syndromes in dogs with chronic spinal transection. The basic types as defined by Martin are a mu or morphine receptor, a kappa or ketocyclazocine receptor, and a sigma or SKF-10047 receptor. In vivo, kappa drugs are distinguished from mu drugs by the absence, after an acute dose, of bradycardia and by failure to either precipitate or suppress withdrawal in morphine-dependent dogs. In contrast, an acute dose of SKF 10047 produces a general activation of the CNS including pupillary dilatation, tachypnea, tachycardia, and mania.

After the development of in vitro receptor binding paradigms, opiate receptor types again began to emerge. The isolation and identification of [Leu]- and [Met]enkephalin revealed an enkephalin or delta receptor and an alkaloid or mu receptor present in both peripheral organs and the central nervous system (Lord et al 1977). Similarly, the isolation of beta-endorphin and its demonstration in brain suggested a beta-endorphin or epsilon receptor in brain and peripheral tissue (Akil et al 1980, Schulz et al 1979). Further binding studies with the sigma receptor suggest that this may not be a typical opiate receptor since it does not show the classic stereospecificity of an opiate receptor (both D and L SKF 10047 are equipotent) and it binds to PCP "receptors." There appear to be species differences in opiate receptor types, with guinea pig and rabbit having more kappa receptors than rat (Maurer 1982). In addition, there are brain regional differences in the distribution of mu and delta receptors in rat, bovine, and human brains (Chang et al 1979, 1981, Bonnet et al 1981, Ninkovic et al 1981).

The controversy over the existence of multiple opiate receptors and the difficulty in demonstrating their existence in in vitro receptor binding assays has led to a search for more selective opiate agonists and antagonists, as well as tissue preparations containing purer receptor types. Using pure receptor type tissues and selective ligands can aid us in characterizing other alkaloids and peptides in terms of their mu, delta, kappa, and epsilon selectivity. Thus far, only a few tissues, such as the rabbit *vas deferens* (Oka et al 1980), a pure kappa preparation, show the receptor selectivity desired.

Pharmacological Relations to Endogenous Ligands

The isolation of various endogenous opioid peptides and their classification into families from common precursor proteins has led to attempts to pair opioid peptides and receptor types. All of the pro-enkephalin-related peptides show delta receptor activity ranging from [Leu]enkephalin, which is primarily delta, to the enkephalin octapeptide, which appears to be equally mu and delta (H. Kosterlitz 1982 unpublished). The dynorphin precursor family shows a similar receptor selectivity range. All dynorphins and neo-endorphins show kappa preference (Chavkin et al 1982), but dynorphin A(1–8) retains delta capability (Corbett et al 1982 and our own data), whereas dynorphin A-(1–13) is very potent at both mu and kappa receptors. The third family, the proopiomelanocortin family, gives rise to only one opioid peptide, beta-endorphin. Beta-endorphin is very potent at both mu and delta receptors with a slight preference for delta. It should be remembered that the preference of an opioid peptide for a receptor type is an in vitro phenomenon, and the true action of any peptide depends upon the type of receptor present in the synapse of that opioid neuron. In addition, multiple peptides are released simultaneously, and the range of receptor preference may have a physiological function to encode slightly different pieces of information. Thus, a more thorough study of the anatomy of pre- and postsynaptic opioid elements is critical.

Anatomical Studies of Opiate Receptors and Their Relationship to Endogenous Opioid Anatomy

Studies of the anatomical distribution of opiate receptors can provide important clues as to the sites of action and, we hope, the functional significance of different opioid peptide systems in brain. Receptor studies using dissected tissues can be useful in these aims, particularly if a complex "region" such as cerebral cortex is dissected into cytoarchitecturally and functionally defined subareas (M. E. Lewis et al 1981). Although microdissection techniques produce rapid, quantitative results, the level of anatomical resolution is intrinsically poor and fails to place the receptors in the context of complexly distributed neuronal systems. To overcome these difficulties, Pert et al (1975, 1976) devised a radiohistochemical procedure for the visualization of opiate receptors labeled by in vivo administration of [³H]diprenorphine. These and subsequent in vivo autoradiographic studies by Atweh & Kuhar (1977) demonstrated a striking distribution of opiate binding sites in brain and spinal cord. The distribution of sites was not only consistent with a number of pharmacological actions of opiates, but also revealed a variety of unexpected anatomical features, such as dense "patches" of sites in caudate-putamen. Because this in vivo method is technically difficult and limited to the use of very high affinity, slowly dissociating opiates that cross the blood-brain barrier, investigators have devised several methods for the in vitro labeling of opiate receptors on cryostat-cut tissue sections (Young & Kuhar 1979, Herkenham & Pert 1980, 1982, Duka et al 1981a,b, M. E. Lewis et al 1983).

Consistent with studies demonstrating different ratios of apparent mu and delta receptors in different dissected brain regions (Chang & Cuatrecasas 1979, Chang et al 1979, 1981, Pert & Taylor 1980, Bonnet et al 1981, M. E. Lewis et al 1981, Ninkovic et al 1981), autoradiographic studies have revealed a different anatomical distribution of sites binding relatively selective mu and delta ligands (Goodman et al 1980, M. E. Lewis et al 1983). Thus, mu binding is predominant in several hypothalamic and thalamic nuclei, the periaqueductal grey, interpeduncular nucleus, inferior colliculus, and the median raphe nucleus, while delta binding appears relatively greater in the amygdala, nucleus accumbens, olfactory tubercle, and the pontine nuclei (Goodman et al 1980). In cerebral cortex, mu sites show striking laminar heterogeneity across regions, and are densest in the limbic cortex, while delta sites show a much more uniform laminar and regional distribution (M. E. Lewis et al 1983). In the caudate-putamen, mu sites show the characteristic "patchy" distribution first described in the in vivo autoradiographic studies, while delta sites show a heavy, diffuse distribution (Goodman et al 1980). However, Bowen et al (1981) have reported incubation conditions under which a delta ligand, [^{3}H]D-Ala^2, D-Leu^5-enkephalin, can also label sites in a patchy distribution within the caudate-putamen. Since the enkephalin binding in this pattern occurs at the expense of [^{3}H]morphine binding, the authors have argued that opiate receptors in striatal patches can interconvert between mu and delta ligand-preferring forms. A third kappa ligand-preferring conformation has also been postulated on the basis of the finding that [^{3}H]ethylketocyclazocine, under reportedly "kappa-selective" incubation conditions, also binds to opiate receptors in patches in rat caudate-putamen (Quirion et al 1982). However, Goodman & Synder (1982) have reported a differential localization of mu and kappa sites in the guinea pig brain; the apparent kappa sites appear to be concentrated in laminae V–VI of the cerebral cortex and in the caudate-putamen.

Interpretation of opiate receptor autoradiographic studies, like ordinary binding assays, is complicated by the use of partially selective ligands to label the receptor and define its pharmacological characteristics. Furthermore, in the absence of autoradiographic displacement studies, it may be difficult to discriminate a high density of low affinity sites from a low density of high affinity sites (James et al 1982). Future autoradiographic studies will very likely employ selective protection procedures (Smith & Simon 1980, James et al 1982) in an attempt to label opiate receptor subtypes more selectively.

Despite the difficulties of interpreting autoradiographic studies, such studies have provided detailed, otherwise unobtainable information on the distribution of opiate receptors in the brain. Given this knowledge, and the wealth of information on the anatomical distribution of the three families of opioid peptides, it is reasonable to ask whether there are specific relationships between the receptors and their endogenous ligands. To circumvent the need for in vivo labeling and between-brain comparisons, M. E. Lewis et al (1982) performed in vitro autoradiography (with [3H]naloxone) and enkephalin immunocytochemistry on adjacent sections from formaldehyde-perfused rat brains. The labeled sites exhibited the pharmacological properties of mu receptors, while delta sites, labeled by [3H]D-Ala^2-, D-Leu^5-enkephalin, were almost totally eliminated by perfusion fixation (M. E. Lewis, H. Khachaturian, and S. J. Watson, in preparation). There was an excellent correlation between the location of enkephalin immunoreactivity and [3H]naloxone binding sites in many regions, e.g. habenula, interpeduncular nuclear complex, nucleus tractus solitarius, parabrachial nuclei, the nucleus of the spinal tract of the trigeminal nerve, and the dorsal horn of the spinal cord. A close association between [3H]naloxone binding sites and enkephalin, β-endorphin, and dynorphin immunoreactivity has also been detected in various regions of the rhesus monkey brain. Further studies will be required to determine whether or not there is a unique relationship between the individual opioid peptide systems and the different receptor subtypes. Although some in vitro studies would suggest such a relationship between the enkephalins and delta receptors, and between dynorphin and kappa receptors, it does not follow that this "segregation" occurs in intact circuits under physiological conditions. The existence of multiple opiate receptors does not logically require unique receptor-ligand relationships, as shown by the finding of multiple receptors for acetylcholine, dopamine, norepinephrine, serotonin, and histamine.

FUNCTION

In thinking about endorphin function, it is difficult to avoid the observation that all three endogenous opioid families are heavily invested in systems that regulate the body's responses to stress. Indeed, the three opioid systems can be found in the hypothalamus, the three lobes of the pituitary, and the adrenal medulla, while the adrenal cortex is exquisitely sensitive to the products of POMC, the ACTH/β-END precursor. Similarly, the autonomic nervous system has been shown to contain endorphins (Schultzberg et al 1978) and opioid receptors (Young et al 1980). Possibly of even greater relevance is that central nuclei that regulate autonomic function, such as the nucleus tractus solitarius, are enriched with endogenous opioids and their receptors (see below). These observations point to a powerful role of the opioids in regulating the organism's response to physiological and environmental demands, including physical and psychological stress. Like the catecholamines, the endorphins may have a basic, multisystem function essential to the maintenance of homeostasis and to the survival of the organism. From this starting point, opioids may have evolved more complex and integrated functions, recruiting higher levels of the neuraxis, serving in the control of affect and mood, drive and reinforcement, or in the process of filtering information and controlling attentional mechanisms.

We focus primarily on the more basic functions related to stress and its sensory and autonomic concomitants. This involves a review of changes in pain responsiveness after stress (stress-induced analgesia), and changes in cardiovascular responses upon opioid administration or after shock (opioid-related shock-induced hypotension). While the more complex paradigms involving learning and memory, mood and affect are intrinsically fascinating, their study is more difficult, and the body of information on their opioid underpinnings is more sketchy. It is often difficult to determine whether the effects of endorphins on complex behaviors are primary or secondary to their ability to modify sensory or autonomic variables.

Pain, Stress, and Endorphins

ACTIVATION OF ENDOGENOUS OPIOIDS TO CONTROL PAIN   Control of pain is possibly the first function that comes to mind when thinking about the role of endorphins. The existence of an endogenous pain-controlling system was first suggested by studies on stimulation-produced analgesia (SPA) whereby stimulation of specific sites in the brain led to a reduction in responsiveness to pain, both in lower animals and in man (Mayer et al 1971, Richardson & Akil 1977). When this analgesia was found to be reversible by naloxone (Akil et al 1972, 1976a,b) and cross-tolerant with morphine (Mayer & Hayes 1975), the critical involvement of an endogenous opioid system was put forth. Although it is evident that nonopioid systems are also involved in SPA (Akil et al 1972, 1976a,b, Basbaum et al 1977), the body of work on SPA gave us the first direct association between a behavioral event and activation of endogenous opioids. In man, it was shown that pain relief is accompanied by an increase in enkephalin-like material, as well as β-END immunoreactivity in the CSF (Akil et al 1978a,b, Amano et al 1982, Hosobuchi et al 1979). Although a cause and effect relationship between the endorphin release and the production of analgesia is far from established, these correlations, along with those uncovered by Terenius and co-workers (cf Terenius 1982), provide a compelling body of evidence for the involvement of endogenous opioids in regulation of responsiveness to pain.

A major theoretical question that arose soon after the discovery of enkephalin concerned the conditions that would engage endogenous opioids to control pain. Opiate antagonists, such as naloxone, were typically thought of as inactive in normal animals, and functioning only to reverse the effects of opiate agonists. The work of Jacob et al (1974) showing a lowering of pain threshold after administration of naloxone was indeed unusual. His findings, along with the naloxone-induced reversal of SPA, were the only shreds of evidence for an endogenous opioid pain inhibitory system. Further, one could argue that brain stimulation is not physiological. We therefore needed to uncover conditions that reliably induced an opioid-mediated inhibition of pain responsiveness. Because it would appear adaptive to become analgesic in the face of stress, we investigated the effect on footshock to pain responsiveness (Akil et al 1976a, Madden et al 1977). We discovered that acute stress induced by intermittent footshock for a period of 15 minutes or more led to a reliable insensitivity to pain, so-called stress-induced analgesia (SIA). This analgesia was comparable in its potency to a large dose of morphine; it was found to be partially reversible by the opioid antagonist naloxone (Akil et al 1976a). Repeated daily exposure to the intermittent footshock led to adaptation of the organism, such that further stress no longer produced analgesia. We showed that SIA was accompanied by an increase in brain endogenous opioids (Madden et al 1977), but the multiplicity of endogenous peptides within the brain was not known at the time, and the exact system or systems being activated, be they synthetic intermediates or final products, are not currently defined. There is evidence that hypothalamic β-END content is decreased (Rossier et al 1977), a finding we have replicated (Akil et al 1981b), possibly due to release of the stored material upon stress. However, a close correlation between the biochemical change and the behavioral time course has not been well established. Other studies, reporting changes in occupancy of opioid receptors after stress, also suffer from the lack of specificity of the measures being used (Chance 1980). More recently, the enkephalinase inhibitor, thiorphan (Roques et al 1980), has been shown to potentiate stress-induced analgesia, and the potentiation was reversible by naloxone (Chipkin et al 1982). While this finding argues for the involvement of enkephalins in SIA, the selectivity of thiorphan in inhibiting enkephalin breakdown, but not the breakdown of other opioids, has not been fully established. Thus, while there is some circumstantial evidence linking various brain endogenous opioids to SIA, no conclusive case has been made for any given family or specific neuronal circuit. This problem is an excellent example of the difficulties encountered in linking a specific opioid to a specific phenomenon, due to the biochemical and anatomic complexities of the systems, and the lack of availability of dynamic measures.

OPIOID AND NONOPIOID FORMS OF STRESS-INDUCED ANALGESIA, NALOXONE, AND CROSS-TOLERANCE WITH MORPHINE   It should be pointed out at this juncture that several stress paradigms beyond footshock have been reported to induce analgesia. These include cold-swim stress (Bodnar et al 1978a), food deprivation (Bodnar et al 1978b, McGivern et al 1979), sexual behavior (Crowley et al 1977), conditioning (Chance 1980, Olivero & Castellano 1982, Watkins et al 1982b), or learned helplessness (Jackson et al 1979), all of which have been shown to produce analgesia that is more or less antagonized by naloxone. A naloxone reversible stress analgesia has also been experimentally demonstrated in man (Willer et al 1981). Part of the complexity resides in the fact that stress-induced analgesia activates both opioid and nonopioid mechanisms. The original report of Akil et al (1976a) described partial anatgonism by naloxone. Mayer and co-workers (Hayes et al 1978) reported no naloxone antagonism in their paradigm involving shorter periods of footshock. J. W. Lewis, Liebeskind, and their collaborators (J. W. Lewis et al 1980) resolved the discrepancy by showing that different durations of stress induce analgesia via different mechanisms. Short durations of stress cause nonopioid analgesia, whereas longer durations appear to induce opioid mechanisms. The conditions yielding opioid analgesia allowed them to establish complete reversibility by naloxone and cross-tolerance to morphine (J. W. Lewis et al 1981b). J. W. Lewis and co-workers have shown that the two stress paradigms lead to very different patterns of change in brain endogenous opioids (Akil et al 1981b). Mayer and co-workers (Watkins et al 1982a) have further shown that the same stressor could lead to either opioid or nonopioid analgesia, depending on the part of the body being shocked. Possibly, pain inhibition induced by shock recruits both local circuits, which may not be opioid in nature, as well as more central circuits that are, in part, endorphinergic.

The multiple mechanisms involved in SIA may also explain the discrepant findings obtained when the issue of cross-tolerance between SIA and morphine is investigated. Whereas J. W. Lewis et al (1981b) and Drugan et al (1981) have found that chronic morphine administration leads to a decrease in stress-induced analgesia, Akil et al (1978c) had reported the converse, i.e. that rats made tolerant to morphine with three daily pellets became more sensitive to stress-induced analgesia. Millan et al (1981a) confirmed this observation by showing that chronic administration of morphine (30 days) leads to an enhancement of SIA. These findings further emphasize the need for defining the exact parameters of testing, including the quantitation of morphine tolerance, since the existence of complex and parallel pain control systems is likely to confound the issues.

PERIPHERAL OPIOIDS AND STRESS-INDUCED ANALGESIA: THE PITUITARY   Although we have implicitly focused on potential brain circuits involved in SIA, there is evidence that the phenomenon depends on peripheral organs. The existence of β-END in anterior pituitary corticotrophs, and its co-secretion with ACTH upon stress (Guillemin et al 1977), are clues suggesting a role of pituitary β-END in stress-induced analgesia. J. W. Lewis et al (1981a) showed that hypophysectomy blocked the opioid-type of stress-induced analgesia, without altering the nonopioid type, further implicating pituitary endorphins in the phenomenon. However, Herz and his co-workers (Millan et al 1980, 1981b) suggested that although the anterior pituitary appeared involved, the critical element did not appear to be β-END. Administration of dexamethasone, a synthetic steroid that leads to the inhibition of the release of ACTH/β-END, did not abolish the naloxone-reversible SIA. Note, however, that dexamethasone is a potent steroid that mimics the natural products of the adrenal cortex. Indeed, MacLennan et al (1982) have shown that stress-induced analgesia can be abolished by hypophysectomy, and reinstated by treatment with corticosterone. Thus, treatment with dexamethasone may have complex effects, causing a reduction in the release of pituitary β-END/ACTH, but compensating for it by its own direct actions.

The case for a critical role of pituitary β-END in the production of SIA can be made, but it is far from perfect. Unquestionably, the conditions of stress used to produce opioid SIA produce a significant depletion of anterior lobe endorphins, and a significant increase into the blood stream (Lim et al 1982, Millan et al 1981a). However, there are stressors which cause equally profound changes in plasma β-END and yet do not produce analgesia (Lim et al 1982). Furthermore, chronic daily stress, which leads to a loss of SIA, still produces significantly elevated levels of β-END in rat plasma (J. Matthews et al, in preparation). While the chronic stress paradigm does lead to an adaptation of anterior lobe POMC, it leads to a simultaneous induction of the intermediate lobe POMC. Thus, we have shown (Shiomi & Akil 1982) that acute stress brings about an increase in the rate of POMC biosynthesis and an acceleration in the conversion of POMC to β-LPH and β-END. These changes do in fact disappear with repeated daily stress. However, the intermediate lobe appears to follow the opposite course, showing mild activation with changes in content, in release of α-MSH and N-acetyl-β-END into plasma, and acceleration in rate of biosynthesis in pulse-labeling experiments. Immediately after a stress session, in an animal with a history of repeated footshock, plasma β-END levels are higher than those seen after the first acute stress. Both the N-acetylated forms and the nonacetylated forms of β-END are elevated. Thus, the animal is no longer analgesic in spite of extremely high opioid levels in plasma. Either the material in plasma plays no significant role in SIA, or significant changes in the target receptors (down regulation?) have taken place, rendering the change in levels functionally irrelevant. It is also conceivable that the great increase in the acetylated forms of β-END seen with chronic stress plays an inhibitory role on the expression of SIA.

We have focused on pituitary β-END; however, it should be remembered that the other two opioid families are present in the pituitary. Enkephalin is present in the neural lobe, while dynorphin is localized in both posterior lobe processes and anterior lobe cells. Millan et al (1981b) have shown that dynorphin-like immunoreactivity was significantly decreased in the anterior lobe following five minutes of acute stress, while posterior lobe immunoreactivity was unchanged and hypothalamic immunoreactivity (the source of posterior lobe material) was elevated.

PERIPHERAL OPIOIDS AND STRESS-INDUCED ANALGESIA: THE ADRENAL   J. W. Lewis and co-workers have implicated the adrenal medulla in SIA. Anatomical studies had first revealed the existence of enkephalins in adrenal chromaffin cells (Schultzberg et al 1978). Work by the groups of Viveros and Costa demonstrated that enkephalin-like peptides are costored and coreleased with catecholamines in response to stress (Viveros et al 1979, Costa et al 1981, Saria et al 1980, Livett et al 1981). Therefore, J. W. Lewis et al (1982) examined the effect of adrenalectomy, adrenal demedullation, or denervation of the medulla via celiac ganglionectomy on opioid and nonopioid SIA. Opioid SIA was selectively prevented by adrenal manipulations. There appeared to be no difference between total adrenalectomy and demedullation in terms of blocking SIA, suggesting to these authors that the adrenal cortex does not contribute to SIA in a significant manner. Further studies examined opioid adrenal content as well as catecholamine content in acutely stressed and chronically stressed (14 days of 30 minute sessions) animals, as well as reserpinized rats, either stressed or unstressed. The amount of opiate-like material in the adrenal medulla was reduced after acute stress, but returned to normal in chronically stressed rats. Rats pretreated with reserpine showed large increases in enkephalin-like material in the adrenal medulla, which again was depleted by an acute stress (J. W. Lewis et al 1982). In similar studies (Alessi et al 1982), using a radioimmunoassay for peptide F, a fragment of adrenal enkephalin, we have found a depletion of peptide F-like immunoreactivity after acute stress in the adrenal medulla. Again, in chronic stress the levels return to normal.

Thus, stress appears to activate the HPA axis of the peripheral sympathetic system, leading to release of endogenous opioid peptides from both the anterior pituitary and the adrenal medulla. Since the release of monoamines and peptides from the adrenal medulla is under peripheral nervous system control rather than pituitary or hormonal control, it would appear that these two systems may be independent in regulation and function.

ENDORPHINS AND PAIN CONTROL IN THE HUMAN   The activation of the endogenous opioid mechanisms to control pain in humans has not been extensively documented. Several studies in humans have implicated endorphins in pain relief or in pain appreciation, either in pathological conditions (cf Terenius

1982, Dubois et al 1981) or in very artificial situations such as electrical stimulation of the brain, or acupuncture (Akil et al 1978a,b, Mayer et al 1976). However, little is known about the natural conditions that activate endorphins in man. The work of Fields and his collaborators implicating endorphins in placebo analgesia comes closest to satisfying a naturalistic setting (Levine et al 1978), though the role of stress perception in placebo-responders versus non-responders and its relation to opioid mechanisms remains to be determined.

Pregnancy and labor can be construed as a natural biological and psychological stress. We, and others (Thomas et al 1982), had shown an elevation of β-END levels during pregnancy and labor. We sought to correlate changes in plasma β-END with pain and anxiety ratings before, during, and after labor (Cahill & Akil 1982). Although we could reliably show an elevation in β-END levels during pregnancy, and a further elevation during labor, followed by a substantial decrease 24 hours after delivery, no correlation could be obtained with pain perception. It appeared to us that the physical stress of labor produced a significant release of β-END, regardless of whether the woman felt in control of her pain (e.g. by using breathing techniques) or even whether her pain was completely blocked with spinal anesthesia. This finding, although somewhat disappointing, does not necessarily mean that the increase in endorphins plays no role in pain modulation, but rather that it may reflect a number of processes including hormonal controls and primary physical stress, and may even play some role in the initiation or maintenance of labor (Thomas et al 1982; our own observations).

OPIOID RECEPTORS MEDIATING STRESS-INDUCED ANALGESIA   There are to date no direct studies investigating which opioid receptor type(s) mediates stress-induced analgesia. However, there are several studies examining the involvement of multiple opioid receptors in the reduction of responsiveness to nociceptive inputs. Based on his use of the irreversible antagonist, naloxazone, Pasternak (1981) suggested that a high affinity receptor, termed mu 1, mediates analgesia whether it is induced by opiate alkaloids, such as morphine, or opioid peptides, such as enkephalin analogues, even of the delta type. This site is distinct from the lower affinity site, which mediates respiratory depression and eventually leads to death. Further evidence for the preferential involvement of mu receptors in analgesia was obtained by Gacel et al (1981) by using a highly selective delta-preferring enkephalin analogue, and showing that its high affinity and potency in a delta system is not seen when one is testing analgesia, presumably because the latter phenomenon is mu-receptor mediated. Consistent with these results is the finding of Holaday and co-workers (1982b) that a selective delta antagonist, M-154, 129, capable of reversing shock-induced hypotension, does not block morphine analgesia (see below).

There are, however, studies implicating other receptor subtypes in opiate analgesia. Tung & Yaksh (1982) studied rat spinal analgesia with intrathecal administration of various opioids and opiates, calculated pA2 values, and examined cross-tolerance between the various agents. They suggested that there were two types of opiate receptors mediating the spinal analgesia, a mu-like receptor, also recruited by kappa ligands such as ethylkeotcyclazocine, and delta receptors recruited by prototypical delta-preferring ligands such as DADL. Animals made tolerant to morphine showed cross-tolerance to EKC, but no cross-tolerance to DADL. These authors could not uncover any selective effects of kappa ligands in their preparation. Similar evidence for the existence of delta involvement in the production of analgesia was obtained by Hynes & Frederickson (1982), who showed that their delta ligand, [Met]enkephalin, does not exhibit cross-tolerance to morphine and produces analgesia in the presence of blockade of the mu 1 sites by naloxazone. Although Tung & Yaksh (1982) failed to detect any selective effects of EKC in the spinal cord, Piercey and co-workers (1982) used a highly selective new kappa ligand, U-50488H, and suggested that it acts primarily at a spinal site via a kappa receptor. They further suggested that dynorphin, an endogenous kappa agonist, may be the endogenous ligand for the analgesia kappa receptor in the cord. However, not all investigators have been able to detect spinal analgesia with the intrathecal injection of dynorphin (cf Tung & Yaksh 1982).

To further complicate the issues, there are reports of interactions between mu and delta receptors in the production of analgesia. For instance, Barrett & Vaught (1982) reported that relatively selective delta ligands not only produce potent analgesia, but, at subanalgesic doses, potentiate morphine-induced analgesia. On the other hand, mu-selective peptide ligands, such as morphiceptin, while capable of producing analgesia, are unable to potentiate morphine analgesia.

Even though the weight of the evidence favors a primary involvement of mu receptors in the mediation of classical opiate analgesia, with a possible secondary role of delta and kappa sites, it should be remembered that what we find in these studies is primarily a function of the selectivity of the agents being used, the neuronal circuit being tested, the relative enrichment of that area or circuit with one or the other receptors, and the species of animal being studied. What matters physiologically is the encounter between a particular ligand and a particular receptor, as a function of the local processing, anatomy, and physiology. Hence, while we can implicate mu receptors in the reduction of pain responses, we cannot eliminate an important role of other receptors in this function.

SUMMARY   The work on stress-induced analgesia has uncovered a role of multiple opioid systems in inhibition of pain responsiveness as a function of environmental variables. Although there is evidence for the involvement of multiple opioids and multiple tissues in this phenomenon, it is unclear whether some of the systems are parallel or redundant, or whether different ones become activated with different durations and magnitude of stress, and different stressors, including physical versus psychological. The involvement of one particular receptor subtype is also unclear, although mu receptors probably play a critical role in some of the paradigms. Finally, the role of other, nonopioid substances that are costored, coreleased, and sometimes cosynthesized with the opioids has not received enough attention.

Opioids and Cardiovascular Control

The role of endogenous opioids in cardiovascular control has been fully reviewed recently by Holaday (1983). We discuss a few salient points here, which we believe are relevant to the issues of multiple opioid systems, multiple opioid receptors, and their relation to understanding opioid function.

ANATOMY OF OPIOID SYSTEMS IN REFERENCE TO CONTROL OF BLOOD PRESSURE   There are a number of circuits and sites that would permit the endogenous opioids to exert significant control over blood pressure regulation. In the central nervous system, the most obvious is the vagal-solitary complex in the medulla. This system mediates control over a variety of visceral functions, including cardiac activity and sensory feedback from that organ. The nucleus of the solitary tract is the only site outside the hypothalamus that contains cell bodies for the ACTH-β-END peptides, the proenkephalin-related peptides, as well as the neo-endorphin-dynorphin family of opioids. This anatomical arrangement suggests an important and complex role of the three families of opioids and their receptors in the control of cardiac function in that area. A second possible site for an opioid-blood pressure interaction is the pituitary. The anterior lobe, by releasing ACTH and gamma-MSH (also β-END) triggers a cascade of endocrine events via the adrenal cortex, resulting in the synthesis and release of corticoids, with their known pressor effects. The activation of the corticotrophs results in the eventual induction of PNMT (phenylethanolamine-N-methyl transferase), the enzyme critical for the synthesis of epinephrine, also a potent pressor agent (cf Van Loon et al 1981). The posterior lobe of the pituitary contains dynorphins and neo-endorphins in the same processes that release vasopressin (Watson et al 1982a). Whether these opioids potentiate or antagonize the actions of vasopressin or whether they have any blood pressure effects on their own remains to be determined. It should be noted, however, that Leander et al (1982) have shown a powerful diuretic effect of kappa agents, which is likely to result in a drop in blood pressure. Since dynorphin and α-neo-endorphin are highly selective kappa ligands, they may play a role in inhibiting vasopressin release (negative feedback control?), thereby causing diuresis and hypotension.

A third possible site of opioid control over blood pressure is via the sympathetic nervous system, known to have opioid components (Schultzberg et al 1978, DiGiulio et al 1978). Dilation of the mesenteric vascular bed can result in systemic hypotension. Further, opioids may have a direct effect on the release of monoamines and enkephalins in the adrenal medulla, since Schultzberg et al (1978) have shown that the preganglionic sympathetic input to the adrenal contains an enkephalinergic component. Whether the net effect of the enkephalins there is to produce hypertension or hypotension is not clear because, aside from their presence in the sympathetic terminals, enkephalins are stored with the catecholamines in the adrenal medulla (Viveros et al 1979). The work of Hanbauer et al (1982) suggests that the enkephalin-like peptides released from chromaffin granules in the anesthetized respernized dog cause a reduction in heart rate and blood pressure. Thus, each of the endogenous opioid families acting at sites within and outside the brain is strategically located to modulate visceral events including the control of blood pressure.

OVERVIEW OF CHANGES IN CARDIOVASCULAR FUNCTION PRODUCED BY OPIATES AND OPIOIDS   The excellent review by Holaday (1983) details the mixed effects on cardiovascular function that have been reported with injections of opioids. Given the anatomy briefly sketched above, it is not surprising that one can detect both pressor and depressor effects of opioids (Feldberg & Wei 1978, Florez & Mediavilla 1977, Laubie et al 1977, Lemaire et al 1978). The effects vary depending on whether the substances are administered centrally or peripherally, whether the studies are carried out in an anesthetized preparation or an awake animal, and whether the ligand is mu or delta preferring (Holaday & Faden 1982). The site of injection in brain is also critical, because the results vary from tissue to tissue, ventricle to ventricle, or even hypothalamic nucleus to nucleus. Based on this overview of the literature, Holaday (1983) has suggested that "mu receptor binding is critical to the decrease in heart rate produced by morphine, whereas both mu and delta actions may be involved in the hypotensive and respiratory depressant actions of this ligand" (p. 555). He goes on to suggest that bradycardia and hypotension, which appear mu-mediated, are probably controlled via central, brainstem sites, which activate parasympathetic outflow. On the other hand, substances, such as DADL, that preferentially activate delta receptors appear to stimulate central sympathetic outflow and elevate blood pressure in awake, unanesthetized experimental animals. While this appears to be a reasonable conclusion to draw from the work on pharmacological administration of various types of opioids, it is not entirely consistent with the opiate receptor selectivity seen with blood pressure changes following shock.

OPIOIDS AND SHOCK   A body of work using a number of shock models has produced convincing evidence that opioids contribute to the cardiovascular changes, particularly hypotension, seen in these conditions (Holaday & Faden 1978, 1980, Faden & Holaday 1979, 1980, 1981). The shock-induced hypotension is reversible by naloxone. This opiate antagonist appears to act via central sites (Janssen & Lutherer 1980, Holaday & Faden 1980). Further, adrenal opioids appear not to be critical for expression of the shock-induced hypotension. On the contrary, adrenalectomy potentiates shock susceptibility and prevents its reversal by naloxone (Holaday 1983). Furthermore, while β-END levels are greatly increased in plasma after shock, as would be expected with such severe stress (our own observations), the functional significance of this elevation is not determined. Thus, Holaday (1983) appears to favor a central mediation of the opioid effects in shock.

Of particular interest are the studies with selective opioid receptor antagonists on endotoxic shock hypotension. Because naloxone, at high enough doses, does not distinguish between mu, kappa, or delta receptors, the use of the newer, more selective antagonists is valuable. Holaday and his co-workers have recently reported that the hypotension is blocked by the putative delta-selective antagonist M154,129, but not by the mu-selective irreversible antagonist, β-FNA (Holaday et al 1982). This picture is the converse of what is observed with these antagonists vis-à-vis morphine-like analgesia.

SUMMARY   The work on shock and cardiovascular regulation has served as a useful probe for studying endorphin functions. It has reinforced the importance of understanding the multiplicity of endogenous opioid systems and their mutual interactions, as well as their interactions with other neuromodulators. As with stress-induced analgesia, we cannot pinpoint a single anatomical circuit, a single endogenous ligand, or a single opioid receptor which, upon activation, can mediate the full range of reported effects. The potential separation of some of the effects on the basis of receptor selectivity is promising, and may prove clinically important. However, the possible interactions between the two functions—control of nociception and control of blood pressure—are equally fascinating. The few reports of correlations between changes in blood pressure and in pain responsiveness (Zamir et al 1980, Zamir & Shuber 1980, Maixner et al 1982) and those linking opioid stress-induced analgesia to respiration (Isom & Elshowihy 1982) or to stress-induced hyperglycemia (Amir & Bernstein 1982) may help us reach a more cohesive picture of opioid functions in controlling the body physiology in both normal and emergency situations.

Multiplicity of Peptides, Active Cores, and Actions

The study of the biology and function of peptidergic systems carries with it a number of unique challenges not encountered previously by the neuroscientist. These systems encompass in their protein code a great deal of biological information that directs their maturation and processing, as well as their effects at the receptors. As we have described, each of the opioid precursors contains multiple active peptides, which are further modified differentially in different brain areas. They also often coexist with other neurotransmitters, such as the catecholamines in the adrenal medulla, or acetylcholine in preganglionic sympathetic nerves, or with other neuropeptides, such as vasopressin, in the magnocellular posterior pituitary pathway. Thus, activation of a certain cell results in the release of many substances that are capable of biological activity. Oftentimes, we have some notions about the actions of one of the actors: ACTH deriving from the anterior lobe, vasopressin deriving from the posterior lobe, catecholamines from the adrenal gland, β-END in the brain. We then tend to ignore the potential effects of the others. We have seen, however, that adrenal enkephalins may have a hypotensive effect, whereas the costored catecholamines may perform the opposite task. Dynorphin, as a putative kappa ligand, may have diuretic and hypotensive actions, whereas vasopressin does the opposite. On the other hand, there is evidence of synergism or potentiation between products derived from POMC. Pedersen et al (1980) have shown that gamma-MSH potentiates the steroidogenesis brought about by ACTH. We (Walker et al 1980a) have shown that in the central gray, ACTH produces analgesia via a nonopioid receptor, that ACTH-like agents and β-END are additive in producing analgesia, and that gamma-MSH, while inert alone, potentiates the analgesia produced by ACTH analogues. In other test systems, however, the synergism is not as evident and there are several reports of antagonism between ACTH and β-END effects (e.g. Belcher et al 1982, Datta et al 1982). A great deal remains to be learned. For example, what is the function of α-MSH in the periphery? Is there a function of the N-acetylated forms of C-terminus-cleaved forms of β-END at nonopiate receptors? Is there a function of other fragments of the opioid precursors, such as peptide F (cf Alessi et al 1982), or the N-terminus pieces of prodynorphin? Can we hope to understand the functions of these circuits without understanding the interactions among the natural products? One can conceive of them actively monitoring each other, synergizing in some cases, setting upper or lower limits in others, or mediating electrical events of varying durations. Studying each one at a time may represent a perturbation of the system, not only because we administer pharmacological doses, but because we may be offsetting a delicate balance critical in orchestrating the final physiological event.

Not only are there multiple active substances within a given precursor, but is now clear that there are several biologically active sites within a single peptide. Schwyzer & Eberle (1977) showed this for ACTH by examining the effects of various fragments of ACTH on a variety of peripheral responses. For typical ACTH effects the core sequence Met-Glu-His-Phe-Arg-Trp-Gly is necessary [i.e. ACTH-(4–10)]. "Core sequences" seem to occur in other systems; for example, the enkephalin sequence appears to be a core for opiate peptides. However, the other segments of ACTH also have biological activity. ACTH-(11–24) acts as an ACTH antagonist in rat adrenal cortical cells and adipocytes, but the addition of N-terminal amino acids to this hormone fragment bestows new biological activity. Thus ACTH-(7–24) stimulates adenylate cyclase in these cells. Further lengthening of the sequence to ACTH-(5–24) is required to observe steroidogenesis and lipolysis.

As with ACTH-related peptides, opiate peptides in the CNS can have several biologically active sites. We recently observed such a phenomenon with the peptide, dynorphin A. The core sequence in dynorphin A is Tyr-Gly-Gly-Phe-Leu, which of course has typical opiate effects in a variety of preparations. The extended 1–17 sequence, however, produces quite a mixture of effects, many of which are not reversed, or are only partially reversed by naloxone (Walker et al 1980b). Many of these latter effects can be reproduced with a fragment of dynorphin that fails to bind to opiate receptors, dynorphin A(2–17) or des-Tyr-dynorphin A (Walker et al 1982). Like the extended ACTH-(11–24) sequence, des-Tyr-dynorphin A produces effects that oppose those of classic opiates such as the inhibition of morphine analgesia and the depression of hippocampal unit firing (Walker et al 1982, 1983).

Similar findings of multiple active cores within one peptide were demonstrated in the work of DeWied and co-workers (e.g. Kovacs et al 1981) on des-tyrosine-gamma-endorphin and des-tyrosine alpha-endorphin, peptides with no opioid activity but with significant and opposing behavioral effects of their own. Opposite effects have also been reported for [Leu] versus [Met]enkephalin (Gibson et al 1980). Of equal interest are the recent findings of Wei and his associates (Mues et al 1982) suggesting that Arg-Phe, the last two residues of [Met]enkephalin heptapeptide (cf Table 1,) produces its own cardiovascular effects, which are not naloxone reversible. Sander & Giles (1982) have shown that the removal of the tyrosine from a [Leu]enkephalin analogue, which renders it inactive at the opioid receptor, does not prevent it from producing pressor effects. Thus, even within the short code of a 7- or 5-residue peptide, there may be multiple messages for effects mediated at multiple receptors.

Finally, we cannot ignore the potential complexity introduced by having multiple opioid receptors in the same vicinity. Lee, Loh and their associates have pioneered work on the possible interactions among multiple opioid receptors, both in the naive and in the morphine-tolerant animal (e.g. Lee et al 1980, Tulunay et al 1981). Whether these results are best explained in terms of multiple active cores of the peptides (Walker et al 1982) or in terms of interactions between multiple opioid receptors merits further investigation. However, the concept of interactions among receptors is very exciting, and presents us with yet another dimension in the study of endogenous opioid function.

CONCLUSIONS

The multiplicity we behold in studying endogenous opioid function is dizzying. Anatomically, biochemically, and pharmacologically, the potential interactions appear infinite. Sometimes, we long for the simple old days, before endogenous opioids, when we were looking for one receptor, and possibly one endogenous ligand. But with our knowledge have come exciting notions that have had impact on all of neuroscience. Many critical and useful tools have also emerged in molecular biology, immunology, biochemistry, and pharmacology. It is now up to the physiologist and the behaviorist to meet the challenge, to ask their questions within this rich biological framework, and to begin evolving the functional principles that appear to govern the actions of endorphins. A tall order, but endorphins have been known to set the pace for neuropeptide research.

ACKNOWLEDGMENTS

This work was supported by NIDA grant #R01DA02265 to H.A. NSF grant #BNS8004512 and NIMH grant #R01MH36168 to S.J.W., and the Theophile Raphael Fund.

Literature Cited

Akil, H. 1982. *Measurement of multiple forms of beta-endorphin and alpha-MSH by HPLC and multiple immunoassays: Logic, strategy, and methods.* Short course presented at Soc. for Neurosci. Ann. Meet., Minneapolis, Minn.

Akil, H., Hewlett, W. A., Barchas, J. D., Li, C. H. 1980. Binding of [³H]beta-endorphin to rat brain membranes: Characterization of opiate properties and interaction with ACTH. *Eur. J. Pharmacol.* 64:1–8

Akil, H., Hughes, J., Richardson, D. E., Barchas, J. D. 1978a. Enkephalin-like material elevated in ventricular cerebrospinal fluid in pain patients after analgetic focal stimulation. *Science* 201:463–65

Akil, H., Madden, J., Patrick, R. L., Barchas, J. D. 1976a. Stress-induced increase in endogenous opiate peptides: Concurrent analgesia and its partial reversal by naloxone. In *Opiates and Endogenous Opioid Peptides*, ed. H. Kosterlitz, pp. 63–70. Amsterdam: Elsevier

Akil, H., Mayer, D. J., Liebeskind, J. C. 1972. Comparaison chez le rat entre l'analgesie iduite par stimulation de la substance grise peri-aqueducale et l'analgesie morphinique. *CR Acad. Sci. Paris* 274:3603–5

Akil, H., Mayer, D. J., Liebeskind, J. C. 1976b. Antagonism of stimulation-produced analgesia by naloxone, a narcotic antagonist. *Science* 191:961–62

Akil, H., Richardson, D. E., Barchas, J. D., Li, C. H. 1978b. Appearance of beta-endorphin-like immunoreactivity in human ventricular cerebrospinal fluid upon analgesic electrical stimulation. *Proc. Natl. Acad. Sci. USA* 75:5170–72

Akil, H., Ueda, Y., Lin, H.-S. 1981a. A sensitive coupled HPLC/RIA technique for separation of endorphins: Multiple forms of beta-endorphin in rat pituitary intermediate vs. anterior lobe. *Neuropeptides* 1:429–46

Akil, H., Ueda, Y., Lin, H.-L., Lewis, J. W., Walker, J. M., Shiomi, H., Liebeskind, J. C., Watson, S. J. 1981b. Multiple forms of beta-endorphin in pituitary and brain: Effect of stress. In *Advances in Endogenous and Exogenous Opioid Peptides*, ed. H. Takagi, E. Simon, pp. 116–18. Tokyo: Kodansha

Akil, H., Watson, S. J., Berger, P. A., Barchas, J. D. 1978c. Endorphins, beta-LPH, and ACTH: Biochemical, and anatomical studies. *Adv. Biochem. Psychopharmacol.: The Endorphins* 18:125–39

Akil, H., Young, E., Watson, S. J., Coy, D. 1981c. Opiate binding properties of naturally occurring N- and C-terminus modified beta-endorphin. *Peptides* 2:289–92

Alessi, N., Taylor, L., Akil, H. 1982. Peptide'

F (proenkephalin fragment): Radioimmunoassay, and stress-induced changes in adrenal. *Life Sci.* 31:1875–78

Amano, K., Tanikawa, T., Kawamura, H., Iseki, H., Notani, M., Kawabatake, H., Shiwaku, T., Suda, T., Demura, H., Kitamura, K. 1982. Endorphins and pain relief. Further observations on electrical stimulation of the lateral part of the periaqueductal gray matter during rostral mesencephalic reticulotomy for pain relief. *Appl. Neurophysiol.* 45:123–35

Amir, S., Berstein, M. 1982. Endogenous opioids interact in stress-induced hyperglycemia in mice. *Physiol. Behav.* 28(3):575–77

Atweh, S., Kuhar, M. J. 1977. Autoradiographic localization of opiate receptors in rat brain. I. Spinal cord and lower medulla. *Brain Res.* 129:1–12

Barrett, R. W., Vaught, J. L. 1982. The effects of receptor selective opioid peptides on morphine-induced analgesia. *Eur. J. Pharmacol.* 80:427–30

Basbaum, A. I., Marley, N. J., O'Keefe, J., Clanton, C. H. 1977. Reversal of morphine and stimulus produced analgesia by subtotal spinal cord lesions. *Pain* 3:43–56

Belcher, G., Smock, T., Fields, H. L. 1982. Effects of intrathecal ACTH on opiate analgesia in the rat. *Brain Res.* 247:373–77

Bloch, B., Bugnon, C., Fellman, D., Lenys, D. 1978. Immunocytochemical evidence that the same neurons in the human infundibular nucleus are stained with antiendorphins and antisera of other related peptides. *Neurosci. Lett.* 10:147–52

Bloom, F. E., Battenberg, E., Rossier, J., Ling, N., Leppaluoto, J., Vargo, T. M., Guillemin, R. 1977. Endorphins are located in the intermediate and anterior lobes of the pituitary gland, not in the neuropophysis. *Life Sci.* 20:43–48

Bloom, F. E., Rossier, J., Battenberg, E., Bayon, A., French, E., Henricksen, S. J., Siggins, G. R., Segal, D., Browne, R., Ling, N., Guillemin, R. 1978. Betaendorphin: Cellular localization, electrophysiological and behavioral effects. *Adv. Biochem. Psychopharmacol.: The Endorphins,* 18:89–109

Bodnar, R. J., Kelly, D. D., Spiaggia, A., Ehrenberg, C., Glusman, M. 1978a. Dose-dependent reductions by naloxone of analgesia induced by cold-water stress. *Pharmacol. Biochem. Behav.*8:661–66

Bodnar, R. J., Kelly, D. D., Spiaggia, A., Glusman, M. 1978b. Biphasic alterations of nociceptive thresholds induced by food deprivation. *Physiol. Psychol.* 6:391–95

Bonnet, K. A., Grotin, J., Gioamnini, T., Cortes, M., Simon, E. J. 1981. Opiate receptor heterogeneity in human brain regions. *Brain Res.* 221:437–40

Bowen, W. D., Gentleman, S., Herkenham, M., Pert, C. B. 1981. Interconverting mu and delta forms of the opiate receptor in rat striatal patches. *Proc. Natl. Acad. Sci. USA* 78:4818–22

Bradbury, A. F., Feldberg, W. F., Smyth, D. G., Snell, C. 1976. Lipotropin C-fragment: An endogenous peptide with potent analgesic activity. See Akil 1976a, pp. 9–17

Cahill, C. A., Akil, H. 1982. Plasma beta-endorphin-like immunoreactivity, self reported pain perception and anxiety levels in women during pregnancy and labor. *Life Sci.* 31:1879–82

Chance, W. T. 1980. Autoanalgesia: Opiate and non-opiate mechanisms. *Neurosci. Biobehav. Rev.* 4:55–67

Chang, K.-J., Cooper, B. R., Hazum, E., Cuatrecasas, P. 1979. Multiple opiate receptors: Different regional distribution in the brain and differential binding of opiates and opioid peptides. *Mol. Pharmacol.* 16:91–104

Chang, K.-J., Cuatrecasas, P. 1979. Multiple opiate receptors: Enkephalins and morphine bind to receptors of different specificity. *J. Biol. Chem.* 254:2610–18

Chang, K.-J., Hazum, E., Cuatrecassas, P. 1981. Novel opiate binding sites for benzomorphan drugs. *Proc. Natl. Acad. Sci. USA* 78:4141–45

Chavkin, C., James, I. F., Goldstein, A. 1982. Dynorphin is a specific endogenous ligand of the kappa opioid receptor. *Science* 215:413–15

Chipkin, R. E., Latranyi, M. B., Iorio, L. C. 1982. Potentiation of stress-induced analgesia (SIA) by thiorphan and its block by naloxone. *Life Sci.* 31(12–13):1189–92

Chretien, M., Benjannet, S., Dragon, N., Seidah, N. G., Lis, M. 1976. Isolation of peptides with opiate activity from sheep and human pituitaries: Relationship to beta-LPH. *Biochem. Biophys. Res. Commun.* 72:472–78

Comb, M., Herbert, E., Crea, R. 1982. Partial characterization of the mRNA that codes for enkephalins in bovine adrenal medulla and human pheochromocytoma. *Proc. Natl. Acad. Sci. USA* 79:360–64

Corbett, A. D., Patterson, S. J., McKnight, A. T., Magnan, J., Kosterlitz, H. W. 1982. Dynorphin-(1–8) and dynorphin-(1–9) are ligands for the kappa subtype of opiate receptor. *Nature* 299:79–81

Costa, E., Guidotti, A., Hanbauer, I., Hexum, T., Saiani, L., Stien, S., Yang, H-Y. T. 1981. Regulation of acetylcholine receptors by endogenous cotransmitters: Studies of adrenal medulla. *Fed. Proc. Fed. Am. Soc. Exp. Biol.* 40:160–65

Crowley, W. R., Rodriguez-Sierra, J. F., Komisaruk, B. R. 1977. Analgesia induced by vaginal stimulation in rats is apparently independent of a morphine-sensitive process. *Psychopharmacology* 54:223–25

Datta, P. C., Sandman, C. A., Hoehler, F. K. 1982. Attentuation of morphine analgesia by alpha-MSH, MIF-I, melatonin and naloxone in the rat. *Peptides* 3:433–37

DiGiulio, A. M., Yang, H-Y. T., Lutold, B., Fratta, W., Hong, J., Costa, E. 1978. Characterization of enkephalin-like material extracted from sympathetic ganglia. *Neuropharmacology* 17:989–92

Drouin, J., Goodman, H. M. 1980. Most of the coding region of rat ACTH beta-LPH precursor gene lacks intervening sequences. *Nature* 288:610–13

Drugan, R. C., Grau, J. W., Maier, S. F., Madden, J., Barchas, J. D. 1981. Cross tolerance between morphine and the long-term analgesic reaction to inescapable shock. *Pharmacol. Biochem. Behav.* 14:677–82

Dubois, M., Pickar, D., Cohen, M. R., Roth, Y. F., MacNamara, T., Bunney, W. E. Jr. 1981. Surgical stress in humans is accompanied by an increase in plasma beta-endorphin immunoreactivity. *Life Sci.* 29:1249–54

Duka, T., Schubert, P., Wuster, M., Stoiber, R., Herz, A. 1981a. A selective distribution pattern of different opiate receptors in certain areas of rat brain as revealed by *in vitro* autoradiography. *Neurosci. Lett.* 21:119–24

Duka, T., Wuster, M., Schubert, P., Stoiber, R., Herz, A. 1981b. Selective localization of different types of opiate receptors in hippocampus as revealed by *in vitro* autoradiography. *Brain Res* 205:181–86

Eipper, B., Mains, R. 1978. Existence of a common precursor to ACTH and endorphin in the anterior and intermediate lobes of the rat pituitary. *J. Supramol. Struct.* 8: 247–62

Eipper, B. A., Mains, R. E. 1980. Structure and function of preadrenocorticotropin/endorphin and related peptides. *Endocr. Rev.* 1:247–62

Elde, R., Hökfelt, T., Johansson, O., Terenius, L. 1976. Immunohistochemical studies using antibodies to leucine enkephalin: Initial observations on the nervous system of the rat. *Neuroscience* 1:349–51

Evans, C. J., Lorenz, R., Weber, E., Barchas, J. D. 1982. Variants of alpha-melanocyte stimulating hormone in rat brain and pituitary: Evidence that acetylated alpha-MSH exists only in the intermediate lobe of the pituitary. *Biochem. Biophys. Res. Commun.* 106:910–19

Faden, A. I., Holaday, J. W. 1979. Opiate

antagonists: A role in the treatment of hypovolemic shock. *Science* 205:317–18

Faden, A. I., Holaday, J. W. 1980. Naloxone treatment of endotoxin shock: Stereospecificity of physiological and pharmacologic effects in rat. *J. Pharmacol. Exp. Ther.* 212:441–47

Faden, A. I., Holaday, J. W. 1981. A role for endorphins in the pathophysiology of spinal cord injury. *Adv. Biochem. Psychopharmacol.* 28:435–46

Feldberg, W., Wei, E. 1978. Central cardiovascular effects of enkephalins and C-fragment of lipotropin. *J. Physiol.* 280:18P

Fischli, W., Goldstein, A., Hunkapiller, M. W., Hood, L. E. 1982. Two "big" dynorphins from procine pituitary. *Life Sci.* 31:1769–72

Florez, J., Mediavilla, A. 1977. Respiratory and cardiovascular effects of [Met]enkephalin applied to the ventral surface of the brain stem. *Brain Res.* 138:585–90

Gacel, G., Fourni'e-Zaluski, M. D., Fellion, E., Roques, B. P. 1981. Evidence of the preferential involvement of mu receptors in analgesia using enkephalins highly selective for peripheral mu or delta receptors. *J. Med. Chem.* 24:1119–24

Gibson, A., Hart, S. L., Shabib, A. 1980. Leucine-enkephalin and methionine-enkephalin produce opposing effects on plasma corticosterone levels in ether-stressed mice. *Br. J. Pharmacol.* 70:509–11

Goldstein, A., Fischli, W., Lowney, L. I., Hunkapiller, M., Hood, L. 1981. porcine pituitary dynorphin: Complete amino acid sequence of the biologically active heptadecapeptide. *Proc. Natl. Acad. Sci. USA* 78:7219–23

Goldstein, A., Tachibana, S., Lowney, L. I., Hunkapiller, M., Hood, L. 1979. Dynorphin-(1–13), an extraordinarily potent opioid peptide. *Proc. Natl. Acad. Sci. USA* 76: 6666–70

Goodman, R. R., Snyder, S. 1982. Opiate receptors localized by autoradiography to deep layers of cerebral cortex: Relation to sedative effects. *Proc. Natl. Acad. Sci. USA* 79: 5703–7

Goodman, R. R., Snyder, S. H., Kuhar, M. J., Young, W. S. III 1980. Differentiation of delta and mu opiate receptor localizations by light microscopic autoradiography. *Proc. Natl. Acad. Sci. USA* 77:6239–43

Gubler, U., Kilpatrick, D. L., Seeburg, P. H., Gage, L. P., Udenfriend, S. 1981. Detection and partial characterization of proenkephalin mRNA. *Proc. Natl. Acad. Sci. USA* 78:5484–87

Guillemin, R., Ling, N., Burgus, R. 1976. Endorphins, peptides d'origine hypothalamique et neurohypophysaire d'activite morphinomimetique. Isolement et structure moleculaire d'alpha-endorphine. *CR Acad. Sci. Ser.* D 282:783–85

Guillemin, R., Vargo, T., Rossier, J., Minick, S., Ling, N., Rivier, C., Vale, W., Bloom, F. 1977. Beta-endorphin and adrenocorticotropin are secreted concommitantly by the pituitary gland. *Science* 197:1367–69

Hanbauer, I., Govoni, F., Majane, E., Yang, H-T. Y., Costa, E. 1982. *In vivo* regulation of the release of [Met]enkephalin-like peptides from dog adrenal medulla. *Adv. Biochem. Psychopharmacol.* 33:63–69

Hayes, R. L., Bennett, G. J., Newlon, P. G., Mayer, D. J. 1978. Behavioral and physiological studies of non-narcotic analgesia in the rat elicited by certain environmental stimuli. *Brain Res.* 166:69–90

Herkenham, M., Pert, C. B. 1980. *In vitro* autoradiography of opiate receptors in rat brain suggests loci of "opiatergic" pathways. *Proc. Natl. Acad. Sci. USA* 77:5532–36

Herkenham, M., Pert, C. B. 1982. Light microscopic localization of brain opiate receptors: A general autoradiographic method which preserves tissue quality. *J. Neurosci.* 8:1129–49

Hökfelt, T., Elde, R., Johansson, O., Terenius, L., Stein, L. 1977a. The distribution of enkephalin-immunoreactive cell bodies in the rat central nervous system. *Neurosci. Lett.* 5:25–31

Hökfelt, T., Ljungdahl, A., Terenius, L., Elde, R., Nilsson, C. 1977b. Immunohistochemical analysis of peptide pathways possibly related to pain and analgesia: Enkephalin and substance P. *Proc. Natl. Acad. Sci. USA* 74:3081–85

Holaday, J. W. 1983. Cardiovascular effects of endogenous opiate systems. *Ann. Rev. Pharmacol. Toxicol.* 23:541–94

Holaday, J. W., Faden, A. I. 1978. Naloxone reversal of endotoxin hypotension suggests role of endorphins in shock. *Nature* 275:450–51

Holaday, J. W., Faden, A. I. 1980. Naloxone acts at central opiate receptors to reverse hypotension, hypothermia and hypoventilation in spinal shock. *Brain Res.* 189:295–99

Holaday, J. W., Faden, A. I. 1982. Selective cardiorespiratory differences between third and fourth ventricular injections of "mu" and "delta" opiate agonists. *Fed. Proc.* 41:1468

Holaday, J. W., Ruvio, B. A., Robles, L. E., Johnson, D. E., D'Amato, R. J. 1982. M154,129, a putative delta antagonist, reverses endotoxic shock without altering morphine analgesia. *Life Sci.* 31:2359–62

Hosobuchi, Y., Rossier, J., Bloom, F., Guillemin, R. 1979. Electrical stimulation of periaqueductal gray for pain relief in humans is accompanied by elevation of immunoreac-

tive beta-endorphin in ventricular fluid. *Science* 203:279–81

Hughes, J., Smith, T. W., Kosterlitz, H. W., Fothergill, L. A., Morgan, B. A., Morris, H. R. 1975. Identification of two related pentapeptides from the brain with potent opiate agonist activity. *Nature* 258:577–79

Hynes, M. D., Frederickson, R. C. 1982. Cross-tolerance studies distinguish morphine- and metkephamid-induced analgesia. *Life Sci.* 31(12–13):1201–4

Isom, G. E., Elshowihy, R. M. 1982. Interaction of acute and chronic stress with respiration: Modification by naloxone. *Pharmacol. Biochem. Behav.* 16(4):599–603

Jackson, R. L., Maier, S. F., Coon, D. J. 1979. Long-term analgesic effects of inescapable shock and learned helplessness. *Science* 206:91–93

Jacob, J. J., Tremblay, E. C., Colombel, M.-C. 1974. Facilitation de reactions nociceptives par la naloxone chez la souris et chez le rat. *Psychopharmacologia* 37:217–23

Jacobowitz, D. M., O'Donohue, T. L. 1978. Alpha-MSH stimulating hormone: Immunohistochemical identificaton and mapping in neurons of rat brain. *Proc. Natl. Acad. Sci. USA* 75:6300–4

James, I. F., Chavkin, C., Goldstein, A. 1982. Preparation of brain membranes containing a single type of opioid receptor highly selective for dynorphin. *Proc. Natl. Acad. Sci. USA* 79:7570–74

Janssen, H. F., Lutherer, L. O. 1980. Ventriculocisternal administration of naloxone protects against severe hypotension during endotoxin shock. *Brain Res.* 194:608–12

Kakidani, H., Furutani, Y., Takahashi, H., Noda, M., Morimoto, Y., Hirose, T., Asai, M., Inayama, S., Nakanishi, S., Numa, S. 1982. Cloning and sequence analysis of cDNA for porcine beta-neo-endorphin/ dynorphin precursor. *Nature* 298:245–49

Kangawa, K., Minamino, N., Chino, N., Sakakibara, S., Matsuo, H. 1981. The complete amino acid sequence of alpha-neoendorphin. *Biochem. Biophys. Res. Commun.* 99:871–78

Khachaturian, H., Lewis, M. E., Hollt, V., Watson, S. J. 1983. Telencephalic enkephalinergic systems in the rat brain. *J. Neurosci.* 3:844–55

Khachaturian, H., Lewis, M. E., Watson, S. J. 1982a. Immunocytochemical studies with antisera against [Leu]enkephalin and an enkephalin-precursor fragment (BAM-22P) in the rat brain. *Life Sci.* 31:1879–82

Khachaturian, H., Watson, S. J., Lewis, M. E., Coy, D., Goldstein, A. 1982b. Dynorphin immunocytochemistry in the rat central nervous system. *Peptides* 3:941–54

Kilpatrick, D. L., Wahlstrom, A., Lahm, H.

W., Blacker, R., Udenfriend, S. 1982. Rimorphin, a unique, naturally occuring [Leu]enkephalin-containing peptide found in association with dynorphin and alpha-neoendorphin. *Proc. Natl. Acad. Sci. USA* 79:6480–83

Kimura, S., Lewis, R. V., Stern, A. S., Rossier, J., Stein, S. Udenfriend, S. 1980. Probable precursors of [Leu] and [Met]enkephalin in adrenal medulla: Peptides of 3–5 kilodaltons. *Proc. Natl. Acad. Sci. USA* 77:1681–85

Kita, T., Inoue, A., Nakanishi, S., Numa, S. 1979. Purificaiton and characterization of the messenger RNA coding for bovine corticotropin/beta-lipotropoin precursor. *Eur. J. Biochem.* 93:213–20

Kovacs, G., Bohus, B., Dewied, D. 1981. Retention of passive avoidance behavior in rats following alpha- and gamma-endorphin administration: Effects of postlearning treatment. *Neurosci. Lett.* 22:79–82

Krieger, D. T., Liotta, A., Brownstein, M. J. 1977. Presence of corticotropin in brain of normal and hypophysectomized rats. *Proc. Natl. Acad. Sci. USA* 74:648–52

Laubie, M., Schmitt, H., Vincent, M., Remond, G. 1977. Central cardiovascular effects of morphinomimetic peptides in dogs. *Eur. J. Pharmacol.* 46:67–71

Leander, J. D. 1982. A kappa opioid effects increased urination in the rat. *J. Pharmacol. Exp. Ther.* 224:89–94

Lee, N. M., Leybin, L., Chang, J. K., Loh, H. H. 1980. Opiate and peptide interaction: Effect of enkephalins on morphine analgesia. *Eur. J. Pharmacol.* 68:181–85

Lemaire, I., Tseng, R., Lemaire, S. 1978. Systematic administration of beta-endorphin: Potent hypotensive effect involving a serotonergic pathway. *Proc. Natl. Acad. Sci. USA* 75:6240–42

Lewis, J. W., Cannon, J. T., Liebeskind, J. C. 1980. Opioid and nonopioid mechanism of stress analgesia. *Science* 208:623–25

Lewis, J. W., Chudler, E. H., Cannon, J. T., Liebeskind, J. C. 1981a. Hypophysectomy differentially affects morphine and stress analgesia. *Proc. West. Pharmacol. Soc.* 24:323–26

Lewis, J. W., Sherman, J. E., Liebeskind, J. C. 1981b. Opioid and non-opioid stress analgesia: Assessment of tolerance and cross-tolerance with morphine. *J. Neurosci.* 1:358–63

Lewis, J. W., Tordoff, M. G., Sherman, J. E., Liebeskind, J. D. 1982. Adrenal medullary enkephalin-like peptides may mediate opioid stress analgesia. *Science* 217:557–59

Lewis, M. E., Khachaturian, H., Watson, S. J. 1982. Visualization of opiate receptors and opioid peptides in sequential brain sections. *Life Sci.* 31:1247–50

Lewis, M. E., Mishkin, M., Bragin, E., Brown, R. M., Pert, C. B., Pert, A. 1981. Opiate receptor gradients in monkey cerebral cortex: Correspondence with sensory processing hierarchies. *Science* 211:1166–69

Lewis, M. E., Pert, A., Pert, C. B., Herkenham, M. 1983. Opiate receptor localization in rat cerebral cortex. *J. Comp. Neurol.* 216:339–58

Li, C. H., Chung, D. 1976. Isolation and structure of a triakontapeptide with opiate activity from camel pituitary glands. *Proc. Natl. Acad. Sci. USA* 73:1145–48

Lim, A. T., Wallace, M., Oei, T. P., Gibson, S., Romas, N., Pappas, W., Clements, J., Funder, J. W. 1982. Footshock analgesia. Lack of correlation with pituitary and plasma immunoreactive-beta-endorphin. *Neuroendocrinology* 35:236–41

Ling, N., Ying, S., Minick, S., Guillamin, R. 1979. Synthesis and biological activity of four gamma-melanotropins derived from the cryptic region of the adrenocorticotropin/beta-lipotropin precursor. *Life Sci.* 25:1773–80

Liston, D. R., Vanderhaeghen, J.-J., Rossier, J. 1983. Presence in brain of synenkephalin, a proenkephalin-immunoreactive protein which does not contain enkephalin. *Nature* 302:62–65

Livett, B. G., Dean, D. M., Whelan, L. G., Udenfriend, S., Rossier, J. 1981. Co-release of enkephalin and catecholamines from cultured arendal chromaffin cells. *Nature* 289:317–19

Loh, Y. P., Eskay, R. L., Brownstein, M. 1980. Alpha-MSH-like peptides in rat brain: Identification and changes during development. *Biochem. Biophys. Res. Commun.* 94:916–23

Lord, J. A. H., Waterfield, A. A., Hughes, J., Kosterlitz, H. W. 1977. Endogenous opioid peptides: Multiple agonists and receptors. *Nature* 267:495

MacLennan, A. J., Drugan, R. C., Hyson, R. L., Maier, S. F., Madden, J., Barchas, J. D. 1982. Corticosterone: A critical factor in an opioid form of stress-induced analgesia. *Science* 215:1530–32

Madden, J., Akil, H., Patrick, R. L., Barchas, J. D. 1977. Stress-induced parallel changes in central opioid levels and pain responsiveness in the rat. *Nature* 265:358–60

Mains, R. E., Eipper, B. A., Ling, N. 1977. Common precursor to corticotropins and endorphins. *Proc. Natl. Acad. Sci. USA* 74:3014–18

Maixner, W., Touw, K. B., Brody, M. J., Gebhart, G. F., Long, J. P. 1982. Factors influencing the altered pain perception in the spontaneously hypertensive rat. *Brain Res.* 237(1):137–45

Martin, W. R., Eades, C. G., Thompson, J. A., Huppler, R. E., Gilbert, P. E. 1976. The effects of morphine and nalorphine-like drugs in nondependent and morphine-dependent chronic spinal dog. *J. Pharmacol. Exp. Ther.* 197:517–32

Mayer, D. J., Hayes, R. L. 1975. Stimulation-produced analgesia: Development of tolerance and cross-tolerance to morphine. *Science* 188:941–43

Mayer, D. J., Price, D. D., Rafii, A., Barber, J. 1976. Acupuncture hypalgesia: Evidence for activation of a central control system as a mechanism of action. *Adv. Pain Res. Ther.* 1:751–54

Mayer, D. J., Wolfle, T. L., Akil, H., Carder, B., Liebeskind, J. C. 1971. Analgesia from electrical stimulation in the brainstem of the rat. *Science* 174:1351–54

Maysinger, D., Hollt, V., Seizinger, B. R., Mehraein, P., Pasi, A., Herz, A. 1982. Parallel distribution of immunoreactive alpha-neo-endorphin and dynorphin in rat and human tissue. *Neuropeptides* 2:211–25

McGivern, R. F., Berka, C., Berntson, G. G., Walker, J. M., Sandman, C. A. 1979. Effect of naloxone on analgesia induced by food deprivation. *Life Sci.* 25:885–88

Millan, M. J., Przewlocki, R., Herz, A. 1980. A non-beta-endorphinergic adenohypophyseal mechanism is essential for an analgetic response to stress. *Pain* 8:343–53

Millan, M. J., Przewlocki, R., Jerlicz, M., Gramsch, C., Hollt, V., Herz, A. 1981a. Stress-induced release of brain and pituitary beta-endorphin: Major role of endorphins in generation of hyperthemia, not analgesia. *Brain Res.* 208:325–38

Millan, M. J., Tsang, Y. F., Przewlocki, R., Hollt, V., Herz, A. 1981b. The influence of footshock stress upon brain, pituitary and spinal cord pools of immunoreactive dynorphin in rats. *Neurosci. Lett.* 24:75–79

Mizuno, K., Minamino, N., Kangawa, K., Matsuo, H. 1980. A new family of endogenous "big" [Met]enkephalins from bovine adrenal medulla: Purification and structure of docosa-(BAM22P) and eicosapeptide (BAM20P) with very potent opiate activity. *Biochem. Biophys. Res. Commun.* 97:1283–90

Moon, H. D., Li, C. H., Jennings, B. M. 1973. Immunohistochemical and histochemical studies of pituitary beta-LPH. *Anat. Rec.* 175:524–38

Mues, G., Fuchs, I., Wei, E. T., Weber, E., Evans, C. J., Barchas, J. D. Chang, J.-K. 1982. Blood pressure elevation in rats by peripheral administration of Tyr-Gly-Gly-Phe-Met-Arg-Phe and the invertebrate

neuropeptide Phe-Met-Arg-Phe amide. *Life Sci.* 31:2555–61

Nakanishi, S., Inoue, A., Kita, T., Nakamura, M., Chang, A. C. Y., Cohen, S. N., Numa, S. 1979. Nucleotide sequence of cloned cDNA for bovine corticotropin-beta-lipotropin precursor. *Nature* 278:423–27

Nakanishi, S., Teranishi, Y., Watanabe, Y., Notake, M., Noda, M., Kakidani, H., Jingami, H., Numa, S. 1981. Isolation and characterization of the bovine corticotropin/beta-lipotropin precursor gene. *Eur. J. Biochem.* 115:429–38

Nilaver, G., Zimmerman, E. A., Defendini, R., Liotta, A., Krieger, D. A., Brownstein, M. 1979. Adrenocortioctropin and beta-LPH in hypothalamus. *J. Cell Biol.* 81:50–58

Ninkovic, M., Hunt, S. P., Emson, P., Iversen, L. L. 1981. The distribution of multiple opiate receptors in bovine brain. *Brain Res.* 214:163–67

Noda, M., Furutani, Y., Takahashi, H., Toyosato, M., Hirose, T., Inayama, S., Nakanishi, S., Numa, S. 1982a. Cloning and sequence analysis of cDNA for bovine adrenal preproenkephalin. *Nature* 295:202–6

Noda, M., Teranishi, Y., Takahashi, H., Toyosato, M., Notake, M., Nakanishi, S., Numa, S. 1982b. Isolation and structural organization of the human preproenkephalin gene. *Nature* 297:431

O'Donohue, T. L., Charlton, C. G., Helke, C. J., Miller, R. L., Jacobowitz, D. M. 1979. Identification of alpha-MSH immunoreactivity in rat and human brain and CSF. In *Peptides Structure and Biological Functioning,* ed. E. Gross, J. Meinhoffer, pp. 897–900. Pierce Chem. Co.

Oka, T., Negishi, K., Suda, M., Matsumiya, T., Inazu, T., Ueki, M. 1980. Rabbit vas deferens: A specific bioassay for opioid kappa-receptor agonists. *Eur. J. Pharmacol.* 73:235–36

Olivero, A., Castellano, C. 1982. Classical conditioning of stress-induced analgesia. *Physiol. Behav.* 29:171–72

Pasternak, G. W. 1981. Opiate, enkephalin, and endorphin analgesia: Relations to a single subpopulation of opiate receptors. *Neurology* 31:1311–15

Pedersen, R. C., Brownie, A. C., Ling, N. 1980. Pro-adrenocorticotropin/endorphin-derived peptides: Coordinate action on adrenal steroidogenesis. *Science* 208:1044–46

Pelletier, G. 1980. Ultrastructural localization of a fragment (16K) of the common precursor for adrenocorticotropin and beta-LPH in the rat hypothalamus. *Neurosci. Lett.* 16:85–90

Pelletier, G., Leclerc, R., LaBrie, F., Cote, J.,

Chretien, M., Lis, M. 1977. Immunohistochemical localization of beta-LPH hormone in the pituitary gland. *Endocrinology* 100:770–76

Pert, C. B., Kuhar, M. J., Snyder, S. H. 1975. Autoradiographic localization of the opiate receptor in the rat brain. *Life Sci.* 16:1849–54

Pert, C. B., Kuhar, M. J., Snyder, S. H. 1976. The opiate receptor: Autoradiographic localization in rat brain. *Proc. Natl. Acad. Sci. USA* 73:3729–33

Pert, C. B., Taylor, D. 1980. Type 1 and type 2 opiate receptors: A subclassification scheme based upon GTP's differential effects on binding. In *Endogenous and Exogenous Opiate Agonists and Antagonists*, ed. E. L. Way, pp. 87–90. New York: Pergamon

Pickel, V. M., Sumal, K. K., Beckley, S. C., Miller, R. J., Reis, D. J. 1980. Immunocytochemical localization of enkephalin in the neostriatum of rat brain: A light and electron microscopic study. *J. Comp. Neurol.* 189:721–40

Piercey, M. F., Lahti, R. A. Schroeder, L. A., Einspahr, F. J., Barsuhn, C. 1982. U-50488H, a pure kappa receptor agonist with spinal analgesic loci in the mouse. *Life Sci.* 31(12–13):1197–1200

Quirion, R., Bowen, W., Herkenham, M., Pert, C. B. 1982. Visualization and solubilization of rat brain opiate receptors with a "kappa" ligand selectivity pattern. *Cell. Mol. Neurobiol.* 2:333–345

Richardson, D. E., Akil, H. 1977. Pain reduction by electrical brain stimulation in man (part 2): Chronic self-administration in the periventricular gray matter. *J. Neurosurg.* 47:184–94

Roberts, J. L., Herbert, H. 1977a. Characterization of a common precursor to corticotropin and beta-lipotropin: Cell-free synthesis of the precursor and identification of corticotropin peptides in the molecule. *Proc. Natl. Acad. Sci. USA* 74:4826–30

Roberts, J. L., Herbert, J. 1977b. Characterization of a common precursor to corticotropin and beta-lipotropin: Identification of beta-lipotropin peptides and their arrangement relative to corticotropin in the precursor synthesized in a cell-free system. *Proc. Natl. Acad. Sci. USA* 74:5300–4

Roberts, J. L., Seeburg, P. H., Shine, J., Herbert, E., Baxter, J. D., Goodman, H. M. 1979. Corticotropin and beta-endorphin: Construction and analysis of recombinant DNA complementary to mRNA for the common precursor. *Proc. Natl. Acad. Sci. USA* 76:2153–57

Roques, B. P., Fourni'e-Zaluski, M. C., Soroca, E., Lecomte, J. M., Malafroy, B., Llorens, C., Schwartz, J. C. 1980. The en-

kephalinase inhibitor thiorphan shown antinociceptive activity in mice. *Nature* 288:286–88

Rossier, J., French, E. D., Rivier, C., Ling, N., Guillemin, R., Bloom, F. E. 1977. Footshock induced stress increases beta-endorphin levels in blood but not brain. *Nature* 270:618–20

Sander, G. E., Giles, T. E. 1982. Des-Tyr¹-d-Ala²-leucine-enkephalinamide elevates arterial blood pressure in conscious dogs. *Biochem. Pharmacol.* 31:2699–2700

Sar, M., Stumpf, W. E., Miller, R. J., Chang, K.-J., Cuatrecasas, P. 1978. Immunohistochemical localization of enkephalin in rat brain and spinal cord. *J. Comp. Neurol.* 182:17–37

Saria, A., Wilson, S. P., Molnar, A., Viveros, O. H., Lembeck, F. 1980. Substance P and opiate-like peptides in human adrenal medulla. *Neurosci. Lett.* 20:195–200

Schulz, R., Faase, E., Wuster, M., Herz, A. 1979. Selective receptors for beta-endorphin on the rat *vas deferens*. *Life Sci.* 24:843–50

Schultzberg, M., Hökfelt, T., Lundberg, J., Terenius, L., Elfirm, L.-G., Elde, R. 1978. Enkephalin-like immunoreactivity in nerve terminals in sympathetic ganglia and adrenal medulla and in adrenal medullary gland cells. *Acta Physiol. Scand.* 103:475–77

Schwartzberg, D. G., Nakone, P. K. 1981. *Pro-ACTH/endorphin antigenicities in medullary neurons of the rat.* Presented at Ann. Meet. Soc. Neurosci., 11th, Los Angeles, Calif.

Schwyzer, R., Eberle, A. 1977. On the molecular mechanism of alpha-MSH receptor interactions. *Front. Hormone Res.* 4:18–25

Seizinger, B. R., Hollt, V., Herz, A. 1981a. Evidence of the occurrence of the opioid octapeptide dynorphin(1–8) in the neurointermediate pituitary of rats. *Bioch. Biophys. Res. Commun.* 102:197–205

Seizinger, B. R., Hollt, V., Herz, A. 1981b. Immunoreactive dynorphin in the rat adrenohypophysis consists exclusively of 6000-dalton species. *Biochem. Biophys. Res. Commun.* 103:256–63

Seizinger, B. R., Hollt, V., Herz, A. 1982. Dynorphin-related opioid peptides in the neurointermediate pituitary of rats are not alpha-N-acetylated. *J. Neurochem.* 39:143–48

Shiomi, H., Akil, H. 1982. Pulse-chase studies of the POMC/beta-endorphin system in the pituitary of acutely and chronically stressed rats. *Life Sci.* 31:2271–73

Smith, J. R., Simon, E. J. 1980. Selective protection of stereospecific enkephalin and opiate binding against inactivation by N-ethylmaleimide: Evidence for two classes of opiate receptors. *Proc. Natl. Acad. Sci. USA* 77:281–84

Suda, T., Tozawa, F., Tachibana, S., Demura, H., Shizume, K. 1982. Multiple forms of immunoreactive dynorphin in rat pituitary and brain. *Life Sci.* 31:51–57

Terenius, L. 1982. Endorphins and modulation of pain. *Adv. Neurol.* 33:59–64

Thomas, T. A., Fletcher, J. E., Hill, R. G. 1982. Influence of medication, pain and progress in labour on plasma beta-endorphin-like immunoreactivity. *Br. J. Anaesthesiol.* 54:401–8

Tsukada, T., Nakai, Y., Jingami, H., Imura, H., Taii, S., Nakanishi, S., Numa, S. 1981. Identification of the mRNA coding for the ACTH-beta-lipotropoin precursor in human ectopic ACTH-producing tumor. *Biochem. Biophys. Res. Commun.* 98:535–40

Tulunay, F. C., Jen, M. F., Chang, J. K., Loh, H. H., Lee, N. M. 1981. Possible regulatory role of dynorphin-(1–13) on narcotic-induced changes in naloxone efficacy. *Eur. J. Pharmacol.* 76:235–39

Tung, A. S., Yaksh, T. L. 1982. In vivo evidence for multiple opiate receptors mediating analgesia in the rat spinal cord. *Brain Res.* 247(1):75–83

Uhl, G. R., Goodman, R. R., Kuhar, M. J., Childers, S. R., Snyder, S. H. 1979. Immunocytochemical mapping of enkephalin cotaining cell bodies, fibers and nerve terminals in the brain stem of the rat. *Brain Res.* 116:75–94

Uhl, G. R., Kuhar, M. J., Snyder, S. H. 1978. Enkephalin-containing pathway: Amygdaloid efferents in the stria terminalis. *Brain Res.* 149:223–28

Van Loon, G. R., Appel, N. M., Ho, D. 1981. Beta-endorphin-induced increases in plasma epinephrine, norepinephrine and dopamine in rats: Inhibition of adrenal medullary response by intracerebral somatostatin. *Brain Res.* 212:207–14

Vincent, S. R., Hökfelt, T., Christensson, I., Terenius, L. 1982. Dynorphin-immunoreactive neurons in the central neurons system of the rat. *Neurosci. Lett.* 35:185–90

Viveros, O. H., Diliberto, E. J. Jr., Hazum, E., Chang, K.-J. 1979. Opiate-like materials in the adrenal medulla: Evidence for storage and secretion with catecholamines. *Mol. Pharmacol.* 16:1101–8

Walker, J. M., Akil, H., Watson, S. J. 1980a. Evidence for homologous actions of pro-opiocortin products. *Science* 210:1247–49

Walker, J. M., Katz, R. J., Akil, H. 1980b. Behavioral effects of dynorphin-(1–13) in the mouse and rat: Initial observations. *Peptides* 1:341–45

Walker, J. M., Moises, H. C., Coy, D. H.,

Baldrighi, G., Akil, H. 1982. Nonopiate effects of dynorphin and des-Tyr-dynorphin. *Science* 218:1136–38

Walker, J. M., Tucker, D. E., Coy, D. H., Walker, B. B., Akil, H. 1983. Des-Tyrosine-dynorphin antagonizes morphine analgesia. *Eur. J. Pharmacol.* 185:121–22

Watkins, L. R., Cobelli, D. A., Faris, P., Aceto, M. D., Mayer, D. J. 1982a. Opiate vs. non-opiate footshock-induced analgesia (FSIA): The body region shocked is a critical factor. *Brain Res.* 242:299–308

Watkins, L. R., Cobelli, D. A., Mayer, D. J. 1982b. Classical conditioning of front paw and hind paw footshock induced analgesia (FSIA): Naloxone reversibility and descending pathways. *Brain Res.* 243:119–32

Watson, S. J., Akil, H. 1979. Presence of two alpha-MSH positive cell groups in rat hypothalamus. *Eur. J. Pharmacol.* 58:101–3

Watson, S. J., Akil, H. 1980. Alpha-MSH in rat brain: Occurrence within and outside brain beta-endorphin neurons. *Brain Res.* 182:217–23

Watson, S. J., Akil, H., Fischli, A., Goldstein, A., Zimmerman, E., Nilaver, G., van Wimersma Greidanus, T. B. 1982a. Dynorphin and vasopressin: Common localization in magnocellular neurons. *Science* 216:85–87

Watson, S. J., Akil, H., Ghazarossian, V. E., Goldstein, A. 1981. Dynorphin immunocytochemical localization in brain and peripheral nervous system: Preliminary studies. *Proc. Natl. Acad. Sci. USA* 78:1260–63

Watson, S. J., Akil, H., Richard, C. W., Barchas, J. D. 1978a. Evidence for two separate opiate peptide neuronal systems and the coexistence of beta-LPH, beta-endorphin and ACTH immunoreactivities in the same hypothalamic neurons. *Nature* 275:226–28

Watson, S. J., Akil, H., Sullivan, S. O., Barchas, J. D. 1977a. Immunocytochemical localization of methionine-enkephalin: Preliminary observations. *Life Sci.* 25:733–38

Watson, S. J., Barchas, J. D., Li, C. H. 1977b. Beta-LPH: Localization of cells and axons in rat brain by immunocytochemistry. *Proc. Natl. Acad. Sci. USA* 74:5155–58

Watson, S. J., Khachaturian, H., Akil, H., Coy, D., Goldstein, A. 1982b. Comparison of the distribution of dynorphin systems and enkephalin systems in brain. *Science* 218:1134–36

Watson, S. J., Khachaturian, H., Taylor, L., Fishli, W., Goldstein, A., Akil, H. 1983. Prodynorphin peptides are found in the same neurons throughout brain: An immunocytochemical study. *Proc. Natl. Acad. Sci. USA* 80:891–94

Watson, S. J., Richard, C. W., Barchas, J. D. 1978b. Adrenocorticotropin in rat brain: Immunocytochemical localization in calls and axons. *Science* 200:1180–82

Weber, E., Evans, C. J., Barchas, J. D. 1982a. Predominance of the amino-terminal hectapeptide fragment of dynorphin in rat brain regions. *Nature* 299:77–79

Weber, E., Roth, K. A., Barchas, J. D. 1982b. Immunocytochemical distribution of alpha-neo-endorphin/dynorphin neuronal systems in rat brain: Evidence for colocalization. *Proc. Natl. Acad. Sci. USA* 79:3062–66

Whitfeld, P. L., Seeburg, P. H., Shine, J. 1982. The human pro-opiomelanocortin gene: Organization, sequence and interspersion with repetitive DNA. *DNA* 1:133–36

Willer, J. C., Dehen, H., Cambier, J. 1981. Stress-induced analgesia in humans: Endogenous oopioids and naloxone-reversible depression of pain reflexes. *Science* 212:689–91

Young, W. S., Kuhar, M. J. 1979. A new method for receptor autoradiography: [³H]opioid receptors in rat brain. *Brain Res.* 179:225–70

Young, W. S., Wamsley, J. K., Zarbin, M. A., Kuhar, M. J. 1980. Opiate receptors undergo axonal flow. *Science* 210:76–77

Zakarian, S., Smyth, D. 1982. Beta-endorphin is processed differently in specific regions of rat pituitary and brain. *Nature* 296:250–52

Zamir, N., Shuber, E. 1980. Altered pain perception in hypertensive humans. *Brain Res.* 201:471–74

Zamir, N., Simantov, R., Segal, M. 1980. Pain sensitivity and opioid activity in genetically and experimentally hypertensive rats. *Brain Res.* 184:299–310

Ann. Rev. Neurosci. 1984. 7:257–78

EFFECTS OF INTRACELLULAR H$^+$ ON THE ELECTRICAL PROPERTIES OF EXCITABLE CELLS

William Moody, Jr.

Department of Zoology, University of Washington, Seattle, Washington 98195

INTRODUCTION

Cytoplasmic pH (pH_i) is a tightly regulated quantity, and changes in pH_i tend to exert profound effects on the properties of cells. Substantial advances have been made in the last decade in understanding how pH_i is regulated, principally through the development of techniques for measuring pH_i accurately in living cells. Nonetheless, information about the roles played by changes in pH_i under normal physiological circumstances, or about the effects of experimentally imposed changes in pH_i on various cell properties, is still not very complete. In this paper, I review a number of experiments concerned with the effects of changes in pH_i on the electrical properties of various excitable cells, such as neurons, muscle fibers, and oocytes. Because this review emphasizes intact cells and the possible physiological actions of pH_i changes in modifying ion conductances rather than on the role of titratable chemical groups in the ion permeation process, I also discuss several aspects of pH_i regulation in excitable cells as well as nonelectrophysiological effects of pH_i changes.

Understanding how excitable cells maintain and regulate cytoplasmic H$^+$ ion activity is, for several reasons, essential for interpreting changes in pH_i and their electrical effects.

1. It is necessary to know what homeostatic systems one is working against in attempting to impose pH_i changes experimentally.

2. If one does succeed in overcoming these systems to some extent, it is important to know what changes have occurred in the cell, other than in pH_i,

257

and to take them into account. For electrophysiological studies, it is particularly important to be aware that other intracellular ion concentration changes may be triggered by alterations in pH_i.

3. It is helpful to know the possible sites of action of endogenous or exogenous stimuli that bring about pH_i changes during the normal life of the cell. I refer below to the action of insulin on muscle glycolysis as an example of such a phenomenon.

For these reasons I discuss in the first section of this review selected aspects of pH_i regulation that are of particular concern to experimentalists working on the electrophysiological effects of pH_i changes. The luxury of being very selective in this section is made possible by the recent comprehensive review of pH_i regulation by Roos & Boron (1981). This section concludes with a brief set of practical examples of how aspects of the pH_i-regulating system could affect electrophysiological experiments.

The second section of this paper reviews experiments concerned with the effects of pH_i changes on various ionic conductances in excitable cells. The emphasis here is to draw relationships among and discuss discrepancies between the work of various authors, to criticize methods, and to suggest alternative interpretations of the data. Some attempt is made to relate these pH_i effects to possible physiological roles of pH_i in the cells in question.

Finally, I discuss several selected examples of the involvement of pH_i in non-electrophysiological cellular functions. Because electrophysiologists interested in ion channel modulation by H^+ or any other cytoplasmic molecule are generally concerned with how membrane electrical properties are integrated into the cell's overall biological functioning, I try to emphasize cases in which changes in several cell properties, including electrical, may occur.

ASPECTS OF pH_i REGULATION IN EXCITABLE CELLS

Two basic facts about pH_i regulation are relevant to the experimentalist wishing to understand the complex cellular responses during imposed or physiologically occurring pH_i changes: (a) H^+ is highly buffered in the cytoplasm; (b) H^+ is actively and rapidly transported out of the cell across the plasma membrane by a system that also transports several other ions. pH_i is maintained more alkaline than its equilibrium value by this system.

Buffering of H^+

The intracellular fluid is a strong pH buffer, and quantifying the cytoplasmic buffering power allows one to know the relationship between the amount of H^+ added to the cytoplasm, by whatever means, and the change in pH_i. Buffering power is usually defined as the amount of H^+ (in millimoles per liter) that needs to be added to a solution to produce a 1 unit change in pH. The unit, millimoles per liter pH, is sometimes referred to as the "Slyke." Buffering power has been measured in a variety of cells using a number of methods, such as microinjection of H$^+$, extracellular application of weak acids, or direct titration of cell homogenates (see Roos & Boron 1981, pp. 395–400, Table 13). In neurons, buffering power has only been measured in invertebrates, where an average value of about 20 mmol/liter-pH has been obtained (Spyropoulous 1960, Boron & de Weer 1976, Thomas 1976, Moody 1981). In both vertebrate and invertebrate muscle, the values tend to be higher, on the order of 40–50 mmol/liter-pH (see e.g. Aickin & Thomas 1975, 1977a). One can get a more intuitive feeling for these figures by noting that a buffering power of 40 mmol/liter-pH is equivalent to 70 mmol HEPES at its pK; with this amount of buffering, roughly 40,000 H$^+$ ions must be injected into the cytoplasm for each free H$^+$ ion appearing intracellularly. Estimates of Ca^{2+} buffering in neurons yield figures on the order of 100 total/free ions, a much lower buffering power (Gorman & Thomas 1980, Plant et al 1983).

Active Transport of H$^+$

Although the cytoplasm is passively buffered against large pH$_i$ changes, the long-term maintenance of resting pH$_i$ and the extrusion of H$^+$ after acid loads is the job of a very efficient H$^+$ transport system found in the plasma membrane. This system has been extensively studied in neurons and muscle fibers.

Roos & Boron (1981) have reviewed pH$_i$ regulating systems in great detail. Several facts are important in the context of this paper:

1. Cytoplasmic pH is more alkaline than expected for an equilibrium distribution of H$^+$ at the resting potential. For example, Aickin & Thomas (1977b; their Figure 3) report in mouse soleus fibers a resting pH$_i$ of 7.2 and a membrane potential of –80 mV. Given an external pH of 7.4, pH$_i$ is 1.13 units more alkaline than the equilibrium pH$_i$ of 6.07. These values are typical of excitable cells, including neurons (Thomas 1977, Moody 1981) and oocytes (Moody & Hagiwara 1982), and indicate that the H$^+$ extrusion mechanism must combat the tendency of the cytoplasm to acidify due to both metabolic acid production and passive H$^+$ influx.

2. The H$^+$ transport system requires external Na$^+$ and transports it into the cell during H$^+$ extrusion. Figure 1 (*top*) shows the scheme proposed for H$^+$ extrusion in snail neurons by Thomas (1977). H$^+$ and Cl$^-$ are exported from the cell in response to an acid load and Na$^+$ and HCO$_3^-$ are taken up. Although some variations in this scheme have been reported in other cells (see e.g. Aickin & Thomas 1977b, Moody 1981), the requirement for external Na$^+$ and its transport during H$^+$ pumping appear universal in nerve and muscle. Figure 1 also shows an experiment that I carried out several years ago in a crayfish neuron to demonstrate that Na$^+$ transport by the H$^+$ extrusion system results in

Figure 1 Top: Diagram of the intracellular pH regulating system proposed by Thomas (1977). In response to a cytoplasmic acid load, the cell extrudes H^+ and Cl^- and takes up Na^+ and bicarbonate. *Bottom:* Experiment in which intracellular pH and Na^+ activity were recorded from the same crayfish neuron during application of an acid load using the ammonium prepulse technique. When ammonium was removed from the Ringer, the cytoplasm became more acidic than prior to NH₄ exposure. During recovery from this acidification, the intracellular Na^+ activity increased.

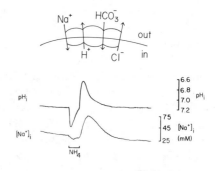

an increase in intracellular Na^+ concentration. The cytoplasm was acidified by the "NH₄Cl prepulse" technique (Boron & de Weer 1976) by which the removal of a small amount of NH₄Cl from the external solution after a brief exposure results in a decrease in pH_i. The $[Na^+]_i$ record below the pH_i trace shows that as pH_i recovers from this acid load, $[Na^+]_i$ rises dramatically. This phenomenon has been demonstrated in a number of cells (e.g. snail neuron, Thomas 1977; mouse muscle, Aickin & Thomas 1977b). Several examples of the possible confusion of the effects of low pH_i and high $[Na^+]_i$ are discussed below.

3. The H^+ extrusion system is relatively insensitive to depletion of metabolic energy stores, though the energy source of H^+ transport is a matter of some controversy. In the squid axon, complete removal of intracellular ATP blocks H^+ extrusion (Russell & Boron 1976). However, in snail neurons, depletion of ATP to levels sufficient to block the Na-K pump does not affect the H^+ pump (Thomas 1977). The energy contained in the Na^+ gradient is sufficient to drive all the ion movements involved in H^+ extrusion in all excitable cells where the relevant quantities have been measured. Moreover, in crayfish neurons, disruption of the balance of the Na^+ and H^+ electrochemical gradients shuts down H^+ extrusion at precisely that value of pH_i at which the two gradients are of equal magnitude (Moody 1981). So depletion of ATP levels may not block H^+ transport, but removal of external Na^+ will.

4. Intracellular pH regulation is related to the maintenance of cytoplasmic Cl^- levels in a complex fashion. The scheme in Figure 1A indicates that in snail neurons, H^+ and Cl^- extrusion are obligatorily coupled. Snail neurons maintain $[Cl^-]_i$ lower than its equilibrium value (see Thomas 1977); hence pH_i and $[Cl^-]_i$ regulation are complementary functions for this system. In crayfish neurons which maintain high levels of intracellular Cl^-, chloride efflux may also occur during acid extrusion; however, a large portion of pH_i regulation occurs via a Na^+-dependent system that does not transport Cl^- out of the cell (Moody 1981). This may allow for more flexibility than the single, four-ion carrier of

Figure 1 in a cell that must balance the conflicting requirements of pH$_i$ and [Cl$^-$]$_i$ regulation. Vaughan-Jones (1982) has studied this question in substantially more detail in sheep cardiac Purkinje fibers. These cells maintain [Cl$^-$]$_i$ at a value roughly four times the equilibrium level. He showed that [Cl$^-$]$_i$ is regulated, at least in part, by a Cl$^-$–HCO$_3^-$ exchange pump running "backwards" (taking up Cl$^-$, extruding HCO$_3^-$); the maintenance of [Cl$^-$]$_i$ occurs at the expense of a standing 0.3 unit acidification of the cytoplasm (see Vaughan-Jones 1982, his Figure 1). The Cl$^-$–HCO$_3^-$ exchange also serves as an active recovery mechanism when the cell is exposed to alkaline loads, a phenomenon not previously demonstrated in other cells. pH$_i$ recovery from acid loads in cardiac muscle may occur via a Na$^+$–H$^+$ pump which does not transport Cl$^-$, as in crayfish neurons (Deitmer & Ellis 1980). So some variation among cells in the pH$_i$ regulating system may occur depending on the normal cytoplasmic Cl$^-$ level; this is important to understand if one wishes to predict the effect of change in pH$_i$ on [Cl$^-$]$_i$. Note that Cl$^-$–HCO$_3$ exchange in barnacle muscle, which operates as a H$^+$ extrusion pump in response to acid loads, can be stimulated markedly by cyclic AMP (Boron et al 1978).

Effects of pH$_i$ Changes on [Ca^{2+}]$_i$

Although Ca^{2+}–H$^+$ exchange seems not to occur during acid extrusion (see Aickin & Thomas 1977b), one might nonetheless predict that pH$_i$ changes would have substantial effects on [Ca^{2+}]$_i$, because H$^+$ competes with Ca^{2+} for binding sites on a variety of molecules (Carvalho et al 1963, Fabiato & Fabiato 1978). A number of workers have measured the effects of an imposed pH$_i$ change on [Ca^{2+}]$_i$. Using Ca^{2+}-sensitive microelectrodes in Xenopus embryonic cells (16 cell stage), Rink et al (1980) recorded only a small increase in Ca^{2+} (from 0.2 to 0.35 μM) when the cells were exposed to 100% CO$_2$, a procedure that produces an intracellular acidification of about 1.3 units. A roughly similar increase in [Ca^{2+}]$_i$ (detected with aequorin) was seen in barnacle muscle fibers exposed to 100% CO$_2$ Ringer (Lea & Ashley 1978). In *Helix* neurons (Alvarez-Leefmans et al 1981), acidification of the cytoplasm with CO$_2$-containing Ringer solutions caused a decrease in [Ca^{2+}]$_i$ of several tenths of a pCa unit. The same effect occurs in Purkinje fibers, in which additional experiments demonstrated that increasing pH$_i$ with NH$_4$Cl increases [Ca^{2+}]$_i$ (Hess & Weingart 1980). In Purkinje fibers, the changes in [Ca^{2+}]$_i$ under cytoplasmic acidification or alkalinization were reflected as decreases or increases in tension, respectively. Thus, the effect on [Ca^{2+}]$_i$ of changing pH$_i$ is not consistent among cells and cannot be predicted with confidence.

The effect on [Ca^{2+}]$_i$ may not necessarily be a good indication of the action of a pH$_i$ change on the cell's various Ca^{2+}-dependent systems. For example, a large decrease in the affinity of a particular binding site for Ca^{2+} could occur during intracellular acidification in the presence of a modest increase in

$[Ca^{2+}]_i$. The phenomenon determined by the binding of Ca^{2+} to that site might act as if it had "seen" a decrease in $[Ca^{2+}]_i$, rather than an increase. Therefore control experiments designed to mimic the change in $[Ca^{2+}]_i$ caused by the imposed pH_i change would not necessarily be appropriate.

The exact mechanism by which changes in pH_i affect $[Ca^{2+}]_i$ is not clear. Active Ca^{2+}-sequestering systems and passive buffers are probably both strongly pH_i-dependent. Alvarez-Leefmans et al (1981) point out that in *Helix* neurons, the decrease in $[Ca^{2+}]_i$ caused by CO_2 long outlasts the expected pH_i decrease (which was not directly measured), thus arguing against a simple effect of pH_i on the binding of Ca^{2+} to passive buffers.

An additional experimental point should be made concerning the effects of imposed pH_i changes on $[Ca^{2+}]_i$. Rink et al (1980) showed that preinjection of EGTA into *Xenopus* embryo cells greatly potentiated the increase in $[Ca^{2+}]_i$ caused by cytoplasmic acidification. This effect is predicted by the very strong pH dependence of the Ca-EGTA dissociation constant (Harufuji & Ogawa 1980). Under most experimental conditions, injection of EGTA would reduce the magnitude of $[Ca]_i$ transients, but when the cause of the transient is a change in pH_i, EGTA could have an effect in either direction, depending on whether cytoplasmic acidification raised or lowered $[Ca^{2+}]_i$ in the absence of EGTA.

Experiments Designed to Change pH_i

The principles of the regulation and maintenance of cytoplasmic pH by cells govern to a great extent the design of experiments to study the electrophysiological effects of pH_i changes.

Three techniques of decreasing pH_i are commonly used in electrophysiological experiments:

1. Direct internal perfusion of buffers in giant axons (Wanke et al 1979), giant crustacean muscle fibers (Hagiwara et al 1968), and, more recently, smaller neuronal somata (Byerly & Moody 1982).

2. The NH_4Cl "rebound" or "prepulse" technique, in which a short exposure of the cell to NH_4Cl in the external Ringer causes a cytoplasmic acidification when normal Ringer is reintroduced (see Boron & de Weer 1976).

3. Exposure of the cell to weak acids, such as CO_2 or acetate, in the external medium. Weak acids cause cytoplasmic acidification during exposure, since the uncharged species can readily cross the membrane and dissociate intracellularly to liberate H^+ (see Figure 3).

Intracellular perfusion requires one to deal directly with the high buffering power of the cytoplasm for H^+. It is hazardous to assume that the cytoplasm near the plasma membrane will attain a pH identical to that of the perfusion fluid, even in the absence of active H^+ transport across the membrane. Two types of control experiments can be carried out in this situation. One can progressively increase the buffer strength of the internal perfusate at constant pH and examine the electrical effect of the pH$_i$ change to see at what buffer strength it no longer increases. Wanke et al (1979) used this method in studying the low pH$_i$ block of delayed K$^+$ currents in the squid axon, and concluded that 45 mM buffer near its pK was required for adequate control of pH$_i$. These experiments were probably carried out in the absence of H$^+$ pumping, since a Cl$^-$-free internal solution, which blocks H$^+$ transport in this cell (Russell & Boron 1976), was used. A second way to confirm the efficacy of internal perfusion is to measure the pH$_i$ change directly with a pH-sensitive microelectrode. This approach has been used by Byerly & Moody (1982) in conjunction with the suction pipet internal dialysis technique (Kostyuk et al 1975, Lee et al 1980, Byerly & Hagiwara 1982). They concluded that even with H$^+$ transport blocked by removal of both external Na$^+$ and internal Cl$^-$, 50–100 mM buffer near its pK was required to achieve control over pH$_i$. Even at these buffer concentrations, changes in pH$_i$ required 10 min to occur, as compared to 1.5–3 min to a complete change in internal K$^+$.

In using either the NH$_4$Cl rebound method or exposure to weak acids, one must first know the exact magnitude and time course of the intracellular acidification produced. For example, exposure to CO$_2$-containing Ringer at constant external pH causes an intracellular acidification that is transient in some cells (snail neuron, Thomas 1977, see his Figure 6; mouse muscle, Aickin & Thomas 1977a; cardiac muscle, Ellis & Thomas 1976), but sustained in others (frog muscle, Bolton & Vaughan-Jones 1977). Similar differences exist in the post-NH$_4$Cl acidification (compare crayfish muscle, Moody 1980, to crayfish neuron, Moody 1981).

When the CO$_2$ or NH$_4$Cl-induced pH$_i$ decreases are transient, several types of experiments can readily be done to separate the electrophysiological effects of low pH$_i$ and high [Na$^+$]$_i$. Since Na$^+$ enters the cell during recovery of pH$_i$, the [Na$^+$]$_i$ transient tends to lag behind the acidification (see Figure 1B). A close correspondence in time between the electrophysiological change and the pH$_i$ transient can thus be taken as evidence against an effect of high [Na$^+$]$_i$. This could be seen, for example, as a lack of hysteresis in a plot of ionic conductances vs pH$_i$ using pH$_i$ values from both the decreasing and recovery limbs of the transient, the same pH$_i$ values being experienced by the cell on the recovery limb in the presence of a higher ambient [Na$^+$]$_i$. An example of such a close correspondence is in the effect of CO$_2$ on light responses in barnacle photoreceptors (Brown & Meech 1979). As shown in Figure 2, the amplitude of the light response closely follows the initial acidification caused by CO$_2$ entry, the recovery of pH$_i$ in the continued presence of CO$_2$, caused by active H$^+$ extrusion, and the longer transient rebound of pH$_i$ when CO$_2$ is removed. Moreover, the [Na$^+$]$_i$ and pH$_i$ transients can be varied independently, so as to

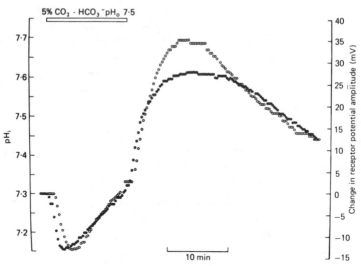

Figure 2 Plot of pH$_i$ (*open circles*) and receptor potential amplitude (*closed circles*) in a barnacle photoreceptor during and after application of an intracellular acid load using CO$_2$ Ringer. See text. Reprinted with permission from Brown & Meech (1979).

determine the electrophysiological effects of each. Removal of external Na$^+$ will, of course, abolish the increase in [Na$^+$]$_i$, while at the same time making the pH$_i$ decrease larger and longer lasting (see e.g. Moody 1981, Figure 2). Similar, though not necessarily identical, effects can be achieved with H$^+$ transport blockers such as SITS (see e.g. Thomas 1977, his Figure 11). With such manipulations, however, one must know the details of the cell's ion transport systems. SITS, for example, blocks H$^+$ transport much more power-fully in snail neurons (Thomas 1977), barnacle muscle (Boron 1977), and squid axon (Russell & Boron 1976) than in cardiac muscle (Ellis & Thomas 1976), mouse muscle (Aickin & Thomas 1977b), or crayfish neurons (Moody 1981). The converse experiment to separate low pH$_i$ and high [Na$^+$]$_i$ effects can also be done. Blocking the Na-K pump with K$^+$-free Ringer, for example, will tend to make the [Na$^+$]$_i$ transient during cytoplasmic acidification larger and longer lasting, while having only a small effect on the acidification itself (see Thomas 1977, his Figure 8).

Weak acids are unfortunately equally good tools to decrease pH$_i$ inadvertant-ly as intentionally. Sharp & Thomas (1981) have examined the effects on pH$_i$ in crab muscle of a series of weak acids, and emphasize that unless one wishes to examine the effects of decreased intracellular pH, these compounds are to be avoided as either low external pH buffers or Cl$^-$ substitutes. A fine example of confusing the electrophysiological effects of external and internal pH by using weak acids in Ringer solutions can be found in the case of the inwardly rectifying K$^+$ conductance in starfish oocytes (compare Hagiwara, Miyazaki, Moody & Patlak 1978 with Moody & Hagiwara 1982) discussed in detail below. Weak acid or base contaminants in the commercial chemicals may also cause unwanted changes in pH$_i$. Zucker (1981a) has reported that commercial tetraethylammonium (TEA) contains about 5% triethylamine, which causes an intracellular alkalinization of about 0.5 pH unit in *Aplysia* neurons exposed to artificial sea water in which all Na$^+$ was replaced by TEA. Purified TEA (ether extraction) had no effect on pH$_i$. Particular caution must be used when replacing either Na$^+$ or Cl$^-$ in a high ionic strength medium such as sea water, because of the high concentrations of substitutes required.

ELECTROPHYSIOLOGICAL EFFECTS OF ALTERED INTRACELLULAR pH

Inwardly Rectifying K$^+$ Conductance

The inwardly rectifying K$^+$ conductance represents the resting K$^+$ permeability of skeletal muscle fibers (Katz 1949, Adrian & Freygang 1962) and many invertebrate oocytes (e.g. see Hagiwara & Takahashi 1974). The biophysical properties of this channel have been extensively characterized in both types of cells (see e.g. Hagiwara et al 1978, Standen & Stanfield 1978, Hagiwara & Yoshii 1979). Inward rectification is blocked by lowering intracellular pH in both frog skeletal muscle (Blatz 1980) and starfish oocytes (Moody & Hagiwara 1982). In muscle, pH$_i$ was lowered either by lowering external pH in the presence of a permeant weak acid (acetate or propionate), or by immersing the cut end of the fiber (using the Hille-Campbell vaseline gap technique) in a low pH solution. In the starfish oocyte, pH$_i$ was lowered with extracellular acetate in low pH artificial sea water.

In the oocyte, Moody & Hagiwara (1982) quantified the effect of internal H$^+$ on the inwardly rectifying K$^+$ conductance by simultaneously measuring the currents under voltage-clamp and monitoring pH$_i$ with a pH-sensitive microelectrode. Figure 3 shows a recording of pH$_i$ in this starfish oocyte as it was sequentially exposed to pH 7.8 sea water, pH 5.0 (biphthalate) sea water, and pH 5.0 (acetate) sea water. Simultaneous voltage-clamp recordings (lower panels) show that only when pH$_i$ falls are the K$^+$ currents blocked. Inward rectification was completely blocked when pH$_i$ was decreased from its normal value of 7.1 to 5.8. Between these values, a smooth titration curve was obtained that could be described quantitatively by assuming that three H$^+$ ions bind to a site with a pK of 6.2 to block the channel (Figure 4). The block showed no voltage-dependence (see Woodhull 1973 for an example of a voltage-dependent H$^+$ block) and H$^+$ did not affect the kinetics of the current. The quantification of this H$^+$ block must be viewed with some skepticism, however, since increased [Na$^+$]$_i$ increases the inwardly rectifying K$^+$ conductance in

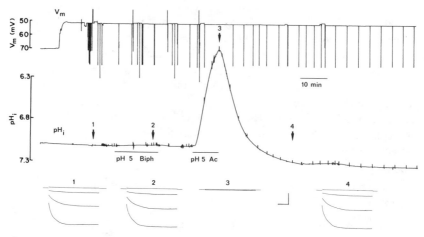

Figure 3   Recording of pH_i and inwardly rectifying K^+ currents from a starfish oocyte during exposure to artificial sea water at low external pH buffered with impermeant (biphthalate) and permeant (acetate) buffers. This experiment demonstrated that inward rectification was blocked by internal and not external pH. Reprinted with permission from Moody & Hagiwara (1982).

starfish oocytes (Hagiwara & Yoshii 1979), and under the low pH_i conditions used, $[Na^+]_i$ increased substantially (see Moody & Hagiwara 1982, their Figure 5). Thus, the titration curve of Figure 4 underestimates the sensitivity of the inward rectifier to cytoplasmic H^+, but by an amount difficult to predict, as it is not known whether H^+ and Na^+ affect this conductance independently. In frog muscle, the inwardly rectifying K^+ conductance is unaffected by cytoplasmic Na^+ (Hestrin 1981).

The history of the low pH_i block of the inward rectifier in starfish oocytes illustrates well the pitfalls of using permeant weak acids as pH buffers in the extracellular solution. Hagiwara, Miyazaki, Moody & Patlak (1978) originally reported that low external pH blocked this conductance. The effect was quantified using acetate-buffered external solutions: three H^+ ions appeared to bind to a site with a pK of 5.3 to block the channel. Only later was it found (Moody & Hagiwara 1982) that when an impermeant buffer, such as biphthalate, was used, a pH 5 external solution had no effect on inward rectification. The original effect had been due entirely to the change in intracellular pH caused by acetate.

In crayfish slow muscle, the inward rectification seen under current clamp is unaffected by lowering pH_i to 6.4 using a NH_4Cl prepulse (see Moody 1980, his Figure 3). This conductance is also unaffected by extracellular Ba^{2+} ion (unpublished observation), a powerful blocker of inwardly rectifying K^+ channels in other cells (Hagiwara et al 1978, Standen & Stanfield 1978).

However, in crayfish muscle, inward rectification may not be due to a K conductance, but rather a voltage-dependent increase in Cl⁻ conductance (see Reuben et al 1962).

The effects of intracellular [Na⁺] and pH on inward rectification in oocytes may have some physiological relevance. The inwardly rectifying K⁺ conductance in starfish is reduced during meiotic maturation (Miyazaki et al 1975, Moody & Lansman 1983) and possibly increased in amplitude after fertilization (Steinhardt et al 1971). The known change in pH_i (Shen & Steinhardt 1978) and $[Na^+]_i$ (Johnson et al 1976) at fertilization and the possibility of their changes during maturation suggest cytoplasmic ionic control of inward rectification during these early developmental stages.

Delayed K⁺ Current

Low intracellular pH has been found to block delayed K⁺ currents in several preparations: squid axon (Wanke et al 1979), *Helix* neurons (Meech 1979), and crayfish muscle (Moody 1980).

This effect has been best quantified in the squid axon, where Wanke et al (1979) used direct internal perfusion of low pH buffers. Their results are explained by assuming that a single H⁺ ion binds to a site with a pK of 6.9 to block the channel, and that when H⁺ is bound to this site, the channel conductance falls to 0.29 of its maximum value (see Figure 4). Neither voltage-dependence of the block nor pH_i effects on the kinetics of the currents were detected. Their results also show that a slight increase in conductance occurs as pH_i is increased from its resting value of 7.32 in the intact axon (Boron & de Weer 1976).

The data are less quantitative in other preparations, but consistent with the squid axon data. Meech (1979) reported some reduction in K⁺ tail currents when a strong (500 mM PIPES) pH 6.8 buffer was microinjected into *Helix* neurons. Moody (1980) reported a substantial reduction of delayed rectification in crayfish slow muscle when pH_i was lowered to 6.4, but the uncertainties introduced by the cable properties of the cells made a quantitative assessment of the effect difficult. However, the block at pH_i 6.4 was substantial enough to convert the normal graded Ca^{2+} responses of the cells to all-or-none action potentials. This action of low pH_i in crayfish muscle may explain the unusual effect of anoxia or metabolic inhibitors in these cells in converting graded electrical responses to all-or-none Ca^{2+} spikes (see Moody 1978, 1983). It is probably also responsible for the effect reported in insect muscle by Fukuda & Kawa (1977), in which low external pH Ringer buffered with acetate caused the generation of large calcium spikes in fibers that under normal conditions are relatively inexcitable.

In contrast to the above reports, decreasing pH_i from 6.8 to 5.9 in *Paramecium* had no effect on delayed K⁺ currents measured under voltage clamp

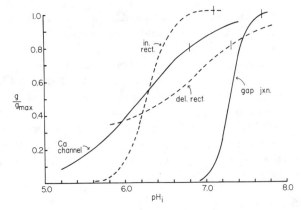

Figure 4 Titration curves for the effects of internal pH on delayed rectification (Wanke et al 1979), inward rectification (Moody & Hagiwara 1982), calcium current (Umbach 1982), and gap junctional conductance (Spray et al 1981). *Curves* are plotted from equations given as best fits of data points in each paper. Each curve is intersected by a *vertical line* to show the resting pH_i of the cell.

(Umbach 1982). However, effects of pH_i on Ca^{2+} currents were found, as discussed below.

Calcium-dependent K^+ Conductance

Almost nothing is known about the pH_i dependence of this conductance. Meech (1979) reported that Ca^2-dependent K^+ tail currents were reduced in amplitude by injection of HCl into *Helix* neurons, but less so than the voltage-dependent (delayed) K^+ currents. Several actions of H^+ could account for this observation. H^+ could compete with Ca^{2+} for the channel binding site. The cytoplasmic acidification could decrease the resting $[Ca^{2+}]_i$ [Alvarez-Leefmans et al (1981) reported this phenomenon in *Helix* neurons; see above], thus causing the $Ca^{2+}]_i$ transient during a voltge-clamp step to attain lower absolute $[Ca^{2+}]_i$ levels. Finally, reduced pH_i might increase cytoplasmic Ca^{2+} buffering and thus reduce the $[Ca^{2+}]_i$ transients during voltage-clamp pulses. This would be complementary to the effects of increased pH_i reported in *Aplysia* neurons by Zucker (1981b).

K^+ Conductance in Sea Urchin Eggs

A large increase in membrane K^+ conductance (Steinhardt et al 1971) immediately follows fertilization in sea urchin eggs. Although the identity of the K^+ conductance has not been established, it could be the same as the inwardly rectifying K^+ conductance, which represents the bulk of the membrane conductance at the resting potential in starfish eggs. This K^+ conductance increase has a close temporal association with the increase in pH_i that accompanies fertilization (Shen & Steinhardt 1978), and a number of experiments have suggested that the increase in K$^+$ conductance is caused by the pH$_i$ change.

Early experiments of Steinhardt & Mazia (1973) showed that eggs exposed to NH$_4^+$-containing sea water underwent an increase in K$^+$ conductance similar to that seen at fertilization. Although the increase in K$^+$ conductance occurs very slowly with NH$_4^+$ exposure (see Steinhardt & Mazia 1973, their Figure 1B), NH$_4^+$ does in fact cause only a very slow increase in pH$_i$ of these eggs (see Shen 1982, Figure 6). However, the effect does not seem to be a simple direct action of cytoplasmic H$^+$ on K$^+$ channels. Shen & Steinhardt (1980), using pH-sensitive microelectrodes in sea urchin eggs, showed that acidification of the cytoplasm with dinitrophenol at long times (25 min) after fertilization was less effective at blocking the new K$^+$ conductance than at short times (5 min). A substantial time lag between the increase in pH$_i$ and the decrease in membrane resistance is also evident in their data (see their Figures 1 and 2 especially). They interpret these data to indicate that the increase in pH$_i$ at fertilization causes the increased K$^+$ conductance, but that soon thereafter the channels lose their pH$_i$ sensitivity. It is still unresolved whether the apparent relation between pH$_i$ and K$^+$ conductance, even at short times after fertilization, represents the direct interaction between H$^+$ and K$^+$ channels. If the interaction were direct, this K$^+$ channel would be substantially more sensitive to cytoplasmic H$^+$ in the physiological pH$_i$ range than the inward rectifier in starfish eggs. The pH$_i$ change at fertilization in sea urchins is from 6.84 to 7.26 (Shen & Steinhardt 1978), which, using the titration curve of Moody & Hagiwara (1982), could not account for the eight-fold conductance increase measured in sea urchins at fertilization (Jaffe & Robinson 1978).

Even assuming a direct interaction, alternative explanations are possible for the apparent change in pH$_i$ sensitivity of the K$^+$ conductance with time after fertilization. Fertilization may well stimulate a membrane Na-H exchange process (see Johnson et al 1976). If the activity of this system increased gradually over 20–30 min after fertilization, the imposed decreases in pH$_i$ at longer times might result in larger increases in [Na$^+$]$_i$. By analogy with the inward rectifier in starfish eggs (Hagiwara & Yoshii 1979, Moody & Hagiwara 1982), the increase in [Na$^+$]$_i$ would increase the K$^+$ conductance and lessen the pH$_i$ effect, without any change in the pH$_i$ sensitivity of the K$^+$ channels. Because the pH$_i$ change is an integral part of the fertilization process, itself causing many metabolic changes in the egg (see Grainger et al 1979), the question of how the membrane electrical properties are integrated into this process is particularly interesting.

Other K$^+$ Conductances

Hagiwara et al (1968) studied the effect of internal pH changes on resting K$^+$ and Cl$^-$ conductances in barnacle giant muscle fibers, using direct internal perfusion with strongly buffered solutions. They found that as pH_i was increased from 4.7 to 9.6, the K^+/Cl^- permeability ratio increased gradually. Below pH_i 5, the membrane was highly permeable to Cl^-. They hypothesized that as pH_i is increased, fixed negatively charged groups in the membrane predominate, and the permeation of cations in the membrane is facilitated.

A decrease in resting K^+ conductance with decreased pH_i has also been reported in crayfish stretch receptors (Moser & Brown 1981) and barnacle photoreceptors (Brown & Meech 1979). In barnacle photoreceptors, the resting K^+ conductance was decreased roughly 50% (estimated from the I–V relations of their Figure 10) for a decrease in pH_i of 0.7 unit. The light response in these cells was considerably more sensitive to decreased pH_i (see below) than the resting conductance.

It is difficult to make productive generalizations from the above data on pH_i and K^+ conductance. The most accurate data are available for the inwardly rectifying K^+ conductance of starfish oocytes and the delayed K^+ conductance of squid axons, whose titration curves are compared in Figure 4. Both are blocked at low pH_i, as are all K^+ channels thus far studied. In both cases, the block occurs without voltage dependence (i.e. the H^+ binding site does not sense the membrane electrical field) and without substantial effects on the kinetics of the currents. Three H^+ ions bind cooperatively to block the inwardly rectifying K^+ channel, whereas only one binds to the delayed K^+ channel. The pK values for the two sites are similar: 6.26 for inward rectification and 6.9 for the delayed K^+ channel. A correction for the Na^+ sensitivity of the inward rectifier might well bring the apparent pK for this conductance even closer to that of the delayed conductance.

Na^+ Conductance

The effects of variations in pH_i on the voltage-dependent Na^+ conductance have been studied in the squid giant axon and in frog skeletal muscle. The effects tend to be more complex than the simple block by internal H^+ of delayed or inward rectification.

In frog muscle, Nonner et al (1980) demonstrated that lowering pH_i blocked inactivation of the Na^+ current. The effect was initially found when low external pH Ringer, buffered with the permeant weak acid acetate, was used on the cells. The study was conducted with direct internal dialysis using the vaseline-gap technique. After long periods (50–100 min) of exposure of the cut end of the fiber to pH 4–5 internal solutions, inactivation was almost completely blocked with little loss of peak current amplitude. The 100 min in pH 5 (50 mM buffer) solution probably was sufficient to bring the fiber interior to near this pH. Exchange of Na^+ into the fiber from the cut end requires about 30 min (Schwarz et al 1977), about ten times the exchange time in internally perfused neuron cell bodies, in which exchange of H$^+$ requires 10 min (Byerly & Moody 1982).

In squid axon, quite the opposite result was obtained. Low pH$_i$ (5.0) depressed peak Na$^+$ currents and enhanced inactivation (Carbone et al 1981), whereas high pH$_i$ (11.0) removed inactivation (Carbone et al 1981, Brodwick & Eaton 1978). The latter authors also reported a substantial depression of peak Na$^+$ currents at high pH$_i$.

The effects of pH$_i$ on both peak Na$^+$ currents and the inactivation process have been analyzed in detail in the squid axon (Wanke et al 1980, Carbone et al 1981). H$^+$ appears to block the channel by binding to two sites, with apparent pK's of 4.6 and 5.8. Both sites are in the membrane field, as indicated by the voltage dependence of the H$^+$ block. Inactivation of Na$^+$ currents was progressively reduced as pH$_i$ was raised from 4.8 to 11.0. The titration curve indicated two sites, pK's 5.4 and 10.4, each binding a single H$^+$ ion (see Carbone et al 1981, their Figure 4). It would be interesting to explore further the opposite effects of low pH$_i$ on Na$^+$ channel inactivation squid axon and frog muscle. Note that, in contrast to the inwardly rectifying and delayed K$^+$ channel (see above), the internal H$^+$ block of the Na$^+$ channel is membrane potential dependent.

Ca Channel

Very little is known about the sensitivity of Ca^{2+} currents to changes in pH$_i$. This is partly because techniques for direct control over the ionic composition of the cytoplasm in cells that have Ca^{2+} currents and whose geometry permits adequate voltage clamp are relatively new. Furthermore, unlike Na$^+$ currents, Ca$^+$ currents tend to be quite labile when the internal dialysis technique is applied (Byerly & Hagiwara 1982).

In *Paramecium*, Umbach (1982) has studied the effects of pH$_i$ changes on Ca^{2+} currents, using CO$_2$ to acidify and NH$_4$Cl to alkalinize the cytoplasm. She found that decreased pH$_i$ blocked and increased pH$_i$ enhanced the Ca^{2+} current. The titration curve for the effect (Umbach 1982, Figure 11) indicated a single titratable group with a pK of 6.2, which binds a single H$^+$ ion. Working with *Aplysia* neurons, Zucker (1981b) reported no effect of 15 mM NH$_4$Cl, which increased pH$_i$ by 0.35 unit, on Ca^{2+} currents. However, the resting pH$_i$ in these cells was 7.17, as opposed to 6.8 for *Paramecium*. The titration curve obtained by Umbach predicts no effect on the Ca^{2+} current as pH$_i$ is raised from 7.17.

Starfish oocytes have two Ca^{2+} currents, which have been reported to be blocked by low external pH (Hagiwara et al 1975). However, acetate was used as a buffer in these experiments, so the results could also be caused by a block by intracellular H$^+$. It would be interesting to explore this point further.

Receptor Potential in Invertebrate Photoreceptors

The effect of changes in pH_i on light responses in lateral ocelli of the barnacle have been studied by Lantz & Mauro (1978) and Brown & Meech (1979). Both studies concluded that the amplitude of the photoresponse was a sensitive function of pH_i, decreasing as pH_i was lowered and increasing as pH_i was raised, with either NH_4^+ (Lantz & Mauro 1978, Figure 4) or CO_2 (Brown & Meech 1979) in the external Ringer. Effects during the application and "rebound" effects after removal were seen in both cases. Brown & Meech (1979) made simultaneous measurements of pH_i with H^+-sensitive microelectrodes, and showed a close temporal correlation between pH_i and the photoresponse amplitude (see Figure 3).

These two studies concentrated on slightly different aspects of the relation between phototransduction and pH_i. Lantz & Mauro (1978) noted the similarity of the effects of anoxia or metabolic inhibition and decreased pH_i on the photoresponse. They concluded that the effect of anoxia in abolishing the photoresponse was caused by a decrease in pH_i. This conclusion is simlar to that reached by Moody (1978, 1980, 1983) for the action of anoxia in inducing hyperexcitability in crayfish muscle. Brown & Meech (1979) compared the effects of light adaptation and decreased pH_i. They recorded decreases in pH_i of up to 0.2 unit during illumination, and concluded that this change was sufficient to explain the reduction in sensitivity of the photoreceptor on exposure to light. This is one of only a few examples of a change in pH_i occurring under physiological circumstances that can be linked to a modulation of the electrical properties of the cell. Where in the transduction process the pH_i change acts is unknown.

Gap Junction Conductance

Considerable attention has focused recently on the block of gap junctional conductance between cells by decreased pH_i. This phenomenon was first reported by Turin & Warner (1977, 1980), who found that CO_2-equilibrated Ringer reversibly abolished current flow between blastomeres of early *Xenopus* embryos. They used pH-sensitive microelectrodes in these cells and concluded that complete abolition of electrical communication occurred at pH_i 6.4–6.85. The complex geometry of the embryos prevented further quantification of the relation between gap junctional conductance and internal pH.

Spray et al (1981, 1982) were able to quantify this relationship exactly using single pairs of electrically coupled *Fundulus* or *Ambystoma* blastomeres, and monitoring pH_i with pH-sensitive microelectrodes or applying the technique of direct internal perfusion. The titration curve for junctional conductance (Figure 4) indicates that 4–5 H^+ ions bind to a site with a pK of 7.3 to block the conductance. The resting pH_i of both types of cells is about 7.7, so junctional conductance falls off steeply when pH_i is lowered more than 0.2 unit from its normal value. Using a modified internal perfusion technique that allowed direct application of solutions to one face of a gap junction, Spray et al (1982) further concluded that in *Fundulus* blastomeres, H$^+$ is about 10,000 times more effective at blocking this conductance than is Ca^{2+}. The steep relationship between pH$_i$ and junctional conductance suggests that pH$_i$ changes might play a physiological role in modifying electrical communication between cells.

H$^+$ Channels

A recent report by Thomas & Meech (1982) indicates that in addition to its effects on other ion conductances, H$^+$ might itself act as a charge carrier through H$^+$-specific channels in excitable cells. They reported slow increases in pH$_i$ in cells that had been previously acidified by HCl injection when the cells were depolarized under voltage-clamp. The efflux of H$^+$ from the cell was associated with an outward membrane current. This current, as well as the pH$_i$ increase, could be inhibited by cadmium ion applied externally. The activation kinetics of the conductance pathway for H$^+$ were not resolvable with the techniques used, but were they sufficiently rapid, H$^+$ could be carried out of the cell during trains of action potentials. The authors suggest that this could lessen the acidification caused by Ca^{2+} entry into the cell during the action potential (see Ahmed & Connor 1980, Meech & Thomas 1980). It is not clear whether this represents a new type of channel, selectively permeable to H$^+$, or the flow of H$^+$ through a previously described voltage-dependent channel.

More recently, L. Byerly, R. W. Meech, and W. Moody (in preparation) have studied the H$^+$ current in greater detail using the suction pipet method of internal dialysis in *Limnaea* neurons. They found that this current does indeed activate rapidly, with activation time constants on the order of a few milliseconds. They also were able to quantify the effects of changes in internal and external pH on the kinetics of the current and the reversal potential of tail currents. Since the H$^+$ current is fairly large (ca 20 nA at +30 mV in *Limnaea*), many of the properties of K$^+$ currents that have been studied in this potential range may well be contaminated by the presence of H$^+$ currents.

INVOLVEMENT OF pH$_i$ IN OTHER CELLULAR FUNCTIONS

Discovering the mechanism by which changes in membrane electrophysiology are integrated with more global alterations in cell properties is the purpose of many investigations of pH$_i$ on channel properties. An example of this is the study of K$^+$ conductance and pH$_i$ in sea urchin eggs, where pH$_i$ also plays a role in bringing about many of the metabolic changes associated with fertilization (see above). In this section I very briefly discuss a few examples in which evidence suggests a role for pH$_i$ changes in nonelectrophysiological cell responses, and in which electrophysiological experiments might prove interesting. A more complete discussion of the role of pH_i in a variety of cell processes can be found in Nuccitelli & Deamer (1982).

Control of the Cell Cycle

Oscillations of cytoplasmic pH have been implicated in the control of cell division in several cell types. In *Tetrahymena,* transient increases in pH_i of ca. 0.3 unit occur just before and after DNA synthesis (Gillies & Deamer 1979). In the plasmodium of the slime mold, *Physarum,* there is also cycling of pH_i with the mitotic cycle, with pH_i at its highest (most alkaline) just at the time of mitosis (Gerson & Burton 1977, Steinhardt & Morisawa 1982). Furthermore, the delay of mitosis by starvation in *Physarum* is correlated with a decrease in pH_i, and the resumption of mitosis with refeeding with a reincrease in pH_i. However, as Steinhardt & Morisawa point out, other experiments on *Xenopus* and sea urchin embryos indicate that pH_i may not be a universal regulator of mitosis. This conclusion is based partly on experiments of Lee & Steinhardt (1981), who found that although small (0.05 unit) oscillations of internal pH occurred during the cleavage cycle of early *Xenopus* embryos, imposed intracellular acidifications of up to 0.3 unit failed to disrupt cleavage.

A related phenomenon occurs in lymphocytes during mitogenic stimulation by agents such as concanavalin A (Gerson et al 1982). Upon stimulation, pH_i rises in two phases, the first with a peak at 6–8hr, the second at ca. 48 hr. This first increase is correlated with the early events of stimulation, such as phospholipid and RNA synthesis, the second with the rise in ^{3}H-thymidine incorporation into DNA. With the advent of the suction pipet technique, electrophysiological recordings from small cells such as lymphocytes are feasible, and detailed voltage-clamp analysis of their ionic conductances possible (see e.g. Fukushima & Hagiwara 1983). It will be especially interesting to examine alterations in the electrical properties of these cells which might be correlated with pH_i changes during activation.

pH_i and the Action of Insulin on Muscle

Moore et al (1982) have presented compelling evidence for the involvement of increased pH_i in the stimulation of glycolysis in frog muscle by insulin. These experiments are a good example of how knowledge of the cell's normal pH_i regulating mechanisms is important for understanding effects of an external stimulus that may act via pH_i. Application of insulin to the isolated muscle caused an increase in pH_i of ca. 0.13 unit, which could be blocked either by the diuretic amiloride or by removal of external Na^+, pointing to Na-H exchange as the site of action. Further, when the external Na^+ concentration was gradually lowered, the pH_i change produced by insulin was gradually reduced, and the effects of glycolysis reduced proportionately (see Moore et al 1982, their Figure 4).

What appears to be a similar activation of Na-H exchange by an external agent occurs during stimulation of neuroblastoma cells in culture by serum. Moolinaar et al (1981) observed a persistent increase in ^{22}Na influx and an amiloride-sensitive stimulation of the Na-K pump in serum-treated cells resuming cell division. They proposed that the primary locus of stimulation was the Na-H mechanism, and that an increase in pH$_i$ might be part of the sequence of intracellular events.

Thus, a number of cell activation phenomena are correlated with increased pH$_i$, in many, if not all, cases caused by a stimulation of Na-H exchange (see above and also Johnson et al 1976, Shen & Steinhardt 1978, 1979, sea urchin eggs; Lee et al 1982, sea urchin sperm). In each of these cases pH$_i$ is probably moving farther away from its equilibrium value, and away from that value set by the pH$_i$ regulating system of the cell. Little is known about what determines the resting pH$_i$ of cells. In cells with a separate Na-H exchange system, the Na$^+$ electrochemical gradient appears to provide the energy for hydrogen ion export in that when the H$^+$ gradient is increased by lowering external pH, pH$_i$ can recover from an acid load only to that value—lower than the normal pH$_i$—at which the Na$^+$ and H$^+$ gradients are equal (Moody 1981). However, the Na$^+$ gradient seems not to set the resting pH$_i$ of the cell. Taking again the crayfish neuron as an example (Moody 1981):

$$[Na^+]_o = 200 \text{ mM, } [Na^+]_i = 20 \text{ mM, } pH_o = 7.4;$$

thus a resting pH$_i$ of 8.40 could be maintained by the Na$^+$ gradient, whereas the resting pH$_i$ of this cell is 7.12. Of course the H$^+$ transport system is working against a steady acid load from H$^+$ influx and metabolic H$^+$ production. However, it seems unlikely that this could result in a standing acidification of over one unit, given the speed of the pH$_i$ regulating system and the slowness of the pH$_i$ decrease when H$^+$ transport is blocked (personal observation; see also Thomas 1977, his Figure 7). Therefore, the H$^+$ transport system may have a "set point," and one might look for a change in this set point, rather than for a simple increase in rate, as a mechanism of causing increased pH$_i$.

CONCLUSION

Although the electrophysiological and cellular effects of changes in pH$_i$ tend to be marked, there are few well-documented cases in which pH$_i$ acts as the mechanism by which changes in the physiological state of the cell are correlated with an alteration of its membrane electrical properties. There are, for that matter, few cases in which a physiologically relevant change in cell electrophysiology alone can be attributed to a change in pH_i. Relatively few studies have addressed this point thus far. What I have emphasized in this review is the necessity for understanding the ways in which pH_i is normally regulated in cells in order to approach this general problem. It is essential to understand the complexity of the response of an intact cell, both during experimental manipulations and under normal physiological circumstances involving intracellular pH, in order to be able to isolate the causes of alterations in the electrophysiological behavior of the membrane.

Literature Cited

Adrian, R. H., Freygang, W. H. 1962. Potassium conductance of frog muscle membrane under controlled voltage. *J. Physiol.* 163:61–103

Ahmed, Z., Connor, J. A. 1980. Intracellular pH changes induced by calcium influx during electrical activity in molluscan neurons. *J. Gen. Physiol.* 75:403–26

Aickin, C. C., Thomas, R. C. 1975. Microelectrode measurement of the internal pH of crab muscle fibres. *J. Physiol.* 252:803–15

Aickin, C. C., Thomas, R. C. 1977a. Microelectrode measurement of the intracellular pH and buffering power of mouse soleus muscle fibres. *J. Physiol.* 267:791–810

Aickin, C. C., Thomas, R. C. 1977b. An investigation of the ionic mechanism of intracellular pH regulation in mouse soleus muscle fibres. *J. Physiol.* 273:295–316

Alvarez-Leefmans, F. J., Rink, T. J., Tsien, R. Y. 1981. Free calcium ions in neurones of *Helix aspersa* measured with ion-selective microelectrodes. *J. Physiol.* 315:531–48

Blatz, A. L. 1980. Chemical modifiers and low internal pH block inward-rectifier K channels. *Fed. Proc.* 39:2073

Bolton, T. B., Vaughan-Jones, R. J. 1977. Continuous direct measurement of intracellular chloride and pH in frog skeletal muscle. *J. Physiol.* 270:801–33

Boron, W. F. 1977. Intracellular pH transients in giant barnacle muscle fibers. *Am. J. Physiol.* 233:C61–C73

Boron, W. F., de Weer, P. 1976. Intracellular pH transients in squid giant axons caused by CO_2, NH_3 and metabolic inhibitors. *J. Gen. Physiol.* 67:91–112

Boron, W. F., Russell, J. M., Brodwick, M. S., Keifer, D. W., Roos, A. 1978. Influence of cyclic AMP on intracellular pH regulation and chloride fluxes in barnacle muscle fibres. *Nature* 276:511–13

Brodwick, M. S., Eaton, D. C. 1978. Sodium channel inactivation in squid axon is removed by high internal pH or tyrosine-specific reagents. *Science* 200:1494–96

Brown, H. M., Meech, R. W. 1979. Light-induced changes of internal pH in a barnacle photoreceptor and the effect of internal pH on the receptor potential. *J. Physiol.* 297:73–93

Byerly, L., Hagiwara, S. 1982. Calcium currents in internally perfused nerve cell bodies of *Limnea stagnalis. J. Physiol.* 322:503–28

Byerly, L., Moody, W. J. 1982. Intracellular pH changes in internally perfused snail neurons. *Biophys. J.* 37:181a.

Carbone, E., Testa, P. L., Wanke, E. 1981. Intracellular pH and ionic channels in the *Loligo vulgaris* giant axon. *Biophys. J.* 35:393–413

Carvalho, A. F., Sanui, H., Pace, N. 1963. Calcium and magnesium binding properties of cell membrane materials. *J. Cell. Comp. Physiol.* 62:311–17

Deitmer, J. W., Ellis, D. 1980. Interactions between the regulation of intracellular pH and sodium activity in sheep cardiac Purkinje fibres. *J. Physiol.* 304:471–88

Ellis, D., Thomas, R. C. 1976. Direct measurement of the intracellular pH of mammalian cardiac muscle. *J. Physiol.* 262:755–71

Fabiato, A., Fabiato, F. 1978. Effects of pH on the myofilaments and the sarcoplasmic reticulum of skinned cells from cardiac and skeletal muscle. *J. Physiol.* 276:233

Fukuda, J., Kawa, K. 1977. Initiation of CA-spikes from an insect muscle fiber immersed in a low pH saline solution containing carboxylic anions. *Life Sci.* 21:981–88

Fukushima, Y., Hagiwara, S. 1983. Voltage-gated calcium channel in mouse myeloma cells. *Proc. Natl. Acad. Sci. USA.* 80:2240–42

Gerson, D. F., Burton, A. C. 1977. The relation of cycling of intracellular pH to mitosis in the acellular slime mold *Physarum polycephalum. J. Cell Physiol.* 91:297–315

Gerson, D. F., Kiefer, H., Eufe, W. 1982. Intracellular pH of mitogen-stimulated lymphocytes. *Science* 216:1009–10

Gillies, R. J., Deamer, D. W. 1979. Intracellular pH changes during the cell cycle in *Tetrahymena*. *J. Cell Physiol.* 100:23–32

Gorman, A. L. F., Thomas, M. V. 1980. Potassium conductance and internal calcium accumulation in a molluscan neurone. *J. Physiol.* 308:287–313

Grainger, J. L., Winkler, M. M., Shen, S. S., Steinhardt, R. A. 1979. Intracellular pH controls protein synthesis rate in the sea urchin egg and early embryo. *Dev. Biol.* 68:396–406

Hagiwara, S., Gruener, R., Hayashi, H., Sakata, H., Grinnell, A. D. 1968. Effect of external and internal pH changes on K and Cl conductances in the muscle fiber membrane of a giant barnacle. *J. Gen. Physiol.* 52:773

Hagiwara, S., Takahashi, K. 1974. The anomalous rectification and cation selectivity of the membrane of a starfish egg cell. *J. Memb. Biol.* 18:61–80

Hagiwara, S., Ozawa, S., Sand, O. 1975. Voltage clamp analysis of two inward current mechanisms in the egg cell membrane of a starfish. *J. Gen. Physiol.* 65:617–44

Hagiwara, S., Miyazaki, S., Moody, W., Patlak, J. 1978. Blocking effects of barium and hydrogen ions on the potassium current during anomalous rectification in the starfish egg. *J. Physiol.* 279:167–85

Hagiwara, S., Yoshii, M. 1979. Effects of internal potassium and sodium on the anomalous rectification of the starfish egg as examined by internal perfusion. *J. Physiol.* 292:251–65

Harufuji, H., Ogawa, Y. 1980. Reexamination of the apparent binding constant of EGTA with calcium around neutral pH. *J. Biochem.* 87:1305–12

Hess, P., Weingart, R. 1980. Intracellular free calcium modified by pH$_i$ in sheep cardiac Purkinje fibres. *J. Physiol.* 307:60P–61P

Hestrin, S. 1981. The interaction of potassium with the activation of anomalous rectification in frog muscle membrane. *J. Physiol.* 317:497–508

Jaffe, L. A., Robinson, K. R. 1978. Membrane potential of the unfertilized sea urchin egg. *Dev. Biol.* 62:215–28

Johnson, J. D., Epel, D., Paul, M. 1976. Intracellular pH and activation of sea urchin eggs after fertilization. *Nature* 262:661–63

Katz, B. 1949. Les constantes electriques de la membrane du muscle. *Archs. Sci. Physiol.* 3:285–99

Kostyuk, P. G., Krishtal, O. A., Pidoplichko, V. I. 1975. Effect of internal fluoride and phosphate on membrane currents during intracellular dialysis of nerve cells. *Nature* 257:691–93

Lantz, R. C., Mauro, A. 1978. Alteration of sensitivity and time scale in inverterate photoreceptors exposed to anoxia, dinitrophenol, and carbon dioxide. *J. Gen. Physiol.* 72:219–31

Lea, T. J., Ashley, C. C. 1978. Increase in free Ca^{2+} in muscle after exposure to CO$_2$. *Nature* 275:236–38

Lee, H. C., Forte, J. G., Epel, D. 1982. The use of fluorescent amines for the measurement of pH. See Nuccitelli & Deamer 1982, pp. 135–60

Lee, K. S., Akaike, N., Brown, A. M. 1980. The suction pipette method for internal perfusion and voltage clamp of small excitable cells. *J. Neurosci. Meth.* 2:51–78

Lee, S. C., Steinhardt, R. A. 1981. Observations on intracellular pH during cleavage of eggs of *Xenopus laevis*. *J. Cell Biol.* 91:414–19

Meech, R. W. 1979. Membrane potential oscillations in molluscan "burster" neurones. *J. Exp. Biol.* 81:93–112

Meech, R. W., Thomas, R. C. 1980. Effect of measured calcium chloride injections on the membrane potential and internal pH of snail neurones. *J. Physiol.* 298:111–29

Miyazaki, S.-I., Ohmori, H., Sasaki, S. 1975. Potassium rectifications of the starfish oocyte and their changes during oocyte maturation. *J. Physiol.* 246:55–78

Moody, W. J. 1978. Gradual increase in the electrical excitability of crayfish slow muscle fibers produced by anoxia or uncouplers of oxidative phosphorylation. *J. Comp. Physiol.* 125:327–34

Moody, W. J. 1980. Appearance of calcium action potentials in crayfish slow muscle fibres under conditions of low intracellular pH. *J. Physiol.* 302:335–46

Moody, W. J. 1981. The ionic mechanism of intracellular pH regulation in crayfish neurones. *J. Physiol.* 316:293–308

Moody, W. J. 1983. Intracellular pH regulation and cell excitability. In basic Mechanisms of Neuronal Hyperexcitability, ed. H. Jasper, N. M. van Gelder. New York: Liss. In press

Moody, W. J., Hagiwara, S. 1982. Block of inward rectification by intracellular H$^+$ in immature oocytes of the starfish *Mediaster aequalis*. *J. Gen. Physiol.* 79:115–30

Moody, W. J., Lansman, J. B. 1983. Developmental regulation of Ca and K currents during hormone-induced maturation of starfish oocytes. *Proc. Natl. Acad. Sci. USA.* 80:3096–3100

Moolinaar, W. H., Mummery, C. L., van der Saag, P. T., de Laat, S. W. 1981. Rapid ionic events and the initiation of growth in serum-stimulated neuroblastoma cells. *Cell* 23:789–98

Moore, R. D., Fidelman, M. L., Hansen, J. C., Otis, J. N. 1982. The role of intracellular pH in insulin action. See Nuccitelli & Deamer 1982

Moser, H., Brown, H. M. 1981. pH effects on the crayfish stretch receptor. *Pflueg. Arch.* 391:R40

Nonner, W., Spalding, B., Hille, B. 1980. Low intracellular pH and chemical agents slow inactivation gating in sodium channels of muscle. *Nature* 284:360–63

Nuccitelli, R., Deamer, D. W. 1982. *Intracellular pH: Its Measurement, Regulation, and Utilization in Cellular Functions.* New York: Liss. 594 pp.

Plant, T. D., Standen, N. B., Ward, T. A. 1983. Calcium injection and calcium channel inactivation. In *The Physiology of Excitable Cells*, ed. A. D. Grinnell, W. J. Moody. New York: Liss. In press.

Reuben, J. P., Girardier, L., Grundfest, H. 1962. The chloride permeability of crayfish muscle fibers. *Biol. Bull.* 123:509–10

Rink, T. J., Tsien, R. Y., Warner, A. E. 1980. Free calcium in *Xenopus* embryos measured with ion-selective microelectrodes. *Nature* 283:658–60

Roos, A., Boron, W. F. 1981. Intracellular pH. *Physiol. Rev.* 61:296–434

Russell, J. M., Boron, W. F. 1976. Role of chloride transport in regulation of intracellar pH. *Nature* 264:73–74

Schwarz, W., Palade, P. T., Hille, B. 1977. Local anesthetics: Effect of pH on use-dependent block of sodium channels in frog muscle. *Biophys. J.* 20:343–68

Sharp, A. P., Thomas, R. C. 1981. The effects of chloride substitution on intracellular pH in crab muscle. *J. Physiol.* 312:71–80

Shen, S. S. 1982. The effect of external ions on pH_i in sea urchin eggs. See Nuccitelli & Deamer 1982, pp. 269–82

Shen, S. S., Steinhardt, R. A. 1978. Direct measurement of intracellular pH during metabolic derepression of the sea urchin egg. *Nature* 272:253–54

Shen, S. S., Steinhardt, R. A. 1979. Intracellular pH and the sodium requirement at fertilization. *Nature* 282:87–89

Shen, S. S., Steinhardt, R. A. 1980. Intracellular pH controls the development of new potassium conductance after fertilization of the sea urchin egg. *Exp. Cell Res.* 125:55–61

Spray, D. C., Harris, A. L., Bennett, M. V. L. 1981. Gap junctional conductance is a simple and sensitive function of intracellular pH. *Science* 211:712–15

Spray, D. C., Stern, J. H., Harris, A. L., Bennett, M. V. L. 1982. Gap junctional conductance: Comparison of sensitivities to H and Ca ions. *Proc. Natl. Acad. Sci. USA* 79:441–45

Spyropoulos, C. S. 1960. Cytoplasmic pH of nerve fibers. *J. Neurochem.* 5:185–94

Standen, N. B., Stanfield, P. R. 1978. A potential and time dependent blockade of inward rectification in frog skeletal muscle fibres by barium and strontium ions. *J. Physiol.* 280:169–91

Steinhardt, R. A., Lundin, L., Mazia, D. 1971. Bioelectric responses of the echinoderm egg to fertilization. *Proc. Natl. Acad. Sci. USA* 68:2426–30

Steinhardt, R. A., Mazia, D. 1973. Development of K conductance and membrane potentials in unfertilized sea urchin eggs after exposure to NH_4OH. *Nature* 241:400–1

Steinhardt, R. A., Morisawa, M. 1982. Changes in intracellular pH of *Physarum* plasmodium during the cell cycle and in response to starvation. See Nuccitelli & Deamer 1982, pp. 361–74

Thomas, R. C. 1976. The effect of carbon dioxide on the intracellular buffering power of snail neurones. *J. Physiol.* 255:715–35

Thomas, R. C. 1977. The role of bicarbonate, chloride, and sodium ions in the regulation of intracellular pH in snail neurones. *J. Physiol.* 273:317–38

Thomas, R. C., Meech, R. W. 1982. Hydrogen ion currents and intracellular pH in depolarized voltage-clamped snail neurones. *Nature* 299:826–28

Turin, L., Warner, A. E. 1977. Carbon dioxide reversibly abolishes ionic communication between cells of early amphibian embryo. *Nature* 270:56–57

Turin, L., Warner, A. E. 1980. Intracellular pH in early *Xenopus* embryos: Its effect on current flow between blastomeres. *J. Physiol.* 300:489–504

Umbach, J. A. 1982. Changes in intracellular pH affect calcium currents in *Paramecium caudatum*. *Proc. R. Soc. London Ser. B* 216:209–24

Vaughan-Jones, R. J. 1982. Chloride-bicarbonate exchange in the sheep cardiac Purkinje fibre. See Nuccitelli & Deamer 1982, pp. 239–52

Wanke, E., Carbone, E., Testa, P. L. 1979. K conductance modified by a titratable group accessible to protons from the intracellular side of the squid axon membrane. *Biophys. J.* 26:319–24

Wanke, E., Carbone, E., Testa, P. L. 1980. The sodium channel and intracellular H^+ blockage in squid axons. *Nature* 287:62–63

Woodhull, A. M. 1973. Ionic blockage of sodium channels in nerve. *J. Gen. Physiol.* 61:687–708

Zucker, R. S. 1981a. Tetraethylammonium contains an impurity which alkalizes cytoplasm and reduces calcium buffering in neurons. *Brain Res.* 208:473–78

Zucker, R. S. 1981b. Cytoplasmic alkalization reduces calcium buffering in molluscan centran neurons. *Brain Res.* 225:155–70

Ann Rev. Neurosci. 1984. 7:279–308

INTRACEREBRAL NEURAL IMPLANTS: Neuronal Replacement and Reconstruction of Damaged Circuitries

Anders Björklund and Ulf Stenevi

Department of Histology, University of Lund, S-22362 Lund, Sweden

INTRODUCTION

Neural grafting has emerged over the last decade as a viable approach to studying the development and regeneration of neuronal connections in the CNS of mammals. In this review we focus on one particular aspect of this technique: the use of intracerebral neural implants for the reestablishment of severed connections, the substitution of lost pathways, and the replacement of tissue defects in the adult mammalian CNS. A survey of the literature shows that grafting of neuronal tissue to the mammalian CNS has been frequently attempted since the end of the last century, but that the results of these earlier studies were generally very poor. Thompson (1890) and Saltykow (1905) were possibly the first to report results from grafts of adult CNS tissue, and Del Conte (1907) the first to try grafts of embryonic tissues to the brains of mammals. Their results were generally unsuccessful and no clear-cut evidence of good long-term survival was obtained. Del Conte, in particular, concluded that the brain was an unfavorable transplantation site. Similar negative results were subsequently published, e.g. by Altobelli (1914), Willis (1935), Glees (1955), Wenzel & Bärlehner (1969), and Frotscher et al (1970). Some of the early investigators were, however, more successful. Ranson (1914) and Tidd (1932) obtained partial survival of grafted sensory ganglia in the cerebral cortex of developing rats, and Dunn (1917) reported survival of four out of 44 grafts of neonatal cerebral cortex, implanted into the cortex of nine to ten day-old

279

0147–006X/84/0301–0279$02.00

rats of the same litter. The most interesting study in the earlier literature is that of LeGros Clark in 1940. He described excellent survival and differentiation in a case of embryonic neocortex grafted to the cortex of a six-week old rabbit. It is remarkable that this paper seemed to have passed relatively unnoticed. In his autobiography, LeGros Clark (1968) does not even mention this study and the findings were never followed up.

During the last decade these earlier positive findings have been greatly substantiated and we know today that all parts of the neuraxis can be transplanted with excellent survival, not only to the CNS of developing animals, as shown by the early workers, but also to the brain and spinal cord of adult or even aged recipients. Technically, the principal restriction is that grafts of CNS tissue have to be taken from developing (embryonic or neonatal) donors. In the most primitive vertebrates, such as urodeles (salamanders and newts) and fishes, transplantation, reimplantation, or transposition of CNS tissue is also possible in adult individuals (see below). The reason for this appears to be, at least in urodeles, that the fully differentiated neuronal elements, which do not survive grafting, can be replaced by newly formed neurons regenerating from undifferentiated neuroepithelial cells (Stone & Zaur 1940, Gaze & Watson 1968). In higher vertebrates there is a clear-cut difference between peripheral ganglionic neurons, which partly survive grafting from adult donors (Stenevi et al 1976, Björklund & Stenevi 1977a), and CNS neurons, which survive grafting only when taken from embryonic or developing donors.

We (Stenevi et al 1976, Björklund et al 1976) were the first to study more systematically the possibilities for good and reproducible survival of grafts of central and peripheral neurons in the *adult* mammalian CNS. In submammalian vertebrates this has been a classic approach for studies of mechanisms of neuronal regeneration, especially all in the retino-tectal system. Matthey (1926), Stone (1944, Stone & Zaur 1940), and Sperry (1945), in particular, showed that transplanted or reimplanted eyes can regenerate a new retino-tectal pathway, with restoration of vision, in adult newts and salamanders or in young postmetamorphic frogs. In these grafts of adult or young adult eyes the retinal ganglion cells initially degenerate, but are subsequently regenerated from proliferating cells at the ciliary margin. Within two to three months the grafted eyes were shown to regenerate a new retino-tectal connection, and experiments with reversed and rotated grafts indicated that the new retino-tectal pathway established a topographically ordered projection over the contralateral tectum (Stone 1944, 1963, Sperry 1945). Stone & Farthing (1942) reported that the same eye could be grafted as many as four times with recovery of vision in each of the new hosts.

More recently, the neural grafting technique has also been successfully applied in the optic tectum of adult goldfish and young postmetamorphic frogs

(Sharma & Gaze 1971, Yoon 1973, 1975, 1977, 1979, Levine & Jacobson 1974, Jacobson & Levine 1975a,b). These studies demonstrate that post-embryonic or adult tectal tissue survives reimplantation or transposition in the brain of the same individual. These grafts become reinnervated by the regenerating retino-tectal axons of the host, and in many cases the new afferents have been observed to be organized in the normal topographic manner according to the original polarity of the graft (see e.g. Yoon 1973, 1975, 1977, Levine & Jacobson 1974).

Another example of functionally successful grafting in submammalian vertebrates, again with a long tradition, is the transplantation of segments of the spinal cord in developing amphibians and chicks (Detwiler 1936, Piatt 1940, Weiss 1950, Székely 1963, 1968, Straznicky 1963). Such grafts establish proper neuromuscular connections and can provide normal coordinated limb movements. In salamander larvae, grafts of CNS tissue also survive well in the tail fin, relatively isolated from the rest of the CNS (Weiss 1950). Of particular interest in the present context are Székely's (1963, 1968) observations that segments of developing spinal cord in such an isolated, ectopic position will retain their motor pattern generating properties. Segments of spinal cord, grafted together with a developing limb, were found not only to reinnervate the muscles of the limb graft, but to move the limb in walking-like coordinated movements, either spontaneously or in response to light touches of the area around the graft. This capacity of the grafted cord to elicit coordinated walking-like movements was specific in the sense that it was seen with segments taken from the brachial portion of the cord, normally innervating the forelimbs, but not with segments taken from the thoracic cord, which does not normally control limb movements.

Although these various experiments in lower vertebrates have revealed a remarkable functional potential of CNS tissue grafts, it is only during the last few years that the possibilities for functional neuronal grafting has been subjected to more systematic investigations in the mammalian CNS. As we try to show in this review, the intracerebral grafting technique, though subjected to greater constraints in higher vertebrates, has opened interesting new possibilities for neuronal reconstruction after CNS damage in mammals.

METHODOLOGICAL CONSIDERATIONS

A more detailed account of the technical aspects of neural grafting in the mammalian brain and spinal cord is beyond the scope of this review. We therefore limit ourselves to a brief overview of some alternative methodological approaches and discuss some basic features and limitations of the intracerebral grafting techniques.

Transplantation Methods Currently in Use

Descriptions of a wide variety of techniques used to implant tissue into the brain or spinal cord can be found in the literature. The simplest approach, and the one that was tried initially, is to insert the graft directly into a slit made in superficial cortical matter with fine forceps or similar instruments. This technique was used with poor results by Saltykow (1905) and Del Conte (1907) in adult recipients, and with more positive results by Ranson (1914), Tidd (1932), and Das & Altman (1971, 1972) in developing rats. Our own experience is that this approach is highly unreliable and gives at best suboptimal results, at least in adult rat recipients.

A variant of this technique, which gives considerably better control over the placement of the graft and also allows the use of larger tissue pieces, is to place the graft in a surgically prepared transplantation cavity. This was first used in the transected spinal cord, mainly for grafts of peripheral nerve, (e.g. see Sugar & Gerard 1940, Kao et al 1970). Later it was adapted for grafting of neural tissues into cortical areas by Stenevi et al (1976, 1980a,b) in the adult rat brain, and by Lewis & Cotman (1980, 1983) and Graziadei & Kaplan (1980) in neonatal animals. The principal advantage of this type of procedure is that graft survival can be greatly improved by preparing the cavity in such a way that the graft can be placed on a richly vascularized surface (e.g. the pia in the choroidal fissure) that can serve as a "culturing bed" for the graft (see Stenevi et al 1976, Møllgaard et al 1978). For areas in which such a surface is not available, Stenevi et al (1980a; see also Björklund et al 1980a) have devised a two-stage delayed grafting procedure: A cavity is first made by a suction pipette, and the wound is closed. After usually four to six weeks, when a new vessel-rich pia has grown over the surfaces of the cavity, the cavity is opened and cleaned and the graft inserted. This technique has been used in our laboratory for the transplantation of the embryonic substantia nigra region onto the dorsal or lateral surfaces of the nc. caudatus-putamen in adult rats (Björklund & Stenevi 1979c, Björklund et al 1980a, 1981, Dunnett et al 1981a–c). The observations made with this technique again emphasize the importance of obtaining rapid and efficient revascularization of the implants, particularly when working with grafts of CNS tissue in adult recipients.

Injection of tissue pieces by means of a syringe or a plunger has been used in various ways. This technique was probably first used by Willis (1935) and subsequently by LeGros Clark (1940), Flerkó & Szentágothai (1957), Halasz et al (1962), and Horvat (1966). More recently this approach has been perfected and used with excellent results in the extensive studies of Das and collaborators, especially in the developing cerebellum (see Das 1974, Das et al 1979, 1980), by Lund and collaborators in the developing rat visual system (see Lund & Hauschka 1976, Jaeger & Lund 1979, 1980a,b, McLoon et al 1981, 1982), and by Sunde & Zimmer (1981, 1983) in the developing hippocampal formation. Although most studies employing this technique have been performed in early neonatal recipients, Das & Hallas (1978), Hallas et al (1980), and Sunde & Zimmer (1983) have shown that the techique can also be applied with good results in brains of adult recipients. In our own studies on grafts of thin sheets of peripheral tissue (such as iris, heart valve, and portal vein) in the adult rat brain, we have used a flat glass rod to insert, or push, the graft into deep brain sites (see Björklund & Stenevi 1971, Svendgaard et al 1975b, 1976, Björklund et al 1975, Emson et al 1977).

A special variation of this method was introduced by Rosenstein & Brightman (1978, 1979) and was later employed by Perlow et al (1979), Freed et al (1980), Gash & Sladek (1980), and Krieger et al (1982). In their experiments the graft is introduced into one of the cerebral ventricles by means of a syringe or a glass rod. Both adult ganglia and embryonic CNS tissues have been shown to survive well in this relatively isolated position, although good volume and more extensive interconnections with the host brain seem to be achieved only if the ependyma is damaged so that the graft can fuse with the underlying tissue.

More recently, we introduced a technique whereby embryonic central neurons can be implanted in the form of a dissociated cell suspension (Björklund et al 1980b, 1983d, Schmidt et al 1981). In this method pieces of embryonic CNS tissue are trypsinized and mechanically dissociated into a milky cell suspension. Small volumes (usually 1–5 µl) are then injected stereotaxically into the desired site of the brain or spinal cord by means of a 10 µl Hamilton syringe. The main advantages of this approach are that it induces minimal destruction of the host tissue, it allows precise and multiple placements of the transplanted cells, and it enables accurate monitoring of the number of cells injected by counting the density of cells in the suspension. This technique has, moreover, the potential advantage of allowing the constituent cells to be manipulated, mixed, or cultured before implantation. A wide variety of central neuronal cells types survive grafting to the adult rat brain (Schmidt et al 1983a). We have observed, however, that the noradrenergic neurons of the developing locus coeruleus are relatively sensitive to trypsin, and that these neurons do not survive unless the trypsin incubation step in the dissociation procedure is omitted (Björklund et al 1983a). Other central neuronal cell types may share this sensitivity to trypsin in the suspension grafting procedure.

Graft Placement

The exact site of implantation and the features of the implantation site seem to be far more critical for long-term survival in the adult rat CNS than in the CNS of neonatal, developing hosts. This is probably because the developing host tissue and the small dimensions of the neonatal rat brain provide generally better support for survival, sources for revascularization, access to the cerebrospinal fluid (CSF), etc. Thus, direct intraparenchymal placement of pieces of embryonic or neonatal CNS tissue has given excellent results in several areas of the neonatal rat brain (Das 1974, Das et al 1979, 1980, Sunde & Zimmer 1981, 1983, Arendash & Gorski 1982). In adult recipients, our experience is that intraparenchymal grafts of pieces of CNS tissue (or ganglia) survive poorly unless they are placed in direct contact with vessel-rich pial surfaces or the CSF-filled ventricular system (Stenevi et al 1976). It seems likely that the conditions of the transplantation site are more critical for older donor tissue, where the constituent neurons have undergone their last cell divisions, than for more undifferentiated donor tissue where continued cell proliferation may be able to compensate for an initially poor cell survival.

The primary prerequisite for optimal survival in adult hosts seems to be a rapid and efficient revascularization from the surrounding tissue. Presumably, the CSF can act as a sufficient nutrient medium during the first few days after grafting, until a new blood supply has been established. This has been estimated in the adult rat brain to take about three to five days (Svendgaard et al 1975b). In adult hosts the best results with solid grafts have therefore been obtained with one-stage or two-stage prepared transplantation cavities, or with grafts placed directly into the cerebral ventricles (see above). The conditions are, however, quite different for implants of suspended cells: with this technique good survival is also obtained in intraparenchymal sites, apparently in any site within the brain or spinal cord of adult recipients.

Age of Donor—Age of Host

The single most important factor for good survival of neural grafts is the developmental stage of the donor tissue and, to a lesser degree, the age of the recipient animal. The general rule for grafts of CNS tissue is that the younger the donor, the better the chances for survival and growth of the graft, and that the tolerance limits with respect to donor age are tighter in more mature recipients.

The age-of-donor factor has been most systematically investigated by Seiger & Olson (1977) and Olson et al (1982) for intraocular grafts in adult rats. Our own observations on intracerebral solid grafts (Stenevi et al 1976, Kromer & Björklund 1980, Kromer et al 1983) are consistent with their findings; this suggests that the same rules are applicable for both intraocular and intracerebral grafts of pieces of nervous tissue, at least in adult recipients.

Although in adult recipients the best survival and growth is consistently seen with embryonic donors for tissue from all regions of the CNS, the optimal embryonic donor age varies for different parts of the neuraxis. In general, the best survival seems to coincide with the period of neuronal proliferation and migration within each area. Thus, regions that differentiate earliest, such as many brain stem and spinal cord nuclei, give best results when taken from relatively young embryos (approximately days 15–17 of gestation).

(Embryonic days are given here with day after mating as day 0.) Regions that exhibit a more protracted neurogenesis, such as cerebral cortex and hippocampus, can be grafted with good results from a wider range of donor ages (up to days 20–22 of gestation), although the final size, the intrinsic architecture, and the survival rate of the grafts can be quite different in grafts taken from early and late gestational stages (see Kromer et al 1983 for examples of hippocampal grafts). Partial exceptions to this rule are grafts of developing cerebellar tissue, which produce well-organized grafts only when taken from a fairly restricted period of age (approximately gestational days 11–15), even though neurogenesis in the cerebellum continues into the postnatal period in the rat (Olson et al 1982, Kromer et al 1983, Alvardo-Mallart & Sotelo 1982). For both hippocampal and cortical tissue the time constraints appear to be less pronounced when tissue is grafted to the brains of neonatal recipients (Das et al 1979, 1980, Sunde & Zimmer 1983).

For grafts of neuronal cell suspensions, donor age seems to be an even more critical parameter (Björklund et al 1980b, Schmidt et al 1983a). Nigral dopaminergic neurons, injected into the adult caudate-putamen, show good survival and axonal outgrowth only when taken from donors up to about day 15 of gestation. With suspensions prepared from 16–17 day-old donors the survival was drastically reduced. Similar observations have been made with cerebellar Purkinje cells, of which large numbers survived in implants prepared from day 15 donors, but not from donors of day 18, or older, of gestation. By contrast, cerebellar microneurons, which continue to be generated up into the postnatal period, survive suspension grafting from the late embryonic stages and the early postnatal period. These observations suggest that central neurons survive the dissociation and grafting procedure only when taken during their period of proliferation and migration, which coincides with gestational days 11–15 for the nigral dopamine neurons (Lauder & Bloom 1974) and days 13–15 for the Purkinje cells (Altman 1969). This conclusion finds additional support from observations by Banker & Cowan (1977) on the survival of different types of hippocampal neurons in dissociated cell cultures. The behavior of non-neuronal elements in the grafts is probably different, but this so far is poorly explored.

Immunological Aspects

Most studies so far on intracerebral neuronal grafts have been carried out between different individuals of the same breeder's stock of rats. Although these rats may originally derive from an inbred strain they have subsequently been outbred to the extent that the individuals are, to some degree at least, genetically different. This degree of difference seems to have very little effect on the long-term survival of both intracerebral and intraocular grafts. Thus, intracerebral grafts of CNS tissue have been seen to survive more than one year without any signs of rejection, necrosis, or regression of either the graft tissue or its neuronal projections into the host. Interestingly, grafts of peripheral tissue (which lack a blood-brain barrier) demonstrate signs of regression and axonal degeneration at long survival times under similar conditions (Svendgaard et al 1975a).

Partial long-term survival (up to six months) has been observed with grafts of mesencephalic dopamine neurons to the striatum, transplanted between species (from mouse to rat) (Björklund et al 1982b). Other investigators have reported survival of septal grafts made between different strains of rats (Low et al 1983), or from rat embryos to adult rabbits (Bragin & Vinogradova 1981). In the study of Low et al (1983), however, the grafts, analyzed at three months after transplantation, showed a lymphocyte infiltration indicative of an on-going immune reaction. In the study of Björklund et al (1982b) the actual grafted tissue piece had degenerated by six months, but groups of dopamine neurons, which apparently had migrated into the host caudate-putamen (as well as some apparently avascular clumps of grafted neurons), had survived. These neurons were sufficient to provide good dopamine fiber ingrowth and functional compensation in the initially denervated host neostriatum.

The brain has been suggested to be an immunologically privileged site, probably partly because of its protective blood-brain barrier (see Barker & Billingham 1977, Raju & Grogan 1977). We (Björklund et al 1982b) proposed that although xenogenic grafts of pieces of neural tissue will eventually undergo rejection, some neurons can escape rejection by migrating behind the protective blood-brain barrier of the host.

DEVELOPMENT AND INTRINSIC ORGANIZATION OF THE GRAFTS

In general, the time-course of development of the grafted tissue in its new host environment seems to follow relatively closely the normal in situ developmental sequence. The time-course studies of Jaeger & Lund (1980b, 1981; neocortex) and Wells & McAllister (1982; cerebellum) indicate that cell proliferation and neurogenesis proceeds with a time-course that is very close to the one in situ, whereas neuronal migration and differentiation and folia formation may be delayed or, in some cases, abnormal. Thus, Wells & McAllister (1982) reported for example that Purkinje cell monolayer formation was delayed by five days and folia formation by about ten days in intracortical grafts of cerebellar primordia. Our own fluorescence histochemical studies on monoaminergic neurons (Stenevi et al 1976, Björklund et al 1979b, 1980a, Kromer et al 1983, and Jaeger & Lund's (1980b) observations on neocortical neurons with ^{3}H-thymidine autoradiography have shown that both proliferating neuroblasts and neurons that have already undergone their final cell division will survive grafting. For the noradrenergic locus coeruleus neurons, taken after their final cell division, it has been estimated that about 10–30% will survive grafting to the cortex of adult recipients (Björklund et al 1979b).

The mature transplants develop intrinsic organizational features that general-ly resemble the ones seen *in situ*. So far this has been studied well only for tissues grafted as intact pieces, but for such specimens it applies also to grafts placed in ectopic sites and in relative isolation from the host brain. Already in 1940 LeGros Clark emphasized the normal-looking lamination of cells de-veloped in an intracortical graft of embryonic neocortex. Relatively normal cellular architecture, intrinsic fiber connections, and cortical foliation has more recently been described in intracerebral grafts of cerebellum (Kromer et al 1979, 1983, Alvarado-Mallart & Sotelo 1982, Wells & McAllister 1982), hippocampus (Kromer et al 1979, 1983, Sunde & Zimmer 1981, 1983), and retina (McLoon & Lund 1980a,b). Each of these regions can develop its characteristic laminar architecture, and in the case of hippocampal and cerebel-lar grafts, also some of its normal foliations. The degree to which this happens seems to depend, however, on (*a*) the age of the donor fetus, (*b*) the dissection, orientation, and integrity of the grafted piece, and (*c*) the space available in which the graft can grow. In cerebellar grafts, Alvarado-Mallart & Sotelo (1982) demonstrated the presence of all five major categories of neurons normally present in the cerebellar cortex, as well as several of the normal intrinsic synaptic connections, such as mossy fiber glomeruli and reciprocal cortico-nucleo-cortical connections. Similarly, Sunde & Zimmer (1981, 1983) observed some of the major intrinsic fiber connections in hippocampal grafts, such as the mossy fiber connections of the granule cells onto CA4 and CA3 pyramidal neurons, and the associational fiber connections from the pyramidal cells to the molecular layer of the dentate gyrus region in the graft. Histochemi-cal studies, finally, have shown that different neuron types within the graft will retain or express their normal neurotransmitter characteristics, such as synth-esis and storage of monoamines or neuropeptides (Björklund et al 1976, 1979a, 1980a, Stenevi et al 1980b, Gash & Sladek 1980, Krieger et al 1982, Schult-zberg et al 1983).

Taken together, these various observations indicate that the organization of the intrinsic circuitries as well as the functional properties of the component neurons in intracerebrally implanted grafts may represent relatively faithful replicas of the region *in situ*. This seems to be the case regardless of whether the graft is well integrated with the host CNS tissue, or whether it occurs in relative isolation, such as in the anterior eye chamber (see e.g. Olson et al 1982).

RECONSTRUCTION OF NEURAL CIRCUITRIES IN ADULT RATS

One of the most interesting features of the intracerebral neural grafts is their ability to form extensive axonal connections with the host brain. Several studies in both developing and adult hosts have demonstrated projections from neurons within the implant to areas within the host brain (Björklund et al 1976,

1979a,b, 1980a, Björklund & Stenevi 1977b, 1979c, Beebe et al 1979, Jaeger & Lund 1979, 1980a, McLoon & Lund 1980a,b, McLoon et al 1981, 1982, Kromer & Björklund 1980, Oblinger et al 1980, Oblinger & Das 1982, Lewis & Cotman 1980, 1983), as well as projections from neurons within the host into the implant (Lund & Hauschka 1976, Oblinger et al 1980, Jaeger & Lund 1980a, Lund & Harvey 1981, Harvey & Lund 1981, Kromer et al 1981a,b, Segal et al 1981, Oblinger & Das 1982, Hallas et al 1980, Sunde & Zimmer 1983). Although the establishment of connections between implant and host may be influenced both by the stage of development of the host brain and by the location of the implant and the surgery involved (see below), it seems clear that the connections established by the intracerebral implants can exhibit a high degree of specificity. In this review we focus on our own studies on the ability of embryonic neural implants to substitute for lesioned pathways and lost innervations in adult rats. Some of the parallel work that has been carried out in rats during postnatal development is discussed in this context, but for a more complete coverage of the developmental studies the reader is referred to the reviews by Lund and co-workers (Lund 1980, Lund et al 1982) and Das (1983).

Reinnervation of the Denervated Hippocampus by Monoaminergic and Cholinergic Neurons

In the studies performed in our laboratory over the last few years we have been particularly concerned with the monoaminergic and cholinergic neuronal systems, taking advantage of the availability of selective and sensitive histochemical methods for tracing of the connections of these systems. The monoaminergic and cholinergic systems have an additional advantage, in that their axonal outgrowths can be monitored biochemically (through assays of their transmitters or transmitter-related enzymes). Moreover, the neurotransmission of these neurons is accessible to pharmacological manipulation, which provides possibilities for testing the functional significance of newly established connections.

In a first series of experiments (Björklund et al 1976, 1979a,b, Björklund & Stenevi 1977b, Beebe et al 1979), we studied the growth of monoaminergic and cholinergic axons from transplants of different embryonic brainstem regions into the hippocampal formation in adult recipient rats. The idea behind these experiments was to test to what extent, and with what degree of precision, neural transplants are able to reestablish normal cholinergic and monoaminergic terminal innervation patterns in a previously denervated brain region. The hippocampal formation is ideally suited for such studies for two reasons. First, the terminal fields of the afferent inputs are, for the most part, discretely laminated; this gives each afferent system a characteristic termination pattern. Second, the major afferent fiber inputs to the hippocampal formation are readily accessible to surgical transection, and this makes possible fairly selective denervations of this region of the brain.

The transplants, obtained from 16–17-day-old rat fetuses, comprised (*a*) the septal-diagonal band area (which contains an extensive system of cholinergic neurons normally innervating the hippocampus and large parts of the neocortex), (*b*) the locus coeruleus region of the pons (which contains the nucleus of origin of the hippocampal noradrenergic innervation), (*c*) the pontine and mesencephalic raphe region (with major serotonergic cell groups, including those which innervate the hippocampus), or (*d*) the dopamine-rich ventral mesencephalon (which does not normally innervate the hippocampus). These different transplants were grafted into the hippocampal circuitry in such a way as to allow the grafted neurons to grow toward the denervated hippocampus along the normal routes of the cholinergic and monoaminergic hippocampal afferents. This was achieved by placing the grafts either in a cavity made through the hippocampal fimbria, or in a cavity of the occipital-retrosplenial cortex (transecting the dorsal part of the entorhinal perforant path system), as illustrated in the inset in Figure 1. In both locations, axons from the grafted neurons were found to invade large parts of the hippocampal formation. The mode of growth and the patterning of the ingrowing axons in the hippocampus differed markedly, however, among the different grafted neuron types, suggesting that the axonal ingrowth is very precisely regulated in the denervated hippocampal target. Indeed, the septal and the locus coeruleus transplants were able to form new "septo-hippocampal" and "coeruleo-hippocampal" pathways in rats whose systems had been damaged before the transplantations were made. The new cholinergic and adrenergic innervations closely mimicked the normal cholinergic and adrenergic innervation patterns, respectively, and in successful cases the entire hippocampus and dentate gyrus were reached by the ingrowing axons. The newly formed innervations have been observed to remain unchanged for more than a year after the operation and are thus probably permanent.

The most remarkable features of the graft-to-hippocampus projections are their reproducible and, at least partly, neuron-specific terminal patterning, and the pronounced influences that are exerted by the presence or absence of specific groups of hippocampal afferents. This impression is mainly derived from the following three observations.

1. Different types of grafted neurons, transplanted in the same location, form distinctly different terminal patterns in the hippocampus. The upper panel of Figure 1 illustrates this in rats whose dentate gyrus had been reinnervated by transplants placed in the occipital-retrosplenial cortex (CS). The noradrenergic axons are seen to avoid, and even to grow through, the denervated terminal zone of the entorhinal perforant path fibers, which normally innervate the outer half of the dentate molecular layer. Instead, the adrenergic axons grow into those areas which normally receive a dense noradrenergic innervation from the locus coeruleus, which in the dentate gyrus is the hilar zone. By contrast, the cholinergic, dopaminergic, and serotonergic axons ramify extensively in the

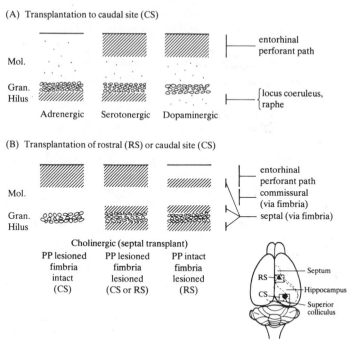

Figure 1   Schematic representation of the patterning of different types of ingrowing axons in the dentate gyrus in rats bearing transplants in the occipital-retrosplenial cortex (caudal site, CS in inset), or in the hippocampal fimbria (rostral site, RS in inset). The *hatched fields* denote the distribution of ingrowing axons in the layers of the dentate (mol.: molecular layer; gran.: granular layer). To the right is given the position of the terminal fields of the normal afferent inputs. (A) The different distribution of axons from locus coeruleus transplants (adrenergic), from raphe transplants (serotonergic), and from ventral mesencephalis transplants (dopaminergic) placed in the caudal site (CS). In all these cases the entorhinal perforant path axons to the outer molecular layer were lesioned, and the monoaminergic afferents were removed by neurotoxin pretreatment. (B) Influence of removal of different afferent inputs on the patterning of cholinergic (AChE positive) fibers growing into the dentate from septal transplants placed in the caudal (CS) or rostral (RS) site. When the normal septal cholinergic input through the fimbria is intact (left-hand side of figure) the transplanted cholinergic axons become restricted to the denervated perforant path (PP) zone, but when the normal septal input is removed (middle of the figure), the ingrowing cholinergic axons also expand into the terminal fields of the septal afferents. Likewise, the extension into the PP zone is inhibited by the presence of an intact entorhinal input (right-hand side of the figure). (From Björklund & Stenevi 1979b.)

denervated perforant path zone to form a dense band of terminals in the outer molecular layer of the dentate. Despite their neurochemical relatedness to the noradrenergic neurons, the transplanted dopaminergic neurons show a very limited growth into the denervated terminal zones of the noradrenergic afferents.

2. When the ingrowing axons are given a choice between different denervated terminal fields, they show in certain cases a clear preference for the zones

denervated of the homologous fiber type. This is illustrated by the above quoted example of noradrenergic axons growing into the dentate gyrus from locus coeruleus grafts (Figure 1). Another example is given by septal grafts in the fimbrial site (lower panel RS in Figure 1): Although the transplantation lesion in the fimbria removes not only the septal cholinergic afferents but also the extensive system of commissural noncholinergic afferents, the ingrowing cholinergic axons from the graft (as monitored by acetylcholine esterase, AChE, histochemistry) become restricted to the normal cholinergic terminal fields. Zones receiving dense commissural, but sparse septal, innervations (such as the stratum radiatum of CA1 and CA3, and the supragranular zone in the temporal parts of the dentate molecular layer; cf Figure 1B) are largely devoid of fibers from the transplant.

3. As seen in experiments with septal transplants (lower panel in Figure 1), the extension of ingrowing cholinergic axons into the terminal fields of the normal cholinergic innervation is markedly modified by the presence or absence of the intrinsic cholinergic innervation. In our first experiment (Björklund & Stenevi 1977b) the AChE staining indicated that in the dentate gyrus the extension of AChE-positive axons from the septal graft (placed in the caudal site in Figure 1) was substantially reduced into the normal terminal zones of the cholinergic afferents if the fimbria, carrying the intrinsic septo-hippocampal cholinergic afferents, were left intact. Instead, the fibers terminated heavily in the entorhinal perforant path zone, denervated by the transplantation lesion (left in lower panel in Figure 1). In follow-up experiments using the suspension grafting technique (F. H. Gage et al, in preparation) we have observed that the ingrowth is markedly reduced but not completely blocked in animals with intact fimbrial inputs. The choline acetyltransferase (ChAT) level in the host hippocampus thus reached about 75% of normal in animals with the fimbria lesioned at the time of transplantation, as compared with about 25% in rats whose fimbria was left intact until one week before sacrifice. The parallel AChE histochemistry suggested that the denervating fibrial lesion, made at the time of transplantation, made the cholinergic axons from the graft grow over greater distances within the host hippocampus and ramify more extensively in some of the normal terminal zones of the intrinsic sept-hippocampal cholinergic projection. Interestingly, McLoon & Lund (1980b) have reported a similar effect of removal of the normal retinal input on the ingrowth of axons from retinal grafts into the superior colliculus in neonatal rats.

These observations suggest that the ingrowth and patterning of axons from implanted neurons are greatly influenced by the target tissue, and that this regulatory influence exhibits a relative specificity with respect to different types of neurons within the graft. The effects of denervating lesions suggest moreover that the proliferation and terminal patterning of the reinnervating axons may in some way be related to the filling of vacated terminal space in the denervated target, analogous to what has been described for lesion-induced collateral sprouting in the septo-hippocampal system (see Raisman & Field 1973, Cotman & Lynch 1976). The monoaminergic and cholinergic neurons in the implants can also reinnervate the adult host hippocampus in a reproducible and orderly manner although the fibers enter from abnormal directions and along routes that are inconspicuous in normal animals. The extension, proliferation, and patterning of the ingrowing fibers can be drastically modified by the removal of intrinsic afferents of both homologous and heterologous kinds. As in the experiments of McLoon & Lund (1980b) in neonatal recipients, the ingrowing axons in the adult hippocampus thus exhibit in some respects a considerable degree of specificity, while being at the same time markedly flexible and modifiable to the extent that abnormal zones of termination can be induced.

Use of Neural Transplants as "Bridges" for Regenerating Axons in the Host

Attempts to use transplants to promote axonal regeneration across tissue defects in the CNS date back to Tello (1911) and Cajal (1928), who introduced the idea of using pieces of peripheral nerve to bridge lesions and stimulate sprouting from lesioned central axons. Numerous studies have subsequently been done along these lines, but it is only recently, with the elegant and important work of Aguayo and collaborators (Richardson et al 1980, 1982, David & Aguayo 1981, Benfey & Aguayo 1982), that the viability of this approach has been well substantiated.

In our own experiments we have asked the questions: To what extent can an embryonic hippocampal implant replace the normal hippocampus in the formation of connections with the host septum, and to what extent can such an implant promote regeneration of the septohippocampal pathway and serve as a bridge for the regeneration of axons from the septal nuclei to the hippocampus? From previous studies (Björklund & Stenevi 1971, Björklund et al 1975, Svendgaard et al 1975b, 1976) we knew that transplants of denervated peripheral tissues, such as iris, heart, or portal vein, promote the regeneration of central axons into the transplanted target. Interestingly, the degree of adrenergic or cholinergic reinnervation of the transplant appeared to be related to the property of the target to possess a normal adrenergic or cholinergic nerve supply. This may suggest that lesioned central neurons are susceptible to trophic mechanisms normally operating during reinnervation of denervated muscle tissue. In the septohippocampal system, in particular, implants of denervated iris have thus been shown to stimulate the cholinergic septal neurons of the host to reinnervate large parts of the graft in an organotypic manner (Svendgaard et al 1976, Emson et al 1977).

Against this background, it seemed reasonable that a graft of embryonic hippocampus, deprived of its normal external afferents, could actively promote the regeneration of axons in the septohippocampal circuitry in order to establish connections with the deafferented graft. In these experiments (Kromer et al 1981a,b, Segal et al 1981) the rostral tip of the hippocampus plus the hippocampal fimbria were removed by suction on one side in young adult female rats, leaving an approximately 3 × 3 mm wide cavity that completely severed the septohippocampal connections. A transplant of embryonic hippocampus was then placed in the cavity, onto the vessel-rich ependyma overlying the anterior thalamus, in contact with the cut fimbria, rostrally, and with the cut hippocampus, caudally.

In most cases the implant was found to have fused with the cut surface of the host hippocampus and formed tissue bridges with both the septum and the caudate-putamen of the host. Over these tissue bridges AChE-positive fibers were seen to invade the implant to give rise to an AChE-positive neuropile that, with time, progressed to cover the entire implant. In parellel, the cholinergic marker enzyme, ChAT, increased progressively to between 40 and 60% of the normal hippocampal level by 6–24 weeks after transplantation. Septal lesions and HRP injections into the graft have provided evidence that the principal origins for the cholinergic afferents were the medial septum and the diagonal band nucleus of the host brain (Kromer et al 1981a).

These results suggest that a piece of deafferented embryonic CNS tissue—similar to denervated peripheral tissue—can indeed promote the regeneration of axotomized central neurons. This effect may be a fairly specific one, providing a stimulus for directed sprouting into the denervated target by axons that normally innervate the implanted tissue. The regenerating septohippocampal axons were not restricted, however, to the implant: they also expanded across the implant-hippocampal border into the dorsal part of the host hippocampus (Kromer et al 1981b). Due to the axotomy caused by the transplantation cavity, the AChE positive innervation of the dorsal parts of the hippocampus was initially totally removed and the ChAT level reduced to 5–10% of normal. By four to six weeks after transplantation, new AChE-positive fibers appeared in parts of the subiculum, hippocampus, and/or dentate gyrus, immediately bordering on the implant. By three months the new fibers had expanded about 2.5–3.5 mm further caudally to cover the entire dorsal hippocampal formation. In parallel, ChAT activity (not seen in control rats without implants) reappeared to an average of 30% of normal in the dorsal hippocampus. In the most successful cases, a recovery of as much as 50% of normal ChAT activity was recorded. As with the regenerated fibers in the implant, the newly formed AChE-positive innervation in the host hippocampus could be removed by a lesion of the ipsilateral medial septum-diagonal band area, and HRP-injections, made into the host hippocampus behind the implant, resulted in labeling of neurons in this area.

These results may have interesting implications for the understanding of the regulation of regenerative responses in the mammalian CNS. The effect of the implant is probably not to induce regeneration *per se,* but to stimulate and promote a regenerative response already present as a result of the lesion. It is tempting to suggest that this effect of the embryonic neural graft is related to its normal property to receive a certain type of innervation (in the present case a cholinergic one), and thus that the very same factors that guide axons to their terminal sites during ontogenesis may in certain cases also operate in the mature CNS to stimulate and guide regenerating axons back to their original sites of termination. This interpretation would imply that at least some types of neurons in the mature CNS retain their responsiveness to factors regulating the formation of neuronal connections during embryonic development. It is conceivable, therefore, that axonal regeneration after lesions in the adult mammalian CNS could be limited by the availability (or access to) factors released by (or present in) the denervated target tissue, rather than by a defective intrinsic regenerative capacity of the mature central neurons themselves.

Results obtained with intracerebral implants in other model systems indicate, however, that the mechanisms underlying the formation of afferents to the implanted grafts are considerably more complex (Harvey & Lund 1981, Lund & Harvey 1981, Jaeger & Lund 1980a, Oblinger et al 1980, Oblinger & Das 1982, David & Aguayo 1981, Benfey & Aguayo 1982). The following factors may influence or determine the formation of afferents to intracerebral grafts:

1. The extent of integration or fusion between the implant and the host brain tissue: Oblinger & Das (1982), in particular, have pointed out that in neocortical grafts in the adult cerebellum the area innervated by afferents from the host was proportional to the extent of tight fusion between graft and host.

2. The availability or proximity of axons in the area of the implant: Observations made on heterotopically placed grafts (i.e. neocortex to cerbellum and peripheral nerve into the brain) suggest that axons in close proximity of the implant would be the ones most likely to invade the implant, regardless of their natural anatomical relationship with the grafted tissue (Oblinger et al 1980, Oblinger & Das 1982, David & Aguayo 1981, Benfey & Aguayo 1982). It should be kept in mind, however, that anatomically distinct systems may share functional properties, e.g. transmitter characteristics (cf Svendgaard et al 1975b, 1976, Lewis & Cotman 1983). Thus, denervated irises will be reinnervated by cholinergic and andrenergic axons regardless of their anatomical identity.

3. In developing rats the stage to which different axonal projections have developed in the host may determine their abiltiy to expand into the implant. This has been proposed, e.g. by Lund et al (1982), as an explanation of differences in the input from cortical and geniculate areas into tectal transplants in neonatal rats.

4. In adult rats the site and extent of axotomy of different pathways may determine their ability to sprout into the implant. This is suggested, for

instance, by the observations of Svendgaard et al (1975b) and Emson et al (1977) that central monoaminergic axons will reinnervate intracerebral implants of iris only when the iris tissue is in close proximity of transected and actively sprouting preterminal axon bundles.

5. Specific interactions between host neurons and grafted target: The studies of Lund and collaborators (1982), in particular, have shown that the implants do not appear as passive receivers of axons, but that the types of afferent connections formed depend on the particular region that is grafted. This is exemplified by their observation that discrete areas of tectal implants, grafted adjacent to the superior colliculus in newborn rats, receive retinal afferents from the host, whereas cortical and retinal implants are devoid of such afferents [cf also the results of Yoon (1979) in adult goldfish]. Specific interactions between host and target are also supported by our own observations on the reinnervation of intracerebral implants of peripheral target tissues by central monoaminergic and cholinergic neurons. Our observations suggest that (a) different monoaminergic neuron types differ in their "affinity" for a denervated iris target, and the appropriate noradrenergic neurons are clearly favored (Svengaard et al 1975b, Björklund & Stenevi 1979a); (b) tissue normally innervated by adrenergic neurons (e.g. iris) will permit or stimulate the ingrowth by regenerating central noradrenergic axons, whereas targets normally lacking such innervation (e.g. uterus muscle) will not do so (Björklund & Stenevi 1971). As during regeneration in the PNS (Olson & Malmfors 1970, Ebendal et al 1980), both the density and the patterning of the ingrowing central monoaminergic and cholinergic afferents seem to be under regulation by the denervated implanted target.

Growth Regulation and Trophic Mechanisms

Three main lines of evidence indicate that the growth response of intracerebrally implanted neurons is due to a fairly specific interaction with the surrounding tissue of the adult host brain.

1. Implanted dopamine neurons have been observed to grow preferentially into areas that are normal targets for the dopamine neurons in the brain (Björklund & Stenevi 1979c, Björklund et al 1980a, Dunnett et al 1981b). Further, dopamine neurons implanted in the form of cell suspension into different regions of the brain exhibit an extensive fiber outgrowth when placed in the striatum, which is a normal target area, but they show no or very little fiber outgrowth into the host tissue when placed in parietal cortex, lateral hypothalamus, or globus pallidus, which are areas that do not belong to the normal primary targets of the mesencephalic dopamine neurons (Björklund et al 1983c). Interestingly, this differential growth response is similar to that observed on dopamine neurons in vitro by Di Porzio et al (1980), Prochiantz et al (1981), Hemmendinger et al (1981), and Appel (1981). They reported a

selective growth stimulatory effect on the cultured dopamine neurons by striatal tissue (i.e. the appropriate target tissue) or by membrane or protein fractions obtained from that tissue. In intracortical mesencephalic dopaminergic grafts innervating the adult neostriatum, Schultzberg et al (1983) have recently reported immunocytochemical observations suggesting that only those dopamine neurons which normally innervate the neostriatum (characterized by a lack of the peptide cholecystokinin) grow extensively into the underlying target. By contrast, the dopamine neurons of the ventral tegmental area (characterized by their concomitant cholecystockinin content), which normally innervate ventral forebrain areas but not the neostriatum, showed very poor ingrowth. This may suggest that the stimulation of fiber outgrowth from the implant is selective with respect to different subpopulations of dopamine neurons, and that the subpopulation normally innervating the neostriatum is favored.

2. As mentioned above, the fiber outgrowth from embryonic neural implants may be substantially modified by the presence or absence of intrinsic afferents, and in some cases the fiber outgrowth can be greatly stimulated by prior deafferentiation of the target. We have studied this in particular with respect to the outgrowth of cholinergic neurons in septal implants grafted to the hippocampal formation. Here, the fiber outgrowth into the host hippocampus (as monitored by measurements of ChAT activity) is about three-fold greater when the intrinsic cholinergic afferents have been removed (F. H. Gage et al, in preparation). Interestingly, the removal of a noncholinergic set of afferents to the hippocampus, i.e. the entorhinal perforant path input, has been shown not only to modify the patterning of ingrowing axons, but also to accelerate reinnervation and increase the magnitude of ingrowth by cholinergic axons (Björklund et al 1979a).

3. In the PNS, target tissues are known to exert trophic influences on neurons that innervate the target (see Hamburger 1977, Hendry 1976, Landmesser & Pilar 1978). Likewise, the growth-stimulating effect of target cell deafferentiaion has been proposed to be mediated by retrogradely acting trophic factors, elaborated by the denervated target (Ebendal et al 1980, Henderson et al 1983, Hill & Bennett 1983). The mechanisms operating during reinnervation in the CNS may for some systems by quite similar to those in the PNS. Thus, for example, a peripheral target tissue normally richly innervated by adrenergic fibers, such as the iris, has been seen to interact with regenerating central adrenergic neurons (i.e. those of the locus coeruleus) in very much the same way as it does with peripheral, ganglionic adrenergic neurons (Björklund & Stenevi 1979a). In fact, the growth-stimulating effect of a denervated iris implant on regenerating central adrenergic neurons is much reduced if the iris is incubated in antiserum to nerve growth factor (NGF) prior to transplantation (Bjerre et al 1974). These effects resemble those exerted by denervated irides

on peripheral adrenergic neurons in vitro (Johnson et al 1972) and in vivo (Ebendal et al 1980).

In a recent study (Björklund & Stenevi 1981) we have attempted to monitor the growth-stimulating effect expressed by the adult rat hippocampus in vivo after denervating lesions by means of superior cervical ganglia implanted adjacent to the hippocampal formation. Grafted adult sympathetic ganglia grow poorly into a hippocampus with intact septohippocampal inputs. A lesion of the septohippocampal pathway (made at the time of grafting or two months after grafting) causes a massive stimulation of axonal growth into the hippocampus. The increase was more than 100-fold by one month after lesion and was still about ten-fold by three months. The denervating lesion, in addition, induced a 3.6-fold increase in the average size of the gangionic cell bodies in the graft, without any clear-cut effect on the number of surviving ganglionic neurons. In a subsequent study (Gage et al 1983c) we have observed that the denervating lesion also has a dramatic effect on the survival of similarly grafted neonatal sympathetic ganglionic neurons. Neonatal sympathetic neurons, which normally are strictly dependent on NGF for their survival (see Hendry 1976), thus did not survive at all on the surface of the host hippocampus unless the hippocampus was deafferented at the time of implantation.

This lesion-induced trophic response (increases in axonal outgrowth, cell body size, and noradrenaline content in adult ganglia, and increase of neuronal survival in neonatal ganglia) is specific for lesions of the septal (probably primarily cholinergic) innervation of the hippocampus, while lesions of other major inputs (entorhinal or commissural) do not have this effect. Lesions of the intrinsic adrenergic afferents (from the locus coeruleus) are also ineffective.

These results speak strongly in favor of the notion that a neuronotrophic factor(s) takes part in the regulation of fiber ingrowth, and perhaps also neuronal survival, in the establishment of connections from intracerebral implants, and that this factor, or factors, is under the control of non-adrenergic (probably cholinergic) afferents originating in the septal-diagonal band area. Interestingly, the effects of this putative factor on the sympathetic neurons resemble that of NGF, and seem to act over relatively large distances. Crutcher & Collins (1982) and Nieto-Sampedro et al (1982) have recently been able to demonstrate, in extracts or fluid from the adult rat hippocampus, the presence of growth factors both similar to and different from NGF, and Barde et al (1982) have purified a factor from mammalian brain tissue with chemical (but not immunological) properties similar to NGF. It seems quite possible, therefore, that diffusible growth factors may operate in a very similar manner, both in the peripheral and central nervous systems to regulate sprouting and reinnervation of denervated targets. Intracerebral neural implants should provide a new powerful tool to monitor and explore such growth-regulating factors in the brain in vivo.

FUNCTIONAL EFFECTS OF INTRACEREBRAL IMPLANTS

A more complete account of the studies carried out on the functional properties of intracerebal neural grafts is outside the scope of the present review. In this section, therefore, we briefly summarize some of the key features that have emerged from experiments carried out in adult recipient rats. For a more complete coverage of this topic, the reader is referred to the recent reviews by Dunnett et al (1982a,b, 1983a,b) and Björklund et al (1983b).

Electron Microscopy

Ultrastructural analysis of transplant-host connections has so far been performed only in a few cases. Thus, axons from implants of brainstem tissue have been shown to make synaptic contacts with dendrites in the dentate molecular layer in adult hosts (Beebe et al 1979), and axons from retinal grafts have been shown to establish normal synaptic contacts in the superior colliculus of developing hosts (McLoon et al 1982). Further, Lund & Hauschka (1976) and Lund & Harvey (1981) observed in neonatal rats synapses in implants of tectal tissue, derived from the ingrowing host retinal afferents.

Biochemistry

Grafted dopamine neurons, reinnervating the initially denervated neostriatum, exhibit transmitter synthesis and turnover rates close to those of the normal nigrostriatal dopamine pathway. This has been recorded in animals with solid grafts of embryonic mesencephalon, placed in an intracortical cavity (Schmidt et al 1982), as well as with suspended dopamine neurons injected into the depth of the neostriatum (Schmidt et al 1983b). Interestingly, these biochemical studies suggest that the activity of the grafted dopaminergic neurons may to some extent be regulated through interactions with the reinnervated target. Thus, the turnover rates of dopamine were inversely related to the magnitude of fiber ingrowth from the graft, and in the more sparsely reinnervated specimens the turnover was considerably above normal. In the study of Schmidt et al (1982), the general metabolic activity of the implant, as measured by ^{14}C-deoxyglucose autoradiography, was found to be close to that of the substantia nigra region in situ.

Peptide Secretion

Grafts of embryonic hypothalamic tissue have been shown to restore endocrine function in animals with congential deficiency of vasopressin (Gash & Sladek 1980, Gash et al 1980) or gonadotrophin-releasing hormone (Krieger et al 1982).

Electrophysiology

There are now several electrophysiological studies confirming that functional connections are established between the graft and the host brain. Electrical stimulation of locus coeruleus implants (Björklund et al 1979b) and of septal implants (Low et al 1982) innervating the host hippocampus, have provided evidence for the establishment of normal inhibitory and excitatory synaptic connections, respectively, with neurons in the host hippocampus. Furthermore, in animals with "bridge" grafts of embryonic hippocampus, placed in a cavity in the hippocampal fimbria (see above), Segal et al (1981) have reported atropin-sensitive evoked responses from neurons within the implant, as well as in the reinnervated host hippocampus, after stimulation of the host medial septum. This suggests that the regenerating cholinergic septo-hippocampal neurons innervating the hippocampal implant and the host hippocampus across the newly formed tissue bridge indeed form functional connections with neurons in the reinnervated areas. Similarly, Harvey et al (1982) have provided electrophysiological evidence for the formation of functional synaptic contacts between host cortical afferents and transplanted tectal neurons, in rats grafted during the neonatal period.

Behavior

Behavioral recovery has been demonstrated on a range of motor and sensorimotor tests in rats with dopamine-rich nigral grafts reinnervating the neostriatum (Björklund & Stenevi 1979c, Perlow et al 1979, Björklund et al 1980a, 1981, Freed et al 1980, Dunnett et al 1981a–c). Nigral grafts have thus been seen to provide a complete, or near-complete, recovery of both spontaneous and drug-induced turning behavior, postural asymmetries, akinesia, as well as in the sensorimotor deficits (so-called sensory neglect) that develop contralateral to a lesion of the intrinsic nigrostriatal dopamine pathway. Interestingly, the recovery of the different behavioral parameters are regionally specific, in that grafts reinnervating the dorsal neostriatum showed pronounced recovery of the turning behavior but no effect on the sensorimotor deficit, whereas grafts reinnervating the ventrolateral parts of the neostriatum had the opposite effects (Dunnett et al 1981b). The behavioral recovery induced by nigral grafts has so far, however, not been complete: in particular, the profound eating and drinking impairments, which develop in rats with bilateral lesions of the mesotelencephalic dopamine pathway, have not been significantly improved with the types of implants hitherto studied.

The ability of intracerebral implants to improve the rats' behavior in more complex "cognitive" tasks has been demonstrated in maze-learning tests in animals with bilateral lesions of the hippocampal fimbria (Low et al 1982, Dunnett et al 1982c). Septal implants, grafted either as solid pieces into the fimbrial cavities or as cell suspensions directly into the initially denervated hippocampus, have thus been found to improve significantly the ability of the lesioned rats to learn spatial memory tasks in eight-armed (Low et al 1982) or three-armed radial mazes (Dunnett et al 1982c). This recovery was only partial, and in the eight-armed maze task the recovery was seen only after administration of physostigmine (which enhances cholinergic transmission by blocking the degrading enzyme, AChE). The histochemical analysis of these animals indicated that cholinergic reinnervation of the host hippocampus from the implants was a necessary but not sufficient prerequisite for the behavioral recovery to occur (Dunnett et al 1982c).

NEURONAL REPLACEMENT IN THE ADULT CNS

Morphological and behavioral studies, performed mainly on animals with grafts of monoamine-, acetylcholine-, or peptide-producing neurons, show that implanted embryonic nerve cells can in some cases substitute quite well for a lost intrinsic neuronal system. The intracerebral implants probably exert their effects in several ways. The functional effects seen with grafts placed into one of the cerebral ventricles, such as in the studies of Perlow et al (1979), Freed et al (1980, 1981), and Gash et al (1980), are thus probably explained on the basis of a diffuse release of the active amine or peptide into the host CSF and adjacent brain tissue. In other instances, such as in animals with dopamine-rich grafts reinnervating the neostriatum or acetylcholine-rich grafts reinnervating the hippocampus, we believe that the available data quite strongly show that the behavioral recovery is caused by the ability of the grafted neurons to reinnervate relevant parts of the host brain. This is illustrated by the studies mentioned above showing that the degree of functional recovery in rats with nigral transplants is directly correlated with the extent of striatal dopamine reinnervation, and that the "profile" of functional recovery is dependent on which area of the neostriatum is reinnervated by the graft. This point is particularly well illustrated in a further study (Fray et al 1983), in which rats with electrodes implanted into the center of intracortical nigral grafts were allowed to "self-stimulate" via the graft. The results show that the graft can indeed sustain self-stimulation behavior, and that the rate of lever pressing is related to the proximity between the electrode tip and the dopamine-containing neurons in the graft. This strongly supports the notion that the implanted dopamine neurons can transmit behaviorally meaningful and temporally organized information to the host brain via their efferent connections.

To what extent the intracerebral implants can be functionally integrated with the host brain is, however, still poorly known and remains therefore in interesting topic for further investigation. The chances for extensive integration may be greatest for neuronal suspension grafts implanted as deposits directly into the depth of the brain. Nevertheless, as discussed above, even solid grafts inserted as whole pieces into the brain have in several cases been seen to become reinnervated from the host brain, both in adult and developing recipients. The host afferents to the graft can, to some degree, be derived from functionally appropriate projections or neuron types within the host and can establish functional synaptic connections.

Available data, derived primarily from studies in monoaminergic and cholinergic systems, indicate that implanted embryonic central neurons can substitute to some degree for a lost set of afferents to a denervated brain region in adult rats, and replace a lost intrinsic neuronal system in normalizing the rat's behavior. This indicates a remarkable plasticity of the mature rat CNS in incorporating new neuronal elements into its already established circuitries. There is now abundant evidence that the adult CNS can reorganize and rebuild itself in reponse to damage (see Cotman & Lynch 1976, Tsukahara 1981), and perhaps in some cases as a way of normal physiological adaptation of the system (Graziadei & Monti Graziadei 1978, Nottebohm 1981, Cotman et al 1981, Tsukahara 1981). Neuronal replacement by neural implants is a striking further example of how the brain can allow new elements to be inserted and linked into its own functional subsystems. Obviously there must be definite limitations as to which types of neurons or functional subsystems can successfully be manipulated in this way. Neural implants would seem most likely to have behaviorally meaningful functional effects with types of neurons that normally do not convey, or link, specific or patterned messages, e.g. in sensoric or motoric input and output systems. Indeed, functional or behavioral recovery in the neuronal replacement paradigm has so far been demonstrated only for neurons of the types that normally appear to act as tonic regulatory or level-setting systems. Such a mode of operation makes it conceivable that implanted neurons can also function well in the absence of some, or perhaps even all, of their normal afferent inputs. Elsewhere (Björklund et al 1981, 1983b) we have discussed in some detail how implants of neurons with a normal level setting or "command" function can operate to reactivate damaged circuitries in the brain or spinal cord.

The use of neuronal implants for neuronal replacement has a particular interest in the context of animal models of neurodegenerative disorders. As summarized above, the most extensive studies have been performed in rats with 6-hydroxydopamine induced lesions of the meso-telencephalic dopamine system, which is a lesion that has been extensively employed as a model for Parkinson's disease. It has also proved feasible, however, to implant embryonic striatal neurons into the striatum of rats whose intrinsic striatal neurons have been destroyed by the neurotoxin, kainic acid (Schmidt et al 1981). Such rats reproduce the neuropathological changes associated with Huntington's disease in man. Moreover, Gage et al (1983b) have recently found that suspensions of embryonic septal or mesencephalic tissue will survive and grow well after implantation into the intact hippocampus and neostriatum, respectively, of aging rats. To the extent that the aging process is associated with a decline in function, or even an actual loss, of selected neuronal elements in the brain, neuronal implantation may provide an interesting new approach for the analysis of the cellular events underlying age-related functional impairments in rats (Gage et al 1983a). In fact, the observations of Gage et al (1983b) have furnished some preliminary evidence that dopamine neurons implanted into the depth of the neostriatum in aging rats will not only establish a new dopamine-containing terminal plexus in the surrounding striatal tissue, but also restore some aspects of the age-related impairment of motor performance in these rats.

Literature Cited

Altman, J. 1969. Autoradiographic and histological studies of postnatal neurogenesis. II. Dating the time of production and onset of differentiation of cerebellar microneurons in rats. *J. Comp. Neurol.* 136:269–94

Altobelli, R. 1914. Innesti cerebrali. *Grazz. Int. Med. Ghir.* 17:25–34

Alvarado-Mallart, R. M., Sotelo, C. 1982. Differentiation of cerebellar anlage heterotypically transplanted to adult rat brain: a light and electron microscopic study. *J. Comp. Neurol.* 212:247–67

Appel, S. H. 1981. A unifying hypothesis for the cause of amyotrophic lateral sclerosis, Parkinsonism, and Alzheimer disease. *Ann. Neurol.* 10:499–505

Arendash, G. W., Gorski, R. A. 1982. Enhancement of sexual behaviour in female rats by neonatal transplantation of brain tissue from males. *Science* 217:1276–78

Banker, G. A., Cowan, W. M. 1977. Rat hippocampal neurons in dispersed cell culture. *Brain Res.* 126:397–425

Barde, Y.-A., Edgar, D., Thoenen, H. 1982. Purification of a new neurotrophic factor from mammalian brain. *EMBO J.* 1:549–53

Barker, C. F., Billingham, R. E. 1977. Immunologically privileged sites. *Adv. Immunol.* 25:1–54

Beebe, B. K., Møllgård, K., Björklund, A., Stenevi, U. 1979. Ultrastructural evidence of synaptogenesis in the adult rat dentate gyrus from brain stem implants. *Brain Res.* 167:391–95

Benfey, M., Aguayo, A. J. 1982. Extensive elongation of axons from rat brain into peripheral nerve grafts. *Nature* 296:150–52

Bjerre, B., Björklund, A., Stenevi, U. 1974. Inhibition of the regenerative growth of central noradrenergic neurons by intracerebrally administered anti-NGF serum. *Brain Res.* 74:1–18

Björklund, A., Dunnett, S. B., Stenevi, U., Lewis, M. E., Iversen, S. D. 1980a. Reinnervation of the denervated striatum by substantia nigra transplants: functional consequences as revealed by pharmacological and sensorimotor testing. *Brain Res.* 199:307–33

Björklund, A., Johansson, B., Stenevi, U., Svendgaard, N.-A. 1975. Reestablishment of functional connections by regenerating central adrenergic and cholinergic axons. *Nature* 253:446–48

Björklund, A., Kromer, L. F., Stenevi, U. 1979a. Cholinergic reinnervation of the rat hippocampus by septal implants is stimulated by peforant path lesion. *Brain Res.* 173:57–64

Björklund, A., Nornes, H., Dunnett, S. B., Glage, F. H., Stenevi, U. 1983a. Intracerebral and intraspinal implants of locus coeruleus cell suspensions: deleterious effect of trypsin in the suspension medium. *Soc. Neurosci. Abstr.* In press

Björklund, A., Schmidt, R. H., Stenevi, U. 1980b. Functional reinnervation of the neostriatum in the adult rat by use of intraparenchymal grafting of dissociated cell suspensions from the substantia nigra. *Cell Tissue Res.* 212:39–45

Björklund A., Segal, M., Stenevi, U. 1979b. Functional reinnervation of rat hippocampus by locus coeruleus implants. *Brain Res.* 170:409–26

Björklund, A., Stenevi, U. 1971. Growth of central catecholamine neurones into smooth muscle grafts in the rat mesencephalon. *Brain Res.* 31:1–20

Björklund, A., Stenevi, U. 1977a. Experimental reinnervation of the rat hippocampus by grafted sympathetic ganglia 1. Axonal regeneration along the hippocampal fimbria. *Brain Res.* 138:259–70

Björklund, A., Stenevi, U. 1977b. Reformation of the severed septohippocampal cholinergic pathway in the adult rat by transplanted septal neurones. *Cell Tissue Res.* 185:289–302

Björklund, A., Stenevi, U. 1979a. Regeneration of monoaminergic and cholinergic neurons in the mammalian central nervous system. *Physiol. Rev.* 59:62–100

Björklund, A., Stenevi, U. 1979b. Reconstruction of brain circuitries by neural transplants. *Trends Neurosci.* 2:301–6

Björklund, A., Stenevi, U. 1979c. Reconstruction of the nigrostriatal dopamine pathway by intracerebral nigral transplants. *Brain Res.* 177:555–60

Björklund, A., Stenevi, U. 1981. In vivo evidence for a hippocampal adrenergic neurotrophic factor specifically released on septal deafferentation. *Brain Res.* 229:403–28

Björklund, A., Stenevi, U., Dunnett, S. B. 1983b. Transplantation of brain stem monoaminergic "command systems": A model for functional reactivation of damaged CNS circuitries. In *Spinal Cord Reconstruction*, ed. C. C. Kao, R. P. Bunge, P. J. Reier, pp. 397–413. New York: Raven

Björklund, A., Stenevi, U., Dunnett, S. B., Gage, F. H. 1982b. Cross-species neural grafting in a rat model of Parkinson's disease. *Nature* 298:652–54

Björklund, A., Stenevi, U., Dunnett, S. B., Iversen, S. D. 1981. Functional reactivation of the deafferented neostriatum by nigral transplants. *Nature* 289:497–99

Björklund, A., Stenevi, U., Schmidt, R. H., Dunnett, S. B., Gage, F. H. 1983c. Intracerebral grafting of neuronal cell suspensions. II. Survival and growth of nigral cells implanted in different brain sites. *Acta Physiol. Scand. Suppl.* 522:11–22

Björklund, A., Stenevi, U., Svendgaard, N.-A. 1976. Growth of transplanted monoaminergic neurones into the adult hippocampus along the perforant path. *Nature* 262:787–90

Bragin, A. G., Vinogradova, O. S. 1981. Homo- and hetero-specific transplantation of embryonal central nervous tissue. *Bull. Exp. Biol. Med.* 10:486–89 (In Russian)

Cajal, S. R. 1928. *Degeneration and Regeneration of the Nervous System*. London: Oxford Univ. Press

Cotman, C. W., Lynch, G. S. 1976. Reactive synaptogenesis in the adult nervous system. In *Neuronal Recognition*, ed. S. H. Barondes, pp. 69–108. New York: Plenum

Cotman, C. W., Nieto-Sampedro, M., Harris E. W. 1981. Synapse replacement in the nervous system of adult vertebrates. *Physiol. Rev.* 61:684–761

Crutcher, K. A., Collins, F. 1982. In vitro evidence for two distinct hippocampal growth factors: Basis of neuronal plasticity? *Science* 217:67–68

Das, G. D. 1974. Transplantation of embryonic neural tissue in the mammalian brain. I. Growth and differentiation of neuroblasts from various regions of the embryonic brain in the cerebellum of neonate rates. *T.I.T.J. Life Sci.* 4:93–124

Das, G. D. 1983. Neural transplantation of the spinal cord of the adult mammal. See Björklund et al 1983b, pp. 367–96

Das, G. D., Altman, J. 1971. Transplanted precursors of nerve cells: Their fate in the cerebellums of young rats. *Science* 173:637–38

Das, G. D., Altman, J. 1972. Studies on the transplantation of developing neural tissue in the mammalian brain. I. Transplantation of cerebellar slabs into the cerebellum of neonate rats. *Brain Res.* 38:233–49

Das, G. D., Hallas, B. H. 1978. Transplantation of brain tissue in the brain of adult rats. *Experientia* 34:1304–6

Das, G. D., Hallas, B. H., Das, K. G. 1979. Transplantation of neural tissue in the brains of laboratory mammals: Technical details and comments. *Experientia* 35:143–53

Das, G. D., Hallas, B. H., Das, K. G. 1980. Transplantation of brain tissue in the brain of rat. I. Growth characteristics of neocortical transplants from embryos of different ages. *Am. J. Anat.* 158:135–45

David, S., Aguayo, A. J. 1981. Axonal elongation into peripheral nervous system "bridges" after central nervous system injury in adult rats. *Science* 214:931–33

Del Conte, G. 1907. Einpflanzungen von embryohalem Gewebe ins Gehirn. *Beitr. Pathol. Anat. Allg. Pathol.* 42:193–202

Detwiler, S. R. 1936. *Neuroembryology: An Experimental Study*. New York: Macmillan

Di Porzio, U., Daguet, M. C., Glowinski, J., Prochiantz, A. 1980. Effect of striatal cells on *in vitro* maturation of mesencephalic dopaminergic neurons grown in serum-free conditions. *Nature* 288:370–73

Dunn, E. H. 1917. Primary and secondary findings in a series of attempts to transplant cerebral cortex in albino rat. *J. Comp. Neurol.* 27:565–82

Dunnett, S. B., Björklund, A., Stenevi, U. 1983a. Dopamine-rich transplants in experimental Parkinsonism. *Trends Neurosci.* 6:266–70

Dunnett, S. B., Björklund, A., Stenevi, U. 1983b. Transplant-induced recovery from brain lesions: A review of the nigrostriatal model. In *Neural Tissue Transplantation Re-*

search, ed. R. B. Wallace, G. D. Das. New York: Springer–Verlag. In press

Dunnett, S. B., Björklund, A., Stenevi, U., Iversen, S. D. 1981a. Behavioural recovery following transplantation of substantia nigra in rats subjected to 6-OHDA lesions of the nigrostriatal pathway. I. Unilateral lesions. *Brain Res.* 215:147-61

Dunnett, S. B., Björklund, A., Stenevi, U., Iversen, S. D. 1981b. Grafts of embryonic substantia nigra reinnervating the ventrolateral striatum aneliorate sensorimotor impairments and akinesia in rats with 6-OHDA lesions of the nigrostriatal pathway. *Brain Res.* 229:209-17

Dunnett, S. B., Björklund, A., Stenevi, U., Iversen, S. D. 1981c. Behavioural recovery following transplantation of substantia nigra in rats subjected to 6-OHDA lesions of the nigrostriatal pathway. II. Bilateral lesions. *Brain Res.* 229:457-70

Dunnett, S. B., Björklund, A., Stenevi, U., Iversen, S. D. 1982a. CNS transplantation: Structural and functional recovery from brain damage. *Prog. Brain Res.* 55:431-44

Dunnett, S. B., Gage, F. H., Björklund, A., Stenevi, U., Low, W. C., Iversen, S. D. 1982b. Hippocampal deafferentation: Transplant-derived reinnervation and functional recovery. *Scand. J. Psychol. Suppl.* 1:104-11

Dunnett, S. B., Low, W. C., Iversen, S. D., Stenevi, U., Björklund, A. 1982c. Septal transplants restore maze learning in rats with fornix-fimbria lesions. *Brain Res.* 251:335-48

Ebendal, T., Olson, L., Seiger, A., Hedlund, K.-O. 1980. Nerve growth factors in the rat iris. *Nature* 286:25-28

Emson, P. C., Björklund, A., Stenevi, U. 1977. Evaluation of the regenerative capacity of central dopaminergic, noradrenergic and cholinergic neurones using iris implants as targets. *Brain Res.* 135:87-105

Flerkó, B., Szentágothai, J. 1957. Oestrogen sensitive nervous structures in the hypothalamus. *Acta Endocrinol.* 26:121-27

Fray, P. J., Dunnett, S. B., Iversen, S. D., Björklund, A., Stenevi, U. 1983. Nigral transplants reinnervating the dopamine-depleted neostriatum can sustain intracranial self-stimulation. *Science* 219:416-19

Freed, W. J., Morihisa, J. M., Spoor, E., Hoffer, B. J., Olson, L., Seiger, Å, Wyatt, R. J. 1981. Transplanted adrenal chromaffin cells in the rat brain reduce lesion-induced rotational behavior. *Nature* 292:351-52

Freed, W. J., Perlow, M. J., Karoum, F., Seiger, A., Olson, L., Hoffer, B. J., Wyatt, R. J. 1980. Restoration of dopaminergic function by grafting of fetal rat substantia nigra to the caudate nucleus: Long-term behavioural,

biochemical and histochemical studies. *Ann. Neurol.* 8:510-19

Frotscher, M., Buck, E., Mannsfeld, B., Wenzel, J. 1970. Zur frage der regeneration des cortex cerebri nach replantation eines cortexabschnittes bei rattus norvegicus B. *J. Hirnforsch.* 33:123-33

Gage, F. H., Björklund, A., Stenevi, U., Dunnett, S. B. 1983a. Intracerebral grafting in the aging brain. In *Aging in the Brain*, ed. Gispen, Taber. In press

Gage, F. H., Dunnett, S. B., Björklund, A., Stenevi, U. 1983b. Aged rats: Recovery of motor coordination impairments by intrastriatal nigral grafts. *Science.* 221:966-69

Gage, F. H., Björklund, A., Stenevi, U. 1983c. Denervation releases a neuronal survival factor in adult rat hippocampus. *Science* Submitted

Gash, D., Sladek, J. R. 1980. Vasopressin neurons grafted into Brattleboro rats: Viability and activity. *Peptides* 1:11-14

Gash, D., Sladek, J. R., Sladek, C. D. 1980. Functional development of grafted vasopressin neurons. *Science* 210:1367-69

Gaze, R. M., Watson, W. E. 1968. Cell division and migration in the brain after optic nerve lesions. In *Growth of the Nervous System*, ed. G. E. W. Wolstenholme, M. O'Connor, pp. 53-67. London: Churchill

Glees, P. 1955. Studies of cortical regeneration with special reference to cerebral implants. In *Regeneration in the Central Nervous System*, ed. W. F. Windle, pp. 94-111. Springfield: Thomas

Graziadei, P. P. C., Kaplan, M. S. 1980. Regrowth of olfactory sensory axons into transplanted neural tissue. 1. development of connections with the occipital cortex. *Brain Res.* 210:39-44

Graziadei, P. P. C., Monti Graziadei, G. A. 1978. Continuous nerve cell renewal in the olfactory system. In *Handbook of Sensory Psychology*, Vol. 9, *Development of Sensory Systems*, ed. M. Jacobson. New York: Springer-Verlag

Halasz, B., Pupp, L., Uhlarik, S. 1962. Hypophysiotrophic area in the hypothalamus. *J. Endocrinol.* 25:147-59

Hallas, B. H., Das, G. D., Das, K. G. 1980. Transplantation of brain tissue in the brain of rat. II. Growth characteristics of neocortical transplants in hosts of different ages. *Am. J. Anat.* 158:147-59

Hallas, B. H., Oblinger, M. M., Das, G. D. 1980. Heterotopic neural transplants in the cerebellum of the rat: Their afferents. *Brain Res.* 196:242-46

Hamburger, V. 1977. The developmental history of the motor neuron. *NRP Bull.* 15:1-37 (Suppl.)

Harvey, A. R., Golden, G. T., Lund, R. D.

1982. Transplantation of tectal tissue in rats. *Exp. Brain Res.* 47:437–45

Harvey, A. R., Lund, R. D. 1981. Transplantation of fetal tissue in rats. II. Distribution of host neurons which project to transplants. *J. Comp. Neurol.* 202:505–20

Hemmendinger, L. M., Garber, B. B., Hoffmann, P. C., Heller, A. 1981. Target neuron-specific process formation by embryonic mesencephalic dopamine neurons *in vitro*. *Proc. Natl. Acad. Sci. USA* 78:1264–68

Henderson, C. E., Huchet, M., Changeux, J.-P. 1983. Denervation increases a neurite-promoting activity in extracts of skeletal muscle. *Nature* 302:609–11

Hendry, I. A. 1976. Control in the development of the vertebrate sympathetic nervous system. *Rev. Neurosci.* 2:149–78

Hill, M. A., Bennett, M. R. 1983. Cholinergic growth factor from skeletal muscle elevated following denervation. *Neurosci. Lett.* 35:31–35

Horvat, J.-C. 1966. Comparison des réactions régénératives provoquées dans le cerveau et dans le cervelet de la souris par des greffes tissulaires intraraciales. *Bull. Assoc. Anat.* 51:487–99

Jacobson, M., Levine, R. L. 1975a. Plasticity in the adult frog brain: Filling the visual scotoma after excision or translocation of parts of the optic tectum. *Brain Res.* 88:339–45

Jacobson, M., Levine, R. 1975b. Stability of implanted duplicate tectal positional markers serving as targets for optic axons in adult frogs. *Brain Res.* 92:468–71

Jaeger, C. B., Lund, R. D. 1979. Efferent fibers from transplanted cerebral cortex of rats. *Brain Res.* 165:338–42

Jaeger, C. B., Lund, R. D. 1980a. Transplantation of embryonic accipital cortex to the tectal region of newborn rats: A light microscopic study of organization and connectivity of the transplants. *J. Comp. Neurol.* 194:571–97

Jaeger, C. B., Lund, R. D. 1980b. Transplantation of embryonic occipital cortex to the brain of newborn rats. An autoradiographic study of transplant histogenesis. *Exp. Brain Res.* 40:265–72

Jaeger, C. B., Lund, R. D. 1981. Transplantation of embryonic occipital cortex to the brain of newborn rats: A golgi study of mature and developing transplants. *J. Comp. Neurol.* 200:213–30

Johnson, D. G., Silberstein, S. D., Hanbauer, I., Kopin, I. J. 1972. The role of nerve growth factor in the ramification of sympathetic nerve fibres into the rat iris in an organ culture. *J. Neurochem.* 19:2025–29

Kao, C. C., Schimizu, Y., Perkins, S. C.,

Freeman, L. W. 1970. Experimental use of cultured cerebellar cortical tissue to inhibit the collagenous scar following spinal cord transection. *J. Neurosurg.* 33:127–39

Krieger, D. T., Perlow, M. J., Gibson, M. J., Dames, T. F., Zimmerman, E. A., Ferin, M., Charlton, H. M. 1982. Brain grafts reverse hypogonadism of gonadotropin-releasing hormone deficiency. *Nature* 298:468–71

Kromer, L. F., Björklund, A. 1980. Embryonic neural transplants provide model systems for studying development and regeneration in the mammalian CNS. In *Multidisciplinary Approach to Brain Development*, ed. C. di Bernadetta, R. Balazs, G. Gombos, G. Porcellati, pp. 409–26. Amsterdam: Elsevier/North Holland

Kromer, L. F., Björklund, A., Stenevi, U. 1979. Intracerebral implants: A technique for studying neuronal interactions. *Science* 204:1117–19

Kromer, L. F., Björklund, A., Stenevi, U. 1981a. Innervation of embryonic hippocampal implants by regenerating axons of cholinergic septal neurons in the adult rat. *Brain Res.* 210:153–71

Kromer, L. F., Björklund, A., Stenevi, U. 1981b. Regeneration of the septohippocampal pathway in adult rats is promoted by utilizing embryonic hippocampal implants as bridges. *Brain Res.* 210:173–200

Kromer, L. F., Björklund, A., Stenevi, U. 1983. Intracephalic neural implants in the adult rat brain. I. Growth and mature organization of brain stem, cerebellar and hippocampal implants. *J. Comp. Neurol.* 218:433–59

Landmesser, L., Pilar, G. 1978. Interactions between neurons and their targets during in vivo synaptogenesis. *Fed. Proc.* 47:2016–22

Lauder, J. M., Bloom, F. E. 1974. Ontogeny of monoamine neurons in the locus coeruleus, raphe nuclei and substantia nigra of the rat. I. Cell differentiation. *J. Comp. Neurol.* 155:469–82

LeGros Clark, W. E. 1940. Neuronal differentiation in implanted foetal cortical tissue. *J. Neurol. Psychiatr.* 3:263–84

LeGros Clark, W. E. 1968. *Chant of Pleasant Exploration*. London: Livingstone

Levine, R., Jacobson, M. 1974. Deployment of optic nerve fibers is determined by positional markers in the frog's tectum. *Exp. Neurol.* 43:527–38

Lewis, E. R., Cotman, C. W. 1980. Mechanisms of septal lamination in the developing hippocampus revealed by outgrowth of fibers from septal implants. I. Positional and temporal factors. *Brain Res.* 196:307–30

Lewis, E. R., Cotman, C. W. 1983. Neurotransmitter characteristics of brain grafts: Striatal and septal tissues form the same laminated input to the hippocampus. *Neuroscience* 8:57–66

Low, W. C., Lewis, P. R., Bunch, S. T. 1983. Embryonic neural transplants across a major histocompatibility barrier: Survival and specificity of innervation. *Brain Res.* 262:328–33

Low, W. C., Lewis, P. R., Bunch, S. T., Dunnett, S. B., Thomas, S.R., Iversen S. D., Björklund, A., Stenevi, U. 1982. Functional recovery following neural transplantation of embryonic septal nuclei in adult rats with septohippocampal lesions. *Nature* 300: 260–62

Lund, R. D. 1980. Tissue transplantation: A useful tool in mammalian neuroembryology. *Trends Neurosci.* 3(11):12–13

Lund, R. D., Harvey, A. R. 1981. Transplantation of tectal tissue in rats. I. Organization of transplants and pattern of distribution of host afferents within them. *J. Comp. Neurol.* 201:191–209

Lund, R. D., Harvey, A. R., Jaeger, C. B., McLoon, S. C. 1982. Transplantation of embryonic neural tissue to the tectal region of newborn rats. In *Changing Concepts of the Nervous System*, ed. A. R. Morrison, P. L. Strick. New York: Academic

Lund, R. D., Hauschka, S. D. 1976. Transplanted neural tissue develops connections with host rat brain. *Science* 193:582–84

Matthey, R. 1926. Lagreffe de l'oeil. Etude experimental de la greffe de l'oeil chez le Triton (Triton cristatur). *Arch. Entwicklungsmech. Org.* 109:326–41

McLoon, S. C., Lund, R. D. 1980a. Identification of cells in retinal transplants which project to host visual centers: A horseradish peroxidase study in rats. *Brain Res.* 197:431–95

McLoon, S. C., Lund, R. D. 1980b. Specific projections of retina transplanted to rat brain. *Exp. Brain Res.* 40:273–82

McLoon, L. K., Lund, R. D., McLoon, S. C. 1982. Transplantation of reaggregats of embryonic neural retinae to neonatal rat brain: Differentiation and formation of connections. *J. Comp. Neurol.* 205:179–89

McLoon, L. K., McLoon, S. C., Lund, R. D. 1981. Cultured embryonic retinae transplanted to rat brain: Differentiation and formation of projections to host superior colliculus. *Brain Res.* 226:15–31

Møllgaard, K., Lundberg, I. J., Beebe, B. K., Björklund, A., Stenevi, U. 1978. The intracerebrally cultured "microbrain": A new tool in developmental neurobiology. *Neurosci. Lett.* 8:295–301

Nieto-Sampedro, M., Lewis, E. R., Cotman, C. W., Manthorpe, M., Skaper, S. D., Barbin, G., Longo, F. M., Varon, S. 1982.

Brain injury causes a time-dependent increase in neuronotrophic activity at the lesion site. *Science* 217:860–61

Nottebohm, F. 1981. A brain for all seasons: Cyclic anatomical changes in song control nuclei of the canary brain. *Science* 214: 1368–70

Oblinger, M. M., Das, G. D. 1982. Connectivity of neural transplants in adult rats: Analysis of afferents and efferents of neocortical transplants in the cerebellar hemisphere. *Brain Res.* 249:31–49

Oblinger, M. M., Hallas, B. H., Das, G. D. 1980. Neocortical transplants in the cerebellum of the rat: Their afferents and efferents. *Brain Res.* 189:228–32

Olson, L., Malmfors, T. 1970. Growth characteristics of adrenergic nerves in the adult rat. *Acta Physiol. Scand. Suppl.* 348:1–112

Olson, L., Seiger, A., Strömberg, I. 1982. Intraocular transplantation in rodents. A detailed account of the procedure and example of its use in neurobiology with special reference to brain tissue grafting. In *Advances in Cellular Neurobiology*, ed. S. Federoff, Vol 4. New York: Academic

Perlow, M. J., Freed, W. J., Hoffer, B. J., Seiger, A., Olson, L., Wyatt, R. J. 1979. Brain grafts reduce motor abnormalities produced by destruction of nigrostriatal dopamine system. *Science* 204:643–47

Piatt, J. 1940. Nerve-muscle specificity in Amblystoma, studied by means of heterotopic cord grafts. *J. Exp. Zool.* 85:211–41

Prochiantz, A., Daguet, M. C., Herbet, A., Glowinski, J. 1981. Specific stimulation of *in vitro* maturation of mesencephalic dopaminergic neurons by striatal membranes. *Nature* 293:570–72

Raisman, G., Field, P. M. 1973. A quantitative investigation of the development of collateral reinnervation after partial deafferentation of the septal nuclei. *Brain Res.* 50:241–64

Raju, S., Grogan, J. B. 1977. Immunologic study of the brain as a privileged site. *Transplant. Proc.* 9:1187–91

Ranson, S. W. 1914. Transplantation of the spinal ganglion, with observations on the significance of the complex types of spinal ganglion cells. *J. Comp. Neurol.* 24:547–58

Richardson, P. M., McGuinness, U. M., Aguayo, A. J. 1980. Axons from CNS neurons regenerate into PNS grafts. *Nature* 284:264–65

Richardson, P. M., McGinness, U. M., Aguayo, A. J. 1982. Peripheral nerve autografts to the rat spinal cord: Studies with axonal tracing methods. *Brain Res.* 237: 147–62

Rosenstein, J. M., Brightman, H. W. 1978. Intact cerebral ventricle as a site for tissue transplantation. *Nature* 276:83–85

Rosenstein, J. M., Brightman, M. W. 1979.

Regeneration and myelination in autonomic ganglia transplanted to intact brain surface. *J. Neurocytol.* 8:359–79

Saltykow, S. 1905. Versuche über Gehirnplantation, zugleich ein Beitrag zur Kenntriss der Vorgänge an den zelligen Gehirneelementen. *Arch. Psychiatr. Nervenkr.* 40: 329–88

Schmidt, R. A., Björklund, A., Stenevi, U. 1981. Intracerebral grafting of dissociated CNS tissue suspensions: A new approach for neuronal transplantation to deep brain sites. *Brain Res.* 218:347–56

Schmidt, R. H., Björklund, A., Stenevi, U., Dunnett, S. B. 1983a. Intracerebral grafting of dissociated CNS tissue suspensions. In *Nerve, Organ and Tissue Regeneration: Research Perspectives,* ed. F. J. Seil. New York: Academic. In press

Schmidt, R. H., Ingvar, M., Lindvall, O., Stenevi, U., Björklund, A. 1982. Functional activity of substantia nigra grafts reinnervating the striatum: Neurotransmitter metabolism and [14C]-2-deoxy-D-glucose autoradiography. *J. Neurochem.* 38:737–48

Schultzberg, M., Dunnett, S. B., Iversen, S. D., Björklund, A., Stenevi, U., Dackray, G., Hökfelt, T. 1983. Dopamine and cholecystokinin immunorective neurones in mesencephalic grafts reinnervatin the neostriatum: Evidence for selective growth regulation. *Neuroscience.* In press

Segal, M., Stenevi, U., Björklund, A. 1981. Reformation in adult rats of functional septohippocampal connections by septal neurons regenerating across an embryonic hippocampal tissue bridge. *Neurosci. Lett.* 27: 7–12

Sieger, A., Olson, L. 1977. Quantitation of fiber growth in transplanted central monoamine neurons. *Cell Tissue Res.* 179: 285–316

Sharma, S. C., Gaze, R. M. 1971. The retinotopic organization of visual responses from tectal reimplants in adult goldfish. *Arch. Ital. Biol.* 109:357–66

Sperry, R. W. 1945. Restoration of vision after uncrossing of optic nerves and after contralateral transposition of the eye. *J. Neurophysiol.* 8:15–28

Stenevi, U., Björklund, A., Dunnett, S. B. 1980a. Functional reinnervation of the denervated neostriatum by nigral transplants. *Peptides* 1(Suppl. 1):111–16

Stenevi, U., Björklund, A., Kromer, L. F., Paden, C. M., Gerlach, J. L., McEwen, B. S., Silverman, A. J. 1980b. Differentiation of embryonic hypothalamic transplants cultured on the choroidal pia in brains of adult rats. *Cell Tissue Res.* 205:217–28

Stenevi, U., Björklund, A., Svendgaard, N.-Aa. 1976. Transplantation of central and peripheral monoamine neurons to the adult rat brain: Techniques and conditions for survival. *Brain Res.* 114:1–20

Stone, L. S. 1944. Functional polarization in retinal development and its reestablishment in regenerated retinae of rotated eyes. *Proc. Soc. Exp. Biol. Med.* 57:13–14

Stone, L. S. 1963. Vision in eyes of several species of adult newts transplanted to adult *Triturus viridescens. J. Exp. Zool.* 153:57–67

Stone, L. S., Farthing, L. S. 1942. Return of vision four times in the same adult salamander eye *(Triturus viridescens)* repeatedly transplanted. *J. Exp. Zool.* 91:265–85

Stone, L. S., Zaur, J. S. 1940. Reimplantation and transplantation of adult eyes in the salamander *(Triturus viridescens)* with return of vision. *J. Exp. Zool.* 85:243–70

Stranznicky, K. 1963. Function of heterotopic spinal cord segments investigated in the chick. *Acta Biol. Acad. Sci. Hung.* 14:143–55

Sugar, O., Gerard, R. W. 1940. Spinal cord regeneration in the rat. *J. Neurophysiol.* 3:1–19

Sunde, N. Aa., Zimmer, J. 1981. Transplantation of central nervous tissue. An introduction with results and implications. *Acta Neurol. Scand.* 63:323–35

Sunde, N. Aa., Zimmer, J. 1983. Cellular histochemical and connective organization of the hippocampus and fascia dentata transplanted to different regions of immature and adult rat brains. *Dev. Brain Res.* 8: 165–91

Svendgaard, N.-Aa., Björklund, A., Hardebo, J.-E., Stenevi U. 1975a. Axonal degeneration associated with a defective blood-brain barrier in cerebral implants. *Nature* 255:334–36

Svendgaard, N.-Aa., Björklund, A., Stenevi, U. 1975b. Regenerative properties of central monoamine neurons as revealed in studies using iris transplants as targets. *Adv. Anat. Embryol. Cell Biol.* 51:1–77

Svendgaard, N.-Aa., Björklund, A., Stenevi, U. 1976. Regeneration of central cholinergic neurons in the adult rat brain. *Brain Res.* 102:1–22

Székely, G. 1963. Functional specificity of spinal cord segments in the control of limb movements. *J. Embryol. Exp. Morphol.* 11:431–44

Székely, G. 1968. Development of limb movements: Embryological, physiological and model studies. In *Growth of the Nervous System,* ed. G. E. W. Wolstenholme, M. O'Connor pp. 77–93. London: Churchill

Tello, F. 1911. La Influencia del neurotropismo en la regeneracion de los centros nerviosos. *Trab. Lab. Invest. Biol.* 9:123–59

Thompson, W. G. 1890. Successful brain grafting. *NY Med. J.* 51:701–2

Tsukahara, N. 1981. Synaptic plasticity in the mammalian central nervous system. *Ann. Rev. Neurosci.* 4:351–79

Tidd, C. W. 1932. The transplantation of spinal ganglia in the white rat. A study of the morphological changes in surviving cells. *J. Comp. Neurol.* 55:531–43

Weiss, P. A. 1950. The deplantation of fragments of nervous system in amphibians. I. Central reorganization and the formation of nerves. *J. Exp. Zool.* 113:397–462

Wells, J. McAllister, J. P. 1982. The development of cerebellar primordia transplanted to the neocortex of the rat. *Dev. Brain Res.* 4:167–79

Wenzel, J., Bärlehner, E. 1969. Zur Regeneration des Cortex cerebri bei Mus musculus. II. Morphologische Befunde regenerativer Vorgänge nach Replantation eines Cortexabschnittes. *Z. Mikrosk. Anat. Forsch.* 81:32–70

Willis, R. A. 1935. Experiments on the intracerebral implantation of embryo tissue in rats. *Proc. R. Soc. B Ser.* 117:400–12

Yoon, M. G. 1973. Retention of the original topographic polarity by the 180° rotated tectal reimplant in young adult goldfish. *J. Physiol* 233:275–88

Yoon, M. G. 1975. Readjustment of retinotectal projection following reimplantation of a rotated or inverted tectal tissue in adult goldfish. *J. Physiol.* 252:137–58

Yoon, M. G. 1977. Induction of compression in the re-established visual projections on to a rotated tectal reimplant that retains its original topographic polarity within the halved optic tectum of adult goldfish. *J. Physiol.* 264:379–410

Yoon, M. 1979. Reciprocal transplantations between the optic tectum and the cerebellum in adult goldfish. *J. Physiol.* 288:211–25

REFERENCES ADDED IN PROOF:

Björklund, A., Stenevi, U., Schmidt, R. H. Dunnett, S. B., Gage, F. H. 1983d. Intracerebral grafting of neuronal cell suspensions. I. Introduction and general methods of preparation. *Acta Physiol. Scand. Suppl.* 522:1–10

Schmidt, R. H., Björklund, A., Stenevi, U., Dunnett, S. B., Gage, F. H. 1983b. Intracerebral grafting of neuronal cell suspensions. III. Activity of intrastriatal nigral suspension implants as assessed by measurements of dopamine synthesis and metabolism. *Acta Physiol. Scand. Suppl.* 522:23–32

Ann. Rev. Neurosci. 1984. 7:309–38
Copyright © 1984 by Annual Reviews Inc. All rights reserved

ENDOGENOUS PAIN CONTROL SYSTEMS: Brainstem Spinal Pathways and Endorphin Circuitry

Allan I. Basbaum and Howard L. Fields

Departments of Anatomy, Physiology and Neurology, University of California at San Francisco, San Francisco, California 94143

INTRODUCTION

In 1978, we published two reviews on pain control mechanisms in the central nervous system. One concentrated on brainstem control of spinal nociceptive neurons (Fields & Basbaum 1978). The other focused on those pain control systems which use endogenous opioid compounds and presumably mediate the analgesic action of exogenous opiate analgesics (Basbaum & Fields 1978). Since those reviews were published, information in both areas has grown rapidly. For example, it is now known that there are at least three families of endogenous opioid peptides (endorphins), each having a different precursor and a differential distribution in the CNS. In addition, knowledge of the immunohistochemistry and pharmacology of the brainstem and spinal neurons involved in pain transmission and modulation has expanded. Finally, much more is known about the circuitry underlying both the transmission and control of pain. These new observations make revisions necessary in the mechanisms proposed to account for the analgesic action of exogenous opiates and electrical brain stimulation. In this paper we review the new information, concentrating on those studies that necessitate changes in the original model.[1]

---

[1]Abbreviations: BAM, bovine adrenal medulla; CSF, cerebrospinal fluid; DLF, dorsal part of the lateral funiculus; GABA, gamma aminobutyric acid; MSH, melanocyte stimulating hormone; NRM, n. raphe magnus; PAG, periaqueductal gray; pCPA, parachlorophenylalanine; POMC, proopiomelanocortin; Pro A, Proenkephalin A; Rgc, n. reticularis gigantocellularis; Rgcα, n. reticularis gigantocellularis, pars alpha; Rmc, n. reticularis magnocellularis; Rpg, n. reticularis paragigantocellularis; Rpgl, n. reticularis paragigantocellularis lateralis; RVM, rostral ventral medulla; SG, substantia gelatinosa; SIA, stress-induced analgesia; SP, Substance P; SPA, stimulation-produced analgesia; TRH, thyrotropin releasing hormone; VIP, vasoactive intestinal polypeptide; II$_i$, inner layer of the substantia gelatinosa; II$_o$, outer layer of the substantia gelatinosa; 5HT, 5-hydroxytryptamine, serotonin.

0147-006X/84/0301-0309$02.00

THE PREVIOUS MODEL OF BULBOSPINAL CONTROL

The original model outlined a three-tiered pain control system (Basbaum & Fields 1978). Its major components included the midbrain periaqueductal gray (PAG), several nuclei of the rostral ventral medulla (RVM), specifically the midline nucleus raphe magnus (NRM) and adjacent reticular nuclei, and the spinal dorsal horn. We proposed the following. The analgesic action of opiates given systemically, or via intracerebral injection into the PAG, results from activation of excitatory connections between the PAG and the raphe. Raphe neurons, in turn, project, via a pathway in the dorsal part of the lateral funiculus (DLF) of the spinal cord (Basbaum et al 1978, Basbaum & Fields 1979) to the region of nociceptors in the spinal dorsal horn, and its trigeminal equivalent, the nucleus caudalis. These raphe-spinal neurons selectively inhibit dorsal horn nociceptive neurons, including interneurons (Fields et al 1977) and a population of rostrally projecting spinothalamic and spinoreticular neurons (Willis et al 1977).

The original model proposed major endorphin links at the level of the PAG and spinal cord. The PAG contains both high affinity opiate binding sites (Atweh & Kuhar 1977b) and significant levels of endogenous opioid peptides (Hökfelt et al 1977, Moss et al 1983). More important, injection of the specific opiate antagonist, naloxone, into the PAG (Tsou & Jang 1964, Yeung & Rudy 1978) or the third ventricle (Yeung & Rudy 1980) reverses the analgesic action of systemic opiates, and microinjection of opiates into the PAG generates analgesia that is reversed by lesions of the spinal dorsolateral funiculus (Murfin et al 1976). Thus, exogenous opiates "substitute" for the endogenous opioid peptide, thereby activating pain control circuits that originate in the PAG. A second endorphin component in this descending pain control model is in the spinal dorsal horn. This region also has high levels of immunoreactive enkephalin (Hökfelt et al 1977, Glazer & Basbaum 1981) and opiate binding sites (Atweh & Kuhar 1977b, LaMotte et al 1976). Furthermore, there is now evidence that descending serotonergic raphe-spinal axons exert their antinociceptive effects, in part, via synapses with opioid peptide-containing neurons of the dorsal horn. First, intrathecal injection of naloxone can antagonize the analgesic action of raphe stimulation (Zorman et al 1982); second, using a method to localize tritiated serotonin and immunoreactive enkephalin simultaneously at the ultrastructural level, we demonstrated that the terminals of descending 5HT axons are presynaptic to enkephalin-containing neurons of the spinal dorsal horn (Basbaum et al 1982, Glazer et al 1981, 1983b).

In addition to 5HT and enkephalin links, the original model included descending catecholamine systems, thought to originate in the dorsolateral pons, and a pharmacologically undefined descending system originating in the nucleus reticularis magnocellularis (Rmc) located lateral to the raphe. As described below, additional bulbospinal systems have now been identified and the contribution of the catecholamines is known to be more complicated than originally proposed.

Although exogenous opiates or electrical brain stimulation can activate this endorphin-mediated pain control system, the factors that naturally activate the system are poorly understood. Taking into account the evidence that "pain inhibits pain" (Melzack 1975) and the consistent observations that noxious stimuli activate raphe spinal neurons (Anderson et al 1977, Guilbaud et al 1980), we proposed that pain, itself, is a critical factor that reliably activates these pain control circuits. It is now clear that a variety of behavioral contingencies affect the operation of this system. These we describe below.

THE ENDORPHINS

Information about the endorphins was limited when our earlier reviews were written. Although leucine and methionine enkephalin and β-endorphin had been isolated from brain and pituitary, respectively, their precursors were unknown. Recently, attention has focused on another class of endorphins, specifically dynorphin and related opioid peptides (Goldstein et al 1979, 1981). Thus, the enkephalins, dynorphin and β-endorphin, represent three distinct families of endogenous opioid peptides (Hollt 1983) (Table 1). Each class is cleaved from a different precursor and each has a distinct anatomical distribution. Our discussion of the endorphins highlights those aspects relevant to pain control. (See also Akil et al 1984, this volume.)

β-Endorphin

β-endorphin, ACTH, and three copies of melanocyte-stimulating hormone (MSH) are derived from a common precursor molecule, proopiomelanocortin (POMC) (Mains et al 1977, Roberts & Herbert 1977). Unlike the enkephalins and dynorphin, which are widely distributed in the brain, POMC-neurons are concentrated in the basal hypothalamus (Bloom et al 1978). Axons of these cells course caudally along the wall of the third ventricle, toward the midbrain PAG and locus coeruleus. The possibility has been raised that stimulation-produced analgesia from electrodes placed in the midbrain PAG, rather than arising from activation of cell somata within the PAG, results from activation of the axons of β-endorphin neurons that pass through the PAG. This possibility has not been ruled out; however, the fact that glutamate injection in the PAG (which would not activate axons of passage) can generate analgesia (Behbehani & Fields 1979) indicates that local cells can activate the system. On the other hand, increases in cerebrospinal fluid (CSF) β-endorphin after brain stimulation in humans are consistent with a contribution of this peptide to pain control (Akil et al 1978, Hosobuchi et al 1979). That the CSF levels reflect changes

TABLE 1   Amino acid sequences of major endogenous opioid peptides

Leucine-enkephalin	Try-gly-gly-phe-leu-oH
Methionine-enkephalin	Try-gly-gly-phe-met-oH
β-Endorphin	Try-gly-gly-phe-met-thr-ser-glu-lys-ser-gln-thr-pro-leu-val-thr-leu-phe-lys-asn-ala-ile-val-lys-asn-ala-his-lys-gly-gln-oH
Dynorphin A	Try-gly-gly-phe-leu -arg-arg-ile-arg-pro-lys-leu-lys-try-asp-asn-gln-oH
Dynorphin B	Try-gly-gly-phe-leu-arg-arg-gln-phe-lys-val-thr
α-Neoendorphin	Try-gly-gly-phe-leu-arg-lys-try-pro-lys

resulting from the "stress" of surgery must, however, also be considered (R. H. Gracely and R. Dubner, personal communication).

Whether pituitary β-endorphin plays any role in descending pain control remains a mystery. While it is possible that pituitary β-endorphin can enter the brain via retrograde flow in the portal system, this has not been demonstrated. Systemically administered β-endorphin has been recovered in spinal CSF (Houghten et al 1980); however, the quantities were exceedingly low and probably insufficient to have significant biological effects. Moreover, since 96% of intermediate lobe β-endorphin is acetylated and thus inactive as an analgesic, it is even more difficult to assess its contribution. On the other hand, the report that hypophysectomy can interfere with certain forms of stress-generated analgesia (see below) indicates that pituitary endorphins may contribute to pain control.

Finally, recent studies have demonstrated interactions between the different peptides derived from POMC. Antagonism between ACTH and β-endorphin is the most frequently reported interaction. Thus, for example, intraventricular (Smock & Fields 1980) or intrathecal (Belcher et al 1982) microinjection of ACTH can antagonize morphine or β-endorphin induced analgesia. Intraventricular injection of ACTH, however, has been reported to generate analgesia (Walker et al 1981). These data indicate that cosynthesis and release of β-endorphin and ACTH may interact physiologically at CNS synapses. Furthermore, the action of ACTH in the cord raises the possibility that CSF β-endorphin and ACTH, released by hypothalamic neurons, may reach the cord via the CSF and influence spinal nociceptors directly.

Enkephalin and Dynorphin

The discovery of the enkephalin precursor molecule Pro-enkephalin A (ProENK A) has generated renewed focus on the first opioid peptides that were described. ProENK A was originally isolated from the adrenal medulla (Kimura et al 1980), but is also found in brain. Probable cleavage products of the ProENK A molecule include six copies of met-enkephalin and one copy of leu-enkephalin (Comb et al 1982, Gubler et al 1982). Two of the met-enkephalin sequences are extended and include the octapeptide, met-enkephalin-arg-gly-leu, and the heptapeptide, met-enkephalin-arg-phe. Other larger cleavage products have also been identified. That the enkephalins and the dynorphin-related peptides are different was unequivocally established with the identification of the prodynorphin molecule, from which dynorphin and two other peptides with N-terminal leucine enkephalin residues, alpha-neo-endorphin and dynorphin-B, are cleaved (Kakidani et al 1982).

The possible functional significance of the different enkephalin fragments to pain control has been recently demonstrated by Hollt et al (1982). They found that while the short enkephalin-containing cleavage products of ProENK A are very weak or inactive as analgesics, significant analgesia is produced by intracerebral injection of the larger products isolated from bovine adrenal medulla, BAM 12 and 22, and Peptides E and F. In fact, on a molar basis, these ProENK A derived peptides are somewhat more potent than morphine, but still significantly less potent than β-endorphin. In contrast to the larger ProENK A cleavage products and β-endorphin, intracerebral injection of prodynorphin-cleavage peptides had no analgesic effect. On the other hand, other studies indicate an analgesic effect of spinally administered dynorphin (see below).

The majority, if not all, of studies that have examined the distribution of immunoreactive enkephalin (leu or met) did not examine for cross-reactivity of antibodies with the prodynorphin peptides (for review see Miller 1981). It is thus possible that much of what was described is actually dynorphin or alpha-neo-endorphin. Fortunately, using antibodies directed against those sequences of proENK A and prodynorphin that differ, it is now possible to stain selectively for immunoreactive enkephalin and dynorphin (Watson et al 1982, Weber et al 1982). The distribution of the two compounds is similar. However, in some areas significant differences are found. For example, the substantia nigra has very high levels of dynorphin and very low enkephalin; the opposite is found in the interpeduncular nucleus (a gift of Dr. E. Weber).

We have also used selective antisera to examine the distribution of dynorphin and enkephalin (Basbaum et al 1983a). One antibody was directed against the met-enk-arg-gly-leu peptide of the proENK A molecule; it has no cross-reactivity with any of the prodynorphin peptides. To define dynorphin-like immunoreactivity, we used an antiserum directed against the prodyn C-terminal leu-enk-containing peptide, i.e. Dynorphin B. This antiserum does not cross-react with proENK A products. We focused our attention on "pain" related areas.

In agreement with previous studies we found significant overlap of immunoreactive enkephalin and dynorphin; however, there are some important differences. Both enkephalin- and dynorphin-positive cells and terminals are found in the PAG. In general, however, the dynorphin-positive cells are located more ventrally. Since β-endorphin terminals are also found in the PAG, these data indicate that the endorphin link in the periaqueductal gray could result from release of any, or all, of the opioid peptides found there.

Analysis of immunoreactive dyn and enk in the rostral medulla also proved interesting. We had previously reported that some 5HT-containing neurons in the raphe magnus, pallidus, and adjacent nucleus reticularis paragigantocellularis lateralis (Rpgl) also contain immunoreactive enk (Glazer et al 1981). In the present study, we examined serial three-micron frozen sections of the medulla of the rat, for dynorphin, enkephalin, and 5HT. We found that not only are there 5HT/enkephalin neurons, but there are separate 5HT/dynorphin neurons. In general, there are more enkephalin than dynorphin immunoreactive neurons in the medulla. While the majority of cells stain for one or the other opioid peptide, in some raphe and neurons, dynorphin and enkephalin coexist. We also found coexistence of dynorphin and enkephalin in some dorsal horn neurons (Mulcahy & Basbaum 1983).

That the different endorphins might have different physiological actions vis-à-vis pain was suggested by our studies in the spinal dorsal horn and its brainstem homolog, the trigeminal nucleus caudalis (Glazer & Basbaum 1981). We found that the densest concentration of immunoreactive enkephalin is in laminae I (the marginal zone), II (the substantia gelatinosa), and in the region of lamina V. Enkephalin-labelled cells are located in both the marginal zone and in the substantia gelatinosa. The distribution of immunoreactive dynorphin, however, is much more limited; the staining is concentrated in the marginal zone. Overall the dynorphin terminal staining is much less than enkephalin, even in colchicine-treated animals; however, the number of immunoreactive cells in the marginal layer far exceeds the enkephalin-positive neurons recorded in an adjacent section.

There is general agreement that intracerebral injection of dynorphin does not produce analgesia. In fact, studies indicate that intracerebral dynorphin can antagonize morphine analgesia (Tulunay et al 1981). Several laboratories, however, report that intrathecal dynorphin generates a prolonged analgesia (Han & Xie 1982, Piercey et al 1982), particularly with tests using noxious heat. We have also found that both dynorphin and α-neoendorphin generate prolonged analgesia when administered intrathecally (Basbaum et al 1983a). A brief paralysis is often produced but it is dissociable from the analgesia. While naloxone did not reverse the analgesia once it was established, pretreatment with naloxone could prevent it, indicating that an opiate receptor is involved.

These data indicate that several endorphins are involved in the descending control of spinal neurons. It is possible that the descending axons contact enkephalin and/or dynorphin neurons. Moreover, different postsynaptic elements may be acted upon by the two putative opioid transmitters. Given its restricted terminal distribution, dynorphin might predominantly influence the projection neurons of the marginal zone, while both projection neurons and interneurons (of the substantia gelatinosa) may be inhibited by enkephalin.

THE COMPONENTS OF AN ENDOGENOUS PAIN CONTROL SYSTEM

The Periaqueductal Gray

The periaqueductal gray was the first region to be implicated in pain modulation (Reynolds 1969, Mayer et al 1971). Although more rostral sites are usually stimulated in humans, there is evidence that the analgesia elicited from these sites is transmitted via the periaqueductal gray (Rhodes 1979). Whether the periaqueductal gray is functionally homogeneous is controversial. Analgesia can be generated from all regions of the PAG; however, several workers have reported that the ventrolateral region is the most effective (Gebhart & Toleikis 1978). Other investigators, however, emphasize that the midline raphe dorsalis (a specialized midbrain region located within the ventral PAG) is the most effective site for stimulation-produced analgesia (SPA) (Oliveras et al 1979).

The fact that microinjection of opiates into (Murfin et al 1976) or electrical stimulation of (Basbaum et al 1977) the PAG generates analgesia (via a pathway in the DLF) and inhibits the firing of dorsal horn neurons (Liebeskind et al 1973) is consistent with the view that opiate analgesia and stimulation-produced analgesia operate via a common neural mechanism (Mayer & Liebeskind 1974, Mayer & Price 1976). The anatomical substrate for opiate and stimulation-produced analgesia may, however, not be completely identical. Some studies found differences in the PAG loci most effective for stimulation-produced and opiate analgesia (Gebhart 1982).

Although its importance to analgesia mechanisms is unquestioned, the intrinsic circuitry of the PAG is largely unknown. Hamilton (1973) characterized three major cytoarchitectural subdivisions of the PAG and implied that they represented functional subdivisions. Other studies could not distinguish these regions, either with Golgi techniques or on the basis of the afferent and efferent connections of the PAG (Mantyh 1982a,b, 1983). Despite this lack of agreement on cytoarchitecture, immunohistochemical studies of the PAG clearly demonstrate its chemical heterogeneity. For example, in the caudal PAG, enkephalin cells and terminals are concentrated ventrolaterally, but their distribution shifts dorsally in the rostral midbrain (Moss et al 1983). As described above, dynorphin cells have a different location. The distribution of immunoreactive Substance P cells is similar to that of enkephalin; however, the terminal fields of the two peptides are not identical (Moss & Basbaum 1983b). In marked contrast is the distribution of immunoreactive vasoactive intestinal polypeptide (VIP), a peptide which, by PAG microinjection, generates a profound, naloxone-insensitive analgesia (Sullivan & Pert 1981). VIP cells are concentrated just ventral to the aqueduct, along the rostral caudal extent of the PAG. Although the raphe dorsalis contains a variety of peptidergic neurons, including enkephalin, that are also found within the PAG (Moss et al 1981), its dense concentration of 5HT-containing neurons readily distinguishes it from the rest of the PAG.

When we proposed our original model, information about inputs to the PAG was sparce. Recent studies have demonstrated that the PAG is pivotally located to transmit cortical and diencephalic inputs to the lower brainstem. Retrograde transport studies have established that the PAG receives significant inputs from the frontal and insular cortex, the amygdala, and the hypothalamus (Beitz 1982b, Mantyh 1983). Because endorphin-mediated analgesia can be conditioned (see below), it is likely that cognitive factors can activate these analgesia systems. Whether the cortical inputs to the PAG are a route by which these cognitive inputs exert their influence is unknown, but they do provide a possible anatomical substrate.

The brainstem inputs to the PAG are also diverse. The majority derive from the nucleus cuneiformis, the pontine reticular formation, and from the locus coeruleus. Taken together with the known direct spinal input to the PAG (Mehler 1962), the former two regions provide a probable relay for the nociceptive input that activates PAG neurons (see Gebhart 1982). The locus coeruleus projection is of interest since it may contribute to the known norepinephrine antagonism of opiate and stimulation-produced analgesia (Akil & Liebeskind 1975).

Details of the PAG connections to the rostral medulla have been demonstrated by both anterograde and retrograde (Mantyh 1983) tracing methods (Gallagher & Pert 1978, Abols & Basbaum 1981). Of particular interest are the studies of Beitz (1982a,c), who examined the PAG-medullary connections with combined retrograde tracing methods and immunocytochemistry. His studies established that both 5HT and neurotensin neurons of the midbrain project to the medulla. Substance P and enkephalin neurons of the PAG did not. This is consistent with the idea that the latter are interneurons that modulate neurotensinergic projection neurons (see PAG circuitry below).

The Rostral Medulla

CYTOARCHITECTURE, CYTOCHEMISTRY, AND PHARMACOLOGY   The rostral medulla, particularly its ventral aspect (RVM), is the major source of axons projecting via the DLF to the spinal cord (Basbaum & Fields 1979, Martin et al 1978, Liechnitz et al 1978), and thus it is a critical link in the descending contral exerted from the PAG. Medullary cells of origin of DLF axons are found in the NRM, and in the adjacent reticular formation; all are located ventral to nucleus reticularis gigantocellularis (Rgc) (Figure 1). In the rat and cat, neurons of at least three different regions, in addition to the raphe, contribute axons to the

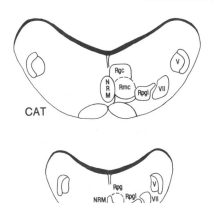

Figure 1 Schematic illustration of the major subnuclei of the rostral medulla of the cat and rat. As indicated in the text, each of these regions contributes to descending control; however, the different regions can be distinguished on cytoarchitectural and cytochemical grounds.

DLF: the n. reticularis paragigantocellularis (Rpg), the n. reticularis gigantocellularis pars α (Rgcα), located ventral to the Rpg, and, more laterally, the n. reticularis paragigantocellularis lateralis (Rpgl). The nucleus reticularis magnocellularis (Rmc) of the cat occupies roughly the region corresponding to the rat Rpg. There is evidence that cells throughout the NRM, Rmc, Rgcα, and Rpgl are involved in pain modulation. All receive projections from the PAG (Beitz 1982c, Mantyh 1983), all send axons to the spinal cord via the DLF, and all produce "analgesia" when electrically stimulated at low intensities (Zorman et al 1981). Finally, in order to block completely the effect of midbrain stimulation, NRM, Rgcα and Rpgl must be simultaneously interrupted, either by lesions (Prieto et al 1983) or by injection of local anesthetic (Sandkuhler et al 1982).

Despite the anatomical and functional similarities, there is evidence that the neighboring regions of the RVM differ. First, the cells of Rpgl and Rgcα are morphologically distinct from those of NRM and Rmc/Rpg; the former are predominantly fusiform, the latter are larger and multipolar. Rmc and Rpgl receive significant afferent projections from the spinal cord (Abols & Basbaum 1981) and project to the cord via both DLF and ventral funiculi (Basbaum et al 1978). In contrast, NRM projects to the cord only via the DLF and receives no direct connections from the spinal cord.

Cytochemical studies also support a parcellation of the region. Thus, NRM, Rgcα, and Rpgl all have significant numbers of 5HT-containing cells (Dahlstrom & Fuxe 1964); Rmc/Rpg does not (Wiklund et al 1982). Although double labeling techniques indicate that many of the 5HT-containing cells in NRM and Rpgl project to the spinal cord (Bowker et al 1981a,b), it has been reported that 5HT-cells in the NRM of the rat do not project to the spinal cord via the DLF (Johannessen et al 1981). This is surprising since the superficial dorsal horn receives its medullary input via axons in the DLF (Basbaum et al 1978) and since 5HT levels in superficial dorsal horn drop after NRM lesions (Oliveras et al 1977).

Although our original model concentrated on the 5HT component in the control of spinal nociceptive neurons (with the exception of the pharmacologically undefined projection from the Rmc), much new information is now available concerning the chemical heterogeneity of neurons of the RVM. For example, many peptidergic neurons have been demonstrated in the NRM, in Rgcα, and in Rpgl. These include enkephalin, Substance P, somatostatin, and thyrotopin-releasing hormone-containing (TRH) neurons. Although some of these neurons are probably local interneurons, double labeling studies have established that many peptidergic neurons of the medulla project to the spinal cord (Hökfelt et al 1979, Bowker et al 1981a).

Perhaps the most significant finding of the past five years and one which has direct bearing on pain control mechanisms is the coexistence of two putative transmitters in a single neuron (Hökfelt et al 1980). It was first demonstrated that in some cells of the NRM, the Rgcα, and the Rpgl, serotonin and Substance P coexist (Hökfelt et al 1978). Our own studies established that 5HT and enkephalin coexist, particularly in neurons of the Rpgl (Glazer et al 1981). A later study revealed neurons in which three putative transmitters—5HT, SP, and TRH—are colocalized (Johansson et al 1981). That the bulbospinal axons (as well as the cell body) contain multiple transmitters was confirmed using the neurotoxin, 5,7-dihydroxytryptamine. This toxin destroys the 5HT terminals in the spinal cord and concomitantly reduces the level of Substance P (Hökfelt et al 1978, Gilbert et al 1982).

The biological significance of transmitter coexistence is only beginning to be examined, but already several intriguing possibilities can be envisioned. For example, it is generally assumed that 5HT inhibits dorsal horn nociceptors; however, it is not known whether the bulbospinal Substance P-containing terminal excites or inhibits spinal neurons. Nor is it known whether the same terminal releases both 5HT and Substance P. A given terminal could have mixed effects, depending on the amount and nature of the transmitter released, and upon on the distribution of postsynaptic receptors. In our description of circuitry (Figure 2), we have considered the possibility that different bulbospinal terminals release 5HT or Substance P. This raises the possibility that NRM neurons produce both inhibitory and excitatory effects at the spinal level. In fact, in animals depleted of 5HT with parachlorophenylalanine (pCPA), raphe stimulation excites dorsal horn nociceptive neurons (Rivot et al 1980). This could result from an unmasking of excitatory effects of Substance P.

In addition to the anatomical heterogeneity of the RVM, there is evidence in rat, albeit controversial, for significant pharmacological differences among its subdivisions. Rpg has been reported to be exquisitely sensitive to opiate microinjection (Akaike et al 1978)—perhaps two orders of magnitude more sensitive than NRM or the PAG. Thus, nanograms doses of morphine injected into the Rpg produces potent behavioral analgesia. Using different methods to assess analgesia, however, Dickenson et al (1979) reported that NRM is the most sensitive site. A third group, using still another method to assess analgesia, agreed that the more lateral sites (Rpg and Rpgl) are most sensitive, although not by orders of magnitude (Azami et al 1982).

Another approach to the analysis of pharmacological differences between classes of spinally projecting RVM neurons is to activate them in the medulla and attempt to block their action at the level of the spinal cord, using specific transmitter antagonists. For example, using medullary microstimulation, Zorman et al (1982) showed that stimulation-produced analgesia from RVM could be antagonized by lumbar intrathecal naloxone. Using a similar approach, bulbospinal 5HT and norepinephrine axons have been implicated. The analgesia elicited from the lateral Rpg is blocked by α-adrenergic antagonists (Kuraishi et al 1979), whereas that elicited from NRM is blocked by serotonergic antagonists (Satoh et al 1980).

In summary, there are several chemically distinct classes of neurons in RVM, each of which may be at the origin of a parallel bulbospinal control system. The descending systems, though parallel, are unlikely to be entirely redundant. As pointed out by previous investigators (Casey 1971, Gebhart 1982), this region of the brainstem, classically considered to be part of the reticular formation, is far from homogeneous; different subpopulations of neurons contribute to diverse functions. Whether each of these chemically distinct classes of neuron has a different physiological role is an important question.

AFFERENT CONNECTIONS   The major afferent connections of the RVM originate in the PAG and the adjacent midbrain nucleus cuneiformis. Because early studies found minimal direct spinal projections from the PAG, the RVM was considered a necessary "relay" between the PAG and the spinal cord. A recent study, however, using very sensitive retrograde tracers, revealed a far more extensive direct PAG-spinal projection (Mantyh & Peschanski 1982). This direct connection may contribute to pain control by influencing neurons of the spinal dorsal horn; however, since simultaneous interruption of the NRM and Rpgl counteracts the effects of PAG stimulation, the medullary link is essential. This concept is strengthened by the observation that activation of PAG by electrical stimulation (Fields & Anderson 1978, Oleson et al 1978), opiate microinjection (Behbehani & Pomeroy 1978), or glutamate injection (Behbehani & Fields 1979) has a predominantly excitatory effect on NRM and Rmc/Rpg neurons. More recently, the introduction of immunocytochemical

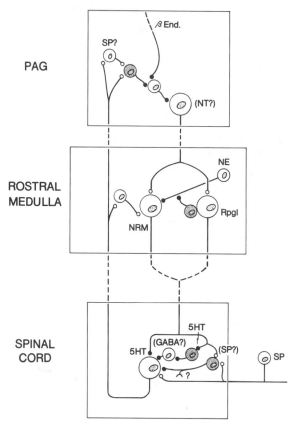

Figure 2 This figure illustrates proposed midbrain (PAG), medullary, and spinal circuitry related to the control of spinal nociceptive neurons. *Unfilled "boutons"* indicate release of an excitatory transmitter, *filled "boutons"* indicate an inhibitory input. In the PAG, the output neuron is depicted as an excitatory neurotensinergic (NT) neuron (Beitz 1982c) that activates cells of the NRM and of the lateral Rpgl of the rostral medulla. An endogenous opioid peptide neuron (*stippled*) in the PAG is presumed to inhibit an inhibitory interneuron that, in turn, controls the PAG output neuron. Input to the opioid interneuron may derive from ascending nociceptive pathways, via a local Substance P-containing (SP) neuron. Inputs from β-endorphin cells of the hypothalamus may also contribute to the opioid link in the PAG. It is not known whether all of the endorphin subtypes act in a similar fashion in the PAG.

At the level of the rostral ventral medulla, we have indicated that there is an inhibitory norepinephrine (NE) input to bulbospinal raphe neurons. We have also included the possibility that local opioid neurons presynaptically control the NE input to raphe-spinal axons. Although not illustrated, it is possible that the noradrenergic neurons that control raphe-spinal neurons are the same as those which exert a direct bulbospinal control. There are a variety of other peptides that must eventually be included in this diagram, but because there is little information as to their local connectivity or possible function in analgesia, these have been omitted.

The greatest complexity is at the level of the dorsal horn. We have only included a few of the synaptic interactions that may be relevant to nociceptive control. While there are numerous other techniques has made it possible to differentiate the chemical nature of the inputs to the RVM. We will highlight a few that appear relevant to pain control circuitry.

Enkephalin   Because microinjection of minute quantities of morphine into RVM generates analgesia, the analysis of opioid peptide inputs is particularly important (Akaike et al 1978, Dickenson et al 1979). Combined retrograde tracing and immunocytochemical double labeling studies in the rat demonstrated several brainstem sources of the enkephalin-like immunoreactivity in or adjacent to the NRM (Beitz 1982a). These include the midbrain nucleus cuneiformis, the nucleus of the solitary tract, and the dorsal parabrachial nucleus of the pons. A smaller enkephalin input originates in the laterally located medullary A5 noradrenergic cell group.

Identifying the extrinsic source of enkephalin inputs is, of course, complicated by the presence of enkephalin neurons within the NRM itself, and particularly by the observation that in some RVM neurons, enkephalin and 5HT coexist. Moreover, the afferent connections originating from dynorphin cells have not been studied; it is likely that these also exist. It is thus possible that the exquisite sensitivity of the RVM to opiate microinjection reflects an extensive, convergent input from both enk- and dyn-containing neurons.

Neurotensin   Although the literature on the contribution of neurotensin (NT) to pain control mechanisms is limited, there is evidence that it is relevant (Clineschmidt et al 1979). Intracisternal neurotensin produces profound analgesia (Kalivas et al 1982), possibly by direct activation of raphe-spinal inhibitory neurons. Thus, studies showing that neurotensin immunoreactive cells of the PAG project to the RVM are particularly important (Beitz 1982c). Other neurotensin inputs to the RVM derive from the dorsolateral pons and from the ventrolateral medulla, in the region corresponding to the chain of catecholamine cells groups, A1 to 5.

---

peptidergic elements that have been defined, there is presently little information about their synaptic relationships. In this diagram the bulbospinal serotonin (5HT) axons are shown to inhibit the projection neurons via two circuits. The simpler is a direct postsynaptic inhibition. Another illustrated possibility is that 5HT exerts its effect through as inhibitory opioid interneuron. As described in the text, it is likely that this latter arrangement (for which there is anatomical evidence) would require a further, interposed, inhibitory interneuron, analogous to the arrangement proposed in the PAG. Based on studies in the hippocampus (Nicoll et al 1980) and given the high concentration of GABA in the superficial dorsal horn, GABAergic interneurons may be involved. Alternatively, the bulbospinal axons may excite the inhibitory endorphin interneurons by releasing its cotransmitter Substance P. The opioid interneuron may, as anatomical studies indicate, directly control the projection neurons, or as biochemical studies indicate, presynaptically control nociceptive primary afferent inputs, some of which may contain Substance P.

Norepinephrine   In our original model, we proposed a bulbospinal nore-pinephrine (NE) pathway synergistic with the inhibitory serotonergic projection that controls spinal nociceptive neurons. (For a recent review see Basbaum et al 1983b.) The norepinephrine inhibits spinal neurons has been confirmed in several studies (Belcher et al 1978, Headley et al 1978); however, the origin of the descending norepinephrine input is still unclear. To complicate matters, recent studies indicate that norepinephrine neurons in the brainstem inhibit the descending 5HT system. For example, microinjection of the norepinephrine antagonist, phentolamine, into the NRM produces a hypoalgesia, which is blocked by intrathecal methysergide (Hammond et al 1980). These data suggest that there is a tonic norepinephrine-mediated inhibition of the bulbospinal 5HT pain-modulatory neurons of the NRM. The observation that iontophoresis of norepinephrine inhibits the firing of raphe-spinal neurons supports this hypothesis (Behbehani et al 1981).

Unfortunately, the connections of norepinephrine neurons within the brainstem are still unclear. Although the origin of the input of norepinephrine to the NRM is not known, it is likely that it derives, in part, from the A3 and A5 norepinephrine cell groups of the lateral medulla (Takagi et al 1981). The contribution of the locus coeruleus is apparently limited (Moss & Basbaum 1979).

Acetylcholine   Iontophoresis of acetylcholine excites NRM neurons (Behbe-hani 1982); more important, injection of carbachol, an acetylcholine mimetic, into the NRM generates analgesia (Brodie & Proudfit 1982). It is important to establish the origin of this input of acetylcholine. Histochemical studies that combine retrograde tracers with markers of choline-acetyltransferase may provide the answer.

SUMMARY   The evidence reviewed above indicates that no one chemical marker will be sufficient to define the location and pharmacology of those neuronal populations which control the firing of NRM neurons. Our original model emphasized the excitatory inputs from the midbrain PAG. There is now evidence for multiple brainstem inputs, some of which are excitatory, others inhibitory. Just as the thalamus is no longer considered a simple relay between receptor and cortex, so the raphe and adjacent reticular formation are not mere relays between the PAG and cord. These medullary neurons receive input from nociceptive afferents and integrate it with inputs from more rostral structures. This integration establishes the level of control that the bulbospinal neurons exert on the transmission of nociceptive messages at the spinal cord.

Spinal Dorsal Horn

Because the descending inhibitory effects are exerted on spinal cord neurons it is important to review briefly some recent observations on the anatomy and physiology of dorsal horn neurons. Ultrastructural examination of physiologically identified and intracellularly filled primary afferent axons and dorsal horn neurons have provided particularly valuable new data. The general anatomical (Rexed 1954) and physiological (Wall 1967) laminar schema of the dorsal horn is still accepted, but some modifications are required, particularly in the substantia gelatinosa (SG), lamina II. (For reviews see Cervero & Iggo 1980, Dubner & Bennett 1983). The substantia gelatinosa can be divided into an outer (IIo) and inner (IIi) layer. The former receives inputs from small diameter high threshold primary afferent fibers (Light & Perl 1979) and contains neurons physiologically similar to those in the marginal zone, lamina I. That is, IIo neurons respond to both noxious and non-noxious inputs (Light et al 1979, Bennett et al 1980). It has been hypothesized (Gobel 1979) and there is evidence that neurons in lamina IIo relay nociceptive inputs from primary afferents to marginal neurons (Price et al 1979). The latter hypothesis is particularly relevant because it raises the possibility that the bulbospinal inhibition of marginal neurons could be indirect, i.e. by inhibition of these putative relay interneurons in IIo. Neurons in IIi, in contrast, receive inputs from small diameter, low threshold mechanoreceptors and contain neurons predominantly responsive to non-noxious inputs. IIi may contain interneurons involved in the segmental, inhibitory control exerted by non-noxious peripheral stimuli.

Based on Golgi studies, the majority of neurons of the substantia gelatinosa can be assigned to either one of two distinct morphological types, the stalk or islet cell (Gobel 1978). Immunohistochemical studies, however, reveal greater complexity. Several peptides have been identified in "islet" cells (Glazer & Basbaum 1981, Hunt et al 1981, Bennett et al 1982, Seybold & Elde 1982). In fact, the list of peptides contained in SG neurons continues to grow and includes Substance P, enkephalin, dynorphin, neurotensin, cholecystokinin, and avian pancreatic polypeptide (Gibson et al 1981, Hunt et al 1981). There is also a large population of GABAergic neurons, some of which may control primary afferents presynaptically (Barber et al 1978, Basbaum et al 1981). The functional properties of these chemically distinct neurons have yet to be established. Which are excitatory and which inhibitory? Which elements receive direct input from nociceptive primary afferents which are contacted by bulbospinal axons? These questions will require anatomical and chemical marking of physiologically identified elements in the superficial dorsal horn.

CIRCUITRY

When our original model was proposed, the relevant synaptic circuitry in the PAG, RVM, and cord was largely unknown. With the advent of EM immunohistochemical analysis, however, much new information is available. In this section we propose to provide further details regarding the circuits that have been defined and discuss the evidence for additional circuits relevant to nociceptive control (Figure 2).

The Periaqueductal Gray

In general, opiate actions on target neurons are inhibitory (Nicoll et al 1980). Thus, the direct action of opiates on the postsynaptic neuron that they contact in the periaqueductal gray is probably inhibition. Since excitation of the PAG output neuron is required to initiate descending control, it follows that morphine, or the endogenous opioid equivalent, does not directly act upon the PAG output neuron. Given that the vast majority of enkephalin-immunoreactive terminals in the PAG are presynaptic to dendrites (Moss & Basbaum 1983a), we propose that endogenous opioid peptides (either enkephalin, dynorphin, β-endorphin, or all three) activate PAG output neurons by inhibiting an inhibitory interneuron. A similar model has been proposed to account for the opiate excitatory effects on hippocampal pyramidal cells (Nicoll et al 1980). In that case, the intervening inhibitory interneuron is probably GABAergic.

In addition to morphine and the endogenous opiates, the nonopioid peptide, Substance P, produces a naloxone-reversible analgesia when injected into the PAG (Fredricksen et al 1978, Mohrland & Gebhart 1979). Antibodies to met-enkephalin also antagonize Substance P analgesia (Naranjo et al 1982). Although there are large numbers of enkephalin cells in the ventrolateral PAG and dorsal raphe (Moss et al 1981, 1983), Substance P cells are not common (Moss et al 1983). In contrast, there is a high concentration of Substance P terminals in these regions and thus enkephalin neurons probably receive a significant Substance P input. These observations indicate that Substance P acts upon local opioid peptide neurons. Figure 2 illustrates one simple circuit through which endogenous opioid peptides or exogenous morphine activate the descending control. Based on Beitz' recent studies, we have indicated the possibility that the PAG-raphe connection, in part, involves neurotensin neurons. It is clearly important to identify the proposed inhibitory interneuron that we hypothesize receives the opioid peptide input.

Rostral Ventral Medulla

Based on our anatomical demonstration of bulbospinal pathways that course in the DLF and terminate in the spinal dorsal horn, we originally proposed at least two descending control systems, one originating in the midline raphe, the second in the adjacent Rpg/Rmc. As described above, a third system has been characterized, namely that deriving from the Rpgl. Although it is not clear whether these systems are activated in parallel, their effects appear comparable (Zorman et al 1981).

Figure 2 illustrates possible medullary circuitry. Local injection of opiates into the RVM generates analgesia. Following the same line of reasoning we used for the PAG, it follows that the bulbospinal projection neurons are disinhibited by local opioid-containing interneurons. As discussed above, there is an inhibitory catecholamine effect on raphe spinal neurons. Because precedence for presynaptic opiate control of catecholamines release has been described (Llorens et al 1978), we propose that an opioid peptide interneuron (either enkephalin or dynorphin) presynaptically inhibits an inhibitory catecholamine input to raphe-spinal axons. Opiates, therefore, would act, at least partially, by disinhibiting the RVM output neuron.

Some of the cells exerting this opioid peptide control in RVM may be in the Rpgl, a region that contains both enkephalin and dynorphin immunoreactive neurons (Basbaum et al 1983a, Watson et al 1982). The Rpgl also contains many neurons in which opioid peptides coexist with 5HT. Such an arrangement might account both for the analgesic action of Rpgl stimulation and for the observation that a lesion of this area disrupts opiate analgesia (Azami et al 1982).

The Spinal Dorsal Horn

In contrast to the PAG and RVM, where the net effect of opiates is apparently activation of descending projection neurons, at the spinal level it is inhibition of the nociceptor that is required. Thus, direct postsynaptic inhibition of the nociceptive projection neurons by opioid peptides is one obvious mechanism of opiate action. The demonstration that enkephalin-immunoreactive terminals contact spinothalamic tract neurons (Ruda 1982) is consistent with such a direct postysynaptic inhibition.

In addition to these postsynaptic actions, there is considerable evidence for an opioid-mediated presynaptic control of primary afferents. Thus, primary afferents are laden with opiate binding sites (LaMotte et al 1976, Atweh & Kuhar 1977a, Hiller et al 1978, Fields et al 1980) and are sensitive to opiates both in vivo (Carstens et al 1979) and in vitro (Hentall & Fields 1983). Inhibition of Substance P release by opiates has also been demonstrated (Jessel & Iversen 1977, Mudge et al 1979, Yaksh et al 1980). Unfortunately, ultrastructural studies of immunoreactive enkephalin terminals in the dorsal horn reveals that they are exclusively presynaptic to dendritic or somatic profiles (Hunt et al 1980, Aronin et al 1981, Sumal et al 1982, Glazer & Basbaum 1983a). As yet there is no anatomical substrate for presynaptic control of primary afferents by enkephalin. However, numerous associations between enkephalin terminals and primary afferents are found; conceivably the control of primary afferents is exerted via a "nonsynaptic" action, in a manner similar to that described for peptides released into the vicinity of the target cell in the bullfrog sympathetic ganglia (Jan & Jan 1982). Another possibility is that other, as yet unstudied, endogenous opioid peptides, e.g. the prodynorphin products, provide presynaptic control of the primary afferents.

Since intrathecal naloxone antagonizes the analgesic action of IVth ventricle morphine injection (Levine et al 1982c), NRM stimulation (Zorman et al 1982), and forepaw shock (Watkins et al 1982a,b), it follows that there is an opiate link between the bulbospinal axons and the spinal nociceptors. We had originally proposed a 5HT-enkephalin synapse in the cord (Basbaum & Fields 1978) and, in fact, have recently demonstrated this connection anatomically (Basbaum et al 1982, Glazer & Basbaum 1983b). Since 5HT is generally inhibitory to dorsal horn neurons, this synaptic arrangement raised a paradox. Inhibition of the enkephalin neurons by 5HT should disinhibit the spinal nociceptor. Conceivably another inhibitory interneuron, possibly GABAergic, is interposed between the opioid peptide neurons and the nociceptor. It is, of course, possible that 5HT neither excites nor inhibits enkephalin neurons, but modulates other inputs to them (for example, see Davies & Roberts 1981). As described above, it is also possible that the Substance P that coexists in many 5HT-containing raphe neurons is the source of an excitatory input to spinal opioid neurons. Figure 2 also illustrates the segmental inputs that activate opioid-mediated inhibition of pain.

PHYSIOLOGICAL ACTIVATION OF ANALGESIA-PRODUCING NEURAL NETWORKS (STRESS-INDUCED ANALGESIA?)

While it is of interest to analyze the anatomy and physiology of endorphin-related pain-control systems, the most important questions concern their normal function. As described in our previous review, noxious peripheral stimuli are the most consistent way to excite cat RVM neurons, including those that project to the spinal cord (Anderson et al 1977). This observation has been confirmed in rats (Guilbaud et al 1980). Since activation of NRM neurons generates analgesia (Oliveras et al 1975), noxious stimuli should produce analgesia. In fact, stimuli that clearly activate nociceptive primary afferents in awake rats are very effective in producing analgesia.

On the other hand, a variety of environmental stimuli, not all of which are obviously pain-producing, may also have an analgesic effect (Hayes et al 1978, Mayer and Watkins 1981). For example, restraint (Amir & Amit 1978, Bhattachary et al 1978) and hypoglycemia (Bodnar et al 1979a,b) consistently produce analgesia. Not all analgesia-producing environmental perturbations are stressful; nevertheless, the analysis of physiological activation of the endorphin-mediated analgesia system has generally emphasized "stressful stimuli."

Comprehensive treatment of "stress-induced analgesia" research is beyond the scope of this review; however, it is important to discuss some of the concepts and problems that have evolved from these studies. The most common method used to elicit stress-induced analgesia (SIA) is to stimulate somatic structures electrically, typically the foot or tail. Because footshock (Watkins & Mayer 1982, Watkins et al 1982b) or tailshock (Woolf et al 1980) can inhibit nociceptor-induced withdrawal reflexes in spinalized rats, where stress is clearly not a factor, perhaps nociceptor-induced analgesia is a more appropriate description of this phenomenon. If this segmental mechanism contributes to footshock-induced analgesia in the intact animal, can footshock analgesia be considered stress-induced? One is clearly faced with a serious semantic question, specifically, "What is stress?" It may not be useful to group all phenomena that have been labeled stress-induced analgesia.

Regardless of what makes a stimulus "stressful," there is general agreement that footshock in the noxious range produces analgesia. The mechanism of this effect is complex and apparently varies with the location of the stimulus, its duration, and whether the animal can escape from it or control it (Mayer & Watkins 1981, Lewis et al 1980). For example, Lewis et al (1983) showed that both brief (three minute) and prolonged (20 minute) inescapable foot shock (3 mA, 50 Hz sine waves) elicit hypoalgesia of a similar magnitude, but that only the hypoalgesia secondary to prolonged footshock was blocked by naloxone and showed cross-tolerance with morphine. Furthermore, the analgesia to prolonged foot shock was attenuated by adrenalectomy or adrenal denervation (Lewis et al 1982). Watkins et al (1982a) reported that much briefer stimuli can produce naloxone-reversible analgesia (90 sec, 60 Hz, 1.6 mA) but only when the shock is restricted to the forepaws. When all four paws are on the grid, the analgesia is not blocked by naloxone. Only the forepaw induced shock-analgesia was abolished by a lesion of the DLF or the RVM (Watkins & Mayer 1982). The hindpaw shock analgesia effect survives T2 cord transection; thus its basis is largely intraspinal. Finally, since the forepaw analgesic effect persists after either adrenalectomy or hypophysectomy (Watkins et al 1982c), the pituitary-adrenal axis was ruled out in this form of stress analgesia. Apparently the footshock analgesia induction used by Watkins & Mayer differs from the prolonged shock approach of Lewis et al 1980.

Under certain conditions, restraining an animal is sufficient to produce a naloxone-reversible hypoalgesia (Amir & Amit 1978). Since restricting the shock to the forepaws requires restraining the rats by suspending them, it may be that restraint (alone or with footshock) is the relevant stressor. Other nonspecific effects of the shock may also contribute to the analgesia. For example, an elevation of blood pressure has been reported to produce analgesia in rats (Dworkin et al 1979, Zamir & Segal 1979 and may be associated with hypoalgesia in man (Zamir & Shuber 1980); this effect is reversed by naloxone (Zamir et al 1980). Furthermore, Maixner et al (1982) reported that spontaneously hypertensive rats are relatively hypoalgesic and that the hypoalgesia can be blocked by naloxone (without changing blood pressure), by lowering the blood pressure with ganglionic blockers, or by cutting the right vagal trunk. Conceivably many of the so-called stress-induced analgesias are secondary to concomitant changes in blood pressure.

Another interesting feature of shock-induced analgesia is the factor of controllability. Maier and colleagues (1982) showed that in identically shocked rats, those that cannot control (escape from) tailshock, develop potent, naloxone-reversible analgesia (Grau et al 1981, Maier et al 1982). This analgesia apparently depends on the rats' learning that they cannot control the shock. Moreover, the effect can be reinstated 24 hours later, by brief shocks that would otherwise not cause analgesia.

It has also been reported that naloxone-sensitive analgesia can be produced by innocuous stimuli under conditions in which such stimuli serve as cues that a noxious stimulus is about to occur (Bolles & Fanselow 1982). Thus, rats repeatedly exposed to aversive shock become hypoalgesic when placed in the same experimental situation, without further shock (Chance et al 1978). It has been proposed that it is fear that causes the analgesia observed in this situation.

In summary, various environmental factors, not all of which are painful, can activate endorphin-mediated analgesia-producing networks in the central nervous system. Pain-producing stimuli can activate this system either directly by activating ascending nociceptive pathways (e.g. spinoreticular or spinothalamic tracts) or indirectly via stress, conditioning, or hypertension. Other stressors, such as restraint and hypoglycemia, may activate similar systems, yet bypass peripheral and central nociceptive pathways. The mechanisms for these phenomena are as yet unknown. Part of the confusion in the field results from the fact that no two workers are studying the system in precisely the same way, so conflicting results may be due to methodological differences. Alternatively, each laboratory may have uncovered separate pain control mechanisms.

The observation that pain, especially when severe enough to be accompanied by "stress," activates an opioid-mediated analgesia system, leads to the prediction that, in man, interruption of analgesia networks would exacerbate preexisting pain. As pointed out in our previous review, the opiate antagonist naloxone does produce hyperalgesia in patients with postoperative pain (Lasagna 1965, Levine et al 1978b). In a subsequent study we showed that the naloxone effect is dose-dependent (Levine et al 1979) and, curiously, that at low doses naloxone (0.4 and 2 mg) tends to produce hypoalgesia. An analgesic effect of low-dose naloxone has also been reported in arthritic rats (Kayser & Guilbaud 1981). The patients in our clinical studies were pretreated with diazepam (5–10 mg i.v.) and had their surgery carried out under nitrous oxide (N_2O) and local block with xylocaine. Thus, the naloxone effect could have resulted from antagonism of the N_2O or diazepam. However, in our experimental model, naloxone has no hyperalgesic effect in patients in whom N_2O was used as the postoperative analgesic (Levine et al 1982a). Furthermore, Gracely et al (1979) have shown naloxone hyperalgesia in patients who had neither N_2O nor diazepam for their postoperative dental pain. It thus is well established that there is a naloxone-sensitive analgesia demonstrable in postoperative patients.

It is not clear, however, what triggers the naloxone-sensitive analgesia or whether it is seen in all patients. It is possible that the pain and stress of surgery are sufficient to release endogenous opioid peptides. Consistent with this hypothesis, we demonstrated that patients reporting higher initial pain (using the visual analog scale) are more likely to have a subsequent reduction in pain (Levine et al 1982b). Furthermore, there appears to be a threshold level of pain that must be crossed before the appearance of a subsequent reduction of pain.

In all of these studies, patients receiving naloxone were compared to patients receiving placebo, under double-blind conditions. Because our preliminary data had indicated that in the absence of treatment, the pain in the model we used increases steadily, it was conceivable that placebo administration was a major factor in generating the analgesic effect. In subsequent work we defined placebo-responders as those patients whose pain either decreased or was unchanged (i.e. did not show the normal increase) following placebo administration. About 35% of our patients fell into this category. We found that the entire naloxone effect could be accounted for by those patients falling into the placebo-responder category and concluded that release of endogenous opioid peptides contributes to the analgesic effect of placebo administration (Levine et al 1978a).

Our experimental design did not permit us to rule out the possibility that the pain reduction following placebo administration was a coincidence, i.e. it would have occurred whether or not a placebo was given. To establish with certainty that a placebo effect had occurred, we should have included a no-treatment group in the protocol. Using such a design, Gracely et al (1982) found naloxone hyperalgesia in patients with dental postoperative pain who had not received a placebo. They showed that a significant placebo effect persists, despite naloxone treatment. On the other hand, Grevert et al (1983), who also used a no-treatment control group, reported a significant, albeit incomplete reversal of placebo analgesia with naloxone.

It thus appears that our original conclusion that placebo analgesia is completely reversed by naloxone does not apply to all situations. Under certain conditions, placeboes do trigger an opioid-mediated analgesia system. However, there is an additional nonopioid component to their action. To some extent, the very concept of a placebo effect is an oversimplification. By extrapolating from the animal studies it is clear that the triggering of pain modulating systems depends on severity of pain, anxiety, blood pressure, and conditioning or expectation of relief. When these factors are completely understood and taken into account by physicians, the placebo may no longer be a useful concept.

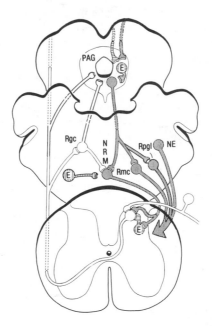

Figure 3 Schematic illustration of the major components of a descending system that contributes to the analgesic action of opiates and of electrical brain stimulation. The basic structure of the original model (Basbaum & Fields 1978) is retained. Highlighted in *stippling* are the connection between the projection neurons of the periaqueductal gray (PAG) and various subregions of the rostral ventral medulla [the nucleus raphe magnus (NRM), the nucleus reticularis magnocellularis (Rmc), and the nucleus reticularis paragigantocellularis lateralis (Rpgl)]. The latter project via the dorsolateral funiculus to the spinal dorsal horn, where they inhibit nociceptive neurons. As indicated in Figure 2, the inhibitory action at the cord may be via direct postsynaptic inhibition, or via an opioid peptide containing endorphin interneuron (indicated by *stripes* and "E"). There are other endorphin links illustrated at the level of the PAG and the rostral medulla; however, their connections are not indicated. Inputs to the PAG (one of which is a hypothalamic β-endorphin pathway) are also illustrated as is the noradrenergic (NE) contribution to bulbospinal control. The NE interactions with the other bulbospinal neurons are better illustrated in Figure 2.

Finally, ascending components of this system are indicated by the *unfilled symbols*. These include afferent inputs (some of which are Substance-P-containing; see Figure 2), projection neurons of the dorsal horn, and their collaterals into the medulla and PAG. The ascending input to the PAG and raphe nuclei is presumed to derive, in part, from collaterals of neurons of the nucleus reticularis gigantocellularis (Rgc).

CONCLUSION

Progress in the field of pain modulation has been rapid and multifaceted. Through a combination of physiological, anatomical, behavioral, and pharmacological approaches there is now much more detailed knowledge of the circuitry involved in pain modulation and the behavioral contingencies that activate pain modulating circuits. Figure 3 illustrates a revised model that incorporates several of the new observations. One of the most significant advances has been the description of multiple, pharmacologically distinct bulbospinal control systems. In addition, although our previous model discussed several enkephalin links, the revised model is more general, in that it implicates three opioid peptide (E) links. These may involve enkephalin or dynorphin neurons. We have also indicated that cortical and diencephalic sites provide significant inputs to the PAG. An important element of this is the β-endorphin from the hypothalamus.

Major questions remain. We do not know how opiates activate output neurons, or how noradrenergic neurons fit into the activation process. What is the functional consequence of coexistence of neurotransmitters/neuromodulators in a single neuron? Is there descending presynaptic control of primary afferents? Do 5HT and norepinephrine terminals interact at the level of the spinal cord? How do the various endogenous opioid peptides interact? Is there an opiate receptor specific for analgesia?

Despite these questions, it does seem clear that pain modulation is a behaviorally significant physiological process, using a discrete CNS network involving release of opioid peptides, biogenic amines, and other transmitters in its operation. With the tools of single cell neurophysiology, ultrastructural immunocytochemistry, and behavioral pharmacology these systems are beginning to yield their secrets.

ACKNOWLEDGMENTS

We thank Ms. Annette Lowe and Dr. Stuart Andersen for artistic help, Ms. Simona Ikeda for photographic assistance, and Ms. Alana Schilling and Ms. Sandy Tsarnas for typing the manuscript. This work was supported by PHS-NS 14627, DA 01949, DE 05369, and NSF-BNS 78–24762. A.I.B. is the recipient of an NIH Research Career Development Award and is a Sloan Foundation Fellow.

Literature Cited

Abols, I. A., Basbaum, A. I. 1981. Afferent connections of the rostral medulla of the cat: A neural substrate for midbrain-medullary interactions in the modulation of pain. *J. Comp. Neurol.* 201:285–97

Akaike, A., Shibata, T., Satoh, M., Takagi, H. 1978. Analgesia induced by microinjection of morphine into and electrical stimulation of the nucleus reticularis paragigantocellularis of rat medulla oblongata. *Neuropharmacology* 17:775–78

Akil, H., Liebeskind, J. C. 1975. Monaminergic mechanisms of stimulation-produced analgesia. *Brain Res.* 84:279–96

Akil, H., Richardson, D. E., Barchas, J. D., Li, C. H. 1978. Appearance of β-endorphin-like immunoreactivity in human ventricular

cerebrospinal fluid upon analgesic electrical stimulation. *Proc. Natl. Acad. Sci. USA* 75:5170–72

Akil, H., Watson, S. J., Young, E., Lewis, M. E., Khachaturian, H., Walker, M. J. 1984. Endogenous opioids: Biology and function. *Ann. Rev. Neurosci.* 7: In press

Amir, S., Amit, Z. 1978. Endogenous opioid ligands may mediate stress-induced changes in the affective properties of pain related behavior in rats. *Life Sci.* 23:1143–52

Anderson, S., Basbaum, A. T., Fields, H. L. 1977. Response of medullary raphe neurons to peripheral stimulation and to system opiates. *Brain Res.* 123:363–68

Aronin, N., Difiglia, M., Liotta, A. S., Martin, J. B. 1981. Ultrastructural localization and biochemical features of immunoreactive leu-enkephalin in monkey dorsal horn. *J. Neurosci.* 1:561–77

Atweh, S. F., Kuhar, M. J. 1977a. Autoradiographic localization of opiate receptors in rat brain. I. Spinal cord and lower medulla. *Brain Res.* 124:53–67

Atweh, S. F., Kuhar, M. J. 1977b. Autoradiographic localization of opiate-receptors in rat brain. II. The brainstem. *Brain Res.* 129:1–12

Azami, J., Llewelyn, M. D., Roberts, M. H. T. 1982. The contribution of nucleus reticularis paragigantocellularis and nucleus raphe magnus to the analgesia produced by systemically administered morphine, investigated with the microinjection technique. *Pain* 12:229–46

Barber, R. P., Vaughn, J. E., Saito, K., McLaughlin, B. J., Roberts, E. 1978. GABAergic terminals are presynaptic to primary afferent terminals in the substantia gelatinosa of the rat spinal cord. *Brain Res.* 141:35–55

Basbaum, A. I., Clanton, C. H., Fields, H. L. 1978. Three bulbospinal pathways from the rostral medulla of the cat: An autoradiographic study of pain modulating systems, *J. Comp. Neurol.* 1:209–24

Basbaum, A. I., Fields, H. L. 1978. Endogenous pain control mechanisms: Review and hypothesis. *Ann. Neurol.* 4:451–62

Basbaum, A. I., Fields, H. L. 1979. The origin of descending pathways in the dorsolateral funiculus of the cord of the cat and rat: Further studies on the anatomy of pain modulation. *J. Comp. Neurol.* 187:513–32

Basbaum, A. I., Glazer, E. J., Lord, B. A. P. 1982. Simultaneous ultrastructural localization of tritiated serotonin and immunoreactive peptides. *J. Histochem. Cytochem.* 30:780–84

Basbaum, A. I., Glazer, E. J, Oertel, W. 1981. A light and EM analysis of immunoreactive glutamic acid decarboxylase (GAD) in the spinal and trigeminal dorsal horn of the cat. *Neurosci. Abstr.* 7:528

Basbaum, A. I., Jacknow, D., Mulcahy, J., Levine, J. 1983a. Studies on the contribution of different endogenous opioid peptides to the control of pain. In *Current Topics in Pain Research and Therapy*, ed. Y. Yokota, R. Dubner, pp. 111–18. Amsterdam: Elsevier

Basbaum, A. I., Marley, N., O'Keefe, J., Clanton, C. H. 1977. Reversal of morphine and stimulus produced analgesia by subtotal spinal cord lesions. *Pain* 3:43–56

Basbaum, A. I., Moss, M. S., Glazer, E. J. 1983b. Opiate and stimulation-produced analgesia: The contribution of the monoamines. *Adv Pain Res. Ther.* 5:323–39

Behbehani, M. M. 1982. The role of acetylcholine in the function of the nucleus raphe magnus and in the interaction of this nucleus with the periaqueductal grey. *Brain Res.* 252:299–307

Behbehani, M. M., Fields, H. L. 1979. Evidence that an excitatory connection between the periaqueductal grey and nucleus raphe magnus mediates stimulation-produced analgesia. *Brain Res.* 170:85–93

Behbehani, M. M., Pomeroy, S. L. 1978. Effect of morphine injected in periaqueductal gray on the activity of single units in nucleus raphe magnus of the rat. *Brain Res.* 149:266–69

Behbehani, M. M., Pomeroy, S. L., Mack, C. E. 1981. Interaction between central gray and nucleus raphe magnus: Role of norepinephrine. *Brain Res.* 6:361–64

Beitz, A. J. 1982a. The nuclei of origin of brainstem enkephalin and Substance P projections to the rodent nucleus raphe magnus. *Neuroscience* 7:2753–68

Beitz, A. J. 1982b. The organization of afferent projections to the midbrain periaqueductal grey of the rat. *Neuroscience* 7:133–59

Beitz, A. J. 1982c. The sites of origin of brainstem neurotensin and serotonin projections to the rodent nucleus raphe magnus. *J. Neurosci.* 2:829–24

Belcher, G., Ryall, R. W., Schaffner, R. 1978. The differential effects of 5-hydroxytryptamine, noradrenaline and raphe interneurons in the cat. *Brain Res.* 151:307–21

Belcher, G., Smock, T., Fields, H. L. 1982. Effects of intrathecal ACTH on opiate analgesia in the rat. *Brain Res.* 247:373–77

Bennett, G. J., Abdelmoumene, M., Hayashi, H., Dubner, R. 1980. Physiology and morphology of substantia gelatinosa neurons intracellularly stained with horseradish peroxidase. *J. Comp. Neurol.* 194:809–27

Bennett, G. J., Ruda, M. A., Gobel, S., Dubner, R. 1982. Enkephalin immunoreactive stalked cells and lamina IIb islet cells in cat substantia gelatinosa. *Brain Res.* 240:162–66

Bhattachary, S. K., Keshary, P. R., Sanyal, A. K. 1978. Immobilization stress-induced antinociception in rats: Possible role of serotonin and prostaglandins. *Eur. J. Pharmacol.* 50:83–85

Bloom, F. E., Battenberg, E., Rossier, J., Ling, N., Guillemin, R. 1978. Neurons containing β-endorphin in rat brain exist separately from those containing enkephalin: Immunocytochemical studies. *Proc. Natl. Acad. Sci. USA* 75:1591–95

Bodnar, R. J., Kelly, D. D., Glusman, M. 1979a. 2-Deoxy-D-glucose analgesia: Influences of opiate and non-opiate factors. *Pharmacol. Biochem. Behav.* 11:297–301

Bodnar, R. J., Kelly, D. D., Mansour, A., Glusman, M. 1979b. Differential effects of hypophysectomy upon analgesia induced by two glucoprivic stressors and morphine. *Pharmacol. Biochem. Behav.* 11:303–8

Bolles, R. C., Fanselow, M. S. 1982. Endorphins and behavior. *Ann. Rev. Psychol.* 33:87–101

Bowker, R. M., Steinbusch, H. W. M., Coulter, J. D. 1981a. Serotonergic and peptidergic projections to the spinal cord demonstrated by a combined retrograde HRP histochemical and immunocytochemical staining method. *Brain Res.* 211:412–17

Bowker, R., Westlund, K. N., Coulter, J. D. 1981b. Origins of serotonergic projections of the spinal cord in rat: An immunocytochemical-retrograde transport study. *Brain Res.* 226:187–99

Brodie M. S., Proudfit, H. K. 1982. The induction of analgesia by the local injection of carbachol into the nucleus raphe magnus. *Neurosci. Abstr.* 8:770

Carstens, E., Tulloch, I., Zieglgansberger, W., Zimmerman, M. 1979. Presynaptic excitability changes induced by morphine in single cutaneous afferent C and A-fibers. *Pflügers Arch.* 379:143–47

Casey, K. L. 1971. Responses of bulboreticular units to somatic stimuli eliciting escape behavior in the cat. *Int. J. Neurosci.* 2:15–28

Cervero, F., Iggo, A. 1980. The substantia gelatinosa of the spinal cord. A critical review. *Brain* 103:717–72

Chance, W. T., White, A. C., Krynock, G. M., Rosecrans, J. A. 1978. Conditional fear-induced decreases in the binding of (^{3}H)-*N*-leu-enkephalin to rat brain. *Brain Res.* 141:371–74

Clineschmidt, B. V., McGuffin, J. C., Bunting. P. B. 1979. Neurotensin: Antinoceptive action in rodents. *Eur. J. Pharmacol.* 54:129–39

Comb, J., Seeburg, P. H., Adelman, J., Eiden, L., Herbert, E. 1982. Primary structure of the human met- and leu-enkephalin precursor and its mRNA. *Nature* 295:663–66

Dahlstrom, A., Fuxe, K. 1964. Evidence for the existence of monoamine neurons in the central nervous system. II. Experimental demonstration of monoamines in the cell bodies of brain stem neurons. *Acta Physiol. Scand.* 62(Suppl. 232):1–55

Davies, J. E., Roberts, M. H. T. 1981. 5-Hydroxytryptamine reduces substance P responses on dorsal horn interneurons: A possible interaction of neurotransmitters. *Brain Res.* 217:399–404

Dickenson, A. H., Oliveras, J. L., Besson, J. M. 1979. Role of the nucleus raphe magnus in opiate analgesia as studied by the microinjection technique in the rat. *Brain Res.* 170:95–111

Dubner, R., Bennett, G. J. 1983. Spinal and trigeminal mechanisms of nociception. *Ann. Rev. Neurosci.* 6:381–418

Dworkin, B. R., Filewich, R. J., Miller, N. E., Craigmyle, N., Pickering, T. G. 1979. Baroreceptor activation reduces reactivity to noxious stimulation: Implications for hypertension. *Science* 205:1299–1301

Fields, H. L., Anderson, S. D. 1978. Evidence that raphe spinal neurons mediate opiate and midbrain stimulation produced analgesia. *Pain* 5:333–49

Fields, H. L., Basbaum, A. I. 1978. Brain stem control of spinal pain transmission neurons. *Ann. Rev. Physiol.* 40:193–221

Fields, H. L., Basbaum, A. I., Clanton, C. H., Anderson, S. D. 1977. Nucleus raphe magnus inhibition of spinal cord dorsal horn neurons. *Brain Res.* 126:441–53

Fields, H. L., Emson, P. C., Leigh, B. K., Gilbert, R. F. T., Iversen, L. L. 1980. Multiple opiate receptor sites on primary afferent fibres. *Nature* 284:351–53

Fredricksen, R. C. A., Burgis, V., Harrell, C. E., Edwards, J. D. 1978. Dual actions of Substance P on nociception: Possible role of endogenous opioids. *Science* 199:1359–62

Gallager, D. W., Pert, A. 1978. Afferents to brain stem nuclei (brain stem raphe, nucleus reticularis pontis caudalis and nucleus gigantocellularis) in the rat as demonstrated by microiontophoretically applied horseradish peroxidase. *Brain Res.* 144:257–76

Gebhart, G. F. 1982. Opiate and opioid peptide effects on brain stem neurons: Relevance to nociception and antinociceptive mechanisms. *Pain* 12:93–140

Gebhart, G. F., Toleikis, J. R. 1978. An evaluation of stimulation-produced analgesia in the cat. *Exp. Neurol.* 62:570–79

Gibson, S. J., Polak, J. M., Bloom, S. R., Wall, P. D. 1981. The distribution of nine peptides in rat spinal cord with special emphasis on the substantia gelatinosa and on the area around the central canal (lamina X). *J. Comp. Neurol.* 201:65–79

Gilbert, R. F. T., Emson, P. C., Hunt, S. P., Bennett, G. W., Marsden, C. A., Sanderg, B. E. B., Steinbusch, H. W. M., Verhofstad, A. A. J. 1982. The effects of monoamine neurotoxins on peptides in the rat spinal cord. *Neuroscience* 7:69–87

Glazer, E. J., Basbaum, A. I. 1981. Immunohistochemical localization of leucine-enkephalin in the spinal cord of the cat: Enkephalin-containing marginal neurons and pain modulation. *J. Comp. Neurol.* 196:377–89

Glazer, E. J., Basbaum, A. I. 1983a. Opioid neurons and pain modulation: An ultrastructural analysis of enkephalin in cat super-ficial dorsal horn. *Neuroscience.* In press

Glazer, E. J., Basbaum, A. I. 1983b. Serotonin axons synapse on enkephalin immunoreactive neurons in cat dorsal horn. Submitted for publication

Glazer, E. J., Steinbusch, H., Verhofstad, A., Basbaum, A. I. 1981. Serotonin neurons in nucleus raphe dorsalis and paragigantocellularis of the cat contain enkephalin. *J. Physiol. Paris* 77:241–45

Gobel, S., 1978. Golgi studies of the neurons in layer II of the dorsal horn of the medulla (trigeminal nucleus caudalis). *J. Comp. Neurol.* 180:395–414

Gobel, S. 1979. Neural circuitry in the substantia gelatinosa of Rolando: Anatomical insights. *Adv. Pain Res. Ther.* 3:175–95

Goldstein, A., Fischi, W., Lowney, L. I., Hunkapiller, M., Hood, L. 1981. Porcine pituitary dynorphin: Complete amino acid sequence of the biologically active hepadecapeptide. *Proc. Natl. Acad. Sci. USA* 78:7219–23

Goldstein, A., Tachibana, S., Lowney, L. I., Hunkapiller, M., Hood, L. 1979. Dynorphin (1–13), an extraordinarily potent opioid peptide. *Proc. Natl. Acad. Sci. USA* 76:6666–70

Gracely, R. H., Deeter, W. R., Wolskee, P. J., et al. 1979. The effect of naloxone on multidimensional scales of post-surgical pain in nonsedated patients. *Neurosci. Abstr.* 5:609

Gracely, R. H., Wolskee, P. J., Deeter, W. R., Dubner, R. 1982. Naloxone and placebo alter postsurgical pain by independent mechanisms. *Neurosci. Abstr.* 8:264

Grau, J. W., Hyson, R. L., Maier, S. F., Madden, J., Barchas, J. D. 1981. Long-term stress-induced analgesia and activation of the opiate systems. *Science* 213:1409–11

Grevert, P., Leonard, H. A., Goldstein, A. 1983. Partial antagonism of placebo analgesia by naloxone. *Pain* 16:129–43

Gubler, U., Seeburg, P., Hoffman, B. J., Gage, L. P., Udenfriend, S. 1982. Molecular cloning established proenkephalin as precursor of enkephalin-containing peptides. *Nature* 295:206–8

Guilbaud, G., Peschanski, M., Gautron, M., Binder, D. 1980. Responses of neurons of the nucleus raphe magnus to noxious stimuli. *Neurosci. Lett.* 17:149–54

Hamilton, B. 1973. Cytoarchitectural subdivisions of the periaqueductal gray matter of the cat. *J. Comp. Neurol.* 149:1–28

Hammond, D. L., Levy, R. A., Proudfit, H. K. 1980. Hypoalgesia induced by microinjection of a norepinephrine antagonist in the raphe magnus: Reversal by intrathecal administration of a serotonin antagonist. *Brain Res.* 201:475–89

Han, J. S., Xie, C. W. 1982. Dynorphin: Po-tent analgesic effect in spinal cord of the rat. *Life Sci.* 31:1781–84

Hayes, R. L., Bennett, G. J., Newlon, P. G., Mayer, D. J. 1978. Behavioral and physiological studies of non-narcotic analgesia in the rat elicited by certain environmental stimuli. *Brain Res.* 155:69–90

Headley, P. M., Duggan, A. W., Griersmith, B. T. 1978. Selective reduction by noradrenaline and 5-hydroxytryptamine of nociceptive responses of cat dorsal horn neurones. *Brain Res.* 145:185–89

Hentall, I. D., Fields, H. L. 1983. Actions of opiates, substance P and serotonin on the excitability of primary afferent terminals and observations on interneuronal activity in the neonatal rat's dorsal horn *in vitro. Neuroscience* 9:521–28

Hiller, J. M., Simon, E. J., Crain, S. M., Peterson, E. R. 1978. Opiate receptors in culture of fetal mouse dorsal root ganglia (DRG) and spinal cord: Predominance in DRG neurites. *Brain Res.* 145:396–400

Hökfelt, T., Johansson, O., Ljungdahl, A., Lundberg, J. M., Schultzberg, M. 1980. Peptidergic neurones. *Nature* 284:515–21

Hökfelt, T., Ljungdahl, A., Steinbusch, H., Verhofstad, A. N., Nilsson, G., Brodin, E., Pernow, B., Goldstein, M. 1978. Immunohistochemical evidence of substance P like immunoreactivity in some 5-hydroxytryptamine containing neurons in the rat central nervous system. *Neuroscience* 3:517–38

Hökfelt, T., Ljungdahl, A., Terenius, L., Elde, R., Nilsson, G. 1977. Immunohistochemical analysis of peptide pathways possibly related to pain and analgesia: Enkephalin and substance P. *Proc. Natl. Acad. Sci. USA* 74:3081–85

Hökfelt, T., Terenius, T., Kuypers, H. G. J. M., Dann, O. 1979. Evidence for enkephalin immunoreactive neurons in the medulla oblongata projecting to the spinal cord. *Neurosci. Lett.* 14:55–60

Hollt, V. 1983. Multiple endogenous opioid peptides. *Trends Neurosci.* 16:24–26

Hollt, V., Tulunay, C., Woo, S. K., Loh, H. H., Herz, A. 1982. Opioid peptides derived from pro-enkephalin A but not that from pro-enkephalin B are substantial analgesics after administration into brain of mice. *Eur. J. Pharmacol.* 83:355–56

Hosobuchi, Y., Rossier, J., Bloom, F. E., Guillemin, R. 1979. Stimulation of human periaqueductal gray for pain relief increases immunoreactive β-endorphin in ventricular fluid. *Science* 203:279–81

Houghten, R. A., Swann, R. W., Li, C. H. 1980. β Endorphin: Stability, clearance behavior and entry into the central nervous system after injection of the tritiated peptide in rats and rabbits. *Proc. Natl. Acad. Sci. USA* 77:4588–91

Hunt, S. P., Kelly, J. S., Emson, P. C. 1980. The electron microscopic localization of methionine-enkephalin within the superficial layers (I and II) of the spinal cord. *Neuroscience* 5:1871–90

Hunt, S. P., Kelly, J. S., Emson, P. C., Kimmel, J. R., Miller, R. J., Wu, J-Y. 1981. An immunohistochemical study of neuronal populations containing neuropeptides or γ-aminobutyrate within the superficial layers of the rat dorsal horn. *Neuroscience* 6:1883–98

Jan, L. Y., Jan, Y. N. 1982. Peptidergic transmission in sympathetic ganglia of the frog. *J. Physiol.* 324:219–46

Jessell, T. M., Iversen, L. L. 1977. Opiate analgesics inhibit substance P release from rat trigeminal nucleus. *Nature* 268:549–51

Johannessen, J. N., Watkins, L. R., Mayer, D. J. 1981. Non-serotonergic cells at the origin of the dorsolateral funiculus (DLF) in rat medulla. *Neurosci. Abstr.* 7:533

Johansson, O., Hökfelt, T., Pernow, B., Jeffcoate, S. L., White, N., Steinbusch, H. W. M., Verhofstad, A. A. J., Emson, P. C., Spindel, E. 1981. Immunohistochemical support for three putative transmitters in one neurons: Coexistence of 5-hydroxytryptamine, substance P and thyrotropin releasing hormone-like immunoreactivity in medullary neurons projecting to the spinal cord. *Neuroscience* 6:1857–81

Kakidani, H., Furutani, Y., Takahashi, H., Noda, M., Morimoto, Y., Hirose, T., Asai, M., Inayama, S., Nakanishi, S., Numa, S. 1982. Cloning and sequence analysis of cDNA for porcine β-neo-endorphin/dynorphin precursor. *Nature* 298:245–49

Kalivas, P. W., Jennes, L., Nemeroff, C. B., Prange, A. J. 1982. Neurotensin: Topographical distribution of brain sites involved in hypothermia and antinociception. *J. Comp. Neurol.* 210:255–38

Kayser, V., Guilbaud, G. 1981. Dose-dependent analgesic and hyperalgesic effects of systemic naloxone in arthritic rats. *Brain Res.* 226:344–48

Kimura, S., Lewis, R. V., Stern, A. S., Rossier, J., Stein, S., Udenfriend, S. 1980. Probable precursors of (Leu) and (Met)enkephalin in adrenal medulla: Peptides of 3–5 kilodaltons. *Proc. Natl. Acad. Sci. USA* 77:1681–85

Kuraishi, Y., Harada, Y., Satoh, M., Takagi, H. 1979. Antagonism by phenoxybenzamine of the analgesic effect of morphine injected into the nucleus reticularis gigantocellularis of the rat. *Neuropharmacology* 18:107–10

LaMotte, C., Pert, C. B., Snyder, S. H. 1976. Opiate receptor binding in primate spinal cord. Distribution and changes after dorsal root section. *Brain Res.* 112:407–12

Lasagna, L. 1965. Drug interaction in the field of analgesic drugs. *Proc. Soc. Exp. Biol. Med.* 58:978–83

Leichnetz, G. R., Watkins, L., Griffin, G., Martin, R., Mayer, D. J. 1978. The projection from nucleus raphe magnus and other brain stem nuclei to the spinal cord in the rat: A study using HRP blue reaction. *Neurosci. Lett.* 8:119–24

Levine, J. D., Gordon, N. C., Fields, H. L. 1978a. The mechanism of placebo analgesia. *Lancet* 2:654–57

Levine, J. D., Gordon, N. C., Fields, H. L. 1979. Naloxone dose dependently produces analgesia and hyperalgesia in postoperative pain. *Nature* 278:740–41

Levine, J. D., Gordon, N. C., Fields, H. L. 1982a. Naloxone fails to antagonize nitrous oxide analgesia for clinical pain. *Pain* 13:165–70

Levine, J. D., Gordon, N. C., Jones, R. T., Fields, H. L. 1978b. The narcotic antagonist naloxone enhances clinical pain. *Nature* 272:826–27

Levine, J. D., Gordon, N. C., Smith, R., Fields, H. L. 1982b. Postoperative pain: Effect of extent of injury and attention. *Brain Res.* 234:500–4

Levine, J. D., Lane, S. R., Gordon, N. C., Fields, H. L. 1982c. A spinal opioid synapse mediates the interaction of spinal and brainstem sites in morphine analgesia. *Brain Res.* 236:85–91

Lewis, J. W., Cannon, J. T., Liebeskind, J. C. 1980. Opioid and non-opioid mechanisms of stress analgesia. *Science* 208:623–25

Lewis, J. W., Terman, G. W., Shavit, Y., Nelson, L. R., Liebeskind, J. C. 1983. Neural, neurochemical and hormonal bases of stress-induced analgesia. *Intra-Sci. Found. Symp., Los Angeles, Calif.*

Lewis, J. W., Tordoff, M. G., Sherman, J. E., Liebeskind, J. C. 1982. Adrenal medullary enkephalin-like peptides may mediate opioid stress analgesia. *Science* 217:557–59

Liebeskind, J. C., Guilbaud, G., Besson, J. M., Oliveras, J. L. 1973. Analgesia from electrical stimulation of the periaqueductal gray matter in the cat: Behavioral observations and inhibitory effects on spinal cord interneurons. *Brain Res* 50:441–46

Light, A. R., Perl, E. R. 1979. Spinal termination of functionally identified primary afferent neurons with slowly conducting myelinated fibers. *J. Comp. Neurol.* 186:133–50

Light, A. R., Trevino, D. L., Perl, E. R. 1979. Morphological features of functionally identified neurons in the marginal zone and substantia gelatinosa of the spinal dorsal horn. *J. Comp. Neurol.* 186:151–71

Llorens, C., Martres, M. P., Baudry, M. 1978. Hypersensitivity to norepinephrine in the cortex after chronic morphine relevant to tolerance and dependence. *Nature* 274: 603–8

Maier, S. F., Drugan, R. C., Grau, J. W. 1982. Controllability, coping behavior, and stress-induced analgesia in the rat. *Pain* 12:47–56

Mains, R. E., Eipper, B. A., Ling, N. 1977. Common precursor to corticotropins and endorphins. *Proc. Natl. Acad. Sci. USA* 74:3014–18

Maixner, W., Touw, K. B., Brody, M. J., Gebhart, G. F., Long, J. P. 1982. Factors influencing the altered pain perception in the spontaneously hypertensive rat. *Brain Res.* 237:137–45

Mantyh, P. W. 1982a. Forebrain projections to the periaqueductal grey in the monkey with observations in the cat and rat. *J. Comp. Neurol.* 204:349–63

Mantyh, P. W. 1982b. The midbrain periaqueductal gray in the rat, cat and monkey: a Nissl, Weil and Golgi analysis. *J. Comp. Neurol.* 204:349–63

Mantyh, P. W. 1983. Connections of midbrain periaqueductal gray in the monkey. II. Descending efferent projections. *J. Neurophysiol.* 49:582–94

Mantyh, P. W., Peschanski, M. 1982. Spinal projections from the periaqueductal grey and dorsal raphe in the rat, cat and monkey. *Neuroscience* 7(11):2769–76

Martin, R. F., Jordan, L. M., Willis, W. D. 1978. Differential projections of cat medullary raphe neurons demonstrated by retrograde labelling following spinal cord lesions. *J. Comp. Neurol.* 182:77–88

Mayer, D. J., Liebeskind, J. C. 1974. Pain reduction by focal electrical stimulation of the brain: An anatomical and behavioral analysis. *Brain Res.* 68:73–93

Mayer, D. J., Price, D. D. 1976. Central nervous system mechanisms of analgesia. *Pain* 2:379–404

Mayer, D. J., Watkins, L. R. 1981. Role of endorphins in endogenous pain control systems. *Mod. Probl. Pharmacopsychiatry* 17:689–96

Mayer, D. J., Wolfe, T. L., Akil, H., Carder, B., Liebeskind, J. C. 1971. Analgesia from electrical stimulation in the brain stem of the rat. *Science* 174:1351–54

Mehler, W. R. 1962. The anatomy of the so-called "pain tract" in man: An analysis of the course and distribution of the ascending fibers of the fasciculus anterolateralis. In *Basic Research in Paraplegia*, ed. J. D. French, R. W. Porter, pp. 26–55. Springfield, Ill: Thomas

Melzack, R. 1975. Prolonged relief of pain by brief, intense transcutaneous somatic stimulation. *Pain* 1:357–73

Miller, R. J. 1981. Peptides as neurotransmitters: Focus on the enkephalins and endorphins. *Pharmacol. Ther.* 12:73–108

Mohrland, J. S., Gebhart, G. F. 1979. Substance P-induced analgesia in the rat. *Brain Res.* 171:556–59

Moss, M. S., Basbaum, A. I. 1979. The efferent projections of the dorsolateral pontine tegmentum of the cat. *Neurosci. Abstr.* 5

Moss, M. S., Basbaum, A. I. 1983a. The fine structure of the caudal periaqueductal grey of the cat: Morphological and synaptic organization of normal and immunoreactive enkephalin-labelled profiles. *Brain Res.* In press

Moss, M. S., Basbaum, A. I. 1983b. The peptidergic organization of the cat periaqueductal grey. II. The distribution of immunoreactive substance P and vasoactive intestinal polypeptide. *J. Neurosci.* 3:1437–49

Moss, M. M., Glazer, E. J., Basbaum, A. I. 1981. Enkephalin-immunoreactive perikarya in the nucleus raphe dorsalis of the cat. *Neurosci. Lett.* 21:33–37

Moss, M. S., Glazer, E. J., Basbaum, A. I. 1983. The peptidergic organization of the cat periaqueductal grey: I. The distribution of enkephalin-containing neurons and terminals. *J. Neurosci.* 3:603–16

Mudge, A. W., Leeman, S. E., Fischbach, G. D. 1979. Enkephalin inhibits release of substance P from sensory neurons in culture and decreases action potential duration. *Proc. Natl. Acad. Sci. USA* 76:526–30

Mulcahy, J., Basbaum, A. I. 1983. Coexistence of immunoreactive enkephalin and dynorphin in single brainstem neurons. Submitted for publication

Murfin, R., Bennett, J., Mayer, D. J. 1976. The effect of dorsolateral spinal cord (DLF) lesions on analgesia from morphine microinjected into the periaqueductal gray matter (PAG) of the rat. *Neurosci. Abstr.* 2:946

Naranjo, J. R., Franco-Sanchez, F., Garzon, J., del Rio, J. 1982. Analgesic activity of Substance P in rats: Apparent mediation by met-enkephalin release. *Life Sci.* 30:441–46

Nicoll, R. A., Alger, B. E., Nicoll, R. A. 1980. Enkephalin blocks inhibitory pathways in the vertebrate CNS. *Nature* 287:22–25

Oleson, T. D., Twombly, D. A., Liebeskind, J. C. 1978. Effects of pain attenuating brain stimulation and morphine on electrical activity in the raphe nuclei of the awake rat. *Pain* 4:211–30

Oliveras, J. L., Bourgoin, S., Hery, F., Besson, J. M., Hamon, M. 1977. The topographical distribution of serotoninergic terminals in the spinal cord of the cat: Biochemical mapping by the combined use of microdissection and microassay procedures. *Brain Res.* 138:393–406

Oliveras, J. L., Guilbaud, G., Besson, J. M. 1979. A map of serotoninergic structures involved in stimulation producing analgesia in unrestrained freely moving cats. *Brain Res.* 164:317–22

Oliveras, J. L., Redjemi, F., Guilbaud, G., Besson, J. M. 1975. Analgesia induced by electrical stimulation of the inferior centralis nucleus of the raphe in the cat. *Pain* 1:139–45

Piercey, M. F., Lahti, R. A., Schroeder, L. A., Einspahr, F. J., Barsuhn, C. 1982. U-50488H, a pure kappa receptor agonist with spinal analgesic loci in the mouse. *Life Sci.* 31:1197–1200

Price, D. D., Hayaski, H., Dubner, R., Ruda, M. A. 1979. Functional relationships between neurons of the marginal and substantia gelatinosa layers of the primate dorsal horn. *J. Neurophysiol.* 42:1590–1608

Prieto, G. J., Cannon, J. T., Liebeskind, J. C. 1983. Nucleus raphe magnus lesions disrupt stimulation-produced analgesia from ventral but not dorsal midbrain areas in the rat. *Brain Res.* 261:53–57

Rexed, B. 1954. A cytoarchitectonic atlas of the spinal cord in the cat. *J. Comp. Neurol.* 100:297–380

Reynolds, D. V. 1969. Surgery in the rat during electrical analgesia induced by focal brain stimulation. *Science* 164:444–45

Rhodes, D. L. 1979. Periventricular system lesions in stimulation-produced analgesia. *Pain* 7:31–51

Rivot, J. P., Chaouch, A., Besson, J. M. 1980. Nucleus raphe magnus modulation of response of rat dorsal horn neurons to unmyelinated fiber inputs: Partial involvement of serotonergic pathways. *J. Neurophysiol.* 44:1039–57

Roberts, J. L., Herbert, E. 1977. Characterization of a common precursor to corticotropin and beta-lipotropin: Identification of beta-lipotropin peptides and their arrangement relative to corticotropin in the precursor synthesized in a cell-free system. *Proc. Natl. Acad. Sci. USA* 74:5300–4

Ruda, M. A. 1982. Opiates and pain pathways: Demonstration of enkephalin synapses on dorsal horn projection neurons. *Science* 215:1523–25

Sandkuhler, J., Thalhammer, J. G., Gebhart, G. F., Zimmerman, M. 1982. Lidocaine microinjected in the NRM does not block the inhibition by stimulation in the PAG of noxious-evoked responses of dorsal horn neurons in cat. *Neurosci. Abstr.* 8:768

Satoh, M., Akaike, A., Nakazawa, T., Takagi, H. 1980. Evidence for involvement of separate mechanisms in the production of analgesia by electrical stimulation of the nucleus reticularis paragigantocellularis and nucleus raphe magnus in the rat. *Brain Res.* 194:424–529

Seybold, V. S., Elde, R. P. 1982. Neurotensin immunoreactivity in the superficial laminae of the dorsal horn of the rat: I. Light microscopic studies of cell bodies and proximal dendrites. *J. Comp. Neurol.* 205:89

Smock, T., Fields, H. L. 1980. $ACTH_{1-24}$ blocks opiate-induced analgesia in the rat. *Brain Res.* 212:202–6

Sullivan, T. L., Pert, A. 1981. Analgesic activity of non-opiate neuropeptides following injections into the rat periaqueductal grey matter. *Neurosci. Abstr.* 7:504

Sumal, K. K., Pickel, V. M., Miller, R. J., Reis, D. J. 1982. Enkephalin-containing neurons in substantia gelatinosa of spinal trigeminal complex: Ultrastructural and synaptic interaction with primary sensory afferents. *Brain Res.* 248:223–36

Takagi, H., Yamamoto, K., Shiosaka, S., Senba, E., Takatsuki, K., Inagaki, S., Sakanaka, M., Tohyama, M. 1981. Morphological study of noradrenaline innervation in the caudal raphe nuclei with special reference to fine structure. *J. Comp. Neurol.* 203:15–22

Tsou, K., Jang, C. S. 1964. Studies on the site of analgesic action of morphine by intracerebral microinjection. *Sci. Sin.* 13:1099–9

Tulunay, F. C., Jen, M. F., Chang, J. K., Loh, H. H., Lee, N. M. 1981. Possible regulatory role of dynorphin on morphine and β-endorphin-induced analgesia. *J. Pharmacol. Exp. Ther.* 219:296

Walker, J. M., Bernston, G. G., Sandman, C. A., Kastin, A. J., Akil, H. 1981. Induction of analgesia by central administration of ORG 2766, an analog of $ACTH_{4-9}$. *Eur. J. Pharmacol.* 69:71–79

Wall, P. D., 1967. The laminar organization of dorsal horn and effects of descending impulses. *J. Physiol.* 188:403–23

Watkins, L. R., Cobelli, D. A., Faris, P., Aceto, M. D., Mayer, D. J. 1982a. Opiate vs. non-opiate footshock induced analgesia (FSIA): The body region shocked is a critical factor. *Brain Res.* 242:299–308

Watkins, L. R., Cobelli, D. A., Mayer, D. J. 1982b. Opiate vs. non-opiate footshock analgesia (FSIA): Descending and intraspinal components. *Brain Res.* 245:97–106

Watkins, L. R., Cobelli, D. A., Newsome, H. H., Mayer, D. J. 1982c. Footshock induced analgesia is dependent neither on pituitary nor sympathetic activation. *Brain Res.* 245:81–96

Watkins, L. R., Mayer, D. J. 1982. Organization of endogenous opiate and non-opiate pain control systems. *Science* 216:1185–92

Watson, S. J., Khachaturian, H., Akil, H. 1982. Comparison of the distribution of dynorphin systems and enkephalin systems in brain. *Science* 218:1134–36

Weber, E., Roth, K. A., Evans, C. J., Chang, J.-K., Barchas, J. D. 1982. Immunohistochemical localization of dynorphin (1–8) in hypothalamic magnocellular neurons; Evidence for absence of proenkephalin. *Life Sci.* 31:1761–64

Wiklund, L., Leger, L., Persson, M. 1982. Monoamine distribution in the cat brainstem: A fluorescence histochemical study with quantification of indolaminergic and locus coeruleus cell groups. *J. Comp. Neurol.* 203:613–47

Willis, W. D., Haber, L. H., Martin, R. F. 1977. Inhibition of spinothalamic tract cells and interneurons by brain stem stimulation in the monkey. *J. Neurophysiol* 40:968–81

Woolf, C. J., Mitchell, D., Barrett, G. D. 1980. Antinociceptive effect of peripheral segmental electrical stimulation in the rat. *Pain* 8:237–52

Yaksh, T. L., Jessell, T. M., Gamse, R., Mudge, A. W., Leeman, S. E. 1980. Intrathecal morphine inhibits substance P release from mammalian spinal cord in vivo. *Nature* 286:155–57

Yeung, J. C., Rudy, T. 1980. Sites of antinociceptive action of systemically injected morphine: Involvement of supraspinal loci as revealed by intracerebroventricular injection of naloxone. *J. Pharmacol. Exp. Ther.* 215:626–32

Zamir, N., Segal, M. 1979. Hypertension induced analgesia: Changes in pain sensitivity in experimental hypertensive rats. *Brain Res.* 160:170–73

Zamir, N., Shuber, E. 1980. Altered pain perception in hypertensive humans. *Brain Res.* 201:471–74

Zamir, N., Simantov, R., Segal, M. 1980. Pain sensitivity and opioid activity in genetically and experimentally hypertensive rats. *Brain Res.* 184:299–310

Zorman, G., Hentall, I. D., Adams, J. E., Fields, H. L. 1981. Naloxone-reversible analgesia produced by microstimulation in the rat medulla. *Brain Res.* 219:137–48

Zorman, G., Belcher, G., Adams, J. E., Fields, H. L. 1982. Lumbar intrathecal naloxone blocks analgesia produced by microstimulation of the ventromedial medulla in the rat. *Brain Res.* 236:77–84

Ann Rev. Neurosci. 1984. 7:339-77

MODULATION OF CELL ADHESION DURING INDUCTION, HISTOGENESIS, AND PERINATAL DEVELOPMENT OF THE NERVOUS SYSTEM

Gerald M. Edelman

The Rockefeller University, New York, New York 10021

INTRODUCTION

From the earliest embryonic events in which neural precursors can be discerned to the adult CNS in which significant molar rearrangements of neuronal connections no longer occur, one must conceive of the nervous system as one in which continuing development takes place. With this in mind, it is meaningful and often convenient to consider neural development as divided into epochs defined by more or less irreversible processes. In terms of significant neural embryonic events one may distinguish (*a*) neural induction and determination with its succeeding period of neurulation or neural tube formation, (*b*) histogenetic rearrangements that are local to an area involving glial and neuronal interactions, (*c*) histogenetic rearrangements involving long range migrations as seen in neural crest cell movements as well as in fiber tract mapping within the CNS, (*d*) continuing perinatal rearrangements in which increasing function determines final structure. The extent and duration of these epochs will of course differ among different species and therefore the definition of an epoch is relative. Indeed, in some cases a simple distinction among early embryonic, organogenetic, and adult epochs will suffice. Nevertheless, even within this crude temporal distinction, clear cut and sometimes irreversible events charac-

339

0147–006X/84/0301–0339$02.00

teristic of different epochs can be seen to occur at the cellular and molecular level. One of the key tasks of developmental biology is to document such events.

With respect to developmental anatomy, particular structures are in every case formed as a result of complex interactions of the primary processes of development: cell division, cell migration, differentiation, cell adhesion, and cell death (Cowan 1978). Each of these processes is in turn the result of a series of complex molecular interactions, most of which have not yet been fully defined. While each of these primary processes can play a dominant and decisive role at a particular time, adhesion occupies a special place in the sense that it is essential at all epochs. Only recently, however, has it become possible to trace some of the mechanisms of cell adhesion in molecular terms and to begin to discern its particular roles in different epochs ranging from the earliest embryonic events to adult life.

In this review, I shall emphasize studies that have led to this molecular view of adhesion in different epochs, particularly as it applies to the nervous system. The discussion shall proceed in critical terms and is guided by the view that progress so far represents only an early beginning in understanding some key problems of neurogenesis. Nevertheless, the picture that emerges is considerably different from that seen just a decade ago and it encourages one to predict that adhesion will be the first of the primary processes to be understood in detail at the molecular level.

The questions to be addressed here are:

1. What are the chemical properties and binding specificities of cell adhesion molecules (CAMs), particularly those of the nervous system?
2. What mechanisms lead to alterations in binding between CAMs as well as between the cells with which they are associated?
3. What are the evolutionary and species relationships among those molecules?
4. What is the connection between the appearance of CAMs and neural induction?
5. What is the role of CAMs in later epochs of neural histogenesis?
6. Are the same CAMs used for similar functions in all epochs of neural development?

Because the answers to the first two questions on structure and binding are critical for answering the more biological questions, a major part of this review is dedicated to a description of the chemical nature of CAMs. Although the main emphasis is on CAMs associated with the nervous system, it becomes apparent that other CAMs must be considered, particularly in analyzing early embryogenesis.

A number of reviews have covered the subject of cell adhesion (Frazier &

Glaser 1979, Lilien et al 1979, Moscona 1974) and reviews pertinent to the discovery and analysis of CAMs have appeared (Edelman 1983) or are in press (Edelman et al 1983a). A number of other general reviews on cell adhesion are also useful as background (Letourneau et al 1980, Garrod & Nicol 1981, Grinnell 1978). They may be consulted for historical details and are only recapitulated here insofar as they bear upon the specific questions about neural development mentioned above.

CONCEPTS AND NOMENCLATURE

Adhesion plays a pivotal role in pattern formation and morphogenesis at least to the extent that the other developmental primary processes depend upon cell-cell interactions. This role is not only direct in the sense that cell-cell interactions are prerequisite for tissue patterns but also regulatory in the sense that these interactions can affect the sequence of other primary processes (Edelman 1983, Edelman 1976). Of particular significance is the expression of genes specifying cell adhesion molecules at particular times of development or in a particular milieu. Regulation of this process by cell-cell interaction is likely to be a primary factor in embryonic induction.

A key theoretical distinction in considering such cell interactions concerns the specificity of cell adhesion molecules as they appear on various forms of differentiated cells making up a tissue. Certain proposals (Moscona 1962, Sperry 1963, Meyer & Sperry 1973) have suggested that the level of specificity is in fact defined almost to the level of pairs of interacting cells; this would imply a large repertoire of specific cell surface markers. At the other extreme are theories suggesting the absence of specificity (Garrod & Nicol 1981), theories asserting the absence of specific cell adhesion molecules, the role of which is replaced by weak forces (Curtis 1967), or theories that deal with adhesion in terms of thermodynamic equilibria (Steinberg 1970).

The picture that is beginning to emerge from the study of CAMs (Edelman 1983) suggests instead the following formulation:

1. There are CAMs of different specificity.
2. This specificity is not expressed in terms of a very large set of complementary gene products but instead appears to be related to major differences in binding mechanisms, e.g. Ca^{2+}-independent or Ca^{2+}-dependent.
3. Interactions of a particular CAM can be altered dynamically (i.e. under constraint of kinetically determined processes) by changes in binding efficiency that result from differences in CAM amount, distribution, or chemical form at the cell surface. Changes of these kinds, collectively called local cell surface modulation, have been proposed as a major means of altering cell recognition (Edelman 1976, Edelman 1983).

Evidence for each of these conclusions is reviewed below; at this point they serve to emphasize two main issues that are important to developing an adequate nomenclature for cell adhesion processes. The first issue is that varying combinations of different primary processes resulting from varying kinetic constraints will lead selectively to different tissue patterns. The selectivity of these processes will be altered by cell surface modulation. These events can in turn lead to the establishment of modally similar tissue patterns in the same anatomical region of two different animals. Such dynamically established patterns, although not completely identical, will appear to be highly specific at the cellular level because of their complexity and overall similarity, tempting one to conclude incorrectly that a large repertoire of predetermined surface markers may be responsible for their order. The second issue to be stressed in later sections of this review is that even if such patterns can arise dynamically, CAMs of different molecular specificity are still necessary, particularly in early development. Nonetheless, CAM specificity, while essential for embryogenesis, is not in itself sufficient to lead to detailed patterns of the kinds seen in highly developed tissues.

With these considerations in mind, a useful nomenclature may be adopted discriminating among (a) adhesion between the same or different cell types, (b) binding between the same or different kinds of CAM molecules, (c) binding between cells and substratum, (d) binding between cells by specialized structures. In this review, the conventions and nomenclature shown in Table 1 are used. Although the major portion of the review is concerned with CAMs, the presence of cell interactions by means of intercellular junctions and via cell substrate interactions must also be briefly touched upon, particularly because the sequence of these three kinds of interactions in any particular developing tissue will almost certainly be a major determinant in pattern formation.

The conventions adopted in Table 1 accommodate the possibility that heterotypic interactions may occur that are mediated by the same CAM. Some unavoidable ambiguity resides in these definitions but specific examples will almost always make the meaning clear. If, for example, Purkinje cells and granule cells are considered to be different types of neurons, then it is already clear (Edelman & Chuong 1982) that their heterotypic binding is mediated by N-CAM in a homophilic mechanism. Another potential ambiguity is related to the suggestion (to be reviewed below) that there may be two sets of CAMs in different epochs: a primitive set employed in not yet fully differentiated tissue areas during early induction events and a secondary set of CAMs that is expressed only in later events of histogenesis. The difficulty in nomenclature is that CAMs in a primitive set (such as N-CAM) may be used in later events as well. Rather than adopt a noncommittal designation for a CAM (such as numbers or molecular weight), an exemplary designation has been adopted. Thus N-CAM, or neural cell adhesion molecule, which was first obtained from

TABLE 1 Nomenclature for cell-cell and cell-substrate interaction

Binding event	Cellular binding mode	Molecule	Molecular binding mechanism
Cell-cell adhesion	Homotypic (same cell type) Heterotypic (different cell type	Cell adhesion molecules; CAMs (e.g. N-CAM, Ng-CAM, L-CAM)	Homophilic: (A-A) Heterophilic: A-B (2nd order) A-B-A (3rd Order)
Cell-substrate adhesion	—	Substrate adhesion molecules; SAMs (e.g. laminin, collagen, fibronectin)	Homophilic Heterophilic
Formation of intercellular junctions (ICJs)	Homotypic Heterotypic	Cell junction molecules; CJMs (e.g. connexin)	Homophilic Heterophilic

differentiated neural tissue and named on that basis, is also found on muscle precursors as well as on very early germ layer derivatives (Edelman et al 1983a). Similarly, L-CAM or liver cell adhesion molecule, is also found in early germ layers and in certain of their later derivatives. As long as these facts are understood, the adoption of a nomenclature for the primitive set CAMs that is related to a major tissue from which they were first obtained is not confusing and appears to be preferable to a noncommittal designation. Moreover, this convention is consistent with that for secondary set CAMs that appear later, such as Ng-CAM (neuron-glial CAM). In this nomenclature, the capital letter refers to the cell of origin of the molecule and the small letter to the cell that is bound in a heterotypic interaction.

IMMUNOLOGICAL IDENTIFICATION OF CAMs AND PERTURBATION ASSAYS

The search for cell adhesion molecules was first directed at embryonic tissues that were well into histogenesis. The methodology was based on several presuppositions (Edelman 1983, Brackenbury et al 1977, Thiery et al 1977):

1. If local cell surface modulation (Edelman 1976) is a major factor in histogenesis, the number of different CAMs in a tissue would not be large.
2. In vitro assays for cell adhesion should be short term to avoid potentially confusing contributions of other primary processes.
3. Antibodies to a predominant tissue CAM raised in a species other than that of CAM origin might yield Fab' fragments capable of blocking adhesion.

4. Specific fractions of cell surface antigens containing CAMs could then be identified by their ability to neutralize the inhibition of adhesion by the anti-CAM Fab' fragments.

Thus, the assay sequence was

(a) $cell_1$ + $cell_2$ ⎯⎯⎯⎯⎯⎯⎯⎯⎯→ adhesion
  inhibition by Fab'

(b) surface antigen fraction containing CAM
  + anti-CAM Fab' ⎯⎯⎯→ neutralization of inhibition.

Iterative application of this procedure allowed the identification of particular antigenic fractions that could then be used for re-immunization; by this means highly specific polyclonal antibodies against various CAMs (Brackenbury et al 1977, Thiery et al 1977, Hoffman et al 1982, Gallin et al 1983) from a variety of species were obtained. Subsequently, monoclonal antibodies were prepared using purified CAM antigens (Hoffman et al 1982, Gallin et al 1983, Chuong et al 1982). With such reagents, CAMs could be found in different species (Chuong et al 1982, McClain & Edelman 1982) and a number of proteins studied in a number of laboratories could be identified definitively or provisionally as related to the CAMs (Table 2).

In addition to their use in direct adhesion assays, antibodies to N-CAM and L-CAM were used in vitro to perturb pattern formation in tissue culture. Anti N-CAM was shown to alter fasciculation patterns of neurites from dorsal root ganglia in culture (Rutishauser et al 1978a,b), to perturb orderly layering in developing chick neural retinal explants (Buskirk et al 1980), and to inhibit retinal cell-myoblast interactions (Grumet et al 1982) and neurite-myotube interactions (Rutishauser et al 1983) in culture. Such perturbation experiments were consistent with the finding that fluorescently labeled anti-N-CAM antibodies stained all central and peripheral neurons (including their processes) as well as myoblasts and myotubes. Similarly, anti-L-CAM antibodies stained hepatocytes and strongly inhibited histotypic organizations of liver cells in culture (Bertolotti et al 1980, Gallin et al 1983).

While the results of perturbation experiments lend support to the view that CAMs present on cell surfaces mediate certain aspects of histogenesis, a deeper understanding of the relation of adhesion to pattern formation and other primary processes of development requires a knowledge of the structure, synthesis, binding mechanisms, expression, and molecular dynamics of various CAMs in various epochs of development. Analyses of the chemical structure of N-CAM and L-CAM have provided major clues to a beginning understanding both of the role of CAMs in very early embryogenesis and of the role of the nervous system CAMs (N-CAM and Ng-CAM) in neurulation and neural histogenesis. Before turning to these key embryonic processes, the

TABLE 2   Immunologically identified CAMs and related molecules

CAM	Tissues used for original isolation	Other proteins			Ion dependence of binding	Animal species so far studied
		Identical or related	(i) (r)	(Ref.)[a]		
N-CAM	Chick neural retina and brain	D_2 BSP-2	(r) (i)	(1–3) (4–6)	Ca^{2+}–independent	Fish, lizard, frog, chicken, mouse, rat, human
Ng-CAM	Chick brain		—		Ca^{2+}–independent	Chick
L-CAM	Chick liver	Cell CAM 120/80	(i)	(7)	Ca^{2+}–dependent	Chick, mouse, human, rat
		Uvomorulin	(?i)	(8)		
		Teratocarcinoma protein	(?r)	(9)		
		Cell-CAM 105	(?r)	(10)		
		LAM	(?i)	(11)		

[a](1) Jørgensen et al 1980; (2) Jørgensen & Moller (1980); (3) Jørgensen 1981; (4) Hirn et al 1983; (5) Rougon et al 1982; (6) Finne et al 1983; (7) Damsky et al 1982; (8) Hyafil et al 1980; (9) Yoshida & Takeichi 1982; (10) Ocklind & Obrink 1981; (11) Nielsen et al 1981.

chemical structure and binding mechanisms and certain time-dependent chemical alterations of CAMs must be reviewed.

STRUCTURE OF CAMs

In this section, the chemistry of three known CAMs is reviewed; although the chemical description is still incomplete, it provides an essential basis for understanding dynamic, morphogenetic, and regulatory roles of CAMs in development.

N-CAM

N-CAM is a large sialoglycoprotein (Hoffman et al 1982, Cunningham et al 1983) present in early embryonic tissues that is synthesized by and expressed on the surface of neurons in various stages of differentiation and is present as well on the surface of myoblasts and myotubes. Carbohydrate moieties are attached to the polypeptide chain of N-CAM in several places; at least one of these sites has a covalently linked set of carbohydrates with a very high content of sialic acid, which appears to be present in a form unusual in higher organisms, polysialic acid (Hoffman et al 1982, Finne et al 1983). Preliminary estimates (Hoffman et al 1982) indicate that N-CAM comprises about 1% of the neuronal cell surface proteins in the embryo; the percentage is somewhat less in the adult. Extraction experiments suggest (Cunningham et al 1983) that it is an integral membrane protein, and fluorescence photobleach recovery measurements (Gall & Edelman 1981) indicate that the molecule is mobile in the cell membrane ($D = 6 \times 10^{-10}$ cm^2/sec). While N-CAM is by no means a rare antigen, large quantities of protein are not readily available and the structural studies carried out so far have required the processing over a decade of about 1/2 million chick embryos.

Careful isolations of N-CAM have revealed two forms of polypeptide chain (Figure 1): one with a molecular weight of 170,000 and one with a molecular weight of 140,000 (Cunningham et al 1983). Experiments on synthesis using tunicamycin (Cunningham et al 1983) have yielded chains lacking carbohydrate and having lower molecular weights corresponding to carbohydrate-free chains. This indicates that the carbohydrate is N-linked and suggests that both chains are synthesized in the cell. Thus, if one chain is a proteolytic product of the other rather than an independent gene product, the cleavage must occur intracellularly. Chemical analysis suggests that the 170 kD and 140 kD chains have similar NH$_2$ terminal sequences and differ only at the COOH terminus.

In addition to these two polypeptides, two cleavage products have been prepared by spontaneous autolysis at 37° (Hoffman et al 1982, Cunningham et al 1983) and by treatment of membrane vesicles containing N-CAM with V8 protease (Cunningham et al 1983). These fragments, known respectively as Fr1 and Fr2 (Figure 1), appear to reflect the organization of the chain into domains;

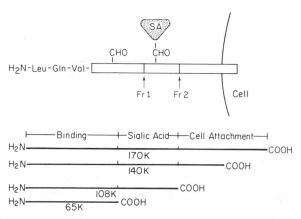

Figure 1 Linear structure of N-CAM. Three structural and functional regions of N-CAM (Cunningham et al 1983) deduced from studies of the intact molecule and a series of fragments: the NH$_2$-terminal region includes a specific binding domain and carbohydrate (CHO) but little, if any, sialic acid; the neighboring region is rich in sialic acid (SA), probably present as polysialic acid; and the COOH-terminal region is associated with the plasma membrane. Fr1 is the 65 kD fragment; Fr2 is the 108 kD fragment. The formation of fragments is consistent with a domain hypothesis. The two chains of 140 kD and 170 kD are isolated from cells.

some evidence supporting this idea has been obtained by electron microscopic studies (C. Cohen, B. Reinhardt, S. Hoffman, B. A. Cunningham, and G. M. Edelman, unpublished observations). A summary of the relationships between the chains and the fragments is shown in Figure 1. This figure also indicates the putative functions of different regions of the chain; those related to binding are discussed more fully below.

Unusual Carbohydrate Structures in N-CAM

As mentioned previously, N-CAM is an unusual glycoprotein in that it contains a very large amount of sialic acid in the presence of otherwise typical amounts of neutral and amino sugars (Table 3). Although some carbohydrate is present in the NH_2 binding domain or Fr1, practically all of the sialic acid is present in the central portion of the N-CAM chain that comprises the COOH terminal portion of Fr2 (Figure 1). As shown in Table 3, N-CAM from embryonic tissues has been found to contain about three times as much sialic acid as that from adult tissues (Rothbard et al 1982). When specific immunoprecipitates of N-CAM are examined by SDS gel electrophoresis, this difference is reflected in a particularly striking way. Embryonic N-CAM is present on the gel as a diffuse band ranging from an apparent molecular weight of 180,000 to a molecular weight of 250,000. In contrast, adult N-CAM shows relatively sharp bands of 180,000 and 140,000 molecular weight in the chick (Rothbard et al 1982), with an additional minor band of 120,000 molecular weight in the mouse (Edelman & Chuong 1982). Inasmuch as the amino acid compositions, known sequences, and peptide maps of the embryonic (E) and various adult (A) forms are similar or identical (Hoffman et al 1982, Cunningham et al 1983, Rothbard et al 1982), these observations indicate that a major change in the carbohydrate structure occurs as a function of development. This must result either from the action of a specific cell surface sialidase or from alteration in the action of an intracellular sialyl transferase, or possibly from both.

TABLE 3   Carbohydrate compositions of embryonic and adult N-CAM

	g/100 g Protein	
	Embryonic	Adult
Fucose	1.3	1.3
Mannose	3.4	2.9
Galactose	1.9	2.5
Glucosee	0.4	0.4
Glucosamine	6.8	7.0
Galactosamine	0.1	0.1
Sialic acid	30.9	10.8

There are two additional aspects of the carbohydrate structure that bear upon N-CAM activity. The first is the presence of some of the sugar as polysialic acid. This was initially hypothesized (Hoffman et al 1982, Rothbard et al 1982) on the basis of the high ratio of sialic acid to galactose (Table 3) and the slow release by neuraminidases. Additional observations on sialic acid-rich glycopeptides (Finne et al 1983) derived from N-CAM suggest that the polysialic acid contains α2–8 linkages. The second aspect of carbohydrate structure relates to the apparent microheterogeneity. In the E form, the electrophoretic evidence is consistent with the presence of mixtures of molecules with different amounts of polysialic acid. Whether this reflects different N-CAMs from different neurons or whether each cell makes the entire spectrum of molecules is not known. It appears likely that the A forms also have polysialic acid (Rothbard et al 1982, Finne et al 1983) with a more uniform structure from molecule to molecule and possibly attached at a different but neighboring site.

As is reviewed below, E-A conversion plays a significant role in the binding behavior of N-CAM even though the sugar does not participate directly in the binding. This alteration of the molecular structure is a clear-cut example of local cell surface modulation (Edelman 1983, Edelman 1976), and it must occur in vivo. Whether the spontaneous cleavage (Hoffman et al 1982, Cunningham et al 1983) of the molecule to yield the Fr1 fragment at 37° also occurs in vivo as another form of chemical modulation is not yet known but the ready occurrence of this cleavage does imply the likelihood of structural organization of the polypeptide chain into domains.

Domains and the Three-Dimensional Organization of N-CAM

The structural studies reviewed above raise the possibility that the N-CAM polypeptide chain is arranged in three domains linked by protease-susceptible regions of polypeptide chain. Evidence to support this hypothesis, which has a bearing upon the assignment of functions to different regions of the molecule (Figure 1) as well as upon the mechanism of binding, has been accumulating. Preliminary investigations by electron micrography of N-CAM that had been rotary shadowed with platinum (C. Cohen, B. Reinhardt, S. Hoffman, B. A. Cunningham, and G. M. Edelman, unpublished observations) show a variety of forms of the molecule. Although many fields show aggregates, some individual regular structures appear as triskelions (Figure 2A). The individual arms in each structure have length dimensions of approximately 400 Å, but detailed interpretation must be guarded until the polarity of the structure and the statistics of the various forms are established. Nonetheless, although the polarity of N-CAM chains comprising each of the arms in this structure and the actual conformation of the molecule in the cell membrane are unknown, the appearance of discrete areas in the micrographs is consistent with notion that the N-CAM monomer contains large protein domains. This is consistent with

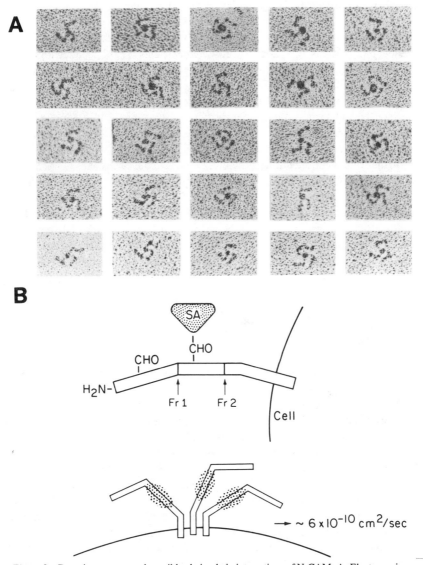

Figure 2   Domain structure and possible chain-chain interactions of N-CAM. A. Electron micrographs of the neural cell adhesion molecule N-CAM. Images of triskelions were selected from pictures of embryonic chicken N-CAM after rotary shadowing with platinum ($\times$ 200,000); the length of each arm is about 400 Å. The precise dimensions of the molecules and the proportion of molecules in the triskelion form remain to be established. Bends in the arms are compatible with inter-domain regions (Edelman et al 1983a). B. Hypothetical placement of a single chain of N-CAM in cell membrane *(top)* and interaction of three such chains to form a triskelion in a trivalent structure *(bottom)*. *Dots* in the second domain represent carbohydrate. The diffusion constant for N-CAM indicates that it is mobile in the plane of the membrane.

the sharp bends seen in the individual arms of the triskelion, each arm presumably representing a single N-CAM chain.

Although N-CAM in these micrographs appears similar to previously published micrographs of clathrin (Crowther & Pearse 1981, Ungewickell & Branton 1982), a combination of immunological analyses, electron microscopic studies, and comparisons of the amino acid compositions suggests that the triskelion observed is not clathrin. The fact that N-CAM is found on the cell surface and clathrin is found beneath the plasma membrane supports this notion. Whether there is an evolutionary relationship between the two molecules, however, remains to be shown.

In the studies carried out so far, views of the A and E forms of N-CAM do not show large differences in the structure; both forms tend to aggregate but the A forms may do so to a much larger degree. If one assumes that the knob at the center represents the confluence of three polypeptide chains of N-CAM, and that the part of each arm beyond the sharp bend represents the NH_2-terminal binding region (Figure 1), a trivalent model emerges. This assumption is consistent with binding data, to be reviewed below, in which Fr1 binding to cells appears to be weaker than that of the whole molecule which may in fact exist in a polyvalent complex.

Inasmuch as these photographs do not represent N-CAM in the membrane, it is not known whether, on the cell surface, the triskelion is only one representative of a population of monomers, dimers, trimers or higher aggregate forms or whether the trimer is the predominant stable form. The possibility that there is a population distribution of higher valence forms in the membrane is by no means excluded; this would be consistent with the fact that, unlike clathrin, N-CAM shows no evidence of a light chain linking its components.

A hypothetical model of the triskelion form in the membrane is shown in Figure 2B to illustrate one of these possibilities. One of the major features of the model is that the interaction of the chains at the cell surface and possibly the orientation of the binding domain (i.e. the degree of bending of the terminal arm) are likely to be influenced by the presence of the highly charged sialic acid residues in the second domain. Such neighboring interactions with charged groups and possibly even the conformation of the binding domain are likely to be influenced by E-A conversion with effects on the binding behavior of the molecule. That there are effects of E-A conversion on binding is now confirmed; the evidence will be considered after comparing N-CAM with CAMs of different specificity.

Ng-CAM, Heterotypic Binding, and the Possibility of an N-CAM Family

Studies of N-CAM in tissue culture, as well as most histologic studies of the distribution of the molecule, are consistent with the view that it mainly mediates neuron-neuron interactions and is present on neurons but not on glia. There is one report, however, that N-CAM can be seen on glia within tissues (Hirn et al 1983). In view of the fact that neuron-glia adhesion is a key feature of early cytoarchitectural events (Rakic 1981), and given the constraints on specificity of the mechanisms of surface modulation in the developing nervous system, at least one other molecule of different specificity would be required to mediate neuron-glia interactions. The search for such a molecule is complicated by the fact that the binding assay for adhesion must measure heterotypic interactions (Table 1) in the presence of potential homotypic interactions. Thus, in the design of the assay, a distinction must be made among neuron-neuron binding, neuron-glia binding, and glia-glia binding. This requires positive identification of each of the participating cell types and a means of suppressing at least one of the homotypic interactions at the molecular level.

To meet these requirements, an assay was developed that allowed quantitative measurement of adhesion between neuronal membrane vesicles and glial cells and permitted the identification of a new CAM (Edelman et al 1983a, Grumet et al 1984). Neuronal membrane vesicles were found to bind to glial cells but not to fibroblasts. This adhesion was inhibited by Fab' fragments prepared from rabbit antisera directed against neuronal cell membranes, but not by Fab' fragments from preimmune sera or by anti-(N-CAM) sera. Extracts of neuronal membranes neutralized the antisera that inhibited the neuron-glia interactions. The assay was used to select fractions rich in neutralization activity during the purification process. Monoclonal antibodies raised against the partially purified activity from brain membranes inhibited the binding of neuronal membrane vesicles to glial cells and allowed subsequent purification of an antigen that turned out to be a new CAM.

The neuronal membrane antigen responsible for neuron-glia binding, named Ng-CAM, has a molecular weight of 135,000 and appears as a single sharp band on gel electrophoresis. Preincubation of Fab' fragments of specific rabbit antibodies to Ng-CAM, or of monoclonal antibody 10F6 (Grumet et al 1984) with affinity purified Ng-CAM, neutralized the inhibition by these antibody fragments of the binding of neuronal membrane vesicles to glial cells. Indirect immunofluorescence staining of cultures of cells from chick brain showed that neurons but not glial cells were recognized by the monoclonal antibody (M. Grumet and G. M. Edelman, unpublished observations). These results suggest that this neuronal cell surface antigen, which is distinct from N-CAM, mediates adhesion between neuronal and glial cells in vitro.

Two points of interest are raised by these findings. The first has to do with whether the heterotypic binding of neurons to glia is also mediated by a heterophilic mechanism. That this may be the case is supported by the fact that Ng-CAM has not been observed in glial cells as well as by the logical argument that homophilic N-CAM binding and simultaneous homophilic Ng-CAM binding would obliterate all necessary distinctions between the two cell types. Were Ng-CAM binding homophilic, the only means by which a particular neuron could distinguish neurons from glia would be alterations in the time and degree of expression on the neuronal surface of N-CAM and Ng-CAM. This would require exquisite regulation of modulation in local regions of histogenesis, a requirement that could be obviated by having a heterophilic mechanism. Examinations of fractions from chick brain tissue have in fact revealed a molecule of molecular weight 200,000 copurifying with Ng-CAM; this may be the glial heterophilic partner (or Gn-CAM) but so far this has not been conclusively demonstrated.

Regardless of this issue, the presence of N-CAM and Ng-CAM on the same cells and some other recent findings raise a second point of interest: N-CAM and Ng-CAM may be evolutionarily related. Recent immunological observations (M. Grumet, S. Hoffman, C.-M. Chuong, and G. M. Edelman, in preparation) lend support to this possibility inasmuch as both polyclonal and monoclonal antibodies have been found that cross-react with N-CAM and Ng-CAM. Although the sites of the cross-reactive antigenic determinants on the different CAMs have not been established, it is conceivable that during evolution a precursor gene may have given rise to molecules that differ in their binding domain, for example, but are similar in other structural regions. This question will be resolved by further structural and molecular genetic analysis of N-CAM and Ng-CAM. Moreover, the co-expression of these two molecules that may arise from a gene family on the cell surface raises the possibility that unusual gene rearrangements may be occurring. It should be noted, however, that as far as can be discerned, the specificity differences between the two molecules do not relate to ion binding, for both operate by a Ca^{2+}-independent mechanism. This distinguishes them from non-neuronal CAMs, such as those isolated from the liver and then also identified in other non-neuronal tissues.

L-CAM

It may seem surprising that a review of cell adhesion in the nervous system would include a description of a non-neural CAM. The reason for its inclusion has to do with some key observations on neural induction. Inasmuch as early neurogenesis involves a sharp and unambiguous segregation of neural precursors from various other cells, it is important to establish that the early embryo contains structurally distinct CAMs that differ sufficiently in specificity to carry out this segregation. Consideration of the limitations of modulation mechanisms (Edelman 1983) as well as the critical nature of early events of neural induction (Ebert & Sussex 1970) favors the view that neural plate formation and the discrimination of neural tissue from other early embryonic tissue depends upon alteration in the expression of at least two molecules of different specificity, one of which is N-CAM. Among the possible candidates, the current evidence already supports the idea that at least one of the other molecules is L-CAM.

L-CAM appears to be widely distributed in several animal species; preliminary evidence suggests that chicken L-CAM resembles molecules detected in mouse embryos (e.g. uvomorulin) and teratocarcinoma cells (Hyafil et al 1980, Yoshida & Takeichi 1982), as well as in human mammary tumor cells (Damsky et al 1982). In addition to differences in structure between N-CAM and L-CAM, a main discriminant between these two CAMs is that L-CAM mediates a calcium-dependent binding process whereas N-CAM interactions are calcium-independent. Anti-L-CAM Fab' fragments inhibit hepatocyte-hepatocyte adhesion and liver colony formation in culture (Bertolotti et al 1980, Gallin et al 1983). Another adhesion molecule has also been isolated from liver; its relation to L-CAM has not yet been established (Ocklind & Obrink 1982).

It is particularly pertinent that L-CAM and N-CAM do not bind to each other, show different cleavage patterns and sugar structures (compare Figure 1 and 3), and are immunologically distinct (Brackenbury et al 1981). The L-CAM molecule, which is absent from brain, is also a glycoprotein, but its carbohydrate moieties contain mostly neutral and amino sugars with relatively little sialic acid. Although its binding mechanism has not been firmly established as homophilic, it is known to be Ca^{2+}-dependent. The molecule has an apparent molecular weight of 124,000 (Figure 3) and in the absence of Ca^{2+} it is subject to rapid proteolytic degradation (Gallin et al 1983). Thus, unlike N-CAM, both its conformation and its binding mechanism depend critically upon a divalent cation. Although certain derivative properties of N-CAM may also depend upon binding of ions particularly to the sialic acid, the discriminant issue is the *essential* requirement for Ca^{2+} in the case of L-CAM binding.

At present, there is no evidence that N-CAM and L-CAM are related or derived evolutionarily from a common precursor, although the possibility can only be ruled out by complete sequence analysis. Moreover, the sugar structure of L-CAM is not unusual and there is no evidence of E-A conversion. Whatever modulation occurs in L-CAM at the cell surface must therefore result from a change in molecular prevalence or distribution, in Ca^{2+} binding, or in proteolytic alteration and destruction. As mentioned above, it is particularly striking that in the absence of Ca^{2+}, L-CAM is rapidly fragmented by proteases, suggesting that its conformation is critically dependent upon this ion.

RELATIONSHIPS BETWEEN CAMs, SAMs, AND CJMs

The mutual transactions between particular primary processes in development, particularly those relating to differentiation, adhesion, and migration, suggest that tissue morphology must depend not only upon cell-cell adhesion but also

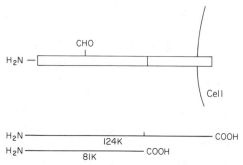

Figure 3 Linear structure of L-CAM obtained by comparing the intact molecule (124 kD) released by detergent extraction with a fragment (81 kD) released by trypsin. The carbohydrate (CHO) is probably attached at several sites. Compare with N-CAM in Figure 1.

upon cell-substrate adhesion and cell communication via intercellular junctions (see Table 1). Complexes of substrate adhesion molecules (SAMs) on cells and in extracellular matrices are currently being studied intensely (Hay 1981); they include various types of collagen, laminin, glycosaminoglycans, and fibronectin. In considering the biological effects of CAMs in embryogenesis, it is well to remember that roles have also been postulated for SAMs and for cell junction molecules (CJMs) in neural induction and neurulation (Toole 1981, Hay 1981), although no causal relationships have been established so far. Similarly, a number of cell junctions have been described (Gilula 1978, Ginzberg & Gilula 1980) and it has been shown that such junctions can appear in neural tissues very early in development (McLachlin et al 1983).

Two problems bearing upon the potential significance of these facts are of direct relevance to the subject of this review. The first is whether CAMs, SAMs, and CJMs are structurally or evolutionarily related. So far, the available evidence suggests no evident structural relationships among the three, although one report (Ocklind et al 1983) has considered the hypothesis that a CAM from liver (cell-CAM 105) may take part in tight junctions. For the rest, however, the chemical, immunological, and histologic evidence suggests that each of the three groups of molecules are structurally distinct and subserve distinct functions. In support of this interpretation, recent studies using methodologies similar to those used for isolating CAMs have succeeded in identifying cell surface molecules that interact with extracellular matrices (Neff et al 1982). While not yet extensively characterized, the molecules appear to be distinct from CAMs.

A second problem of great importance relates to the developmental sequences of the appearance and intermodulation among CAMs, SAMs, and CJMs and the relation of these events to morphogenesis and the regulation of primary developmental processes. With one exception (the neural crest cell to be reviewed below), this important issue remains to be studied in detail in an appropriate developmental system.

BINDING ACTIVITY OF N-CAM IN RELATIONSHIP TO MODULATION MECHANISMS

It is clear from the evidence reviewed above that different CAMs have different specificities as well as different ion dependencies for both conformation and binding. Within any given specificity, it is important to know the mechanism of binding and the effect of modulation events on binding efficacy. N-CAM presently provides the best example for exploration of these issues because most is known about its structure and about the nature of E-A conversion and its occurrence in different neural tissues.

N-CAM in solution tends to aggregate (Hoffman et al 1982), complicating attempts to measure binding constants for N-CAM to N-CAM interactions by thermodynamic methods in solution. To circumvent this difficulty, N-CAM has been incorporated into artificial lipid vesicles. The aggregation of these vesicles and of native brain vesicles has been used to study the specificity and kinetics of N-CAM to N-CAM binding. It was found that free N-CAM and its fragments can bind to cells and inhibit binding of soluble radioidinated N-CAM to various extents (Cunningham et al 1983). It is pertinent that Fr1 binds very weakly to cells in contrast to the whole molecule, a finding that should be considered in terms of the possibility discussed above that trivalent structures with higher apparent binding affinities may be formed in the membrane. Artificial lipid vesicles containing N-CAM can strongly inhibit cell-cell aggregation (Rutishauser et al 1982), and N-CAM and N-CAM fragments inhibit vesicle-vesicle aggregation (Hoffman & Edelman 1983). These data are consistent with a second-order homophilic binding mechanism (Edelman 1983) and they confirm the assignment of the region of binding to the NH_2-terminal domain. Thus, N-CAM from one cell binds directly to N-CAM from another via a region in the domain corresponding to Fr1 (Figure 1).

Of particular significance is the relationship of both the amount and the location of the sialic acid to binding. Is the carbohydrate portion of the molecule directly and specifically involved in the binding? What is the significance of the E-A conversion for the binding event? In order to answer these questions, the various forms of N-CAM were incorporated into lipid vesicles and the kinetics of vesicle-vesicle binding were examined (Table 4) as a function of the amount of sialic acid in the N-CAM and the concentration of N-CAM in each vesicle (Hoffman & Edelman 1983). The results indicate that early (initial) binding approximates a second order process when expressed in terms of vesicle concentration, that the rate of vesicle binding is inversely

TABLE 4 Aggregation of reconstituted vesicles containing adult or embryonic forms of N-CAM

Vesicles (form of N-CAM)	N-CAM/lipid (μg/mg)	k_{agg} (units)[a]
E	17	3.5
A	17	12.2
E	14	1.5
E	17	3.5
E	19	6.4
E	28	54.0
E(100%)	14	1.4
E(75%) + A(25%)		3.3
E(50%) + A(50%)		6.3
E(25%) + A(75%)		9.3
A(100%)	17	12.2

[a]One unit is the rate of aggregation (measured in nanoliters of newly formed vesicle aggregates per millileter per minute) divided by the square of the concentration of vesicles (measured in milligrams of lipid/millileter) (Hoffman & Edelman 1983).

related to the sialic acid content of the N-CAM molecules in the vesicles (Table 4), and that there is a highly nonlinear (fifth order) dependence of binding on the concentration of N-CAM in the vesicles (Table 4). Inasmuch as molecules completely devoid of sialic acid can bind, it is clear that this sugar is not directly involved in the binding site.

It is also clear, however, that the presence of the sialic acid plays a key role in altering rates of binding by an indirect mechanism. Indeed, the results in Table 4 and those obtained by mixing E form and A form vesicles confirm the prediction (Edelman 1983) that the order of increasing binding rates for the forms of N-CAM is E-E<E-A<A-A. Furthermore, neuraminidase treatment of E form vesicles increased their binding rate to the level observed in A form vesicles, while similar treatment of A form brain vesicles had almost no effect on their binding (Hoffman & Edelman 1983). This indicated that the functional difference observed between the E and A form vesicles is, in fact, due to the different sialic acid concentrations of their constituent N-CAM. The data indicate that binding rates are greater in the absence of sialic acid and that no basic change in mechanism occurs upon mixed (E-A) vesicle binding (Table 4). Experiments using native brain vesicles (Hoffman & Edelman 1983) have also shown that the rate of vesicle aggregation is a function of both the sialic acid content of the N-CAM molecules and the relative N-CAM content of the vesicles. This shows that the data obtained with artificial vesicles are not an artifact resulting from the presence of N-CAM alone in the membrane; con-

versely, the presence of other surface proteins in native brain vesicles had no major effect on the phenomenology of binding.

That alteration of the amount of sialic acid leads to changes in binding kinetics even though this sugar does not participate directly in binding constrains the number of possible binding models. A reasonable model based on charge perturbation and domain structure has been proposed (Edelman 1983). The fundamental premises underlying this model are (*a*) binding takes place between the NH_2-terminal domains and involves protein-protein interactions, and (*b*) the presence in the neighboring domain of sialic acid in extraordinarily large amounts (Hoffman et al 1982) perturbs N-CAM to N-CAM interactions. This perturbation could occur by interaction ipsilaterally with charged amino acids to alter conformation of the N-CAM binding region, by competition for the binding of counter-ions, or by direct charge repulsion across the N-CAM to N-CAM bond. An examination of this model and comparison with the model shown in Figure 2B indicates that there are a rich number of possibilities that are related not only to these variables but also to the existence of higher valence states and mobility of the N-CAM in the lipid bilayer. A recent finding upon ultracentrifugation of Fr1 is consistent with the idea that each binding domain (Figure 1, Figure 2B) forms bonds with only one other domain: Fr1 dimerizes but forms no higher aggregation species (W. E. Gall and G. M. Edelman, unpublished observations). Further analysis of this equilibrium binding under different relaxation constraints may eventually permit an estimate of thermodynamic constants for the protein-protein interaction; as already pointed out, these are difficult to measure on the intact protein in aqueous solution.

Although one must proceed cautiously in relating all of these in vitro results to in vivo events, the data are consistent with the idea that there are at least two forms of modulation that can alter the kinetics of binding: one involving a structural change in the molecule itself, and the other involving a change in the surface density or aggregation state of N-CAM monomers in the membrane. The most striking result is the strong dependence on surface density (Table 4), which is reflected in the greater than 32-fold change in rate for a two-fold change in the amount of N-CAM in the vesicles. This is consistent both with increased probabilities of clustering of the mobile N-CAM molecules upon cell-cell binding as well as with changes in its valence state. Further physicochemical studies should clarify the exact mechanisms.

As is shown below, modulation by alteration in prevalence of N-CAM (and L-CAM) plays a signal role particularly in early embryonic events in vivo, and it is perhaps no accident that such strong nonlinear changes occur in binding behavior with changes in CAM surface density. As far as is known, modulation by E-A conversion is a later event concerned particularly with the period in which fiber tracts are formed. While not as dramatic in its effects in vitro as changes in CAM prevalence, E-A conversion still leads to a four- to five-fold increase in binding rates. It is likely that both forms of modulation interact in vivo and that the resultant modifies cellular behavior.

Two other related issues are highlighted by the kinetic and structural data: (a) there is still a considerable amount of sialic acid present in the A forms of the molecule; (b) removal of this residual carbohydrate does not greatly alter rates of vesicle binding. It is possible that the role of the residual sugar in vivo is related to the need for electrostatic repulsion, particularly to guarantee effective cell-cell separation. A measurement of average cell separation in a tissue before and after E-A conversion might in fact reveal smaller separations in the adult state.

General theoretical analyses of the physical chemistry of cell adhesion have been carried out and have stressed the importance of such factors as electrostatic repulsion and various physical variables (Bell 1978). These analyses have been valuable in showing that for membrane bound molecules both forward and reverse rate constants for binding of adhesion molecules will be fairly close to their diffusion limits, and that equilibrium constants will not depend on the diffusion constants in the membrane. Nevertheless, what is needed now is a detailed and specific physicochemical analysis of a realistic example such as N-CAM with known molecular shape and functional relevance. An obvious example of the usefulness of information on the actual molecular structure is the hint that the binding domain of N-CAM may be flexibly hinged. This would allow proper orientation of binding domains from CAMs on adjoining cells without major need for local membrane perturbation. Binding domain orientation may be altered by the charge and distribution of sialic acid on the neighboring domain (Figure 2B), resulting in greater or lesser rates of binding. The time is close approaching that a detailed physicochemical analysis of cell adhesion can be carried out on molecules of known structure.

ROLE OF CAMs IN NEURAL INDUCTION AND NEURULATION

The structural evidence reviewed above provides the necessary background for considering the role of cell adhesion during embryonic development of the nervous system. There are two major stages in the formation of neural structures:

1. Embryogenesis, with induction and segregation of the neural plate coeval with the formation of the neural axis; this is followed by neural tube formation, by cephalocaudal specialization (Jacobson 1981), and by secondary inductions such as those seen for placodes and neural crest cells.

2. Organogenesis, in which detailed neural specializations occur, establishing the connectional structures upon which later functional activity depends (see for example, Rakic 1981, Fraser & Hunt 1980).

Although these two epochs overlap in time, they are distinct in the order of their primary processes.

Insofar as neurulation depends upon a contrasting commitment of cells to neural and non-neural tissues, it is likely to depend upon modulation of more than one kind of CAM. If the conjecture (Edelman 1983) that only a small number of CAMs are involved in early embryonic events is correct, it is likely that once such a cellular commitment takes place, each CAM (e.g. N-CAM in neural tissue) will then be utilized later in further developmental processes within the particular organ in which it is found. This has in fact been found to be the case for N-CAM, the modulation of which first occurs by alteration of prevalence during early embryogenesis and then by E-A conversion in the perinatal period. The different kinds of modulation involved in this sequence are discussed in order in this and subsequent sections. It should be emphasized that the methods used to determine this sequence were qualitative in the early period (fluorescence labeled anti-CAMs) and somewhat more quantitative during later periods (immunoprecipitation of N-CAM forms followed by optical scanning of gels). Although the results of these methods are reasonably clear-cut, it will be necessary to develop more quantitative micromethods of CAM estimation in tissues and cells if detailed interpretations in terms of physical chemistry are contemplated.

Several studies have been carried out to determine the distribution of CAMs in early development of the chick (Thiery et al 1982, Edelman et al 1983). Both N-CAM and L-CAM appear very early in epiblast, hypoblast, and endophyll. While it has not been conclusively proven whether single cells have both CAMs, the staining distribution for each is widespread at this stage and is consistent with this interpretation. A direct stage-by-stage comparison indicated that within a very short time period, N-CAM increased in amount in the presumptive neural plate region while L-CAM disappeared from this region but remained in non-neural ectoderm and endoderm. This is the first indication that during development N-CAM will become mainly (but not exclusively) expressed in several neural structures, whereas L-CAM will be expressed mainly in precursors of epithelial organs (Table 5). Although the staining of neural plate, groove, fold, and tube by anti-N-CAM reagents has been shown to persist, in the surrounding regions and in associated structures an extraordinary dynamism has been seen in N-CAM expression. For example, N-CAM appears in the notochord during neurulation and then disappears; during somitigenesis, each somite stains for N-CAM, and the molecule also appears in the heart region. All of these regions are classically associated with embryonic induction.

TABLE 5   Distribution of L-CAM and N-CAM in three epochs

0–3 Day embryo	5–13 Day embryo	Adult
	L-CAM	
Ectoderm	Epidermis	Skin: stratum germi-
Upper layer	Extraembryonic ectoderm	nativum
Epiblast		
Presumptive		
epidermis		
Placodes		
Mesoderm	Wolffian duct	Epithelium of:
Wolffian duct	Ureter	Urogenital tract
	Most meso- and	
	metanephric epithelium	
Endoderm	Epithelium of:	Epithelium of:
Endophyll	Digestive tract	Digestive tract
Hypoblast	Respiratory tract	Respiratory tract
Gut primordium	Lymphoid organs	Lymphoid organs
and buddings	Secretory glands	Secretory glands
	Extraembryonic	
	membrane	
	N-CAM	
Ectoderm	Nervous system	Nervous system
Upper layer		
Epiblast		
Neural plate		
Placodes		
Mesoderm	Striated muscle	[a]
Notochord	Adrenal cortex	
Somites	Gonad cortex	Testis
Dermomyotome	Some mesonephric	
Somato- and	and metanephric	
splanchno-pleural	epithelia	
mesoderm	Somato- and	
Heart	splanchno-pleural	
Mesonephric primordium	elements	
	Heart	Cardiac muscle

[a]Distribution of N-CAM in adult tissues derived from mesoderm has not yet been rigorously established.

In placodes, the process first seen in the neural plate is echoed; L-CAM and N-CAM appear together but then L-CAM disappears leaving N-CAM behind. This pattern has been seen for structures as varied as the otic placode and the apical ectodermal ridge of the limb bud.

One of the most striking examples of prevalence modulation of CAMs is seen in the case of migrating cells whose surface CAMs disappear during movement. This has been observed for middle layer cells during early gastrulation (Edelman et al 1983) but most strikingly for N-CAM in migrating neural crest cells. These cells stain intensely for N-CAM, but as soon as they start to migrate, they lose their staining, while becoming associated with a fibronectin-rich matrix. After reaching their destination, the fibronectin disappears or is diminished in amount and N-CAM reappears just prior to ganglion formation (Thiery et al 1982, Thiery 1983). In view of the demonstrated strong dependence of rates of binding upon N-CAM surface density, it is perhaps not surprising that effective cell migration requires a depletion in N-CAM surface density below a critical threshhold. The conjugate appearances of N-CAM and fibronectin point up the need to study the possible coordination of appearance of CAMs, SAMs, and CJMs in appropriate model systems. This in turn requires an understanding of the cellular mechanisms of CAM expression and transport to the cell surface. As shown in the last section of this review, transformed cell lines offer a model system for studying these mechanisms in vitro.

All of the examples discussed so far are consistent with the view that differing amounts of one CAM or different ratios of different CAMs on the cell surface may act to regulate pattern. It is clear that for neurulation, the distinction made between N-CAM rich and N-CAM poor regions is accompanied by the occurrence of reciprocal changes in surface modulation of L-CAM. Once such segregation occurs, the homophilic N-CAM binding mechanism with the additional effects of prevalence modulation on a dynamic basis would be expected to regulate other primary developmental processes and to play a major role in pattern formation and morphogenesis of the nervous system. Moreover, as this theme is repeated during secondary induction (suggested for example, by the evidence on placodes), a distinct relationship should emerge between the fate of different early precursor cell lineages and the distribution of the CAMs. It is also clear that a sharp distinction in specificity and binding mechanism is present for each of the CAMs involved in such early segregation mechanisms. The combination of all of these factors should result in a spatiotemporal sequence of appearance of CAMs that is intimately connected with morphogenesis.

The stronger hypothesis that CAM modulation is actually essential for morphogenesis may be tested by preparing a fate map spanning two epochs in which CAM distributions are also mapped. Such a map, covering the epoch of early embryogenesis and the first part of the epoch of organogenesis, has been prepared in the chicken with the help of the cellular fate map constructed by Vakaet (Edelman et al 1983). The procedure was to map N-CAM and L-CAM distributions found in successive stages of embryogenesis back on to those blastoderm cells in the fate map which were known to be precursors to cells present in the later structures on which the CAMs were found. The resulting map (Figure 4) has a number of striking features; these are perforce topological

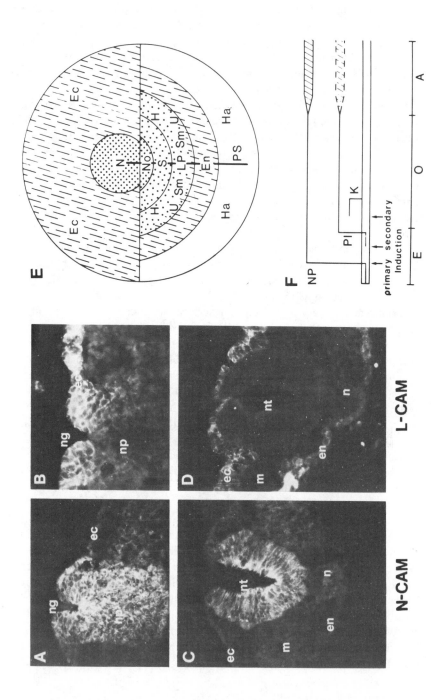

N-CAM   L-CAM

rather than topographic, inasmuch as the four-dimensional distributions of CAMs are mapped onto a two-dimensional sheet of precursor cells.

The key topologic features of this map are the following:

1. The central areas are committed to neural structures, striated muscle precursors, and some kidney elements via N-CAM with its calcium-independent binding mechanism. N-CAM distribution shows a coarse gradient of staining that decreases in a cephalocaudal direction.
2. Completely surrounding this region of calcium-independent binding is a ring of tissue including ectoderm and endoderm that is fated to express the calcium-dependent L-CAM. Future kidney elements stain both with sera to N-CAM and to L-CAM.
3. Together the two CAMs account for more than two-thirds of future embryonic structures. Gaps in the map appear for certain somatopleural and splanchnopleural elements, including smooth muscle and hemangioblastic areas, suggesting (Edelman et al 1983) that at least one more primary CAM will be found.

It should be stressed that this map must be interpreted in terms of dynamic changes in CAM prevalence and not static gradients; a good example is the notochord in which N-CAM appears and then disappears. The early dynamic changes in cell surface ratios of N-CAM and L-CAM, each with separate specificities and modulation mechanisms, could generate a rich set of patterns.

←_____

Figure 4   Distribution of N-CAM and L-CAM during neurulation and later development. A. Cross-section of neural plate (np) as the neural groove (ng) forms. N-CAM is present in large amounts in the chordomesoderm and neural ectoderm and in small amounts in ectoderm (ec) and adjoining mesoderm. B. L-CAM staining disappears from neural ectoderm in the neural groove (ng) and becomes restricted to non-neural ectoderm (ec). C. As the neural tube (nt) closes, N-CAM becomes restricted to the neural ectoderm. Non-neural ectoderm (ec), mesoderm (m), endoderm (en), and notochord (n). D. When the neural tube (nt) has closed, L-CAM is restricted to endoderm (en) and ectoderm (ec). E. A fate map of the blastodisc (stage 0; see Rudnick 1948; also L. Vakaet, personal communication) indicating areas of cells that will give rise to differentiated tissues (indicated by letters). The distribution of N-CAM *(stippled)* or L-CAM *(slashed)* on those tissues at 5–14 days (stage 26–40) as determined by immunofluorescence staining is mapped back onto the blastodisc fate map. Cells that will give rise to the urinary tract (U) express both L-CAM and N-CAM. Smooth muscle (Sm) and hemangioblastic (Ha) tissues express neither N-CAM nor L-CAM, and areas giving rise to these tissues are blank on this map. The *vertical bar* represents the primitive streak (PS); Ec, intraembryonic and extraembryonic ectoderm; En, endoderm; H, heart; LP, lateral plate (splanchno-somatopleural mesoderm); N, nervous system; No, prechordal and chordomesoderm; S, somite. F. Dynamics of expression of L-CAM *(horizontal bar)* and N-CAM *(thin lines)* on several structures during the successive developmental epochs indicated by the *bottom line:* E, early embryogenesis; O, organogenesis; A, adult. Neural plate (NP) loses all detectable L-CAM as the level of N-CAM increases during neurulation. N-CAM subsequently undergoes E-A conversion, represented by the broadening of the line. Placodes (Pl) that will give rise to ganglia contain both N-CAM and L-CAM, then lose L-CAM as N-CAM increases and probably undergo E-A conversion *(broken, cross-hatched wide line)* at later times. Kidney (K): mesonephric tubules express N-CAM as they first condense, then lose N-CAM and express L-CAM as the tubules expand.

This is clearly the case for gross boundary regions in the map such as those between ectoderm and neural tube as well as for local areas such as the kidney in which a complex local set of reciprocal inductions occurs (Edelman et al 1983).

A main issue posed by the topology of the map concerns the events that lead to the segregation of the Ca^{2+}-dependent from the Ca^{2+}-independent CAMs. This is particularly important in connection with the mechanism of formation of the primitive streak and the neural axis. It is difficult to imagine how a geometrically true neural axis is established except through biophysical means, particularly because genes expressing CAMs in each cell obviously do not possess long range spatial cues. One hypothesis is that distributions of calcium currents over such a potential neural axis could act as signals to CAM-containing cells, and a consequent differentiation event could first account for the intense appearance of L-CAM on Hensen's node and then account for loss of the L-CAM, leaving N-CAM only on the cells that will form the future neural plate. Thus, both the larger scale geometry and the early expression of CAMs may relate to early stages of electrical activity and ion fluxes following fertilization (Harris 1981). Whatever the case may be for these very early events, it is amusing that the CAM segregations and interactions seen in the map (Figure 4) are consistent with the folding of topologically continuous regions to form tubes (neural tube; intestine; body wall) by homophilic interactions and to fuse primary germ layers by heterotypic homophilic interactions (as in the pharynx where ectoderm meets endoderm). Apparently, both the expression of CAM genes and the transport of CAMs to the cell surface are under regulation and are subject to a number of strict contingencies that lead in time to these orderly morphogenetic events.

While hardly sufficient for a firm theoretical formulation at this time, the data on CAM distribution during neural induction do hint that pattern formation (and later morphogenesis) depend upon the differential expression of CAM genes in a particular order. Supplemented by intracellular CAM transport mechanisms, such differential expression could result in (a) the segregation of different CAM specificities in different regions; (b) the dynamic increases or losses of surface CAMs; (c) the complete loss of CAMs from migrating cells and to form mesenchyme; (d) the persistence of CAMs in sheets that will self-assemble into three-dimensional tubes (e.g. the neural tube). Further analysis of CAM gene expression and transport should provide a sound molecular basis for understanding these events in terms of concurrent changes in other primary developmental processes. This will be greatly expedited by cloning of CAM genes and study of transformed cell lines in which CAM expression is under facultative control (see below). Later events of chemical modulation can be presently understood in terms of the chemistry already reviewed here; we turn now to their consideration.

N-CAM MODULATION IN NEURAL HISTOGENESIS AND IN THE PERINATAL PERIOD

Following neurulation, a number of histogenetic rearrangements occur involving local interactions of neurons and glia as well as long range interactions that lead to mapping of fiber tracts. In organisms such as the mouse these interactions can continue well into the perinatal period, during which time anatomical arrangements can be modified by behavior and function. If the modulation hypothesis is correct, one would expect all of these events to be reflected by particular and specific alterations of CAM expression and CAM structure:

1. Ng-CAM should be expressed at or prior to the development of glial structures.
2. E-A conversion of N-CAM should occur according to different schedules in different regions of the nervous system.
3. These schedules should be under genetic control.

This section reviews evidence supporting each of these points.

Recent experiments on the distribution of Ng-CAM in the nervous system of chick embryos indicate that it appears at four to five days of development (J.-P. Thiery, A. Delouvée, M. Grumet, and G. M. Edelman, in preparation). Within the sensitivity of the fluorescent antibody method, there was no evidence of Ng-CAM at earlier times. Thus, unlike L-CAM and N-CAM, Ng-CAM is not part of the primitive set of CAMs responsible for the CAM fate map (Figure 4). It is instead the first example of a molecule in a secondary set of CAMs that may appear later during histogenesis to carry out specific differentiated cell-cell interactions within a given tissue. The anti-Ng-CAM stains intensely on neural fibers in the CNS and periphery but also stains cell bodies somewhat more sparsely. In view of the fact that glial cells arise subsequently in this period, it appears that the prevalence of Ng-CAM is the result of gene expression at just the appropriate time. Until Gn-CAM (the postulated heterophilic partner on glia) is isolated, however, the exact sequence of events cannot be worked out with certainty.

A comparison of all of the findings on CAM distribution during the different epochs of embryogenesis suggests that at least some primitive CAMs used in earliest embryogenesis (N-CAM, L-CAM) are used in later histogenesis, but are supplemented by a secondary set of CAMs in specific areas. Moreover, the fact that N-CAM and Ng-CAM are related molecules raises the possibility that specification of certain and possibly all secondary set molecules may have occurred in evolution by duplication of a gene for a primitive set molecule. In each case examined so far, the binding specificity of each CAM molecule is different, as one would expect with segregation based on prevalence modulation.

In the case of N-CAM, the additional mechanism of E-A conversion seems to be called into play during the period of histogenesis. While not changing the specificity of the molecule, this conversion changes the binding regime as shown by the kinetic differences discussed previously. This change would be expected to alter the interaction of other primary processes and to be reflected differently in different parts of the CNS. In accord with this hypothesis, it has been found (C.-M. Chuong and G. M. Edelman, unpublished observations; Edelman & Chuong 1982) that E-A conversion occurs at different rates in different parts of the brain (Figure 5). The gross rates appear to be cerebellum > spinal cord > cerebral cortex > hippocampus > olfactory bulb, tectum, and diencephalon. Supporting evidence for such differences in maturational events in different regions was also obtained using monoclonal antibodies specific for the embryonic form of the N-CAM molecule (C.-M. Chuong and G. M. Edelman, unpublished observations).

All of these observations suggest that different cells of the nervous system process N-CAM on different schedules. These schedules may be complex, differing in time of initiation of the conversion or in the rate at which the conversion occurs. What is critically required to relate such findings to the details of morphogenesis and mechanisms of neural mapping are assays sensitive enough to detect CAM types and CAM distribution on individual neurons. ELISA techniques and radioimmunoassays using a wide array of specific immunoreagents are promising avenues. In this connection, several monoclonal antibodies have been recently obtained that appear to distinguish variations in conformation of N-CAM from different parts of the brain.

Although these results nicely place the observations concerning the effects of E-A conversion upon kinetics of binding in the context of modulation theories, they do not indicate the cause or mechanism of E-A conversion. There are at least two potential mechanisms for E-A conversion: (a) cleavage of sialic acid from N-CAM at the cell surface by surface sialidases; or (b) turnover of N-CAM at the cell surface at different rates with temporal alteration of intracellular sialyl transferase levels. The expression of such enzymatic activity would be expected to be under genetic control; evidence from cerebellar mutants to be reviewed below suggests that this is indeed the case.

Failure to establish E-A conversion is strongly correlated with connectional defects in *staggerer* mutants of the mouse. This mutant is a homozygous recessive, phenotypically expressed in homozygotes by a failure to form synapses between parallel fibers and the tertiary dendrites of Purkinje cells (Sidman 1974, Caviness & Rakic 1978). In addition, there is extensive granule cell death and connectional disarray that lead eventually to ataxia and death. There is no apparent involvement of glial fibers as is seen in the cerebellar

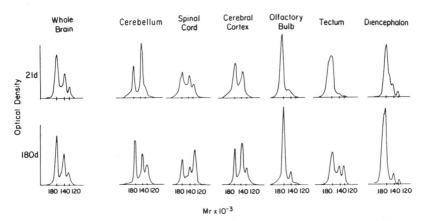

Figure 5 Differential expression of N-CAM forms in different brain regions of 21- and 180-day old mice. Each panel shows a densitometric scan of autoradiographs of anti-N-CAM immunoblots of brain extracts fractionated by SDS-PAGE. Ordinate, optical density; abscissa, relative molecular weight. The total area under each tracing has been normalized to the same value. The three A-forms of Mr = 180,000, 140,000 and 120,000 are marked. Different brain regions are presented from left to right in order of decreasing rates of E-A conversion.

mutants *reeler* and *weaver*. Biochemical analysis has revealed (Edelman & Chuong 1982) several alterations in *staggerer* indicative of a pathology in the modulation of N-CAM:

1. In the mutant, the cerebellum showed extensive delay in E-A conversion (Figure 6).
2. The cerebral cortex showed a slight delay that is apparently not large enough to cause gross functional or connectional aberrations (Figure 6).
3. The heterozygote cerebellum also showed a slight delay, suggesting that the mutant is a semidominant with alterations in penetrance or threshold effects.
4. Other cerebellar mutants, such as *reeler* and *weaver,* appear to be normal in their schedules of conversion.
5. The N-CAM itself appears to be normal in *staggerer,* suggesting that the defect accompanying the disease is related to a failure in structure, function or expression of the enzymes responsible for the conversion.

While these results do not necessarily imply that failure in E-A conversion is etiologic in *staggerer,* they do indicate that the conversion delay is correlated with the pathogenesis of the disease. Indeed, it is tempting to speculate that failure of E-A conversion may result in faulty coordination of other primary processes of development, specifically, the termination of normal migration at the proper times. The resultant failure to establish appropriate connections may in turn lead to excessive granule cell death and other disasters. Whatever the actual relationship of E-A conversion delay to the etiology, it is clear from these

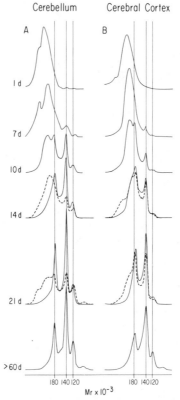

Figure 6 Forms of N-CAM present in the cerebellum and cerebral cortex in *staggerer* (- - -) and normal (–) mouse brain at different ages. The *panels* show densitometric scans of ^{125}I-N-CAM immunoprecipitated from extracts of the indicated brain regions and separated by SDS-PAGE. A. N-CAM from cerebellum; B. N-CAM from cerebral cortex. N-CAM from *staggerer* is shown at 14 and 21 days.

findings that conversion is under genetic control and that different degrees of expression (as in heterozygotes or tissues other than cerebellum in homozygotes) are strongly correlated with different degrees of disorder in connections. These findings suggest that it would be profitable to reexamine classic systems for study of mapping, such as the retinotectal system, in terms of N-CAM modulation and dynamics. Moreover, they point to the need to study factors regulating CAM gene expression and that of the enzymes responsible for chemical modulation.

CAM EXPRESSION: TRANSFORMED NEURONAL CELLS AS A MODEL SYSTEM

The data on the order of binding rates, the data on CAM expression in early embryogenesis and histogenesis, and the results suggesting that expression of enzymes regulates N-CAM carbohydrate structure during histogenesis indicate that it is necessary to study both gene expression and transport of CAMs to the cell surface in order to understand the causal basis of modulation events. Although the eventual understanding of these problems will be greatly expedited by the cloning of CAM genes, there is also a strong need for a model system in which to carry out appropriate biochemical studies. One useful approach to the analysis of N-CAM gene expression and transport in vitro is to attempt to regulate and deregulate these events by transforming viruses. This approach has the additional promise of clarifying the possible role of CAMs in the processes of neoplasia and metastasis.

Recent work (Edelman et al 1983a, M. Greenberg, R. Brackenbury, and G. M. Edelman, in preparation) indicates that transformation of neurons can be correlated both with alterations in N-CAM expression and with cellular binding functions. Dividing retinal cells from 6-day chicken embryos in culture can be infected by Rous sarcoma virus (RSV), yielding a population of morphologically altered cells (Calothy et al 1979). These cells synthesize the src gene product pp60^{src} and express high levels of pp60^{src} tyrosine kinase activity. It has been found that both control cells and transformed retinal cells express adult forms of N-CAM, but transformed cells expressed much less N-CAM. Moreover, such transformed cells aggregated much more slowly and to a much lower extent than non-transformed cells. Infection with transformation-defective viruses had no effect on N-CAM expression, indicating that actual transformation of the cells by pp60^{src} is essential for the effects observed.

To show conclusively that the transformed cells arose from N-CAM-positive neural precursor cells rather than from cells that lack N-CAM, a pure population of N-CAM positive cells was selected with a fluorescence-activated cell sorter. This population transformed as rapidly and extensively as an unfractionated cell population, with the accompanying loss of N-CAM. Several cell lines containing a temperature-sensitive transformation-defective mutant of RSV that were derived from rat cerebellum were examined in similar fashion (Giotta & Cohn 1982). One line, R2, when grown at the nonpermissive temperature, expressed significant amounts of N-CAM and aggregated by an N-CAM-mediated mechanism. In contrast, phenotypically transformed cells that were grown at the permissive temperature expressed substantially less N-CAM and aggregated poorly.

An analysis of some possible dynamic consequences of the decrease in surface N-CAM were obtained by time-lapse cinematographic comparisons of the normal and transformed chick retinal cells in culture. Whereas the cells in control cultures were in compact aggregates that were relatively static and stable in form, RSV-transformed cells formed looser associations and showed a dynamic flattening and spreading out of the cellular clusters. The larger transformed cells were highly motile, migrating rapidly between and underneath neighboring cells.

These experiments provide the first demonstration that transformation alters the expression of an identified cell-cell adhesion molecule with direct consequences on cell aggregation. While most brain tumors are not metastatic, the findings have significant implications for an understanding of the behavior of invasive neural tumors and perhaps for tumors in other tissues that express N-CAM. They also provide a basis for similar explorations of the role of L-CAM and other non-neural CAMs in metastasis.

It is pertinent that the transformed cells so far examined expressed the adult form of N-CAM when they expressed the molecule at all. There are observations that mutant cell lines with reduced cell surface sialic acid tend to bind more tightly to substrates such as collagen type IV and fibronectin (Dennis et al 1982). But it should be emphasized that in the case of the neuronal lines, the main effect on aggregation is attributable to a sharp decrease in expression of the specific adhesion molecules; the relationship of the sialic acid content of transformed cells to CAM-SAM interactions remains to be worked out. Whether or not the tumor lines infected with *ts* mutants prove to be ideal as models for CAM gene expression and transport to the cell surface, they do provide yet another (and particularly dramatic) case showing the effect of surface modulation on a cell-cell adhesion.

CAM DISTRIBUTION IN ADULT TISSUES

As shown in Table 5, the CAMs persist in definite distributions in adult life. N-CAM is present on all neural tissues and possibly in certain mesodermal derivatives such as the testis and cardiac muscle. L-CAM is present in various epithelia of endodermal, ectodermal, and mesodermal derivatives. In some locations, L-CAM is found in a polar distribution, e.g. it is present on the apicolateral surfaces of pancreatic exocrine cells. Such polarity modulation may serve a structural role in glandular epithelia. In any case, it is clear that by the time adult life is reached, the development of cell substrate interactions and of cell junctions has variegated the morphological picture seen in earlier epochs and the known CAMs are not expressed on certain types of cells in various organs. This raises the question whether additional CAMs will account for the lacunae. On the basis of the CAM fate map, at least one other primary set CAM (for smooth muscle and hemangioblastic cells) has been predicted. But the possibility of secondary set CAMs (of which an example is Ng-CAM) appearing in a variety of organs cannot be excluded.

Another question raised by the findings in adult tissue concerns the function of the CAMs in adult life. At present one can only speculate based on indirect evidence. N-CAM appears to have a domain structure and only the binding domain has been assigned a well-defined function; the possibility of functions for the other domains must to be explored. On the basis of rather sparse evidence, the so-called D2 protein isolated from synaptosomal preparations has been located in presynaptic complexes (Jørgensen 1981). This protein has been shown to be immunochemically related to N-Cam (Jørgensen et al 1980) and thus it is possible that N-CAM or a protein related to it is present in the synapse, serving structural or other functions.

Finally, the generalized distribution of the CAMs and the findings on transformed cells reviewed above raise the question of the roles of these molecules in various diseases. It would be particularly valuable to examine N-CAM and Ng-CAM in a number of neurological disorders. More constructively, the potential role of these molecules in neural regeneration is a subject of great fundamental as well as practical importance. Once the structure and genetics of N-CAM, Ng-CAM, and the postulated Gn-CAM are better understood, for example, it may be possible to regulate their expression. It is not inconceivable that appropriate suppression of Ng-CAM or Gn-CAM would allow neurite extension past an area of gliosis or potential glial scarring.

CONCLUSIONS AND PROSPECTS

While the subject of cell adhesion still remains to be explored, particularly at the level of molecular mechanisms, recent progress stemming from the discovery of cell adhesion molecules has clarified several long standing issues and has provided a basis for new ideas. Perhaps the major methodological illumination arising from the study of this difficult subject has come from the realization that while adhesion molecules must be identified with clear-cut assays, such assays are not enough. It is also necessary to explore functional, genetic, histologic, and developmental features of a putative CAM before its role can confidently be assigned. With actual CAM molecules now in hand, this exploration has become feasible. Several other major realizations are worth stressing, for they finally dispose of several controversial issues inherited from the past. They may be listed as follows:

1. CAMs of different specificity, structure, and ion dependence do exist but do not appear to represent a very large repertoire of prespecified gene products determining the positions of cells exactly, or acting as absolute cell specific markers.

2. Instead, a variety of mechanisms of local cell surface modulation (Figure 7A) are responsible for dynamic alterations, either in CAM binding regimes or in spatial segregation at the cell surface. The fact that prevalence modulation is intimately linked to neural plate formation, placode development, and inductive events suggests that (as for pigments in color vision in relation to millions of possible hues) just a small number of primary set CAMs can account for very large numbers of patterns.

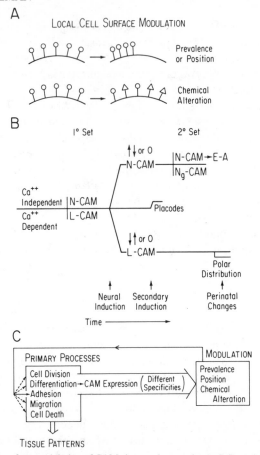

Figure 7 Local surface modulation of CAMs in morphogenesis. A. Schematic representation of local cell surface modulation. The various elements represent specific glycoprotein (e.g. N-CAM) on the cell surface. The upper panel shows modulation by both alteration of the prevalence of a particular molecule and its distribution on the cell surface. The lower panel shows modulation by chemical modification resulting in the appearance of new or related forms *(triangles)* of the molecule with altered activities. Local modulation is distinct from global modulation, which refers to alteration in the whole membrane, affecting a variety of different receptors (Edelman 1976). B. Summary schematic showing temporal sequence of CAM expression and different modulation mechanisms. After an initial differentiation event N-CAM and L-CAM diverge in cellular distribution (see Figure 4) and then are modulated in prevalence (↓ ↑) within various regions of inductions or actually disappear (o) when mesenchyme appears or cell migration occurs. Note that placodes, which have both CAMs, echo the events seen for neural induction. Just before appearance of glia, a 2° set CAM (Ng-CAM) emerges; this CAM is not found in the map shown in Figure 4. In the perinatal period, a series of epigenetic modulations occur: E-A conversion for N-CAM and polar redistribution for L-CAM. The diagrammed events are based mainly on work on the chick. C. Regulatory effects of CAM expression and modulation. While the major effect is on the adhesion process, CAM expression and modulation may alter the sequence and extent of each of the other primary processes either indirectly or (as suggested by the *dotted arrows*) directly. These effects lead in turn to different tissue patterns.

3. While a primary set of CAMs may be used in all epochs, secondary late-appearing CAMs such as Ng-CAM may also be necessary for further refinement of tissue patterns (Figure 7B). There is a suggestion that the one known instance of a secondary set molecule, Ng-CAM, is evolutionarily related to N-CAM. Despite the presence of secondary set CAMs, CAMs in the primary set are also used in histogenesis and also persist into adult life. The evidence so far suggests that primary set CAMs of different specificity occur at fundamental borders, such as that between ectoderm and neurectoderm and in inductions; different secondary specificities occur in late-appearing structures during histogenesis, such as glia.

4. The expression of CAMs at the cell surface in the proper locations is a key event in primary induction and neurulation as well as in secondary inductions. In harmonious combination with other primary processes of development (Figure 7C), cell surface modulation of CAMs appears to regulate kinetically constrained events that are far from equilibrium such as migration, retraction, and death. Such regulation of prevalence can be abrogated by transforming viruses leading to a loss of N-CAM from the surface of affected cells. Moreover, chemical modulation such as the E-A conversion can be altered by mutations such as *sg/sg*. The enzymatic basis of E-A conversion remains to be worked out but clearly represents one of the most important epigenetic variables regulating neurogenesis. Of equal importance, but with less basis in fact, is the possibility that interacting CAMs may provide a means for signaling changes in cell metabolism or differentiation state, either directly or indirectly by ensuring cell proximity and the transfer of other factors.

The picture that emerges from these observations is not consistent with purely thermodynamic models of cell-cell adhesion. Instead, it implies a complex, readjusting set of interactions and regulations both among the results of expression of CAM genes and the relative contributions of the various primary processes at any given time (Figure 7C). This picture is consistent with a large body of phenomenological observations on the plasticity, rearrangements, and variability that occur during the formation of developmental patterns as well as during neural formation of maps.

5. Cell-cell adhesion and cell-substrate adhesion appear to be mediated by completely different families of molecules, CAMs and SAMs. Yet a third group of molecules (CJMs) mediates the assembly of classical cell contact systems such as gap junctions and desmosomes that are related to cell communication and epithelial segregation. Of these systems, CAMs appear to be the most cell-specific and are found in a variety of vertebrate species. SAMs appear more important for migration, for connective tissue networks, and for hard tissues. As shown in the analysis of neural crest cell migration, there is coordination in the expression of N-CAM and a SAM such as fibronectin. It is particularly tempting to suggest that the modulation of the surface prevalence of CAM (as seen for example both in neural crest cells and in middle layer cells

during early embryogenesis) may be the basis of the fundamental epithelial-to-mesenchymal transition, a particularly important process in many morphogenetic events during development.

A final word about evolutionary origins may be in order. So far, the CAMs have been defined only in vertebrate species and appear to be evolutionarily related. It would appear likely that similar molecules and principles exist in invertebrates in which very determinate mapping by pathfinding neurons occurs (Bentley & Keshishian 1983). None of the extant descriptions of such mapping is inconsistent with the presence of CAMs and local modulation. The search for precursors of CAM families would have particular relevance for the connection between evolutionary and developmental events (Bonner 1982). While that search undoubtedly would have to reach below organisms with complex nervous systems, it is in the nervous system that CAM expression and regulation has achieved its most sophisticated form.

ACKNOWLEDGMENTS

The work from the author's laboratory cited in this review was supported by USPHS grants HD-09635, HD-16550, AM-04256, and AI-11378. The author is grateful for the use of the facilities of The Neurosciences Institute of the Neurosciences Research Program during the preparation of this review.

Literature Cited

Bell, G. I. 1978. Models for the specific adhesion of cells to cells. *Science* 200:618–27

Bentley, D., Keshishian, H. 1983. Pathfinding by peripheral pioneer neurons in grasshoppers. *Science* 218:1082–88

Bertolotti, R., Rutishauser, U., Edelman, G. M. 1980. A cell surface molecule involved in aggregation of embryonic liver cells. *Proc. Natl. Acad. Sci. USA* 77:4831–35

Bonner, J. T., ed. 1982. *Evolution and Development: Report of the Dahlem Workshop, Berlin 1981, May 10–15*, (Life Sci. Res. Rep. 22). New York: Springer-Verlag. 357 pp.

Brackenbury, R., Rutishauser, U., Edelman, G. M. 1981. Distinct calcium-independent and calcium-dependent adhesion systems of chicken embryo cells. *Proc. Natl. Acad. Sci. USA* 78:387–91

Brackenbury, R., Thiery, J.-P., Rutishauser, U., Edelman, G. M. 1977. Adhesion among neural cells of the chick embryo. I. An immunological assay for molecules involved in cell-cell binding. *J. Biol. Chem.* 252:6835–40

Buskirk, D. R., Thiery, J.-P., Rutishauser, U., Edelman, G. M. 1980. Antibodies to a neural cell adhesion molecule disrupt histogenesis in cultured chick retinae. *Nature* 285:488–89

Calothy, G., Poirier, G., Dambrine, P., Mignatti, P., Combes, P., Pessac, B. 1979. Expression of viral oncogenes in differentiating chick embryo neuroretinal cells infected with avian tumor viruses. *Cold Spring Harbor Symp. Quant. Biol.* 44:983–90

Caviness, V. S. Jr., Rakic, P. 1978. Mechanisms of cortical development. A view from mutations in mice. *Ann. Rev. Neurosci.* 1:297–326

Chuong, C. -M., McClain, D. A., Streit, P., Edelman, G. M. 1982. Neural cell adhesion molecules in rodent brains isolated by monoclonal antibodies with cross-species reactivity. *Proc. Natl. Acad. Sci. USA* 79:4234–38

Cowan, W. M. 1978. Aspects of neural development. *Int. Rev. Physiol. Neurophysiol. III* 17:149–91

Crowther, R. A., Pearse, B. M. F. 1981. Assembly and packing of clathrin into coats. *J. Cell Biol.* 91:790–97

Cunningham, B. A., Hoffman, S., Rutishauser, U., Hemperly, J. J., Edelman, G. M. 1983. Molecular topography of the neural cell adhesion molecule N-CAM: Surface orientation and location of sialic acid-rich and binding regions. *Proc. Natl. Acad. Sci. USA* 80:3116–20

Curtis, A. S. G. 1967. *The Cell Surface: Its Molecular Role in Morphogenesis.* New York: Academic. 405 pp.

Damsky, C. H., Richa, J., Knudsen, K., Solter, D., Buck, C. A. 1982. Identification of a cell-cell adhesion glycoprotein from mammary tumor epithelium. *J. Cell Biol.* 95:22 (Abstr.)

Dennis, J., Waller, C., Timpl, R., Schirnmaker, V. 1982. Surface sialic acid reduces attachment of metastatic tumor cell to collagen type IV and fibronectin. *Nature* 300:274–76

Ebert, J. D., Sussex, I. M. 1970. *Interacting Systems in Development.* New York: Holt, Rhinehart & Winston. 338 pp. 2nd ed.

Edelman, G. M. 1976. Surface modulation in cell recognition and cell growth. *Science* 192:218–26

Edelman, G. M. 1983. Cell adhesion molecules. *Science* 219:450–57

Edelman, G. M., Chuong, C.-M. 1982. Embryonic to adult conversion of neural cell adhesion molecules in normal and *staggerer* mice. *Proc. Natl. Acad. Sci. USA* 79:7036–40

Edelman, G. M., Gallin, W. J., Delouvée, A., Cunningham, B. A., Thiery, J.-P. 1983. Early epochal maps of two different cell adhesion molecules. *Proc. Natl. Acad. Sci. USA* 80:4384–88

Edelman, G. M., Hoffman, S., Chuong, C.-M., Thiery, J.-P., Brackenbury, R., Gallin, W. J., Grumet, M., Greenberg, M., Hemperly, J. J., Cohen, C., Cunningham, B. A. 1983a. Structure and modulation of neural cell adhesion molecules in early and late embryogenesis. *Cold Spring Harbor Symp. Quant. Biol. Vol.* 48:In press

Finne, J., Finne, U., Deagostinibazin, H., Goridis, C. 1983. Occurrence of α2–8 linked polysialosyl units in a neural cell adhesion molecule. *Biochem. Biophys. Res. Commun.* 112:482–87

Fraser, S. E., Hunt, R. K. 1980. Retinotectal specificity: Models and experiments in search of a mapping function. *Ann. Rev. Neurosci.* 3:319–52

Frazier, W., Glaser, L. 1979. Surface components and cell recognition. *Ann. Rev. Biochem.* 48:491–523

Gall, W. E., Edelman, G. M. 1981. Lateral diffusion of surface molecules in animal cells and tissues. *Science* 213:903–5

Gallin, W. J., Edelman, G. M., Cunningham, B. A. 1983. Characterization of L-CAM, a major cell adhesion molecule from embryonic liver cells. *Proc. Natl. Acad. Sci. USA* 80:1038–42

Garrod, D. R., Nicol, A. 1981. Cell behavior and molecular mechanisms of cell-cell adhesion. *Biol. Rev.* 56:199–242

Gilula, N. B. 1978. Structure of intercellular junctions. In *Intercellular Junctions and Synapses,* Ser. B, ed. J. Feldman, N. B. Gilula, J. D. Pitts, 2:3–22. London: Chapman & Hall. 246 pp.

Ginzberg, R. D., Gilula, N. B. 1980. Synaptogenesis in the vestibular sensory epithelium of the chick embryo. *J. Neurocytol.* 9:405–24

Giotta, G. J., Cohn, M. 1982. Derivation of neural cell lines with Rous sarcoma virus. In *Neuroscience Approached through Cell Culture,* ed. S. E. Pfeiffer, 1:203–25. Boca Raton, FL:CRC Press. 234 pp.

Grinnell, A. 1978. Cellular adhesiveness and extracellular substrata. *Int. Rev. Cytol.* 53:63–144

Grumet, M., Hoffman, S., Edelman, G. M. 1984. Two antigenically related neuronal CAMs of different specificities mediate neuron-neuron and neuron-glia adhesion. *PNAS.* In press

Grumet, M., Rutishauser, U., Edelman, G. M. 1982. Neural cell adhesion molecule is on embryonic muscle cells and mediates adhesion to nerve cells *in vitro. Nature* 295:693–95

Harris, W. A. 1981. Neural activity and development. *Ann. Rev. Physiol.* 43:689–710

Hay, E. D. 1981. Collagen and embryonic development. In *Cell Biology of Extracellular Matrix,* ed. E. D. Hay, pp. 379–409. New York: Plenum. 417 pp.

Hirn, M., Ghandour, M. S., Deagostinibazin, H., Goridis, C. 1983. Molecular heterogeneity and structural evolution during cerebellar ontogeny detected by monoclonal antibody of the mouse cell surface antigen BSP-2. *Brain Res.* 265:87–100

Hoffman, S., Edelman, G. M. 1983. Kinetics of homophilic binding by E and A forms of the neural cell adhesion molecule. *Proc. Natl. Acad. Sci. USA* 80:5762–66

Hoffman, S., Sorkin, B. C., White, P. C., Brackenbury, R., Mailhammer, R., Rutishauser, U., Cunningham, B. A., Edelman, G. M. 1982. Chemical characterization of a neural cell adhesion molecule purified from embryonic brain membranes. *J. Biol. Chem.* 257:7720–29

Hyafil, F., Morello, D., Babinet, C., Jacob, F. 1980. A cell surface glycoprotein involved in the compaction of embryonal carcinoma cells and cleavage stage embryos. *Cell* 21:927–34

Jacobson, A. G. 1981. Morphogenesis of the neural plate and tube. In *Morphogenesis and Pattern Formation,* ed. T. G. Connelly, L. L. Brinkley, B. M. Carlson, pp. 233–64. New York: Raven. 302 pp.

Jørgensen, O. S. 1981. Neuronal membrane D2-protein during rat brain ontogeny. *J. Neurochem.* 37:939–46

Jørgensen, O. S., Delouvée, A., Thiery, J.-P., Edelman, G. M. 1980. The nervous system specific protein D2 is involved in adhesion among neurites from cultured rat ganglia. *FEBS Lett.* 111:39–42

Jørgensen, O. S., Moller, M. 1980. Immunocytochemical demonstration of the D2 protein in the presynaptic complex. *Brain Res.* 194:419–29

Letourneau, P. C., Ray, P. N., Bernfield, M. R. 1980. The regulation of cell behavior by cell adhesion. In *Biological Regulation and Development,* ed. R. Goldberger, 2:339–76. New York: Plenum. 655 pp.

Lilien, J., Balsamo, J., McDonough, J., Hermolin, J., Cook, J., Rutz, R. 1979. Adhesive specificity among embryonic cells. In *Surfaces of Normal and Malignant Cells,* ed. R. O. Hynes, pp. 389–418. New York: Wiley. 471 pp.

McClain, D. A., Edelman, G. M. 1982. A neural cell adhesion molecule from human brain. *Proc. Natl. Acad. Sci. USA* 79:6380–84

McLachlin, J. R., Caveney, S., Kidder, G. M. 1983. Control of gap junction formation in early mouse embryos. *Dev. Biol.* 98:155–64

Meyer, R. L., Sperry, R. W. 1973. Tests for neuroplasticity in the anuran retinotectal system. *Exp. Neurol.* 40:525–39

Moscona, A. A. 1962. Analysis of cell recombinations in experimental synthesis of tissues in vitro. *J. Cell. Comp. Physiol.* 60 (Suppl. 1):65–80

Moscona, A. A. 1974. Surface specification of embryonic cells: Lectin receptors, cell recognition, and specific ligands. In *The Cell Surface in Development,* ed. A. A. Moscona, pp. 67–99. New York: Wiley. 334 pp.

Neff, N. T., Lowrey, C., Decker, C., Tovar, A., Damsky, C., Buck, C., Horwitz, A. F. 1982. A monoclonal antibody detaches embryonic skeletal muscle from extracellular matrices. *J. Cell Biol.* 95:654–66

Nielsen, L. D., Pitts, M., Grady, S. R., McGuire, E. J. 1981. Cell-cell adhesion in the embryonic chick: Partial purification of liver adhesion molecules from liver membranes. *Dev. Biol.* 86:315–26

Ocklind, C., Forsum, U., Obrink, B. 1983. Cell surface localization and tissue distribution of a hepatocyte cell-cell adhesion glycoprotein (cell-CAM 105). *J. Cell Biol.* 96:1168–71

Ocklind, C., Obrink, B. 1982. Intercellular adhesion of rat hepatocytes. Identification of a cell surface glycoprotein involved in the initial adhesion process. *J. Biol. Chem.* 257:6788–95

Rakic, P. 1981. Developmental events leading to laminar and areal organization of the neocortex. In *The Organization of the Cerebral Cortex,* ed. F. O. Schmitt, F. G. Worden, G. Adelman, S. G. Dennis, pp. 7–28. Cambridge: MIT Press. 592 pp.

Rothbard, J. B., Brackenbury, R., Cunningham, B. A., Edelman, G. M. 1982. Differences in the carbohydrate structures of neural cell-adhesion molecules from adult and embryonic chicken brains. *J. Biol. Chem.* 257:11064–69

Rougon, G., Deagostinibazin, H., Hirn, M., Goridis, C. 1982. Tissue and developmental stage-specific forms of a neural cell surface antigen linked to differences in glycosylation of a common polypeptide. *EMBO J.* 1:1239–44

Rudnick, D. 1948. *Ann. NY Acad. Sci.* 49:761–72

Rutishauser, U., Gall, W. E., Edelman, G. M. 1978a. Adhesion among neural cells of the chick embryo. IV. Role of the cell surface molecule CAM in the formation of neurite bundles in cultures of spinal ganglia. *J. Cell Biol.* 79:382–93

Rutishauser, U., Grumet, M., Edelman, G. M. 1983. N-CAM mediates initial interactions between spinal cord neurons and muscle cells in culture. *J. Cell Biol.* 97:145–52

Rutishauser, U., Hoffman, S., Edelman, G. M. 1982. Binding properties of a cell adhesion molecule from neural tissue. *Proc. Natl. Acad. Sci. USA* 79:685–89

Rutishauser, U., Thiery, J.-P., Brackenbury, R., Edelman, G. M. 1978b. Adhesion among neural cells of the chick embryo. III. Relationship of the surface molecule CAM to cell adhesion and the development of histotypic patterns. *J. Cell Biol.* 79:371–81

Sidman, R. L. 1974. Contact interaction among developing mammalian brain cells. In *The Cell Surface in Development,* ed. A. A. Moscona, pp. 221–53. New York: Wiley. 334 pp.

Sperry, R. W. 1963. Chemoaffinity in the orderly growth of nerve fiber patterns and connections. *Proc. Natl. Acad. Sci. USA* 50:703–10

Steinberg, M. S. 1970. Does differential adhesion govern self-assembly processes in histogen? Equilibrium configurations and the emergence of a hierarchy among populations of embryonic cells. *J. Exp. Zool.* 173:395–433

Thiery, J.-P. 1983. Roles of fibronectin in embryogenesis. In *Fibronectin,* ed. D. Mosher, E. Ruoslahti. New York: Academic. In press

Thiery, J.-P., Brackenbury, R., Rutishauser, U., Edelman, G. M. 1977. Adhesion among neural cells of the chick embryo. II. Purification and characterization of a cell adhesion molecule from neural retina. *J. Biol. Chem.* 252:6841–45

Thiery, J.-P., Duband, J.-L., Rutishauser, U., Edelman, G. M. 1982. Cell adhesion molecules in early chicken embryogenesis. *Proc. Natl. Acad. Sci. USA* 79:6737–41

Toole, B. P. 1981. Glycoaminoglycans in morphogenesis. In *Cell Biology of Extracellular Matrix,* ed. E. D. Hay, pp. 259–94. New York: Plenum. 417 pp.

Ungewickell, E., Branton, P. 1982. Triskelions: The building blocks of clathrin coats. *Trends Biochem. Sci.* 7:358–61

Yoshida, C., Takeichi, M. 1982. Teratocarcinoma cell adhesion: Identification of a cell-surface protein involved in calcium-dependent cell aggregation. *Cell* 28:217–24

Ann. Rev. Neurosci. 1984. 7:379–412

THE ANALYSIS OF STEREOPSIS

Gian F. Poggio

The Philip Bard Laboratories of Neurophysiology, Department of Neuroscience, Johns Hopkins University School of Medicine, Baltimore, Maryland 21205

Tomaso Poggio

Department of Psychology and Artificial Intelligence Laboratory, Massachusetts Institute of Technology, Cambridge, Massachusetts 02139

INTRODUCTION

Our eyes capture different 2-D images of the objects around us and our brain uses these images to recover a description of the 3-D structure of the environment. Stereopsis is the process responsible for this reconstruction of the depth dimension in our visual world. Since Wheatstone invented the stereoscope in 1838, the processes underlying primate stereo vision have been intensively studied, first with psychophysical techniques, and more recently in terms of the underlying physiological mechanisms in the visual cortex. During the last few years, the computational aspects of the problem of stereo vision have received increased attention. As with so many other visual tasks that humans perform easily and effortlessly, the development of automatic systems of stereoscopic vision, which would yield immediate and important applications, has proven surprisingly difficult.

It is now clear that the problem of stereopsis is not only a problem in the area of psychophysics and physiology, but also a complex problem in information processing. Because of its knowledge-free, low-level character, the solution of the stereopsis problem may thus uncover some fundamental principles that apply equally well to artificial and natural vision systems.

This somewhat new perspective is reflected in the organization of this paper. We first review the computational problems posed by stereopsis and describe briefly the main theories and models that have been proposed in the context of human and computer vision. The structure of these algorithms and their

379

0147-006X/84/0301-0379$02.00

performance and limitations suggest a number of critical questions about human stereopsis. For each of these areas of investigation, we review relevant psychophysical data and try to put in focus the main open questions. The final two parts of the paper review the physiology of stereopsis and attempt to relate it to both the psychophysical and computational levels. This review is by no means exhaustive; we have concentrated on a few recent developments and aspects of the analysis of stereopsis that we consider of particular importance in providing new information.

INFORMATION PROCESSING IN STEREOPSIS

What is stereopsis? Because our two eyes are located in different positions in the head, their views of a 3-D scene are slightly disparate. One can easily experience directly this binocular disparity by looking at objects not too distant and noting their different relative positions when closing each eye in turn. The disparity of each "point" depends on its distance from the fixation point of the two eyes. Our brain is capable of measuring this disparity and using it to produce the sensation of depth that is the subjective estimate of relative distance. This is stereopsis and its sole basis is the horizontal disparity between the two retinal images. There are, of course, several cues to depth, like perspective, texture gradients, and shading, that are used in our everyday vision. Binocular disparity is one of the most important and accurate of them. We are concerned here only with stereopsis, considered as the information processing module in our brain that is responsible for measuring and using binocular disparity.

The Computational Problems of Stereopsis

From the experience of everyday life, it is not clear how to separate stereoscopic processing from the monocular analysis of each image. Line stereograms (like the ones of a chemical structure) already show that stereopsis only needs disparity information: shading, perspective and texture gradients are not critical. An early step in stereo processing is to compute disparity between the two images. With knowledge of disparity, an estimate of distance can then be recovered from the geometry of the situation. Thus, the processing involved in stereopsis includes matching corresponding points of the images in the two eyes, measuring their disparity, and from this information recovering the 3-D structure of the objects seen by the viewer.

THE CORRESPONDENCE PROBLEM   If one could identify with certainty a location in the two images, the first two steps could be avoided and the problem would be easy. In practice, one cannot mark spots in the scene and the difficult part of the computation is solving the correspondence problem. Before Julesz's work, it was thought that this problem was in fact avoided by first recognizing objects and their components, and then performing an unequivocal correspondence between the same recognized detail in the left and in the right image. In this way, there is never any real problem in deciding what should match what. This comes at the cost of placing the correspondence process underlying stereopsis quite late in the processing of visual information, certainly after object recognition. Julesz (1960) demonstrated that it is not so by inventing the random-dot stereogram, in which there is no information whatever about visible surfaces except for disparity.

Thus, experiments with random-dot stereograms proved that binocular combination need not happen after object recognition. They strongly suggest, in fact, that the correspondence problem preliminary to stereopsis is solved early on, independently from higher level processing. This fits well with our present knowledge of visual physiology. Binocularity appears very early in the visual pathway, before any complex recognition process has taken place.

Julesz's observation is critical because it allows one (a) to consider the processing module involved in decoding stereopsis independent from other processes (at least to a first approximation), and (b) to formulate the computational goal of human stereopsis as the extraction of disparity information from a pair of images, without the need of monocular cues. The main problem that human stereovision has to solve is what has been called the correspondence problem—how to find corresponding points in the two images without recognizing objects or their parts. Random-dot stereograms (Figure 1, top) would seem to confront our brain with an immense number of possible matches between the two images. After all, dots in one image are all the same, of the same size and contrast: Any given dot in one image could in theory be matched with any one of a large number of dots in the other image. And yet our brain solves the false target problem and comes up with the right answer. How does the brain know what corresponds to what?

There is an additional problem preliminary to the matching operation at the level of a computational theory of stereopsis. We have to specify what has to be put in correspondence between the two images. To be sure, the basic information available to a visual system is an array of measurements of light intensity at different image points. In our eye, these measurements are taken by the photoreceptors. Each point in an image is the projection along a line of sight of a point on a physical surface (unless the camera is looking at the sky!). If a point is visible by both eyes or cameras, then its projections are corresponding points in the image and should be matched. Attempts to use directly the intensity values at each pixel as the elements to be matched have had little success. It is not difficult to see why a scheme solely based on intensity is doomed to failure. Typically, corresponding points in the two images do not have exactly the same shade of gray, because of differences in the vantage points of the observer's eyes and noise intrinsic to every imaging and sensing system like our eye or a camera. Furthermore, numerous points in a surface might fortuitously be the same shade of gray, especially in the neighborhood of the correct match. Finally, it is easy to show that images of a stereopair that have rather different contrasts can be easily fused by our brain (Julesz 1971). On the other hand, we have already seen that a high level representation of recognized objects is not used by the stereomodule. More primitive measurements taken on the intensity arrays are used by the correspondence process and a critical problem in stereopsis is the nature of this representation. In brief, for understanding the stereo computation we have to answer two questions: what to match and how to match.

The matching primitives   Gray levels themselves are the most elementary form of "features" that could be used for the correspondence problem. They are, however, as we mentioned above, unreliable. Moreover, the intensity values measured by the photoreceptors are not directly transmitted to the cerebral cortex, where binocular combination occurs. An alternative approach is to match higher level features, typically oriented edges, extracted from each image. In 1967, Barlow, Blakemore & Pettigrew suggested just this solution to the correspondence problem, by pointing out that "the number of identical trigger features lying in the same appropriate region of each eye can safely be assumed to belong to the same object." Most recent computer algorithms and models of human stereovision that rely on some form of edges as the basic primitives perform far better than gray level correlation techniques on most images. This, of course, seems to be in encouraging agreement with the physiological evidence about the "edge detection properties" (without for now specifying what is meant exactly) of binocular neurons in the visual cortex.

Yet, one should not overlook the possibility that the primitive measurements used for stereo matching may be far more dense and specific than just location and parameters of edges. Many different measurements at each point of the image, for instance various types of derivatives, may provide a rich and robust description of each image, suitable for matching. One can easily see how this may work: Almost every point in each image of a stereopair may be characterized in terms of several measures of local variations of light intensity; corresponding locations in the other image can be found by identifying the point characterized by the most similar set of measurements (see Marr & Poggio 1976). An algorithm of this type for solving the correspondence problem has been recently developed and tested with encouraging results (Kass 1983). It is based on the binocular conjunction of a large collection of independent linear measurements, specifically partial derivatives of images smoothed via a few oriented filters of different size.

The other possibility is that a simpler set of measurements on the image intensity allows satisfactory matches to be assigned. Several recent theories of stereopsis are indeed mainly based on one set of primitives—called zero-crossings—which are specific relatives of edges. The hypothesis that a zero-crossing map could be one of the main products of the first phase of visual processing was suggested by a combination of psychophysical and physiological data on human vision (Marr & Poggio 1979). The first step in the scheme was to filter the image through a low-pass filter in order to cut out the high spatial frequency components of the signal. The next step was to take the second derivative of this filtered intensity array and to detect the zero-crossings in it. These zero-crossings then correspond to the inflection points of intensity changes in the filtered array, i.e. the points at which intensity is changing most rapidly. Originally, these steps were conceived as the operation of convolving the image with a particular function, the difference of two Gaussians of opposite sign (DOG function), which effectively performs both operations simultaneously. These filters are not tuned to orientation, but may be quite similar to a center-surround type of receptive field, as suggested by psychophysical observations (Mayhew & Frisby 1978a, 1979a) and computational arguments (Marr & Hildreth 1980). Spatial filters of this type are band-pass: They respond optimally to a certain range of spatial frequencies in the image. In summary, for a given resolution, the process of finding intensity changes consists of convolving the image with a center-surround filter whose spatial dimensions reflect the scale over which the changes have to be detected, and then of locating the zero-crossings in the filtered image. This process is performed in parallel at several different resolutions, that is, with filters of several different dimensions. A theorem by Logan (1977), which does not strictly apply to images filtered through the DOG filters, suggests nevertheless that the relatively sparse number of "discrete" symbols provided by zero-crossing are very rich in information about the filtered image and represent, therefore, one of the candidates for an optimal encoding scheme used for later processes. Marr & Poggio (1979) have suggested that they are the most important (but not the only!) primitives to be matched between the two images.

The constraints of matching   Even for quite specific matching primitives, the false matches problem cannot be completely avoided. Ambiguous matches have to be solved, taking into account constraints that the world and the geometry of the imaging process dictate. Some of these constraints are very simple and straightforward and some are somewhat more subtle. Existing algorithms differ in the emphasis given to these constraints and in their implementation, that is, in their matching rules. The most obvious constraints to be taken into account are the following:

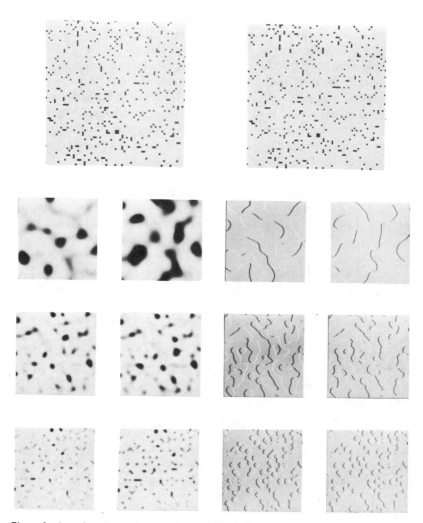

Figure 1 A random-dot stereogram (density 10%) is shown at the top. Its convolutions with center-surround masks of three different sizes are indicated in the *two left columns*. The convolution values are positive and negative: they are represented here in terms of shades of gray, positive values black and negative ones white. The coarse structure of the image is captured by the large filters; increasingly finer details are seen by the smaller masks. If the size of the dots in the stereogram is taken to be $4' \times 4'$ (at the corresponding distance the stereogram subtends slightly more than $4°$ of visual angle) the diameter of the center region of the three masks is $35'$, $17'$, and $9'$ respectively, from top to bottom.

The *two right columns* show the zero-crossings obtained from the convolutions in the left two columns. If positive values in the convolution were associated with activity of ON-center retinal ganglion cells and negative values with the activity of OFF-center cells, then the zero-crossings would correspond to the ON-OFF transitions between the activity of the two types of cells.

(These images are courtesy of E. Grimson of the Artificial Intelligence Laboratory at M.I.T.)

1. The uniqueness constraint: A given point on a physical surface has only one 3-D location at any given time. This translates into the rule that an item in one eye should usually be matched with only one item in the other eye (Marr & Poggio 1976).
2. The continuity and ordering constraint: Variations in the distance of surfaces from the viewer are generally smooth, with discontinuous changes being encountered at object boundaries (Marr & Poggio 1976). Thus, except at object boundaries, the disparity gradient should not be too high and, in particular, ordering of edges in the two images should be preserved (Baker 1982, Mayhew 1983).
3. Trigonometric constraint: This determines the orientation of the epipolar lines, that is, the pairs of straight lines in the two images that match point by point. When the observer fixates a distant point (ideally, at infinity), the epipolar lines are horizontal and parallel: In general, they depend on the direction of gaze. If the epipolar lines are known, the matching problem is essentially one-dimensional: The search for matches is limited to corresponding rasters in the images in the two eyes. If the epipolar lines are not known, the matching problem, now truly two-dimensional, becomes dramatically more difficult, because of the increase in number of false matches.

Other constraints have also been proposed by various authors to help solve the correspondence problem. Two of the most interesting ones are:

4. Connectivity in 3-D space of matched edges (Baker & Binford 1981).
5. Relative orientation of corresponding edges in the two images can be slightly different, depending on the viewing geometry. For the typical geometry and distance characteristic of human vision, the expected half-width of orientation difference is about nine degrees (Arnold & Binford 1980).

In conclusion, most of the recent computational approaches to the matching problem emphasize the use of suitable constraints to eliminate false matches among simple primitives. For "unspecific" primitives, the question of how to match is, of course, critical. An alternative approach is to rely on very specific primitives (as a large set of independent measurements at each point of the image): In this way, the false matches problem may be avoided in most situations, without any need for global constraints. Further on we discuss again this dichotomy of approaches to the matching problem from the perspective of the physiology of stereoscopic vision.

THE PROBLEM OF "STRUCTURE FROM STEREOPSIS" The human visual system can recover the three-dimensional shape of objects and estimate their absolute distance from binocular disparities. Horizontal disparities, which are the carrier of stereoscopic depth perception, are themselves uninterpretable without information about eye position. Knowledge of the interocular separation and of the convergence angle of the eyes is needed to recover the quantitative 3-D structure of a scene from horizontal disparity alone. In artificial systems, the distance between the two "eyes" and their angle of convergence are known precisely (camera model). In biological systems, some absolute depth information has generally been assumed to be extraretinal in origin, most likely derived from the convergence angle of the eyes (Foley 1980). Although this information may indeed be useful, its effective availability and its precision are unknown.

In principle, however, a binocular system that takes into account vertical as well as horizontal disparities can solve the interpretation problem without recourse to extraretinal sources of information. This has been well known to photogrammetrists for several years, but its relevance to human stereopsis was pointed out only recently (Longuet-Higgins 1981). The main result is that if the correspondence problem has been solved for seven points, the 3-D structure can be recovered uniquely (unless the seven points are transversed by two planes with one plane containing the origin or by a cone containing the origin; see Tsai & Huang 1981). Furthermore, there exists a remarkably simple approximate method of deriving the 3-D parameters of a planar surface from the horizontal and vertical disparities of a small number of corresponding points (Mayhew & Longuet-Higgins 1982). The question of whether the human visual system does measure vertical disparities and uses them for estimating distance (instead of simply correcting for them with eye movement) is still open. On the other hand, since vertical disparities are quite small even at short fixation distances and eccentric locations, their accurate measurement may turn out to be unattainable; in addition, the main psychophysical support for the idea—the Ogle-induced effect—has been criticized (Westheimer 1978, Gillam et al 1983) as unreliable and quantitatively in disagreement with the prediction of the Mayhew & Longuet-Higgins (1982) hypothesis. The question of whether vertical disparities are used to recover the 3-D structure is important also for its potential implications for the correspondence problem. If vertical disparities are to be measured precisely, matching may become a genuine two-dimensional problem, instead of a one-dimensional process along scan lines.

Theories of Stereoscopic Matching

The following outline of a few of the current computational approaches to stereopsis takes into account only algorithms that are more directly related to human stereoscopic vision. There is no theory encompassing all aspects of stereopsis; the algorithms described here mainly focus on the specific problem of stereomatching.

1. First, we mention a model proposed by Sperling (1970) based on correlation between two gray-level images. The model was never implemented, but

performance of gray-level correlation schemes is well known to be quite limited. Sperling's work does, however, make an interesting point of the connection between stereopsis and vergence movements. His neurophysiological theory of stereopsis includes several interesting notions, such as inhibition among cells tuned to different disparities and detectors for fine and coarse stereopsis.

2. Another model that has not yet been cast in the form of a precise algorithm is Julesz's dipole model (1971). Each position on each retina is associated with a magnetic dipole whose polarity is determined (in the case of random-dot stereograms) by the retinal intensity value. Spring coupling between the tips of adjacent dipoles implements the continuity rule and, of course, the ordering constraint. The orientation of a dipole represents a disparity value, and the fact that each dipole can have only one orientation at a time provides an implementation of the uniqueness rule. Taken literally, this model would correspond to a scheme in which disparity at each position is signaled by the rate of firing of a single neuron. The model exemplifies a matching scheme based on nonspecific primitives (intensity values). False matches are therefore a real problem, and to cope with them, "global" constraints are enforced (via the magnets and the springs).

3. We consider now a cooperative algorithm that was devised by Marr & Poggio (1976) (the approach was first tried by Dev 1975; see also Nelson 1975, 1977, Marr et al 1978), and which is successful at solving random-dot stereograms. The algorithm, which embodies some of the features of Julesz's dipole model, requires a 3-D network of nodes or "neurons," each of which lies at the intersection of a line of sight from each image. The algorithm implements the uniqueness constraint in that the nodes lying along a given line of sight strictly inhibit each other. To implement the continuity constraint, each node excites its immediate neighbors at the same disparity. In the unnatural case of a random-dot stereogram, each pixel of the stereogram can be made to correspond to one line of sight and will have one of two distinct intensities (black or white). The algorithm does not specify the type of matching primitives to be used, whether values in images filtered through center-surround receptive fields, or the more complex features envisioned by the "primal sketch" (Marr 1976), or a large set of measurements at each point in the image. The algorithm is only concerned with a simple cooperative implementation of the basic matching rules.

The range of effectiveness of this algorithm can be extended to natural images by transforming the images to obtain better primitives than the intensity values themselves. By convolving the image with a center-surround filter and representing the sign of the resulting values, it is possible to convert natural scenes into patterns that bear a striking resemblance to Julesz's binary random-dot stereograms (the sign of this convolution array is equivalent to the zero-crossing map). The cooperative algorithm can now operate on this binary image exactly as on random-dot stereograms and extract the correct disparities.

4. Marr & Poggio (1979) later proposed a rather different algorithm specifically motivated by human psychophysics. It exploits the intrinsic possibility offered by sparse matching primitives like zero crossings at different spatial resolutions. At low spatial resolutions, zero-crossings of a given sign are quite rare and never too close. Consequently, false matches, that is, matches between noncorresponding zero-crossings, are essentially absent over a relatively large disparity range. From a mathematical analysis of the probability of occurrence of zero-crossings in bandpass images, it turns out that if the disparity range that is considered is on the order of the size of the receptive field used to filter the image (more precisely, the diameter of the field's center), false matches are virtually absent. These observations led to a different algorithm for solving the correspondence problem. In its simplest version, the algorithm matches zero-crossings of the same sign in image pairs filtered with receptive fields of three or more different sizes. These ideas lead to several algorithms with somewhat different properties: for instance, matching could take place between the binary function of the filtered images over a certain area (H. K. Nishihara, in preparation).

The main aspect of this theoretical framework is its avoidance of the false matches problem by trading off resolution with disparity range in a coarse-to-fine strategy. [Moravec (1980) was the first to use a matching scheme based on a search from the lowest to the highest resolution]. A computer implementation of this algorithm by Grimson (1981) performs satisfactorily on random-dot stereograms as well as on natural images, showing some of the properties of human depth perception. Psychophysical data against some of the detailed predictions made by Marr & Poggio (1979) have been extensively discussed by Mayhew & Frisby (1981). For instance, vergence movements are not driven only by matches in the largest channels; simpler measurements of disparity are used as well (e.g. based on monocular contours). In addition, the role of eye movements and the independence of the matching at the various resolutions are apparently inconsistent with some of the results reported by Mayhew & Frisby (1979b) and by Julesz & Schumer (1981), for instance, concerning the size of Panum's area for filtered stereograms in the absence of eye movements. Though the theoretical framework may still be correct in its main lines, it is already clear that the scheme of Marr & Poggio will have to be modified in several respects.

One of the main weaknesses of the algorithm is its sensitivity to vertical disparities. Vertical disparities can be well tolerated up to a fraction, about 1/4 of the filter size. Interestingly, human performance deteriorates quite rapidly with vertical disparities in the absence of eye movements, almost as much as predicted by the model (Nielsen & Poggio 1983).

5. An important extension of the Marr-Poggio (1979) theory has been proposed by Mayhew & Frisby (1981). The main modifications they suggest

are (*a*) concerning the matching primitives: in addition to zero-crossings, Mayhew & Frisby advocate the use of peaks in the convolution values, both because of psychophysical evidence and for computational reasons; (*b*) concerning the matching rules: instead of binocular matching performed independently on each spatial frequency channel, Mayhew & Frisby propose a matching rule which takes into account simultaneous correspondences in all channels. In addition, continuity of zero-crossing segments to be matched is also explicitly used and thus the matching primitives are oriented, continuous zero-crossing segments (as in the neural implementation suggested by Marr & Poggio 1979). Computational experiments support the usefulness of these modifications for solving correspondence problems in random-dot stereograms. Extended tests on natural images are not yet available and would be highly desirable.

6. Several other algorithms of stereopsis have been developed without a direct concern with human stereopsis (e.g. Arnold 1982, Gennery 1980). Among them, the most interesting one has been developed by Baker (1980, 1982) and Baker & Binford (1981). Its matching primitives are "edges" (similar to zero-crossings) at various spatial resolutions. In addition to position, contrast orientation and intensity to either side of the edges are used as matching primitives, and provide a weighting factor for the correspondence process. The matching process takes place on epipolar lines (assumed known and parallel) and exploits the ordering constraint (corresponding edges in the two images have the same order). It starts with low resolution edges to bring the two images into rough correspondence. A successive refinement in resolution brings finer details into the analysis. The matching performs the optimal correspondence on each epipolar line. A global constraint is then applied, checking for edge connectivity: a connected sequence of edges in one image should be seen as a connected sequence of edges in the other. Finally, intensity values are matched between matched edges.

7. To recapitulate: The main matching rules proposed in the cooperative algorithm and used by the Marr-Poggio theory (1979) have been refined and extended by Mayhew & Frisby (1981) and by Baker & Binford (1981). Cross channel correspondence, figural continuity, and edge connectivity are the main additions to continuity and uniqueness of matching. Computer simulations suggest that these rules may be advantageous but experiments must be conducted to find out whether they are used by our own visual system.

All these theories agree that "edges" are the main matching primitives. This is consistent with the neurophysiological observations that binocular cortical cells respond to oriented edges. The thrust of the Marr-Poggio theory (1979) is that simple primitives of the zero-crossing type, together with a coarse to fine strategy, already go a long way in avoiding the false matches problem, without disambiguation processes. More complex and specific primitives, as pointed out by Barlow et al (1967), would clearly show an even simpler false matches problem and allow a larger Panum's fusional area. A step further, one may consider whether a larger set of independent measurements on each image point, possibly including zero-crossings, directional derivatives, etc, may be used by the visual system to avoid almost completely the problem of incorrect matches.

All theories discussed above consider the problem of full stereo matching, for the extraction of a precise disparity map. Simpler and more primitive schemes could be used, and are probably used by the visual system, when the only information that is needed is a rough estimate of convergent or divergent depth (see section on Vergence).

PSYCHOPHYSICS OF STEREOPSIS

Primitives and Constraints of Matching

So far, psychophysics does not tell much about the nature of the primitives used in stereoscopic matching. We know they are not intensity values themselves or complex micropatterns of intensity values (Julesz 1971). Available evidence is consistent with the idea that zero-crossings, as described above, are among the matching primitives. In particular: (a) stereograms in which the two images have different contrast (of the same sign) can be easily fused (Julesz 1971), (b) if the contrast of one half-pair is reversed, fusion is impossible (Julesz 1960), and (c) fusion survives blurring of both or just one of the two images. The experiments of Mayhew & Frisby (1981), on the other hand, suggest that zero-crossings alone cannot explain the perception of stereograms composed of a triangle ramp grating paired with a phase-varying ramp grating. They have proposed that peaks in the channel convolutions are also used by the process of stereo matching. The idea, presented above, that several independent measurements, possibly including or being in part equivalent to zero-crossings and peaks, are used in the matching process has still to be tested.

The constraints that the matching process obeys are only indirectly explored with psychophysical techniques. The continuity and ordering constraint is consistent with the "nail illusion" (Kroll & van der Grind 1980) and with the disparity gradient limit found by Burt & Julesz (1980a,b). Convincing evidence is, however, lacking. On the whole, the question of the nature of the primitives and constraints of matching waits for new psychophysical experiments.

Vertical Disparity, Registration, and Eye Movements

Vertical disparity arises naturally because of eye misalignments and because of the stereo geometry. In the latter case, vertical disparity contains valuable information about the parameters of gaze. Can the visual system measure the corresponding small vertical disparity and use it to compute the "camera" parameters? As a separate question, how does the matching process cope with perspective deformation, and in particular, with vertical disparity?

One may expect that eye movements would perform the kind of registration process that must precede full, high-resolution stereomatching. The psychophysical evidence is consistent with this view. In particular, Duwaer (1981) and Duwaer & van der Brink (1981, 1982) have argued that the sensorimotor system eliminates vertical disparities with high accuracy: Vergence eye movements can be initiated by vertical disparities that are much too small to induce diplopia. Residual vertical disparities that can be fused are in the range of 8'–15'. The stimuli Duwaer used did not pose, however, any false match problem. The interesting question concerns the maximum range of vertical disparities that allows fusion without eye movements in the presence of potential false matches. Recently, Nielsen & Poggio (1983) found that for briefly flashed random-dot stereograms about 4'–7' of vertical disparity represent the fusion limit. Because of the small value, this finding may be consistent with a simple one-dimensional matching scheme having a built-in tolerance to small vertical disparities. These data have a clear implication: A registration process based on eye movements must precede the stereo matching process to remove vertical disparities larger than about 4'–7'. Some questions immediately arise about this registration stage; for instance, which measurements on the image are used to control the registration process? These measurements must be simpler and coarser than what is involved in the full matching process. Otherwise, in order to register and control the eyes the visual system would have to solve a 2-D correspondence problem.

Vergence

Given the obvious importance of eye movements in stereopsis, it is surprising that so little is known about the role of vergence. In most computational schemes for stereo, the false matches problem is simplified by considering only those potential left/right fusions with disparities under a certain limiting size (defined by Panum's area). This disparity limit requires that when disparities greater than Panum's area are presented, a suitable vergence change must be initiated to bring corresponding features within the allowed range for fusion. As we mentioned, it is critical to know the nature of the measurements made on the image that guide these vergence movements. Furthermore, one would like to know how vergence is controlled and maintained.

It is long known that vergence movements are independent of conjunctive eye movements, are smooth rather than saccadic, are accurate within 2' of disparity, and are continuously controlled by a disparity signal (Jones & Kerr 1972, Rashbash & Westheimer 1961(a,b), Riggs & Niehl 1960, Westheimer & Mitchell 1969). Marr & Poggio (1979) conjectured that matches between large receptive fields (that can cope with disparities of the same order of the size of the receptive field) initiate vergence movements bringing into correspondence higher spatial frequency channels mediating smaller disparities. Recently, it has become clear that this is not the only mechanism that can trigger correct vergence movements. Kidd, Frisby & Mayhew (1979) have shown that texture contours can initiate correct vergence eye movements even for disparities larger than the ones predicted by Marr & Poggio. Later, Mowforth, Mayhew & Frisby (1981) claimed that high-frequency filtered stereograms could initiate vergence movements well outside the range of the largest receptive field presumably active. The conclusion is inescapable that not only matching of zero-crossings but some other more primitive measurements of disparity control at least the direction of vergence movements. Possibly, prominent monocular cues (like strong edges in some, or all, of the channels) could drive eye movements (Mowforth, Mayhew & Frisby 1981; see also Marr & Poggio 1980). For isolated and diplopic cues, there is no false matches problem and no reason that vergence movement and rough sensation of depth could not be provided even for large disparities.

For dense textures, there are two basic possibilities: (a) Either a similar mechanism works on prominent monocular features of the images (for instance, a few sparse strong zero-crossings, possibly coincident in all channels), or (b) all convergent matches are weighted against all divergent matches (false and correct) over relatively large areas. This scheme may correctly find the overall sign of disparity, but fails to identify the shape of a stereo figure. Moreover, the disparity limit in this case may depend on the size of the figure (see Marr & Poggio 1980). Interestingly, Tyler & Julesz (1980) have reported that detection ability depends on the square root of the area.

Spatial Frequency Channels in Stereopsis

Several data suggest that monocular detection of targets relies upon a set of independent channels of different coarseness—probably corresponding to receptive fields of different sizes. That similar independent channels may be used in stereopsis was demonstrated by Julesz & Miller (1975), who found that masking noise disrupts stereopsis only if it has a spatial spectrum overlapping that of the stereogram. These findings, as well as the work of Frisby & Mayhew (1977), provide the basis for the assumption that binocular matches are made on independently filtered images. The natural consequence of the assumption is that coarse channels encode large disparities whereas fine channels can match only small disparities (without eye movements). Psychophysical observations by Felton et al (1972), Kulikowski (1978), and Levinson & Blake (1979) support this view and are consistent with the physiological observation that neurons with small receptive fields often have a narrower range of disparities than neurons with large receptive fields (see below).

Mayhew & Frisby (1978b), however, found that the contrast threshold for stereopsis does not depend on the independent contrast of different spatial frequency components. The contrast threshold for stereopsis is higher than for monocular detection of narrow-band filtered, random-dot noise at all spatial frequencies by 0.3 to 0.4 log units (Frisby & Mayhew 1978). This may reflect processes at the level of 2 1/2-D sketch of Marr & Poggio (1979), the dynamic buffer where successful matches are stored. More likely, however, it supports the idea that different channels are combined in the matching process (the cross-channel correspondence rules of Mayhew & Frisby 1980). More evidence in this direction has been provided by the same authors (with the "missing fundamental" experiment and with spatial frequency filtered stereograms portraying corrugated surfaces). It seems, therefore, that the strictly independent channel model of Marr & Poggio has to be modified, allowing for some interactions between the channels at the level of disambiguating ambiguous matches. An alternative possibility is the coupling of the channels, not simply via eye movement, but also via a hypothetical cortical shift mechanism.

An important property of the binocular spatial frequency channels is that according to Mayhew & Frisby (1978a, 1979a) they are not orientation selective, a conclusion consistent with the notion that the matching primitives are extracted from the images filtered through center-surround receptive fields. This is not in disagreement with the orientation tuning of disparity selective cells in the cortex since the matching primitives themselves, such as the zero-crossings, can be locally oriented. Oriented receptive fields, however, cannot be completely ruled out on the basis of these experiments alone.

Panum's Area

There is a vast literature, and a correspondingly large amount of data, on Panum's fusional area, and almost as much disagreement on its properties and even its precise definition.

Panum's area represents the total range of disparities that can be fused about the fixation distance without the help of eye movements. The values reported for line stereograms are typically around 10' for the maximum amount of convergent or divergent disparity without diplopia; the extent of Panum's fusional area is, therefore, twice this (Mitchell 1966). Comparable values have been reported for vertical disparities (Schor & Tyler 1980).

While these data refer to tachistoscopic presentations, Fender & Julesz (1967) measured the limits of fusion under stabilized image conditions, and found that fusion occurred between line targets at a maximum disparity of 40'. The corresponding value for random-dot stereograms was 8' for horizontal disparities and 10' for vertical disparities (for the onset of fusion). Fusion, once established, could be maintained over a much larger range. Diner (1978) replicated the experiments of Fender & Julesz but failed to obtain foveal diplopic thresholds beyond the classical limit of 20′ for binocularly stabilized bars.

All these data show a relatively small Panum's area with a size comparable to the maximum size of the receptive fields operating in foveal vision (around 20′ in human according to Wilson & Bergen 1979). However, recent data obtained with dynamic random-dot stereograms imply that Panum's area can be larger. In particular, Schumer & Julesz (1980; see also Julesz & Schumer 1981) have reported that plus or minus 50′ disparities can be successfully fused. Mayhew & Frisby (1979b) observed that convergent disparities could be fused in a flash up to 18′ of disparity: At this value, the latency for discrimination (longer than 180 msec) indicated the need of vergence movements. With filtered stereograms, disparities larger than twice the values predicted by the Marr-Poggio (1979) independent channel algorithm can be successfully fused (Mayhew & Frisby 1979b, Schumer & Julesz 1982). A critical prediction of this theory is that the maximum fusable disparity should scale with the spatial frequency of the stimulus, since the lower spatial frequencies will be detected only by the larger channels. Julesz & Schumer (1981) concluded that there is probably an association between large disparities and coarse channels, but that the complementary association between small disparities and fine channels lacks experimental support.

The presence of Panum's area seems to be the simplest explanation for the persistence of single binocular vision, even though there are constant changes in retinal image disparity caused by ocular drifts and saccades (2′–4′; St. Cyr & Fender 1969). In this view, the disparities for which Panum's area compensates are dynamic. Schor & Tyler (1981) studied the temporal properties of Panum's area and found that for slow changes in disparity (0.2 Hz), the area increasing horizontally by as much as a factor of ten, but remaining basically unchanged vertically, compared with fast disparity variations (5 Hz). They were, however, unable to obtain Panum's areas much greater than 20′ and thus to approach the large extended fusional ranges reported by Fender & Julesz (1967) under stabilized image conditions.

Tyler (1975) and Schor & Tyler (1981) found that Panum's area was reduced with increasing spatial frequency variations of disparity (both for horizontal and vertical directions). Interestingly, Panum's area dependency on temporal modulation of disparity holds only for low spatial frequency.

In order to interpret correctly measurements of Panum's area obtained with line stereograms and with random-dot stereograms, it should be kept in mind that the mechanisms of rivalry and fusion may not coincide. Thus, different measured ranges for diplopia and for fusion may reflect two different underlying processes (Duwaer 1981). Furthermore, successful depth discrimination, especially in a forced-choice paradigm, may not need a solution of the correspondence problem for all matches; a high enough proportion of correct, possibly sparse, matches could be sufficient.

Stereoacuity

The impressive quality of stereoacuity would seem to impose quite restrictive requirements on the mechanisms subserving stereopsis. The threshold for depth discrimination is only a few seconds of arc, which is much less than the size and the separation of foveal photoreceptors. In addition, stereoscopic acuity does not suffer when targets move laterally or in depth by five or more minutes of arc during a short 200-msec exposure (Westheimer & McKee 1978, 1980a). Stereoacuity has also its limitations, however: Synchrony of presentations of left and right image and absence of a standing disparity (of even one minute of arc) are necessary requirements for small discrimination thresholds (Westheimer 1979). In addition, blur of the images decreases stereoacuity more than ordinary visual acuity: For optimal stereoacuity, the full representation of the spatial frequency range admitted by the eye's optics is needed (Westheimer & McKee 1980b).

All these results refer to tachistoscopic presentations of line stereograms. One may ask how stereoacuity is affected by the potential presence of false targets as in a random-dot stereogram, since Westheimer reports that "crowding" of features decreases stereoacuity even with line stereograms. The available data are contradictory on this point. Harwerth & Rawlings (1977) report relatively high acuity thresholds for random-dot stereograms presented in a flash. Uttal et al (1975) find comparable values at 75% correct answer threshold. Lower thresholds have been measured for dynamic random-dot stereograms by Schumer & Ganz (1979). Recently, K. R. K. Nielsen (in preparation) has also measured thresholds as low as 30" for dense random-dot stereograms presented in a short flash of 150 msec. Thus, the potential problem of false targets does not seem to affect dramatically stereoacuity. Interestingly, stereoacuity thresholds depend on exposure duration and are higher for discriminating the form of the (cyclopean) figure than for simply discriminating depth (Harwerth & Rawlings 1977).

Westheimer (1979) has suggested that a differencing mechanism measuring relative disparities is at the basis of stereoacuity. For two points in a stereo target, depth discrimination would be accomplished in terms of the difference of the individual disparities, a value that is independent of ocular stability. More recent experiments by G. Westheimer and G. Mitchison (personal communication) suggest that in a stereogram of several lines, the perceived disparity of a given line is determined by an average of its disparities' differences with respect to the flanking lines. As a consequence, if the lines lie exactly on a tilted plane, any given line (not at the borders) is perceived at the same depth as the left and right flanking lines. It is clearly important to check whether this computation of relative disparity takes place only for small disparities (in the range of stereoacuity) or also for large ones. The underlying computational reasons are still unclear, although evaluation of relative dispari- ties—actually changes in disparities—would give the stereosystem a useful tolerance to eye movements. The precision of stereoacuity requires that the monocular matching primitives are localized with the same level of relative accuracy in the image delivered by each eye. It is possible that the interpolation process underlying monocular hyperacuity—as it is revealed by vernier detec- tion tasks—is used as an input to stereoacuity. In this view, the patterns of activity transmitted by the left and the right retinae would first be interpolated independently, localizing features such as zero-crossing with precision of a few seconds, much less than the spacing and the diameter of the foveal cones. These features would then be matched between the two eyes. Absolute disparities could not be measured reliably because of involuntary eye movements, but differences in disparity between neighboring features could be evaluated with an accuracy of a few seconds of arc.

On the other hand, it has been recently shown that vernier cues need not be input to the stereo matching process (Nishihara & Poggio 1982), as proposed earlier by Julesz & Spivack (1967). The successful fusion of random line stereograms with vernier breaks between the lines as small as $15''$ can be explained in terms of a matching process operating on the coarse structure of these stereograms as extracted by relatively large receptive fields, without recourse to interpolation and detection of the vernier breaks.

Disparity Gradient

Panum's area is an upper limit to the disparities that can be fused without eye movements. Recently, evidence has been accumulating that in addition to disparity magnitude, the rate of change in disparity (the disparity gradient) also sets a limit to stereo fusion. There is in most situations a physical upper limit to the maximum disparity gradient. When one continuous surface changes in depth, one of the two eyes may not see all of the surface: this starts to happen when one line of sight grazes the surface. In situations of this type (which correspond to the Panum limiting case) human vision has difficulties and stereopsis is usually diplopic.

Tyler (1973, 1974, 1975a,b) was the first to observe that there is a limit in the rate of change of disparity across the retina both for line stereograms and for random-dot stereograms. Burt & Julesz (1980a,b) measured the disparity gradient limit of pairs of points at different orientations defined as their total disparity divided by their "binocular" separation. This definition implies that the disparity gradient is $d' = 2$ for the Panum case. They find a value for d'

around 1, roughly independent of orientation for the limit between fusion and diplopia.

The critical question, of course, is the nature of this constraint. Is it a constraint used in matching or is it a property of later processes, for instance, of the way depth is represented? As argued by Baker (1982) and by Mayhew (1983), stereo projection almost always preserves the order of successive pairs of matches in each eye and thus it seems natural to implement this constraint during the matching process. The ordering constraint, however, would correspond to d' less than or equal to 2 and is therefore only indirectly implemented by the human visual system.

There may be additional, nonexclusive reasons for the disparity gradient limit. A large disparity gradient locally compresses and rotates one of the two images relative to the other: thus, the same line segments will differ in orientation and length in the two images and these differences may by themselves make the matching impossible. The results of filtering the two images for extracting the matching primitives (for instance, by ganglion cells' receptive fields) may give different results for high disparity gradients, depending on the size of the operator and of course, on the pattern itself. In other words, the matching features extracted from the monocular patterns may be affected by the disparity gradient. Braddick (1979) has found that the geometrical relations of different parts of the pattern presented to each eye may influence binocular single vision.

Stereopsis by Binocular Delay

Morgan (1975, 1976), Ross & Hogben (1975), and Tyler (1977) studied a Pulfrich-like effect in stereopsis that has some fascinating implications about the spatiotemporal interpolation performed by the visual system. The new observation was that a Pulfrich effect—dimming the view by one eye—still occurs for an object in apparent motion produced by flashing it stroboscopically at a sequence of positions. Burr & Ross (1979) were then able to show that binocular delay of a spot of light moving stroboscopically against a background of dynamic noise yields a vivid stereoscopic depth. Temporal delay of one image seems equivalent to binocular disparity and has an equivalent threshold of a few seconds of arc. The simplest explanation is provided by another (monocular) experiment by Burr (1979): If two line segments are displayed sequentially at a series of stations, an illusory displacement occurs if the line segments are aligned in space but are displayed with a slight delay in one sequence relative to the other. Not the actual portions of the line segments, but the positions estimated by a temporal interpolation process are displaced relative to each other. From this experiment, it seems that the spatial pattern of activity between the flashes is actually reconstituted. This monocular reconstituted pattern of activity could be used by stereopsis and could satisfactorily explain the results of Burr & Ross. An apparent implication of this view is that at least some binocular cells should have access to interpolated monocular activity, reconstructed between the positions at which the stimulus was presented stroboscopically.

Motion and Color in Stereopsis

It is known that contours of discontinuities in a motion field can be matched stereoscopically. Thus, primitives more complex than simple edges or zero-crossings can be used in stereomatching. A more general question is whether motion can interact with stereopsis at the level of matching (helping to disambiguate false matches) or at the level of structure from stereopsis.

On the whole, the interaction of motion with stereopsis still requires an analysis at both the computational and the experimental levels. Similarly, not much is known about color information and its use in the process of stereoscopic matching and stereofusion (but see Julesz 1978).

Development of Stereopsis

Recently, a number of investigators in different laboratories have been able to measure the onset of binocular vision and the development of stereoacuity in infants. These data and their comparison with developmental studies of binocular cortical mechanisms in cats and monkeys promise to provide some important information about the neural organization of stereopsis. Coarse stereopsis appears around the fourth month of age; at this time stereoacuity (fine stereopsis) begins to increase very rapidly, approaching adult levels around the fifth month after birth (Fox et al 1979, Held et al 1980, Petrig et al 1981). There is some evidence that stereoacuity for crossed disparity develops earlier than, but approximately at the same rate as, the uncrossed disparities (Birch et al 1982). This finding is consistent with the notion that at least two stereoscopic mechanisms exist in the adult (Richards 1970, 1971). The rapid development of stereoacuity contrasts sharply with the development of monocular visual acuity for gratings, which increases rapidly during the first six months of age and then slowly over the first few years (Teller 1981). Held (1983) has conjectured that segregation of the ocular dominance columns in layer IVc of the primary visual cortex is responsible for the onset and the development of stereoacuity.

NEUROPHYSIOLOGY OF STEREOPSIS

While psychology and psychophysics provide knowledge of its functional properties and computational capacities of stereoscopic vision, neurophysiology must attempt to unravel the extent and nature of the "cyclopean brain," the processing neural network upon which depends the vivid and rich three-dimensional visual world we may enjoy. The term "cyclopean," after the one-eyed giants, foes of Ulysses, aptly suggests the egocentric one-image visual world that we usually perceive, even though the hapless, monocular Cyclops never experienced stereoscopic depth.

In preceding sections of this article, we have described the psychophysical evidence showing that binocular positional disparity provides the most important and accurate information, indeed the only necessary one, for stereopsis. While it may be argued that "stereopsis constitutes only a fraction of the information upon which depth perception is based" (DeValois & DeValois 1980), this fraction is undeniably of some importance, for it alone makes the monocular and binocular experiences of depth so dramatically and fundamentally different.

The analysis of the neural mechanisms of stereopsis must begin with an attempt to discover where and how binocular disparity is derived and measured by the cells in the brain, and how it is signaled in neural code. In primates, and generally in higher mammals, the region of brain where the inputs from the two eyes converge in the region of cerebral cortex is known as "area striata," area 17 of Brodman, the primary visual cortex, or V1. In the macaque monkey, nearly all visual cortical neurons are binocular. Binocular neurons have two receptive fields, one for each eye, and their activity reflects the dynamic interaction of excitatory and inhibitory influences from each eye; thus, the responses of these neurons to binocularly viewed patterns may vary markedly depending on the receptive field characteristics and on the spatio-temporal configuration of the stimulus in the two eyes.

Neuron Sensitivity to Positional Disparity: "Local" Stereopsis

Barlow, Blakemore & Pettigrew (1967) and Nikara, Bishop & Pettigrew (1968) first presented evidence that neuronal stereoscopic processing can be identified in the primary visual cortex of the cat based on the phenomenon of receptive field disparity. These studies, conducted with isolated line/bar stimuli, showed that the receptive fields of binocular cortical neurons subserving central vision could be in exact spatial correspondence in the two eyes, or could have different relative positions, some field pairs having convergent disparities, others, divergent disparities. These disparity sensitive cells might well play an important role in stereoscopic depth perception because, with normal convergent fixation of binocular vision, they would be selectively activated by objects at different relative depths.

The properties of disparity sensitive neurons in the visual cortex of the cat, and more generally of the neuronal characteristics of binocular interaction, were analyzed in some detail by Bishop and his collaborators (Pettigrew et al 1968, Joshua & Bishop 1970, Bishop & Henry 1971, Bishop et al 1971). These studies, together with more recent ones by von der Heydt et al (1978), Fischer

& Krueger (1979), and Ferster (1981), have confirmed and extended the original observations. Disparity sensitive neurons have been found also in the visual cortex (V1 and V2) of the sheep (Clarke et al 1976), and in the Wulst of the owl (Pettigrew & Konishi 1976).

The stereoscopic properties of cortical neurons of the macaque monkey, an animal whose discriminatory capacities for depth are very similar to those of humans (Bough 1970, Sarmiento 1975), were first studied by Hubel & Wiesel (1970) in anesthetized and paralyzed preparations. These authors were unable to identify disparity sensitive neurons in area 17, but did observe numerous neurons in prestriate area 18, or V2, which were termed "binocular depth cells" for they were uniquely sensitive to simultaneous stimulation of the two eyes, some neurons having receptive fields in exact binocular correspondence, others disparate fields. On the basis of these observations and of similar ones they made in the cat, Hubel & Wiesel (1970, 1973) suggested that the elaboration of stereoscopic mechanisms occurs outside the primary visual cortex.

Studies in alert and visually attentive macaque monkeys under conditions of normal binocular vision, unequivocally demonstrated that a large number (60–70%) of neurons in striate cortex, and an even higher proportion in prestriate cortex, are sensitive to horizontal disparity, and that different types of depth neurons exist, many with the same binocular properties as those described by Hubel & Wiesel (1970) in area 18 (Poggio & Fischer 1977, Poggio & Talbot 1981, Poggio 1984). An account of the findings of those studies is given in that which follows.

Response profiles of single cells in area 17 and 18 were obtained for binocular stimulation with isolated line/bar patterns of optimal size and orientation for the cell under study, and presented in real (Poggio & Fischer 1977), or simulated (dichoptic) depth (Poggio & Talbot 1981). Under these stimulus conditions, in which false matches are not present, two main types of stereoscopic neurons may be recognized: 1. A group of neurons are disparity selective over a limited and often narrow range; excitatory binocular facilitation (*Tuned excitatory neurons*) and, less frequently, suppressive interaction (*Tuned inhibitory neurons*) may be observed. 2. A second type of stereoscopic neurons includes cells with a reciprocal selectivity for crossed and uncrossed disparities: One set of neurons (*Far neurons*) gives excitatory responses to objects farther than the points of fixation and inhibitory responses to nearer objects. Other neurons (*Near neurons*) have the opposite behavior: excitation for nearer objects and inhibition for farther ones. About one third of the neurons in A17, and fewer still in A18, are not disparity selective and give similar responses to stimuli of different disparities within the disparity range of binocular interaction (*Flat neurons*).

TUNED NEURONS   Approximately one half of the disparity selective neurons are Tuned excitatory (Figure 2, left). In the area of striate cortex subserving central vision, peak disparity sensitivity of these neurons is very seldom found outside 12' of crossed or uncrossed disparity, with most neurons maximally excited by disparities within ± 6'. The width of the disparity sensitivity curve for a group of these neurons (taken at the response level of maximal response ÷ $\sqrt{2}$; Schiller et al 1976) was 10' ±4 '(SD) of disparity. (Poggio, 1980, 1984). Frequently, tuned excitatory neurons are directionally selective and many are strictly unidirectional. Most characteristics of depth tuned cells are neurons, found both in striate and prestriate cortex, whose properties appear to be the same as those of the "binocular depth cell" of Hubel & Wiesel (1970). These cells do not respond at all, or only minimally, to monocular stimulation over a very narrow range of disparities at or about zero disparity. No monocularly silent neurons were ever found with other types of stereoscopic properties.

Tuned inhibitory neurons have disparity response profiles opposite to those of the tuned excitatory neurons, in that their binocular responses are suppressed within the same narrow range of disparities near the horopter over which tuned excitatory neurons are binocularly facilitated. For foveal neurons, maximal suppression occurs within ± 6' of disparity and response facilitation at larger crossed and uncrossed disparities is often observed.

RECIPROCAL NEURONS   Neurons with the second main type of stereoscopic properties, the Near and the Far neurons (Figure 2, right), give excitatory responses over a range of disparities of one sign and inhibitory responses over a similar range of opposite sign. For many of these neurons, disparity sensitivity for bar stimuli extends over a range of one degree or more on either side of the horopter; other reciprocal neurons have a narrower, S-shaped depth response profile. Both groups, however, have common properties characterized by a steep response gradient from maximal excitation to maximal inhibition with the mid-point of response activity at, or very close to, zero disparity.

The distribution of stereoscopic types among binocularly "simple" and "complex" cells appears to be remarkably similar, and no functional relationship was observed between the simple/complex classification and the tuned/reciprocal types of depth sensitivity for bar patterns (Poggio 1984). Similar conclusions have been reached by Ferster (1981) for the visual cortex of the cat. Stereoscopic neurons typically signal depth with little response uncertainty; the response to the preferred excitatory or inhibitory disparity is consistently stronger than the response to any other disparity, irrespective of any intrinsic response variability. Thus, under normal binocular vision, the mechanisms of binocular interaction are remarkably secure and capable of compensating dynamically for the small movements of fixation (Motter & Poggio 1981, 1982, Poggio 1984).

The neurophysiological evidence, both from cats and from monkeys, indicates that the characteristics of the binocular response to isolated line patterns are determined by the spatial organization of the component regions of the

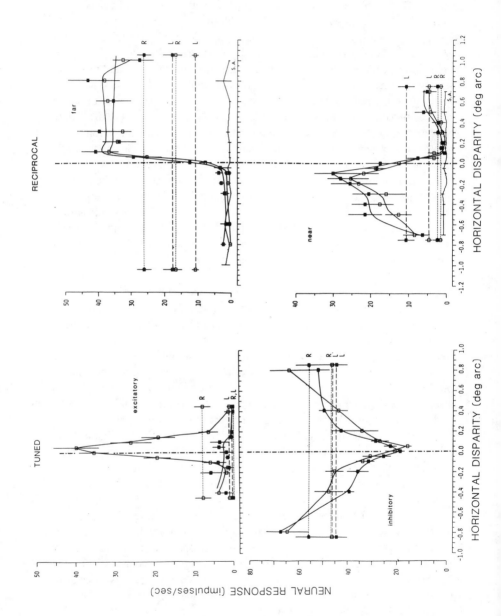

neuron's receptive fields in left and right eyes, and the dynamic processing of the signals reaching the cortical cell. Maximal responses occur when synergistic regions of the two monocular receptive fields are concurrently engaged, and minimal responses when antagonistic regions are stimulated (Barlow et al 1967, Bishop et al 1971, Ferster 1981). Both simple and complex cells measure retinal image disparity by receptive field disparity and may operate in mechanisms of "local" stereopsis appropriate for detecting unambiguous disparities in sparse and simple contour patterns.

The psychophysics of stereoacuity is not easily interpreted in terms of the known properties of binocular cortical neurons. The threshold of stereoacuity is more than one order of magnitude smaller than the width of tuning of disparity-sensitive cells. Perhaps sets of stereoscopic neurons provide the desired precision by means of interpolation process similar and subsequent to the one postulated to occur in layer IVc-beta for monocular hyperacuity (Barlow 1979, Crick et al 1981).

The existence of different types of depth neurons, Tuned neurons and Near and Far neurons, gives support to the suggestion of Richards (1971) that normal stereopsis is based on the activity of three populations of neurons preferentially activated by crossed, near zero, and uncrossed disparity. This suggestion was based on studies of stereoanomalies (Richards 1970, 1971) showing that some individuals are unable to localize stimuli presented with large ($>0.5°$) uncrossed disparities (Far), while other individuals are unable to utilize large crossed disparities (Near). Jones (1977) demonstrated that these forms of stereoblindness may be present in subjects who have normal fine stereopsis. The proposition may be advanced that these persons lack either functionally normal Far or Near neurons but possess a normal Tuned stereosystem (Fischer & Poggio 1979).

Neuron Responses to Random-Dot Stereograms: "Global" Stereopsis

The experimental results described above provide evidence that cortical visual neurons signal and measure horizontal binocular disparity in two distinct ways

←————————————————————————————————

Figure 2 Positional disparity sensitivity profiles for line stereograms of *Tuned* (excitatory and inhibitory) and *Reciprocal* (far and near) foveal cortical neurons of the macaque. The graphs were constructed by plotting mean response magnitude vs. horizontal binocular disparity of stimuli. For each neuron, the response profiles for the two diametrically opposite directions of stimulus motion perpendicular to stimulus orientation are shown (*open* and *filled square symbols*). The *vertical lines* at each point indicate ±1 SE of mean. The magnitude of the response to monocular stimulation of each eye for each of the two directions are shown by the *horizontal lines* across the graphs, identified at each end by the appropriate directional *square symbols*. *Broken line* = left eye (L); *dotted line* = right eye (R).

(tuned and reciprocal), but do not give information on how the brain handles the problem of false matches that may occur during normal binocular vision.

An attempt to understand this aspect of the analysis of stereopsis was made by studying the response of cortical neurons to dynamic random-dot stereograms. Using this powerful form of "cyclopean" stimulation[1], Julesz (1960) conclusively demonstrated that correct stereoscopic matches occur quite early in the processing of visual information, and that basic stereoscopic operation probably precedes form recognition. Thus, it was reasonable to speculate that the neuronal mechanisms for the resolution of ambiguities and for determining the correct correspondence associated with depth should be present at early stages of binocular interaction. In 20% of a sample of 230 neurons whose depth sensitivity was examined both with line/bar stereograms and with (dynamic) cyclopean random-dot stereograms, the latter evoked evident responses, both from disparity selective neurons as well as from neurons responding equally well to all disparities. In most (but not all) instances, the depth response profiles for line patterns and random-dot patterns were qualitatively the same (Poggio 1980, 1984).

The proportion of cells responding to cyclopean random-dot stereograms is not significantly different among the tuned (excitatory and inhibitory) or reciprocal (near and far) neurons. Certain functional properties of the cell, such as eye dominance and directionality, are qualitatively the same with line patterns as with random-dot patterns. For some cells, however, other properties, stimulus orientation and size selectivities in particular, are different for the two types of stimuli in that the response to cyclopean random-dot patterns may show little or no selectivity for the orientation of the monocularly invisible figure and/or its size.

At variance with what obtains with disparity unambiguous line/bar stimuli, an evident correlation exists between sensitivity to random-dot stimulation and simple/complex receptive field organization in that essentially all cells that responded to cyclopean stereograms are complex cells (Poggio 1984). This finding is particularly significant for it assigns to the complex cell the unique capacity of solving the correspondence problem by responding to the "correct" binocular matches over the receptive fields in the two eyes.

CONCLUSION: TOWARD AN INTEGRATED VIEW

The foregoing account of the results of computational, psychophysical, and physiological analyses of stereopsis shows that in recent years, these three

[1]The term *cyclopean* identifies that form of visual stimulation with patterns such as random-dot stereo pairs that are visible only when the brain combines the inputs from the two eyes, patterns that do not exist "physically" under monocular vision (Julesz 1971).

approaches have converged on similar problems. It is now possible to identify a common ground, and for the information gained with one approach to contribute to an explanation of the findings of another. The processing involved in stereopsis includes matching corresponding points of the images in the two eyes, measuring their disparity, and recovering the 3-D structure of the objects seen by the viewer. In what follows, we attempt to discuss some aspects of this processing that may be considered from an integrated perspective.

The correspondence problem has dominated much of the recent computational and psychophysical investigations. As unequivocally shown by the psychophysics with random-dot stereograms, during everyday, normal binocular vision, the brain solves the correspondence problem of stereopsis with amazing ease, precision, and immunity to noise. Potentially numerous false matches do not disturb the performance of our visual system. Existing algorithms are still far from solving the correspondence problem as effectively as our visual system, but suggest ways of how it may be solved at all. Recently, cells have been found in the visual cortex of the macaque that signal the correct disparity of dynamic random-dot stereograms in which there are many possible false matches. These neurons reflect the results of the correspondence process; they are at least one of the neural correlates of "global" stereopsis (Julesz 1971). The computational approach strongly suggests that these neurons must play a critical role in the perception of depth, not only from random-dot stereograms, but most importantly, in natural images. Of singular importance for the understanding of the organization of the visual system, these neurons are found at the earliest stages of binocular interaction in cortical area 17. This discovery suggest new insights about the brain mechanisms of stereoscopic matching.

In terms of the correspondence problem, we ask first what are the matching primitives for the stereoscopic neurons or, in other words, what are the significant "measurements" of the visual images on which the neuronal matching process operates. Second, we ask whether these neurons exploit global matching constraints of the type discussed above such as continuity and uniqueness.

The results of neurophysiological experiments show that neurons in monkey striate and prestriate cortex are sensitive to binocular disparity. When presented with disparity unambiguous stimuli, such as isolated line contours of optimal size and orientation, the majority of cortical neurons, both simple and complex, display disparity selectivity. The evidence suggests that under these conditions cortical neurons measure binocular disparity by receptive field disparity. For each cell a response profile can be constructed along the disparity domain over a range centered on the functionally optimal spatial superposition of the two monocular fields in space, and extending over a range of disparities that depends on the extent of field superposition and on receptive field size (Poggio

& Talbot 1981). In the cat, and presumably in the macaque as well, receptive field width and the width of the disparity tuning curve correlate strongly (Pettigrew et al 1968, Ferster 1981), as expected by some of the models of stereopsis described above. These stereoscopic cells then may be thought to use matching primitives that correspond to isolated oriented edges.

When, on the other hand, the cortical binocular neuron is presented with disparate figures embedded in or as part of a textured pattern, such as occurring in random-dot stereograms (and commonly in natural images as well), only about one out of five of the binocular cells in primary visual cortex, A17, is capable of detecting disparity. A major functional difference is that the simple cells do not and the complex cells do respond to texture patterns without sharp, oriented monocular edges (Hammond & MacKay 1977, Burr et al 1981, Poggio 1984). It seems unlikely that the complex cells require for activation isolated "edges" or zero-crossings of a certain orientation in their receptive field. Indeed, neurons have been found early in the cortical stereoprocessor (in area 17) that are orientation insensitive for cyclopean random-dot stereograms. Under these stimulus conditions, simple positional disparity between the receptive fields in the two eyes does not appear to be a sufficient mechanism for disparity detection of the correct matches. At these cortical neurons, binocular matching appears to be spatially very precise, for these cells are capable of signaling the correct disparity of a few dots whose positions over the cell's receptive fields change 100 times a second (Poggio 1984). A possible arrangement for these neurons could be based on sets of discrete and numerous receptive sites, or subfields, in the receptive field in one eye, "gated" with a similar set of appropriate positional disparity in the other eye. False matches would be avoided by having cells with subfield disparity of the order of the associated precortical (or IVc) monocular receptive field's size, and not larger. This scheme is computationally similar to the correlation of some function of the filtered image over the area defined by the cortical receptive field. Another conjecture on the neural organization for recognizing correct correspondence would regard the stereoscopic neuron matching a large set of local monocular measurements on the image such as, for example, derivatives of image intensities. In this case, one would expect afferents to the cell to carry in parallel the different types of measurement taken at each location in the visual field. The disparity range may easily be larger than in the neural network hypothetized previously (see section on the Correspondence Problem).

In both cases, a key question is how the activity of the binocularly interacting subfields, or measurements, is "integrated" over the area of the cortical receptive field. Candidate operations range from an approximately linear summation to an almost logical combination of the binocular matches. Data from responses of cortical cells to random-dot patterns with different degrees of binocular correlation could provide information of the characteristics of the integration.

Difficulties with matching schemes involving linear summation over relatively large areas (like the receptive fields of complex cells) are well known (Baker 1982). Disparity discontinuities—like the border of a figure in a random-dot stereogram—are neither precisely nor easily detected in this way. Thus, one wonders whether the complex cells' sensitivity to random-dot stereograms may also represent the substrate for the psychophysical percept of sharp boundaries or whether other stereo mechanisms are required.

Little can be said about the neurophysiological correlates of "global" matching constraints. Constraints like continuity and uniqueness are enforced by the mechanisms suggested above. Moreover, a direct implementation of the constraints would also be possible, for example in terms of inhibitory interactions among cells tuned to different disparities, and with overlapping receptive field locations, but there is no physiological evidence either in favor or against it.

A set of complex cells, therefore, seems to represent at least an initial solution to the correspondence problem. But what, then, is the role of disparity sensitive simple cells? The receptive field of simple cells suggests that oriented edges or bars (possibly oriented zero-crossings or peaks) are their matching primitives. Simple cells, however, do not respond to random-dot stereograms. Is it possible that their disparity range and their input organization would make them particularly susceptible to false matches? And that this is the reason they are shut off in situations in which there is a large number of potential false matches, as in random-dot stereograms? One may think that in the presence of textured patterns that pose a severe correspondence problem, simple cells are suppressed monocularly (Burr et al 1981) and binocularly, possibly by complex cells. What would then be their function? Psychophysical as well as computational findings suggest that monocular contours play an important role in driving eye movements for registering the two images. Isolated, oriented, sharp zero-crossing (or edges) are in addition useful matching primitives, as shown by Mayhew & Frisby (1981), both computationally and psychophysically. Simple cells may represent and match monocularly strong isolated borders with a given orientation and their output may be used in the matching process and in the control of vergence movements (see also Marr & Poggio 1979, figure 6).

Other problems also have been brought into sharper focus by the interaction between different approaches, for instance, the role of vertical disparity in stereopsis and the control of eye movements. From the computational perspective, vertical disparity and more generally geometric distortions pose some of the hardest problems for stereoscopic matching. It would be interesting to know the extent to which the complex cells responding to random-dot stereograms can tolerate vertical disparity, and to compare it with psychophysical data and with the performance of different algorithms. Since registration of the two eyes seems so important, it would be natural to expect cells tuned to vertical disparities, especially Near/Far with simple receptive field organization (because monocular contours should play a prominent role in registration), as well as Tuned cells with peak at zero disparity, to have an important role in controlling eye movements.

In addition to the solution of the correspondence problem, the computation of stereoscopic information entails the recovery of the 3-D structure of the visual scene from binocular disparity and knowledge of the fixation distance (for example, from angle of vergence and gaze) or assessment of vertical disparity (Mayhew & Longuet-Higgins 1982). Psychophysical observations have shown that brains do perform this computation, though approximately, but nothing is known of the physiological mechanisms underlying it. Neurons or, more likely, neuronal networks sensitive to viewer-centered coordinates and distances (as opposed to retino-centered coordinates) are expected to be involved. This aspect of stereoscopic analysis is likely to be beyond the initial stages of visual processing and is possibly organized within a nontopographic representation of the visual field.

Many other questions are readily suggested by any attempt to connect the computational point of view with the physiological and psychophysical data. Among the most obvious ones, a general problem is whether cells represent disparity directly or rather the relative disparity with respect to another point in the image. The question is motivated by the apparent stability of the cortical neuron's response in the presence of small involuntary fixation disparities (Poggio 1984) as well as by the psychophysical findings of Westheimer & Mitchison on stereoacuity (see section on Stereoacuity). An associated question concerns the possible existence of a cortical shift mechanism similar to vergence movements over a small range of disparities (less than ten degrees); in computer algorithms, these two possibilities are indistinguishable.

In conclusion: The interaction of the different approaches to stereopsis promises to be very fruitful for understanding both the mechanisms and the information processing aspects of binocular vision. The first effect of an integrated approach is a number of new questions and a recasting of older ones in a different and more meaningful form. Work on algorithms for stereopsis is helping to clarify what the central issues are in terms of information processing; physiology and psychophysics will tell us how and where these problems are solved. Stereopsis is an early but very difficult state of vision; it provides, we believe, an unique opportunity for a combined understanding of information processing and of neural mechanisms. We expect that future research on stereopsis will show that the attempt of understanding human vision and of developing computer vision can fruitfully interact.

ACKNOWLEDGMENTS

We wish to thank the Neuroscience Research Institute of the Neuroscience Research Program for the kind hospitality and for the facilities provided to us during the writing of this review.

We are grateful to E. Grimson, M. Kass, and especially to K. Nishihara and K. Nielsen for many helpful discussions.

This work was supported by NIH grant EYO2966 (G. F. P.), by the Advanced Research Project Agency of the Department of Defense under Office of Naval Research contract N00014–75–C–0643 to the Artificial Intelligence Laboratory of the Massachusetts Institute of Technology (T. P.), and by NIH grant EYO4206 (T. P.).

Literature Cited

Arnold, R. D. 1982. *Automated stereo perception*. PhD thesis. Stanford Univ.

Arnold, R. D., Binford, T. O. 1980. Geometric constraints in stereo vision. *SPIE J.* 238:281–92

Baker, H. H. 1980. Edge-based stereo correlation. *Proc. ARPA Image Understanding Workshop*, Univ. Md., pp. 168–75

Baker, H. H. 1982. *Depth from edge- and intensity-based stereo*. PhD thesis. Univ. Ill., Urbana, IL

Baker, H. H. Binford, T. O. 1981. Depth from edge- and intensity-based stereo. *Proc. 7th Intern. Joint Conf. on Artif. Intell.* Vancouver, British Columbia, pp. 631–36

Barlow, H. 1979. Reconstructing the visual image in space and time. *Nature* 279:189–90

Barlow, H., Blakemore, C., Pettigrew, J. D. 1967. The neural mechanism of binocular depth discrimination. *J. Physiol.* 193:327–42

Birch, E. E., Gwiazda, J., Held, R. 1982. Stereoacuity development for crossed and uncrossed disparities in human infants. *Vision Res.* 22:507–13

Bishop, P. O., Henry, G. H., 1971. Spatial vision. *Ann. Rev. Psychol.* 22:119–61

Bishop, P. Q., Henry, G. H., Smith, C. J. 1971. Binocular interaction fields of single units in the cat striate cortex. *J. Physiol.* 216:39–68

Bough, E. W. 1970. Stereoscopic vision in the macaque monkey: A behavoural demonstration. *Nature* 225:42–44

Braddick, O. J. 1979. Binocular single vision and perceptual processing. *Proc. R. Soc. London Ser. B* 201:503–12

Burr, D. C. 1979. Acuity for apparent Vernier offset. *Vision Res.* 19:835–37

Burr, D. C., Ross, J. 1979. How does binocular delay give information about depth? *Vision Res.* 19:523–32

Burr, D. C., Morrone, C., Maffei, L. 1981. Intra-cortical inhibition prevents simple cells from responding to textured visual patterns. *Exp. Brain Res.* 43:455–58

Burt, P., Julesz, B. 1980a. A disparity gradient limit for binocular fusion. *Science* 208:615–17

Burt, P., Julesz, B. 1980b. Modifications of the classical notion of Panum's fusional area. *Perception* 9:671–82

Clarke, P. G. H., Donaldson, I. M. L., Whitteridge, D. 1976. Binocular visual mechanisms in cortical areas I and II of the sheep. *J. Physiol.* 256:509–26

Crick, F., Marr, D., Poggio, T. 1981. An information-processing approach to understanding the visual cortex. In *The Organization of the Cerebral Cortex*, ed. F. O. Schmitt, F. G. Worden, G. Edelman, S. G. Dennis, pp. 505–33. Cambridge, Mass: MIT Press

Dev, P. 1975. Perception of depth surfaces in random-dot stereograms: A neural model. *Int. J. Man-Machine Studies* 7:511–14

DeValois, R. L., DeValois, K. K. 1980. Spatial vision, *Ann. Rev. Psychol.* 31:309–41

Diner, D. 1978. *Hysteresis in human binocular fusion: A second look*. PhD thesis. Calif. Inst. Technol., Pasadena

Duwaer, A. L. 1981. *Binocular Single Vision: Psychophysical Studies on Underlying Sensory and Motor Processes*. Rotterdam: Delfta Univ. Press. 173 pp.

Duwaer, A. L., van den Brink, G. 1981. Diplopia thresholds and the initiation of vergence eye movements. *Vision Res.* 21:1727–37

Duwaer, A. L., van den Brink, G. 1982. Detection of vertical disparities. *Vision Res.* 22:467–78

Felton, B., Richards, W., Smith, A. Jr. 1972. Disparity processing of spatial frequencies in man. *J. Physiol.* 225:319–62

Fender, D., Julesz, B. 1967. Extension of Panum's fusional area in binocularly stabilized vision. *J. Opt. Soc. Am.* 57:819–30

Ferster, D. 1981. A comparison of binocular depth mechanisms in areas 17 and 18 of the cat visual cortex. *J. Physiol.* 311:623–55

Fischer, B., Krueger, J. 1979. Disparity tuning and binocularity of single neurons in the cat visual cortex. *Exp. Brain Res.* 35:1–8

Fischer, B., Poggio, G. F. 1979. Depth sensitivity of binocular cortical neurons of behaving monkeys. *Proc. R. Soc. London Ser. B* 204:409–14

Foley, J. M. 1980. Binocular distance perception. *Psych. Rev.* 87(5):411–34

Fox, R., Aslin, R. N., Shea, S. L., Dumais, S. T. 1979. Stereopsis in human infants. *Science* 207:323–24

Frisby, J. P., Mayhew, J. E. W. 1977. Global processes in stereopsis: Some comments on Ramachandran and Nelson (1976). *Perception* 6:195–206

Frisby, J. P., Mayhew, J. E. W. 1978. Contrast sensitivity function for stereopsis. *Perception* 7:423–29

Gennery, D. B. 1980. Modeling the environment of an exploring vehicle by means of stereo vision. *Stanford Artif. Intell. Lab. Memo* 339

Gillam, B., Lawergren, B. 1983. The induced effect, vertical disparity, and stereoscopic theory. *Percept. Psychophys.* 34:121–30

Grimson, W. E. L. 1981. *From Images to Surfaces: A Computational Study of the Human Early Visual System.* Cambridge, Mass: MIT Press. 274 pp.

Hammond, P., MacKay, D. M. 1977. Differential responsiveness of simple and complex cells in cat striate cortex to visual texture. *Exp. Brain Res.* 30:275–96

Harwerth, R. S., Rawlings, S. C. 1977. Viewing time and stereoscopic threshold with random-dot stereograms. *Am. J. Optom. Phys. Opt.* 54(7):452–57

Held, R. 1983. Binocular Vision . . . behavioral and neuronal development. In *Neonate Cognition: Beyond the Blooming, Buzzing Confusion,* ed. J. Mehler, R. R. Fox. Hillsdale, N.J.: Erlbaum. In Press

Held, R., Birch, E., Gwiazda, J. 1980. Stereoacuity of human infants. *Proc. Nat. Acad. Sci. USA* 77:5572–74

Hubel, D. H., Wiesel, T. N. 1970. Cells sensitive to binocular depth in area 18 of the macaque monkey cortex. *Nature* 225:41–42

Hubel, D. H., Wiesel, T. N. 1973. A reexamination of stereoscopic mechanisms in the cat. *J. Physiol.* 232:29–30P

Jones, R. 1977. Anomalies of disparity in the human visual system. *J. Phys.* 264:621–40

Jones, R., Kerr, K. E. 1972. Vergence eye movements to pairs of disparity stimuli with shape selection cues. *Vision Res.* 12:1425–30

Joshua, D. E., Bishop, P. O. 1970. Binocular single vision and depth discrimination. Receptive field disparities for central and peripheral vision and binocular interaction on peripheral units in cat striate cortex. *Exp. Brain Res.* 10:389–416

Julesz, B. 1960. Binocular depth perception of computer-generated patterns. *Bell System Tech. J.* 39:1125–62

Julesz, B. 1971. *Foundations of Cyclopean Perception.* Chicago: Univ. Chicago Press. 406 pp.

Julesz, B. 1978. Global stereopsis: Cooperative phenomena in stereoscopic depth perception. In *Hand. Sensory Physiol.* 8:215–52

Julesz, B., Miller, J. 1975. Independent spatial-frequency-tuned channels in binocular fusion and rivalry. *Perception* 4:125–43

Julesz, B. Schumer, R. A. 1981. Early visual perception. *Ann. Rev. Psychol.* 32:575–627

Julesz, B., Spivack, G. J. 1967. Stereopsis based on vernier acuity cues alone. *Science* 157:563–65

Kass, M. 1983. A computational framework for the visual correspondence problem. *Proc. ARPA Image Understanding Workshop, Washington DC*

Kidd, A. L., Frisby, J. P., Mayhew, J. E. W. 1979. Texture contours can facilitate stereopsis by initiating vergence eye movements. *Nature* 280:829–32

Kroll, J. D., van der Grind, A. 1980. The double-nail illusion: Experiments on binocular vision with nails, needles and pins. *Perception* 9:651–69

Kulikowski, J. J. 1978. Limit of single vision in stereopsis depends on contour sharpness. *Nature* 275:126–27

Levinson, E., Blake, R. 1979. Stereopsis by harmonic analysis. *Vision Res.* 19:73–78

Logan, B. F. Jr. 1977. Information in the zero crossings of bandpass signals. *Bell System Tech. J.* 66:487–510

Longuet-Higgins, H. C. 1981. A computer algorithm for reconstructing a scene from two projections. *Nature* 293:4133–35

Marr, D. 1976. Early processing of visual information. *Philos. Trans. R. Soc. London Ser. B* 275:483–524

Marr, D., Hildreth, E. 1980. Theory of edge detection. *Proc. R. Soc. London Ser. B* 207:187–217

Marr, D., Palm, G., Poggio, T. 1978. Analysis of a cooperative stereo algorithm. *Biol. Cyber.* 28:223–39

Marr, D., Poggio, T. 1976. Cooperative computation of stereo disparity. *Science* 194:283–87

Marr, D., Poggio, T. 1979. A computational theory of human stereo vision. *Proc. R. Soc. London Ser. B* 204:301–28

Marr, D., Poggio, T. 1980. Some comments on a recent theory of stereopsis. *MIT Artif. Intell. Memo* 558

Mayhew, J. E. W. 1983. Stereopsis. In *Physiological and Biological Processing of Images,* ed. O. J. Braddick, A. C. Sleigh. Berlin: Springer-Verlag

Mayhew, J. E. W., Frisby, J. P. 1978a. Stereopsis masking in humans is not orientationally tuned. *Perception* 7:431–36

Mayhew, J. E. W., Frisby, J. P. 1978b. Contrast summation effects and stereopsis. *Perception* 7:537–50

Mayhew, J. E. W., Frisby, J. P. 1979a. Surfaces with steep variations in depth pose difficulties for orientationally tuned disparity filters. *Perception* 8:691–98

Mayhew, J. E. W., Frisby, J. P. 1979b. Convergent disparity discriminations in narrow-band-filtered random-dot stereograms. *Vision Res.* 19:63–71

Mayhew, J. E. W., Frisby, J. P. 1980. The computation of binocular edges. *Perception* 9:69–86

Mayhew, J. E. W., Frisby, J. P. 1981. Psychophysical and computational studies towards a theory of human stereopsis. *Artif. Intell.* 16:349–85

Mayhew, J. E. W., Longuet-Higgins, C. 1982. A computational model of binocular depth perception. *Nature* 297:376–78

Mitchell, D. E. 1966. Retinal disparity and diplopia. *Vision Res.* 6:441–51

Moravec, H. P. 1980. Obstacle avoidance and navigation in the real world by seeing a robot rover. *Stanford Artif. Intell. Lab. Memo* 340

Morgan, M. J. 1975. Stereoillusion based on visual persistence. *Nature* 256:639–40

Morgan, M. J. 1976. Pulfrich effect and the filling-in of apparent motion. *Perception* 5:187–95

Motter, B. C., Poggio, G. F. 1981. The three-dimensional locus of fixation of the rhesus monkey. *Invest. Ophthalmol. Vis. Sci. Suppl.* 20:27

Motter, B. C., Poggio, G. F. 1982. Spatial invariance of receptive field location in the presence of eye movements of fixation for neurons in monkey striate cortex. *Soc. Neurosci. Abstr.* 8:707

Mowforth, P., Mayhew, J. E. W., Frisby, J. P. 1981. Vergence eye movements made in response to spatial-frequency-filtered random-dot stereograms. *Perception* 10:299–304

Nelson, J. I. 1975. Globality and stereoscopic fusion in binocular vision. *J. Theor. Biol.* 49:1–88

Nelson, J. I. 1977. The plasticity of correspondence: After-effects, illusion and horopter shifts in depth perception. *J. Theor. Biol.* 66:203–66

Nielson, K. R. K., Poggio, T. 1983. Vertical image registration in human stereopsis. *MIT Artif. Intell. Memo* 743

Nikara, T., Bishop, P. O., Pettigrew, J. D. 1968. Analysis of retinal correspondence by studying receptive fields of binocular signal units in cat striate cortex. *Exp. Brain Res.* 6:353–72

Nishihara, H. K., Poggio, T. 1982. Hidden cues in random line stereograms. *Nature* 300:347–49

Petrig, B., Julesz, B., Kropfl, W., Baumgartner, G., Anliker, M. 1981. Development of stereopsis and cortical binocularity in human infants: Electrophysiological evidence. *Science* 213:1402–5

Pettigrew, J. D., Konishi, M. 1976. Neurons selective for orientation and binocular disparity in the visual Wulst of the barn owl (*Tyto alba*). *Science* 193:675–78

Pettigrew, J. D., Nikara, T., Bishop, P. O. 1968. Binocular interaction on single units in cat striate cortex: Simultaneous stimulation by single moving slits with receptive fields in correspondence. *Exp. Brain Res.* 6:391–410

Poggio, G. F. 1980. Neurons sensitive to dynamic random-dot stereograms in areas 17 and 18 of the rhesus monkey cortex. *Soc. Neurosci. Abstr.* 6:672

Poggio, G. F. 1984. Processing of stereoscopic information in primate visual cortex. In *Dynamic Aspects of Neocortical Function,* ed. G. Edelman, W. M. Cowan, W. E. Gall. New York: Wiley. In press

Poggio, G. F., Fisher, B. 1977. Binocular interaction and depth sensitivity of striate and prestriate cortical neurons of the behaving rhesus monkey. *J. Neurophysiol.* 40:1392–405

Poggio, G. F., Talbot, W. H. 1981. Mechanisms of static and dynamic stereopsis in foveal cortex of the rhesus monkey. *J. Physiol.* 315:469–92

Rashbass, C., Westheimer, G. 1961a. Disjunctive eye movements. *J. Physiol.* 159:339–60

Rashbass, C., Westheimer, G. 1961b. Independence of conjunctive and disjunctive eye movements. *J. Physiol.* 159:361–64

Richards, W. 1970. Stereopsis and stereoblindness. *Exp. Brain Res.* 10:380–88

Richards, W. 1971. Anomalous stereoscopic depth perception. *J. Opt. Soc. Am.* 61:410–14

Riggs, L. A., Niehl, E. W. 1960. Eye movements recorded during convergence and divergence. *J. Opt. Soc. Am.* 50:913–20

Ross, J., Hogben, J. H. 1975. Short-term memory in stereopsis. *Vision Res.* 14:1195–1201

Sarmiento, R. F. 1975. The stereoacuity of macaque monkey. *Vision Res.* 15:493–98

Schiller, P. H., Finlay, B. L., Volman, S. 1976. Quantitative studies of single-cell properties in monkey striate cortex. II. Orientation specificity and ocular dominance. *J. Neurophysiol.* 39:1320–33

Schor, C. M., Tyler, C. W. 1981. Spatiotemporal properties of Panum's fusional area. *Vision Res.* 21:683–92

Schumer, R. A., Ganz, L. 1979. Independent stereoscopic channels for different extents of spatial pooling. *Vision Res.* 19:1303–44

Schumer, R. A., Julesz, B. 1980. Maximum disparity limit for detailed depth resolution. *Invest. Ophthalmol. Visual Sci.* 19:106–7 (Suppl.)

Schumer, R. A., Julesz, B. 1982. Disparity limits in bandpass random-grating sterograms. *Invest Ophthalmol. Visual Sci.* 22:272 (Suppl.)

Sperling, G. 1970. Binocular vision: A physical and a neural theory. *Am. J. Psychol.* 83:461–534

St. Cyr, G., Fender, D. H. 1969. The interplay of drifts and flicks in binocular fixation. *Vision Res.* 9:245–65

Teller, D. Y. 1981. The development of visual acuity in human and monkey infants. *Trends NeuroSci.* 4:21–24

Tsai, R. Y., Huang, T. S. 1981. Uniqueness and estimation of three-dimensional motion parameters of rigid objects with curved surfaces. *Tech. Rep. R921, Coordinated Sci. Lab., Univ. Ill., Urbana.* 60 pp.

Tyler, C. W. 1973. Stereoscopic vision: Cortical limitations and a disparity scaling effect. *Science* 181:276–78

Tyler, C. W. 1974. Depth perception in disparity gratings. *Nature* 251:140–42

Tyler, C. W. 1975a. Spatial organizations of binocular disparity sensitivity. *Vision Res.* 15:583–90

Tyler, C. W. 1975b. Spatial limitations of human stereoscopic vision. *SPIE J.* 120:36–42

Tyler, C. W. 1977. Stereomovement from interocular delay in dynamic visual noise: A random spatial disparity hypothesis. *Am. J. Optom. Phys. Opt.* 54(6):374–86

Tyler, C. W., Julesz, B. 1980. On the depth of the cyclopean retina. *Exp. Brain Sci.* 40:192–202

Uttal, W. R., Fitzgerald, J., Eskin, T. E. 1975. Parameters of tachistoscopic stereopsis. *Vision Res.* 15:705–23

von der Heydt, R., Adorjani, C., Hanny, P., Baumgartner, G. 1978. Disparity sensitivity and receptive field incongruity of units in the cat striate cortex. *Exp. Brain Res.* 31:423–545

Westheimer, G. 1978. Vertical disparity detection: Is there an induced size effect? *Invest. Ophthalmol. Visual Sci.* 17:545–51

Westheimer, G. 1979. Cooperative neural processes involved in stereoscopic acuity. *Exp. Brain Res.* 36:585–97

Westheimer, G., McKee, S. P. 1978. Stereoscopic acuity for moving retinal images. *J. Opt. Soc. Am.* 68:450–55

Westheimer, G., McKee, S. P. 1980a. Stereogram design for testing local stereopsis. *Invest. Ophthalmol. Visual Sci.* 19:802–9

Westheimer, G., McKee, S. P. 1980b. Stereoscopic acuity with defocused and spatially filtered retinal images. *J. Opt. Soc. Am.* 70(7):772–78

Westheimer, G., Mitchell, D. E. 1969. The sensory stimulus for disjunctive eye movements. *Vision Res.* 9:749–55

Wilson, H. R., Bergen, J. R. 1979. A four mechanism model for threshold spatial vision. *Vision Res.* 19:19–32

Ann. Rev. Neurosci. 1984. 7:413–42
Copyright © 1984 by Annual Reviews Inc. All rights reserved

GONADAL STEROID INDUCTION OF STRUCTURAL SEX DIFFERENCES IN THE CENTRAL NERVOUS SYSTEM

Arthur P. Arnold

Department of Psychology and Laboratory of Neuroendocrinology, Brain Research Institute, University of California, Los Angeles, California 90024

Roger A. Gorski

Department of Anatomy and Laboratory of Neuroendocrinology, Brain Research Institute, University of California, School of Medicine, Los Angeles, California 90024

INTRODUCTION

In the last decade, we have experienced a small revolution in the study of the neural and hormonal bases of sex differences in behavior and reproductive function. The previous view had been that these sex differences might have an anatomical basis, but anatomical differences in neural circuits were likely to be small and subtle. This view stemmed from the observation that many behaviors typical of one sex (e.g. lordosis behavior in female rodents) could also be observed to a lesser extent in the opposite sex, as long as the proper hormonal milieu was present. Thus, both sexes must possess the neural circuitry needed for the behavior, and the quantitative (not qualitative) differences between the sexes seemed potentially explained by subtle differences in the microtopography of synaptic relations among neurons, in the metabolism of neurotransmitters, or in steroid receptor mechanisms in neurons. Today, the idea of significant structural sex differences in the CNS is accepted. We now recognize that gonadal steroids can exert profound effects on the morphogenesis and survival of specific neurons resulting in marked sex differences in CNS structure.

We begin by reviewing briefly the history of this field since 1930. We then discuss at length three anatomical dimorphisms that have played a dominant

413

0147-006X/84/0301-0413$02.00

role in shaping our thinking about sexual differentiation of the CNS. We compare these systems, and discuss the implications of work on each. Finally, we suggest that gonadal steroids can be used profitably as tools to perturb certain neural systems in known ways, and thus aid in our understanding of how steroids influence certain developmental processes, and increase our understanding of those processes *per se*.

HISTORICAL PERSPECTIVE

The study of sexual differentiation of the brain can be divided into four phases.

1. The first, or "phenomenological" period, was characterized by a few seminal studies that set the stage for the next phase. In the late 1930s and early 1940s, the results of studies by Pfeiffer (1936), Wilson et al (1940, 1941), Wilson (1943), and Danchakoff (see Young et al 1964) implied that when female rats were exposed to testicular secretions during development, they would fail to ovulate as adults, and would have an increased tendency to show masculine copulatory behaviors and a decreased tendency to show feminine copulatory behaviors, compared to unexposed females. Similarly, although ovaries transplanted into the anterior chamber of the eyes of adult male rats would develop mature follicles, these would not ovulate unless such a male was first deprived of his gonads on the day of birth. Pfeiffer (1936) concluded that the pituitary, which controls ovulation, differs between males and females, and that this difference is induced by gonadal secretions postnatally. With the realization in the late 1940s and 1950s that gonadotropin secretion of the pituitary is under neural control, there developed the idea that the brain may be the site of sex differences in control of behavior and gonadotropin secretion. The studies of Phoenix et al (1959), Barraclough & Gorski (1961), and Harris & Levine (1962) were the first to suggest clearly that gonadal steroids act directly on the brain. Phoenix et al (1959) proposed that steroids can act in two fundamentally different modes: organizational and activational. Sex steroids act early in development to *organize* neural pathways responsible for reproductive behavior, and these differentiating effects of steroids are permanent. Later in adulthood, steroids act on the differentiated pathways to *activate* behaviors or functions thus previously organized. The activational effects are reversible and transient. The organizational effects of steroids lead to sex differences in the brain and behavior, and the activational effects often regulate behaviors and neural circuits that are sexually dimorphic.

2. The shift in interest to direct neural effects of steroids ushered in the second phase in the study of sexual differentiation of brain and sexual function. During the 1960s and 1970s, a great number of studies clarified when gonadal steroids act to organize sexual function, which steroids are effective, and how steroid hormones act in terms of molecular mechanisms. These studies con-

firmed that testicular androgens control the development of masculine and feminine copulatory behaviors. If testosterone is present in the blood during an early critical period (either from an endogenous testis in a normal male, or from injections into a female), the animal will be permanently masculinized and defeminized. Masculinization refers to the organization (i.e. permanent facilitation) of male qualities such as the capacity to show high levels of masculine copulatory behaviors or any other characteristics that are typical of males. Defeminization refers to the lack of development of feminine characteristics such as the tendency to show high levels of feminine copulatory behaviors (e.g. lordosis) in adulthood, and the capacity to show ovulatory surges of gonadotropin.

The timing of steroid action is critical. For example, in rats androgens act prenatally to masculinize sexual behavior without exerting a defeminizing effect (Ward 1969). In contrast, steroid effects on patterns of gonadotropin secretion and feminine sexual behavior are primarily postnatal, although lower doses of steroids disrupt gonadotropin secretion compared with those doses needed to defeminize copulatory behavior (Gorski 1971). Although the timing of the critical periods for masculinization and defeminization depend on the doses used (Gorski 1968), in general permanent steroid effects are not typically seen after about the tenth postnatal day in rats, even when implanted intracerebrally (Lobl & Gorski 1974). There are some exceptions to this rule (Kawashima 1960, Brown-Grant 1975, Harlan & Gorski 1978). One important conclusion drawn from the studies on the timing of permanent steroid effects is that there is more than one critical period. The independence of steroid effects on different differentiating processes (masculinization, defeminization) suggests that steroids independently influence different developmental processes in the brain (Whalen 1981). This conclusion is echoed below in the context of neuroanatomical studies.

Of the conclusions drawn from studies in the second phase of this history, perhaps the most surprising is that estradiol mimics the differentiating effects of testosterone. In fact, estradiol is more potent than testosterone in preventing development of feminine patterns of gonadotropin secretion and behavior (Gorski & Wagner 1965, Sutherland & Gorski 1972). The hypothesis that prevails at present to explain this estradiol effect is that in the neonatal male, testosterone of testicular origin is metabolized (aromatized) in the brain to estradiol, which is at least one of the hormone species responsible for the differentiating effects of testosterone (Lieberburg et al 1977, Goy & McEwen 1980). However, this aromatization hypothesis challenged the general concept of steroid-controlled sexual differentiation of the brain when it was discovered that plasma levels of estrogen are remarkably high during the postnatal period in both male and female rats (Weisz & Gunsalus 1973, Ojeda et al 1975). Why should exogenous estrogen defeminize females, yet normal neonatal females are not defeminized by estradiol secreted by their ovaries? This paradox was in turn apparently resolved by the suggestion that alpha feto-protein, an estrogen-binding plasma protein present in neonatal rats, binds plasma estrogen and prevents its entry into the brain (Nunez et al 1971, Raynaud et al 1971, Plapinger et al 1973). In contrast, testosterone passes into the brain, where it is subsequently aromatized to estradiol, which is necessary for some defeminizing and masculinizing effects of testosterone. However, not all differentiating effects of androgen can be explained by aromatization (e.g. Baum et al 1982; see below), and androgens themselves may play an important role.

Finally, a wealth of studies has clarified the molecular mechanisms of steroid action (McEwen 1976). In general, gonadal steroid action is thought to require entry of the steroid into neurons or other cells in the brain. Specific brain regions contain cells that have cytoplasmic receptors specific for androgens, estrogens, or progestins. These cytoplasmic proteins bind the appropriate steroid, and this steroid-receptor complex then moves to the nucleus, where the steroid is thought to interact with DNA to regulate the production of specific RNA and protein species that alter cell function. It is far from clear precisely which cell functions are modulated developmentally. However, the recognition of the importance of steroid receptors has led to a great many studies that have established the biochemical properties of these receptors, their specificity and topography. This has focused our attention in developmental studies on brain regions that contain high levels of steroid receptors, or that accumulate steroids as determined in autoradiographic studies.

3. The second phase in the study of sexual differentiation is not yet over; many important studies have yet to be performed on the timing and nature of steroid effects on behavior or reproductive function. Yet we have already begun the third phase, involving anatomical measures of the action of steroids in the developing CNS. The beginning of phase 3 is perhaps most accurately marked by the study of Raisman & Field (1973), who found a sex difference in the types of dendritic synapses in the rat preoptic area—a difference that could be reversed appropriately by neonatal manipulations of androgen levels. This study may well have influenced the attitude of the scientific community by suggesting the possibility of morphological sex differences in the brain. In the late 1970s came several reports, in rapid succession, of large anatomical dimorphisms in the brain. Nottebohm & Arnold (1976) found a large sex difference in the volume of brain regions controlling vocal behavior in two passerine bird species. In these species, males sing and females do not, and the males possess certain vocal nuclei that are as much as six times larger than the females'. Gorski et al (1978) found that a distinct nucleus within the preoptic area of the rat (now called the sexually dimorphic nucleus of the preoptic area, SDN-POA) was much larger in males than females. Breedlove & Arnold (1980) discovered sex differences in the number of motor neurons innervating perineal muscles in rats. These studies have allowed investigators to focus their

attention on dimorphic brain regions to measure at a cellular level the perinatal influences of steroid hormones. To date, most such studies have used light microscopic techniques to determine what cellular features are controlled by steroids at different times of life. Perhaps because these discoveries have come at a time characterized by a great proliferation in the kinds of light microscopic analyses available, the use of the light microscope gives considerable power in analyzing such properties as the number of dimorphic neurons, their histochemical characteristics, projections, dendritic morphology, connectivity, and steroid accumulation. Consideration of these light microscopic studies forms the bulk of our discussion below.

4. The fourth phase in the study of sexual differentiation of the brain will begin in the next few years as a direct outgrowth of the present anatomical studies. It can be termed the "molecular phase," since we believe new dependent measures will be increasingly used to understand at the molecular level what genes and gene products are regulated by steroid hormones at various stages of life. In particular, new recombinant DNA techniques will be used to isolate and characterize these gene products, and to understand how they determine the form, connectivity, and function of neurons under the influence (or lack) of steroid hormones.

ANATOMICAL SEX DIMORPHISMS AND THEIR ONTOGENY

Our review focuses on three neural systems in which there are large differences in cell size and cell number.[1] We have chosen this emphasis because we have investigated these systems in our laboratories, and because they are the most thoroughly studied model systems to date. We do not wish to ignore the growing number of studies in other systems that show steroid-induced sex dimorphisms in neural connectivity (Raisman & Field 1971, 1973), neuron or cell size (Dörner & Staudt 1968, Wright & Smollen 1982), or neural sprouting (Milner & Loy 1982), nor do we overlook a variety of studies that document sex differences which may or may not be induced by steroids (e.g. Pfaff 1966, Greenough et al 1977, Gregory 1975, Bubenik & Brown 1973, Henry & Calaresu 1972, Hannigan & Kelley 1981, Diamond et al 1980).

---

[1]The following abbreviations are used: BC, bulbocavernosus; DHT, dihydrotestosterone; DHTP, dihydrotestosterone propionate; DM, dorsomedial nucleus of ICo; EB, estradiol benzoate; E2, estradiol; HRP, horseradish peroxidase; HVc, caudal nucleus of the hyperstriatum ventrale; ICo, intercollicular nucleus; LH, luteinizing hormone; LHRH, luteinizing hormone releasing hormone; MAN, magnocellular nucleus of the anterior neostriatum; MPOA, medial preoptic area; nXIIts, tracheosyringeal portion of the hypoglossal motor nucleus; RA, robust nucleus of the archistriatum; SDN-POA, sexually dimorphic nucleus of the preoptic area; SNB, spinal nucleus of the bulbocavernosus; T, testosterone; TP, testosterone propionate.

Vocal Control Regions in the Song Bird Brain

In the canary *(Serinus canarius)* and the zebra finch *(Poephila guttata)*, there exists an interconnected series of brain regions that project to and control the muscles of the vocal organ, the syrinx (Figure 1). Lesions of some of these brain regions disrupt vocal behavior (Nottebohm et al 1976). Song is a reproductive courtship or territorial behavior typical of males, and is under the control of gonadal androgens. Castration reduces the occurrence of singing, and androgen therapy reverses this decline (e.g. Arnold 1975). Female canaries sing only if given exogenous androgen, and even then sing simpler songs than males. Female zebra finches never sing. All major brain regions considered part of the main song circuit are significantly larger in males than females (Nottebohm & Arnold 1976) (Figure 2). Androgen treatment of adult female zebra finches has no effect on the volume of these brain regions, except for the nucleus containing syringeal motor neurons (Arnold 1980b). When adult female canaries are given androgen, two important vocal control regions (HVc and RA) increase in volume but do not attain male size (Nottebohm 1980b). This correlates well with the behavioral effects of androgen.

The sex difference in volume of various brain regions could be explained by differences in size or number of cells in these regions. Gurney (1981) has shown that cells in RA of the male zebra finch have larger somas, larger dendritic trees, and are more numerous than in RA of females. The sex difference in soma size is clear from casual observation of nuclei HVc, MAN, RA and nXIIts, but this difference has been published only for RA. In canaries, there is also a sex difference in extent of dendritic branching of neurons in RA (DeVoogd & Nottebohm 1981).

Nearly all of the major vocal control regions contain cells that accumulate gonadal steroids. Neurons in HVc, MAN, RA, ICo, and nXIIts accumulate radioactivity after injection of tritiated testosterone or dihydrotestosterone

Figure 1   A schematic parasagittal drawing of the vocal control regions in passerine birds such as canaries and zebra finches, showing some important anatomical pathways connecting the regions. The HVc (hyperstriatum ventrale pars caudale) projects caudally to the robust nucleus of the archistriatum (RA), which projects to nucleus intercollicularis (ICo) and to the tracheosyringeal portion of the hypoglossal motor nucleus (nXIIts), which innervates the vocal organ, the syrinx. Two other regions connect with HVC and RA: the magnocellular nucleus of the archistriatum, MAN, and Area X of the lobus parolfactorus (Nottebohm 1980a, Arnold 1982).

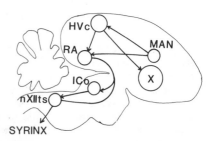

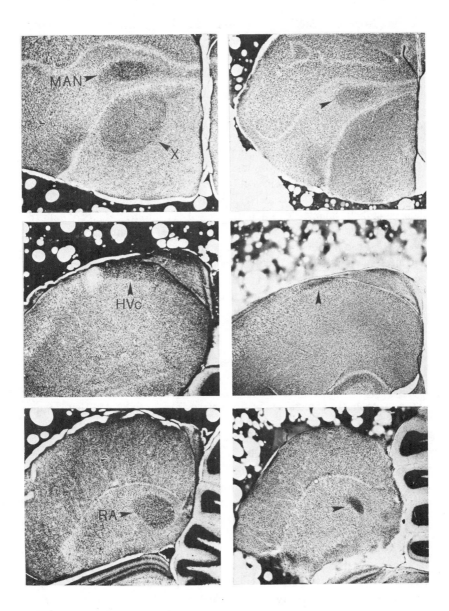

Figure 2 Photomicrographs of cresyl-violet stained coronal sections through the brain of the zebra finch, showing several sexually dimorphic regions. On the *left,* from top to bottom are male MAN and X, HVc and RA. On the *right* are the corresponding regions in females. The boundaries of MAN are much more diffuse in females and Area X cannot be seen. HVc and RA contain fewer, smaller cells. From Arnold (1982).

(Arnold, Nottebohm & Pfaff 1976; Arnold 1980a). Of these regions, only ICo accumulates estradiol or its metabolites. Adult male zebra finches have a higher proportion of steroid-accumulating cells in HVc and MAN than females (Arnold & Saltiel 1979, Arnold 1980a). We return below to this differential accumulation of steroids and its implication for the process of sexual differentiation.

The sex differences in neuron number and size appear to be the result of neonatal effects of sex steroids. Gurney & Konishi (1980) implanted female zebra finches on the day after hatching with silastic pellets containing estradiol (E2), testosterone (T), or dihydrotestosterone (DHT). The E2 treatment masculinized (increased) the volumes of two telencephalic brain regions, HVc and RA, about five-fold. DHT was less effective in masculinizing HVc and RA, increasing their volumes less than two-fold. DHT did increase the volume of the midbrain nucleus DM (dorsomedial nucleus of ICo) by a factor of 1.2, whereas E2 had no effect on this nucleus. The differential effectiveness of E2 and DHT can be understood better by considering their effects on cell number and morphology. Gurney (1981) measured the relative effects of neonatal E2, DHT, or T on three parameters in nucleus RA: neuronal number, neuronal somal size, and the spacing of neurons (inverse of neuronal density). E2 given on the day of hatching doubled the mean somal diameter, quadrupled the spacing of neurons, and caused a modest increase (17%) in the number of neurons. In contrast, DHT caused a modest increase in somal diameter (13%), had no effect on spacing, and doubled neuron number. Since T is thought to be metabolized to both E2 and DHT, T's effects were greatest. T doubled neuron size and number, and increased neuron spacing almost five-fold. These results suggest that regulation of cell number is more sensitive to DHT, and regulation of cell size and spacing is more sensitive to estrogenic metabolites of T. Thus, these two processes are regulated independently.

There is also evidence that neonatal E2 treatment alters the sensitivity of the song system to subsequent androgen treatment. In normal adult females, TP or DHT treatment alters only the volume of the hypoglossal nucleus (Arnold 1980b, Gurney 1982). However, if such adult females were previously given E2 as neonates, then the T implants significantly increase the size of HVc, RA, MAN, Area X, and DM (Gurney 1982). Pilot data suggest that DHT is even more effective than T. This T treatment in adults slightly increases neuron size and spacing in RA but does not increase neuron number (Gurney 1981). The results are interesting in several ways. First, they suggest that neonatal treatment with estradiol regulates the sensitivity of these tissues to testosterone. Second, the brain of females given estradiol at hatching is surprisingly plastic in adulthood, since testosterone can cause significant morphological changes, which imply significant functional alterations. We point this out in passing here, but return to this observation below when discussing the difference between neonatal and adult androgen effects. Finally, Gurney reports that females given E2 neonatally do not sing unless given T or DHT as adults. However, under some conditions, females treated only with E2 neonatally do sing (E. J. Nordeen and A. P. Arnold, in preparation) although it is unclear at this time whether adult androgen treatment improves the complexity or quality of song. In any case, there are behavioral correlates of the morphological masculinizations.

Historically, great emphasis has been placed on the idea that permanent (organizational) effects of steroids result only from steroid actions during an early critical period. It is important, therefore, to ask when steroids can masculinize the song system. Since E2 is potent in masculinizing day-old female zebra finches but fails to masculinize adult females (Gurney 1982), there must be an end to a critical period. Gurney (1980) gave females of various ages the same does of E2 (50 μg mixed with silastic polymer). Thus, the older birds received less estradiol per body weight. Nevertheless, the results are interesting. In 30- or 40-day old birds (who are close to adult weight), estradiol appeared to masculinize (increase soma size) of a fraction of RA neurons. It is surprising that this sort of masculinization can occur so late in life, at least when viewing these data from the perspective of data on mammals, for whom these types of critical periods end within a week or two of birth (see below). A second surprise is that estradiol appears to masculinize only some of the RA cells, increasing their somal size fully to masculine values, and leaving other RA cells unaffected. This suggests that individual RA neurons do not gradually lose their sensitivity to estradiol, but lose it relatively rapidly. However, different RA cells lose sensitivity at different ages, so that by age 40 days RA consists of cells that are either fully sensitive or insensitive. While these preliminary data fall short of proving this theoretical point, they are compatible with it.

The Sexually Dimorphic Nucleus of the Preoptic Area in Rats

The second sexually dimorphic model system is found within the medial preoptic area (MPOA) of the rat hypothalmus. This region had been considered a probable site of sexual differentiation because of its importance in the regulation of several sexually dimorphic functions, including the control of masculine sexual behavior and the cyclic release of gonadotropin necessary for ovulation (Gorski 1966, 1971). In fact, one of the earliest demonstrations that the sex difference in pituitary function suggested by Pfeiffer (1936) was actually due to an alteration in the brain was the demonstration that the medial preoptic area was relatively refractory to electrical stimulation in terms of the induction of ovulation in the masculinized female rat (Barraclough & Gorski 1961). The results of subsequent lesion studies, as well as the action of direct implants of crystalline steroids in the medial preoptic area of the neonatal female in masculinizing brain function, all supported the medial preoptic area as at least one site of androgen action perinatally (see Gorski 1966, 1971, Christensen & Gorski 1978).

This hypothesis has been strongly reinforced by the discovery of a marked structural sex difference within the medial preoptic area, the development of which is (perhaps totally) determined by the hormonal environment perinatally. Within the medial preoptic area is an area of intensely staining neurons (as determined at the level of the light microscope), which is some three- to seven-fold greater in volume in the male rat (Figure 3 A and B) (Gorski et al 1978). The neuronal density within this region is several-fold higher than in a control region of the medial preoptic area just lateral to it. On this basis, Gorski et al (1980) defined this area as a nucleus, the sexually dimorphic nucleus of the preoptic area (SDN-POA). It is noteworthy that there is no apparent difference in neuronal density within the SDN-POA of the male and female. However, because of the large sex difference in SDN-POA volume, it is clear that in the male this nucleus includes more neurons than it does in the female.

The existence of a sex difference, even a structural one, does not prove steroid regulation of sexual differentiation, since such differences could be caused by nonhormonal mechanisms. Thus, it was necessary to document that SDN-POA volume is influenced, if not determined, by the hormonal environment during development. Castration of the newborn male rat produces a significant reduction (about 50%) in the volume of the SDN-POA in adulthood (Gorski et al 1978), which can be completely prevented by the administration of exogenous androgen one day later (Jacobson et al 1981a). Similarly, the single subcutaneous administration of exogenous testosterone propionate (TP) to the newborn female significantly increases the volume of the SDN-POA in the adult (Gorski et al 1978, Jacobson et al 1981a). Note that in both cases, such

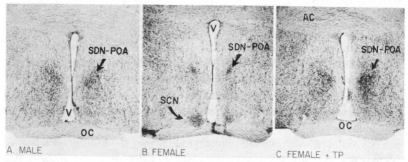

Figure 3 Photomicrographs of coronal sections through the rat brain at the center of the sexually dimorphic nucleus of the preoptic area (SDN-POA) of the female (A), male (B), and the "structurally sex-reversed" female exposed for a prolonged period perinatally to testosterone propionate (C). All photomicrographs at the same power. Abbreviations: AC, anterior commissure; OC, optic chaism; V, third ventricle. From Gorski (1983).

manipulation of the hormonal environment, although changing SDN-POA volume, does not lead to a sex-reversal of this structure. This could be due to (*a*) possible nonhormonal factors influencing SDN-POA differentiation, (*b*) a requirement for a greater, earlier or more prolonged exposure to androgen in the female, or (*c*) for earlier (prenatal) castration in the case of the male. However, prolonged treatment with TP (2 mg daily to the pregnant rat from gestational day 16, and 100 μg daily to individual pups until day 10 of postnatal life) does completely sex reverse the brain of the female in terms of SDN-POA volume (see Figure 3C) (Döhler et al 1982a). Although this observation does not exclude the possibility of a nonhormonal component, it does indicate that the steroid hormonal environment alone can fully determine final SDN-POA volume. It is clear that this sex-reversal is accompanied by a dramatic increase in the number of neurons that comprise the SDN-POA of the genetic female. More recently F. C. Davis and R. A. Gorski (unpublished observation) have found that prolonged treatment may not be necessary; an intrafetal injection of 0.5 mg TP on day 18 of gestation followed by an injection of 1 mg TP on day 2 of postnatal life also sex reverses the SDN-POA of the genetic female.

In agreement with the concept that the aromatization of testosterone to estrogen is necessary for masculinization of the brain, a single injection of estradiol benzoate (EB) postnatally masculinizes the volume of the SDN-POA of the female (Gorski et al 1981). Moreover, prolonged treatment with diethylstilbestrol also completely sex reverses the female brain in terms of SDN-POA volume (Döhler et al 1982b). In addition, in male rats with the testicular mutation *(tfm)*, the SDN-POA has a normal masculine volume (Gorski et al 1981). These males possess 10–15% of the normal level of androgen receptors (Naess et al 1976), but do have normal estrogen receptors (Wieland & Fox 1981). In the same *tfm* animals, other CNS regions are feminine (see below).

Presently, the volume of the SDN-POA does not appear to be sensitive to gonadal steroids in the adult (Gorski et al 1978), even though its neurons accumulate labeled steroids more readily than surrounding neurons (Jacobson et al 1982). Thus, treatment of males and females gonadectomized as adults, with hormonal regimes effective in restoring masculine or feminine copulatory behavior, does not influence SDN-POA volume (Gorski et al 1978). However, since steroids modulate neuronal morphology in adult birds and in rat spinal cord (discussed below), and because in the gerbil the apparent equivalent of the SDN-POA is modified by hormones during adulthood (Commins & Yahr 1982), it may be premature to conclude that in the adult rat there are no morphological effects of gonadal hormones. SDN-POA volume is a relatively crude measure and does not permit conclusions about individual neurons, their processes, or glia. Additional studies utilizing hormonal implants in silastic capsules, and studies at the level of the electron microscope are currently underway. Whether or not there are morphological effects on SDN-POA neurons in adults, it is clear that gonadal hormones exert profound effects on the development of this nucleus.

The SDN-POA is an important potential model of sexual differentiation in general because the sex difference in SDN-POA volume is established during the first week or so of postnatal life, almost exactly parallel to the period of functional sexual differentiation of the brain. Although the nucleus is recognizable as early as day 20 post-fertilization, there is no significant sex difference in its volume until the day of birth (three days later). During the next ten days there is a gradual five-fold increase in SDN-POA volume in the male. Although mean SDN-POA volume almost doubles in the female, there is no statistically significant change in the female from the day of birth through postnatal day 10 (Jacobson et al 1980). This growth differential is presumably due to testicular hormones.

The SDN-POA appears to reach its adult volume by postnatal day 10, but the maturation of neurons in the medial preoptic area in terms of dendritic connections extends well into the third week of postnatal development (Reier et al 19.77, Lawrence & Raisman 1980). Presumably the connections of SDN-POA neurons also continue to form after the termination of the critical period of steroid-induced organization of this nucleus. This is not necessarily incompatible with the assumption that steroids act during the critical period to promote the development of neuronal connections necessary for neuronal survival. Although steroid-induced synaptogenesis may contribute to SDN-POA neuronal survival during the critical period, the final sculpting of connectivity after postnatal day 10 may be independent of steroid hormones.

In contrast to the two other anatomical models currently under active study, little is known of the functional significance of the SDN-POA. This is due to the complex role of the medial preoptic area in many integrative processes. Although this area has been well studied functionally and anatomically by many investigators, most of these studies were performed before the existence of the SDN-POA was recognized. [In spite of the obvious structural sex difference in the hypothalamus (i.e. the SDN-POA), it escaped detection, perhaps because such dramatic sex differences in brain structure were not expected.] Given the complex functional activity of the medial preoptic area it would be dangerous to assign a specific function(s) to the SDN-POA retrospectively. Two examples illustrate this point. The medial preoptic area is generally considered to regulate masculine sexual behavior in the rat, yet when small electrolytic lesions were placed in the general area of the SDN-POA, specific lesions of the nucleus were ineffective, whereas small lesions just dorsal to it did disrupt masculine behavior (Arendash & Gorski 1983). Similarly, the medial preoptic area is thought to regulate the cyclic release of luteinizing hormone (LH) (Gorski 1966, 1971); however, luteinizing-hormone-releasing-hormone (LHRH) producing neurons are clearly localized anterior to the SDN-POA (Silverman et al 1979).

Although there is a need to identify the function(s) of the neurons of the SDN-POA, and to identify their connections and neurochemical specificity, this nucleus can currently be viewed as a valuable model to study the mechanisms of hypothalamic differentiation. Whatever the function(s) of the neurons of the SDN-POA, this nucleus remains a clear morphological signature of the action of gonadal steroids during the development of the brain.

The Spinal Nucleus of the Bulbocavernosus in Rats

The third sexually dimorphic model system is the spinal nucleus of the bulbo-cavernosus (SNB) and its connections. The SNB is a discrete nucleus in the lumbar spinal cord of rats, composed of motor neurons innervating two perineal muscles, the bulbocavernosus (BC) and levator ani (LA). The levator ani/bulbocavernosus complex attaches exclusively to the penis of the male, serves a sexual function, and is absent in adult female rats (Breedlove & Arnold 1980). In males, the SNB is a compact nucleus in the ventral horn near the midline (Figure 4). In females, only one-third the number of motor neurons found in males can be detected in this region using morphological criteria, and those present are one-half as large. Considering the absence in females of the target muscles for SNB neurons, it is not clear what function is served by the few neurons counted as motor neurons in the SNB region in females.

Autoradiographic studies demonstrate that in adult male rats, SNB motor neurons accumulate hormone after injection of tritiated testosterone or dihy-drotestosterone, but not after estradiol injection. Virtually every SNB motor neuron is labeled in autoradiograms, and the SNB is more heavily labeled than other motor neuron pools in the lumbar spinal cord (Breedlove & Arnold 1983c). The androgen accumulation by SNB motor neurons implies that they are direct targets for androgen action in adulthood. Indeed, the size of SNB neurons depends on androgen levels in adult rats (Breedlove & Arnold 1981). The levator ani/bulbocavernosus muscle complex is also sensitive to androgens and contains specific androgen receptors (Wainman & Shipounoff 1941, Dube et al 1976). Castration of adult males markedly reduces the size but not number of levator ani muscle fibers, and androgens reverse this effect (Venable 1966). Thus, the SNB system is composed of motor neurons and muscle targets that are sensitive to androgens. The possible steroid sensitivity of afferents to the SNB has not been studied.

The sex difference in number and size of SNB neurons is clearly regulated by gonadal androgens, and estrogens play little or no role. The first clue confirming this conclusion is the absence of the SNB in genetic males who possess the *tfm* mutation (Breedlove & Arnold 1980, 1981), who are estrogen-sensitive but

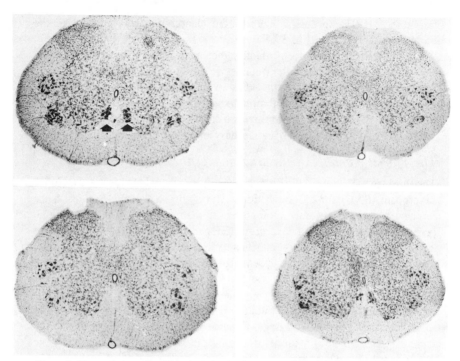

Figure 4   Photomicrographs of thionin-stained transverse sections of the rat spinal cord, showing androgen-induction of sex differences in the SNB. *Above left, arrows* point to the SNB in a control male sham castrated at birth. *Above right,* a control female injected with oil vehicle pre- and postnatally, and ovariectomized at birth. The SNB region is comparable to an untreated female. *Below left,* a male whose mother was injected with the antiandrogen flutamide during the last week of gestation. This male was castrated at birth and received control injections of oil vehicle postnatally. The SNB nucleus is absent as are the levitor ani/bulbocavernosus muscles. *Below right,* a female treated prenatally with testosterone propionate possesses the SNB and its target muscles. From Breedlove & Arnold (1983a,b).

androgen-insensitive (Naess et al 1976, Wieland & Fox 1981). These males lack the bulbocavernosus and levator ani muscles, and they have a feminine paucity of SNB cells. This implies that androgens act to masculinize the SNB system. Furthermore, a single injection of TP but not EB on day 2 after birth will masculinize the SNB system in female rats (Breedlove et al 1982). Such females retain the levator ani/bulbocavernosus muscle complex, which is present at birth in normal females but atrophies unless maintained by androgen treatment (Cihak et al 1970). These androgen-treated females also possess more and larger SNB neurons than normal. Finally, if genetic male rats are treated perinatally with flutamide, a potent antiandrogen, and then castrated at birth, they possess a feminine number of SNB neurons, and they lack the levator ani/bulbocavernosus muscles (Figure 4) (Breedlove & Arnold 1983a).

Pre- or post-natal exposure to androgens can masculinize the SNB and its targets in males or females (Figure 4) (Breedlove & Arnold 1983a,b).

Breedlove & Arnold (1983b) administered androgens to female rats during various periods before and after birth, in an effort to define the limits of any critical period for androgen action, and as an initial step in analyzing what cellular processes are modified by androgens to regulate neuron size and number. The rationale was that if one knows when androgens act to determine cell number or size, this information may place constraints on the possible mechanisms affected. Two androgens were injected during one of three injection periods. Testosterone propionate (TP) or dihydrotestosterone propionate (DHTP) was injected during days 16–22 of gestation (late prenatal group), during days 1–5 postnatally (early postnatal group) or during postnatal days 7–11 (late postnatal group). All females were ovariectomized at birth. In adulthood, these females received TP injections from days 68–102 to equate their adult hormonal level, and then were sacrificed. In general, TP was more effective than DHTP in regulating SNB number, and it was more effective prenatally than postnatally (Figure 5). No injections after postnatal day 6 masculinized SNB number. These results imply that postnatal day 6 is the approximate end to androgen regulation of SNB number.

In contrast, the period for regulation of SNB size extends beyond postnatal day 11 (Figure 5). TP injections in all three periods effectively masculinized SNB size, and DHTP was effective only postnatally. Since all of these females receive injections of TP as adults, it is clear that adult androgen treatment is ineffective in eliminating any difference between control females and females treated neonatally with androgen. Thus, there must be an end to a critical period for neonatal regulation of cell size, but it extends beyond day 11.

Because TP can influence SNB size without influencing SNB number (e.g. late posnatal TP group, Figure 5), it is clear that regulation of these two processes can be independent. This finding confirms a similar independent determination of neuron number and size in the song system of the zebra finch, discussed above.

COMPARISON OF THE THREE MODEL SYSTEMS

Several specific questions or themes recur in our analyses of steroid effects on the ontogeny of the three model systems. First, we wish to determine what cellular morphological parameters are influenced by steroids. These parameters have been discussed above and include neuron number, neuron size, extent of the dendritic tree, and possibly sensitivity to steroids. Second, we wish to determine the mechanisms by which these specific processes are altered. Third, we wish to determine where steroids act in these systems, which steroid effects can be accounted for by direct steroid actions on specific

Figure 5A   A. Perinatal androgens TP (testosterone propionate) and DHTP (dihydrotestosterone propionate) regulate neuronal
number in the SNB. The graph shows mean
number of SNB neurons ± standard errors for
six groups of adult females given TP or DHTP
during one of three periods pre- or postnatally. The late prenatal period is the last week of
gestation, the early postnatal period comprises days 1–5 after birth, and the late postnatal
period comprises days 7–11 after birth. The
prenatal TP group and the early postnatal
DHTP group are the only groups significantly
different from control females. All females
(including control females) were ovariectomized at birth.

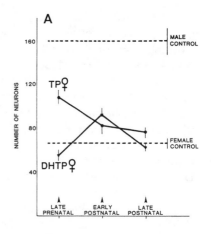

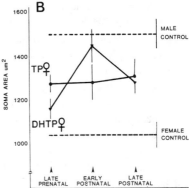

Figure 5B   These graphs show perinatal
androgen regulation of SNB soma size (mean
± standard errors) in the same female groups
shown in A. TP significantly masculinized
(increased) SNB soma size at all three time
periods. DHTP's effects were significant
only in the two postnatal periods.

Figure 5C   These graphs demonstrate perinatal androgen effects on the number of levator ani muscle fibers in the same groups as A
and B. Normal females lack the levator ani,
and the androgen-induced increase in number
of levator ani muscle fibers was significant in
all groups except the late postnatal DHTP
group. From Breedlove & Arnold (1983a,b).

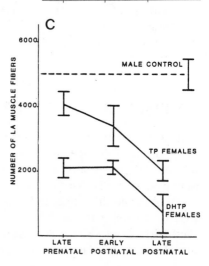

elements in the dimorphic systems, and which effects might be caused indirect-ly. For example, does masculinization of one element in a neural pathway lead, via transsynaptic effects, to masculinization of other elements in the circuit? Finally, what is the nature of the critical periods for steroid action? Why do steroids influence the ontogeny of the system only during specific developmental stages? What factors determine the onset and close of critical periods?

Mechanisms for Steroid Regulation of Neuronal Number

In all three systems, males possess more neurons of specific morphological types, and steroids regulate neuronal number. This could be accomplished in several ways, which are not mutually exclusive: (a) steroids might stimulate neuronal proliferation in males, or (b) steroids might prevent neuronal death, which typically occurs early in development in many neuronal systems (e.g. Cowan 1978, Oppenheim 1981). Finally, we must consider the possibility (c) that the sex differences in neuronal number are only apparent, not real. According to this view, the neurons present in males are also present in females, but are not recognized because they have migrated to other brain regions, or have been specified to play different functional roles in the two sexes. In this case, steroids might then alter neuronal specification, migration, or morphological development (recognizability) of neurons.

To date, the answers to these questions have been addressed most completely in the two rat dimorphic systems. In song birds, we do not have any data on timing of neuronal proliferation and death. However, we can say presently that it is unlikely that female song birds possess a masculine number of HVc neurons, which are somehow hidden or unrecognizable. We do not, for example, find ectopically placed "HVc" neurons in females (i.e. neurons with the same projections and histochemical properties of male HVc neurons) (Gurney 1981, Ryan & Arnold 1981) whose aberrant position would prevent their classification as HVc homologs. If female and male zebra finches possess an equal number of neurons, then it is safe to conclude that homologous neurons play different functional roles in the two sexes.

REGULATION OF NEURONAL NUMBER IN THE SDN-POA   In the case of the SDN-POA, the steroid environment clearly influences, if not determines, the number of neurons that comprise this nucleus (see above). Although data from the literature indicated that the neurons of the medial preoptic area of the rat become post-mitotic by about day 16 of gestation (Ifft 1972, Altman & Bayer 1978, Anderson 1978), fortunately it was considered important to verify that the neurons specifically destined to form the SDN-POA behaved as typical medial preoptic area neurons. Thus, Jacobson & Gorski (1981) administered tritiated thymidine intravenously to pregnant rats bearing an intraatrial cannula on day 14, 15, 16, 17, or 18 of gestation and sacrificed the offspring of these females at 30 days postnatally. Although this experiment confirmed data in the literature with respect to the medial preoptic area, the neurons of the SDN-POA turned out to be unique in two respects. First, the period of neuroblast division of those neurons which form the SDN-POA is markedly prolonged. Second, there are significant sex differences in the labeling index of SDN-POA neurons following tritiated thymidine exposure at these prenatal ages. These two findings are documented in the following paragraphs.

Figure 6 shows the labeling index of neurons from a control region of the medial preoptic area and the SDN-POA following exposure to tritiated thymidine prenatally. Although neurons of the medial preoptic area are essentially born by day 16 of gestation, exposure to tritiated thymidine as late as day 18

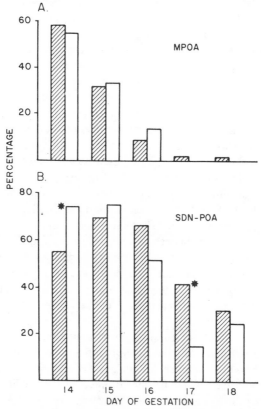

Figure 6   The influence of gestational age at the time of injection of tritiated thymidine on the mean total percentage of labeled neurons in (*a*) the control region of the medial preoptic area (MPOA) immediately lateral to (*b*), the sexually dimorphic nucleus of the preoptic area (SDN-POA) for male *(shaded bars)* and female *(open bars)* rats sacrificed on day 30 postnatally. * Significantly different from value for other sex. Data from Jacobson & Gorski (1981) reprinted with permission from Gorski et al (1981).

labels approximately one-third of the neurons of the SDN-POA. The data from subsequent, still incomplete, studies suggest that mitotic activity ceases by day 20 of gestation for the cells that form the neurons of the SDN-POA (F. C. Davis and R. A. Gorski, unpublished observations). As shown in Figure 7B, the net effect of this prolongation in neurogenesis is that exposure to tritiated thymidine on day 18 produces a highly specific labeling of the neurons of the SDN-POA, which persists in the adult (Jacobson & Gorski 1981). As described below, this offers a unique opportunity to identify the origin of these neurons and their pathway of migration to the SDN-POA.

As shown in Figure 6, statistical analysis revealed a significant effect of sex on the labeling index within the SDN-POA at 30 days of age. When fetal rats are exposed to tritiated thymidine on day 14 of gestation, the labeling index in the SDN-POA of the female rat at day 30 is significantly greater than that of the male. For animals exposed to tritiated thymidine on day 17 of gestation, the sex difference is reversed. No hormonal factor is presently known that explains the sex difference on day 14 and 17 of gestation. [Note that Weisz & Ward (1980) have reported a male-specific surge of testosterone on day 18 of gestation.] One interpretation of these results is that there is a temporal sex difference in

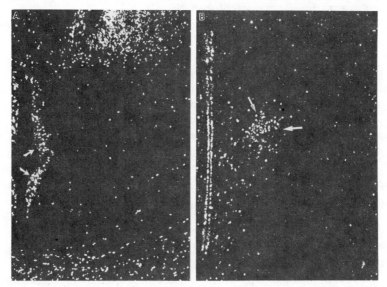

Figure 7 A. Dark field photoradiomicrographs of the preoptic area of the fetal rat sacrificed two hours after the intravenous injection of tritiated thymidine to the pregnant rat on day 18 of gestation. *Arrows* indicate the heavily labeled ependymal lining of the third ventricle.

B. Dark field photoradiomicrograph of the preoptic area of a rat similarly exposed to tritiated thymidine as in A, but not sacrificed until day 32 postfertilization. Although the ependyma is still labeled, other labeled cells are now concentrated almost exclusively in the sexually dimorphic nucleus of the preoptic area *(arrows)*.

neurogenesis, with this process accelerated in the female. However, these data may also suggest that in the male testosterone prolongs or increases the period of neurogenesis. Although a testosterone-dependent stimulation of neurogenesis would provide a simple explanation of the sex difference in the number of neurons that ultimately comprise the SDN-POA, there are three critical caveats against this view.

1. No sex difference in the prenatal hormonal environment is currently known that could explain the sex differences in labeling index on days 14 and 17.
2. Exposure to gonadal hormones clearly after neurogenesis has ceased, i.e. postnatally, still appears to alter neuronal number in the SDN-POA; some other mechanism(s) must be involved.
3. We have evaluated the labeling index at 30 days of age postnatally. Thus, these results are confounded by any hormone-dependent process that takes place after the birth of the neurons of the SDN-POA.

It may be that the simplest explanation that androgen exposure promotes neurogenesis is the most unlikely. As discussed above, hormonal effects on neuronal migration, survival, or specification may be more likely to contribute to the structural differentiation of the CNS.

We have a unique opportunity to follow the migration of SDN-POA neurons from their origin to their adult location, since exposure to tritiated thymidine on day 18 of gestation leads to a relatively specific and permanent labeling of the neurons of the SDN-POA. Figure 7A shows a section through the preoptic-anterior hypothalamic area of a rat fetus killed two hours after exposure to tritiated thymidine on day 18 of gestation. Note the concentration of label over the lining of the walls of the third ventricle. By sacrificing animals similarly exposed to tritiated thymidine on day 18 at different ages, Jacobson et al (1981b) have been able to trace the movement of labeled cells over time. Although the statistical analyses of these data are still in progress, it now appears that with time, labeled cells accumulate at the floor of the third ventricle and in the surrounding tissue. Labeled cells then migrate upward and laterally to the SDN-POA. Over a period of six days, the number of labeled cells increases in the SDN-POA in both sexes, while there is a dramatic decrease in the number of labeled cells in the ependymal lining of the third ventricle and in the region of the brain between the base of the ventricle and the SDN-POA. It is not presently known whether there are significant sex differences in the quantitative aspects of this migration. It is clear, however, that more neurons are labeled by exposure to tritiated thymidine in day 18 of gestation than ultimately come to reside in the SDN-POA. The fate of the other cells, e.g. migration to other regions of the brain or their death, remains to be determined. The original interpretation that a large proportion of the neurons of the SDN-POA are born much later than those of surrounding regions of the medial preoptic area still appears to be true (see Figure 7B), but the SDN-POA neurons appear to be part of a larger group that are still undergoing cell division on day 18. The fate of these other cells (i.e. their migration elsewhere or their death), or the reasons that the late arising neurons of the SDN-POA persist in this nucleus, are important questions, which, when understood, may provide critical insight into the formation of the SDN-POA and its sexually dimorphic characteristics. If we can assume that these late arising neurons of the SDN-POA are typical of all of its neurons, that they can be specifically labeled provides a unique tool to elucidate the development of this nucleus and the mechanism(s) of steroid-induced modification of this process.

In this regard, it is important to stress that the SDN-POA itself may represent only one component of a sexually dimorphic neural system. For example, on the basis of cytoarchitecture, the SDN-POA appears to be one component of a much larger tri-component medial preoptic nucleus (Simerly et al 1983). Moreover, recent studies of the guinea pig reveal that although an SDN-POA also exists in this species (Hines et al 1982), it may be only one component of a larger neural system that includes the bed nucleus of the stria terminalis, which is also sexually dimorphic in volume (M. Hines and R. A. Gorski, unpublished observation). As the late-arising neurons of the SDN-POA may provide a tool to elucidate the mechanisms of the formation and differentiation of the SDN-POA, this nucleus itself may serve as a model of the process of the structural sexual differentiation of other regions of the mammalian brain. Just as in the songbird, the rodent brain may include entire neural systems that are sexually dimorphic.

REGULATION OF NEURONAL NUMBER IN THE SPINAL NUCLEUS OF THE BULBOCAVERNOSUS   Most neurons of the male or female SNB undergo their last mitotic division on day 12 of gestation, with a few divisions occurring as late as day 14 of gestation (Breedlove et al 1983). Since TP injections into females augment adult SNB number when the injections are begun on day 16 of gestation or even postnatally, it is unlikely that these injections alter SNB proliferation. This conclusion is strengthened considerably by the observation from thymidine autoradiographic studies that females exposed to tritiated thymidine on day 12 of gestation and then injected with TP postnatally possess a masculinized SNB whose neurons are labeled with thymidine (S. M. Breedlove and A. P. Arnold, in preparation). This indicates that the neurons present in the SNB proliferated on day 12, before the androgen injections (Breedlove et al 1983). Although we cannot totally exclude the possibility of androgen effects on proliferation, clearly androgens can dramatically increase SNB number after proliferation has ceased.

It is not yet known precisely when SNB motor neurons undergo cell death. In the rat brachial spinal cord, motor neuron death seems to be mostly in the first six postnatal days (Nurcombe et al 1981). We would expect that lumbar motor neuronal death would occur at about the same time or slightly later. Since androgen injections six days or more after birth fail to increase SNB number, it may be that the end of the critical period for androgen effects on SNB number is coincident with the end of the period for SNB death.

We cannot yet exclude the third possibility: androgen alteration of motor neuronal specification. If in the fetal female, presumptive SNB motor neurons fail to establish synaptic connections in the periphery because of atrophy of the target levator ani/bulbocavernosus complex, they might be induced to make "anomalous" connections not found in males. Such motor neurons might well fail to be recognized as SNB neurons because their morphology or spinal location is dependent on their target, which differs from that of the SNB neurons of the male. Although there is no direct evidence that this sequence of events occurs normally in females, recent evidence suggests that lumbar motor neuronal specification can be influenced by androgens. Breedlove (1983) compared adult females that had been exposed during the last week of their gestation to either TP or DHTP. Using horseradish peroxidase (HRP) injected into the bulbocavernosus muscles to locate spinal motor neurons innervating these muscles, he found a majority of motor neurons in TP females in the SNB region as usual, but in DHTP females the majority was in an abnormal lateral position. This suggests that lateral motor neurons, which normally do not innervate the bulbocavernosus, were induced to do so in the DHTP treatment group, and that specific neuron-to-muscle connections may depend on the prenatal hormonal milieu.

Direct versus Indirect Effects

If steroids act at one target to increase the number of neurons that survive to adulthood, this should have profound effects both on neurons projecting to this neuronal pool, and on neurons receiving projections from the pool. It is well known that the normal ontogeny of neurons depends on the characteristics of their targets as well as the nature of afferents to the neurons (Jacobson 1978). Thus, in sexually dimorphic systems that involve two or more sexually dimorphic elements, one can potentially explain the sex differences in one neural population (e.g. Area X in song birds) as a direct effect on that area, or as an indirect masculinizing effect exerted by dimorphic afferents from another area (for example HVc). As a second example, it is possible that androgens act directly on the levator ani/bulbocavernosus muscle complex to prevent its involution in fetal males, and this stabilization of the target for SNB neurons in turn prevents their death. Alternatively, stabilization of SNB neurons may indirectly cause retention of the levator ani/bulbocavernosus complex.

This question is largely unresolved, but certain observations bear on the answer. In the song system, estradiol treatment of day-old female zebra finches masculinizes nucleus RA by increasing its volume and doubling RA neuron soma size (Gurney 1981, 1982). In the same females, E2 had no detectable effect on the volume of nucleus DM, which receives a projection from RA. In contrast, DHT treatment of neonatal females doubled RA neuron number and increased DM volume by 20%. Thus, the independent masculinization of RA and DM suggests that masculinization of one nucleus does not necessarily masculinize its target, and that steroids act at multiple sites, or influence multiple processes to control sexual differentiation. This conclusion is tentative, because it rests on relatively crude measures of masculinization, such as gross volume of DM.

In the rat, estradiol seems to be most responsible for masculinization of the SDN-POA, yet non-aromatized androgens are primarily responsible for masculinization of the SNB system. In *tfm* male rats, the SDN-POA is masculine because of the presence of estradiol receptors in the brain, and the SNB system is absent because of the lack of androgen receptors (Jacobson et al 1981, Breedlove & Arnold 1981). Thus masculinization of one nucleus does not necessarily result in masculinization of the other. In this case, we do not yet know if there are any anatomical connections between the two nuclei.

In the SNB system, late prenatal DHTP treatment partially masculinizes the levator ani/bulbocavernosus complex without increasing SNB number (Breedlove & Arnold 1983b) (Figure 5). This result can be explained in a number of ways.

1. The muscle fibers are uninnervated. This is unlikely, since HRP injections into such females do label motor neurons in the spinal cord (S. M. Breedlove and A. P. Arnold, unpublished).
2. Androgens regulate motor unit size, and the number of SNB neurons present in these females innervate an abnormally large number of muscle fibers.
3. There is a larger than normal number of motor neurons in these DHTP females, but the cell bodies are located outside of the SNB region. If this were true, it would suggest androgen effects on specification or migration of neurons discussed above.

The study of Breedlove (1983) discussed above specifically tested this possibility and found an abnormally large number of bulbocavernosus motor neurons outside the SNB region. This finding is exceptionally exciting and calls for further study defining the conditions under which steroids can influence neuronal specification. To date, no androgen treatment has succeeded in masculinizing the SNB without also masculinizing the levator ani/bulbocavernosus complex.

Steroid Influences on Multiple Developmental Processes

The process of sexual differentiation clearly involves both estrogenic and androgenic influences on several different developmental processes. In both song birds and rats, the developmental effects of estrogens and androgens both contribute independently to masculinization of the CNS. In the rat, androgens can increase SNB neuronal size without increasing their number. These results indicate that there are likely to be multiple sites of action or multiple periods during which steroids must act, and that there are independent developmental processes affected by steroid hormones. This conclusion at a cellular level of analysis corresponds closely with those made previously at a functional/ behavioral level. Androgens can defeminize rats behaviorally without altering their potential to show masculine behavior, and can render a female anovulatory at low doses that leave behavioral capacity unaffected (Whalen 1981). It must also be emphasized that the process of sexual differentiation of the brain applies to functional systems involved in a wide variety of behaviors other than sexual behavior (e.g. social, aggressive behaviors), and these may also be independently regulated (Goy & McEwen 1980).

The Concept of the Critical Period

The recent studies at the anatomical level have confirmed the existence of critical periods discovered in investigations of the organizational (i.e. permanent) effects of steroids on behavior and neuroendocrine function. Certain parameters (e.g. SNB number, SDN-POA volume) are modulated by steroids only at early developmental stages. Part of the excitement generated by these anatomical studies is that they now offer us the chance to evaluate specific hypotheses to define the nature of these critical periods. For example, the period for regulation of SNB number may be coincident with the period of SNB death. In the case of the SDN-POA, the end of the critical period may coincide with loss of the potential to make afferent or efferent connections. Another hypothesis is that the end of the period for estradiol regulation of RA soma size is gradual, not because of a decrease in response of all RA cells to estrogen, but because of a gradual reduction in the number of RA cells that are sensitive to estrogen (Gurney 1981). We can anticipate experimental analysis of these hypotheses in the near future.

The existence of critical periods implies the absence in adulthood of the capacity to respond to steroids in the same ways as in the perinatal period. Although some recent studies confirm this absence of plasticity in adults [e.g. inability of steroids to regulate SNB number in the adulthood (Breedlove & Arnold 1981)], certain studies on song birds reveal a surprising degree of plasticity. One example has already been discussed briefly above. If androgens are given to adult female zebra finches previously treated neonatally with estradiol, the size of Area X increases by about 70% (Gurney 1982). In adult female canaries who have been ovariectomized early in life, treatment with testosterone increases the volume of HVc and RA by 90% and 53%, respectively (Nottebohm 1980b). The same treatment induces growth of dendrites in one specific neuron type in RA, so that their total length is 48% longer (DeVoogd & Nottebohm 1981). Since the density of synapses in RA is largely unaltered by this treatment, DeVoogd et al (1982) suggest that testosterone stimulates the production of new synapses onto these RA neurons with elongated dendrites. The androgen-treated adult females sing unlike untreated adult females, and the androgen induces substantial reorganization of connections onto RA neurons, which may in part be responsible for the change in behavior. This degree of plasticity is unexpected in the adult brain, suggesting that some morphological responses to steroids do not have critical periods.

Preliminary evidence indicates that similar androgen-sensitive synaptic plasticity conceivably occurs in mammals. The soma size of SNB neurons depends on the circulating levels of androgen in adulthood (Breedlove & Arnold 1981). Castration reduces SNB size and androgen treatment of females or castrated males increases soma diameter by 17–22%. Theoretically this means a 17–22% increase in soma membrane area available for axo-somatic synapses. This finding implies either a steroid-induced change in size or number of synapses onto adult SNB neurons, or a change in the percentage of soma membrane contacted by synapses. Either process suggests a steroid-induced reorganization of inputs into the SNB motor neurons. The adult androgen regulation of SNB size should not be confused with an indpendent neonatal regulation of SNB soma size, which occurs during a critical period, and thus probably involves androgen regulation of a different mechanism.

The Concept of Cascading Steroid Influences During Development

Although we have emphasized the multiplicity of processes influenced by steroids during development, it is important to consider that the opposite view has some theoretical merits. According to this alternative view, steroids act on one developmental process, which then exerts pervasive effects on other processes in a cascading manner. Because it is thought that steroids do modify major developmental processes, the cascade theory is attractive. For example, Toran-Allerand (1976) has demonstrated that estrogens or androgens enhance neurite outgrowth from hypothalamic explants from neonatal mice. The neurite growth occurs from regions of the explant that contain steroid-accumulating cells (Toran-Allerand et al 1980). Even if this were the only effect of steroids on neurons, it is conceivable that this enhanced growth could lead to such diverse consequences as sex differences in neuron size, number, and steroid accumulation. For example, if steroids stimulate outgrowth of specific neuron types such as androgen-accumulating neurons in the SDN-POA, or androgen-accumulating neurons in the lumbar spinal cord, such neurons might be in a better competitive position relative to other, steroid-insensitive neurons. Hence the stimulated neurons might make more effective synaptic contacts, which might promote their subsequent survival during a period of cell death. A large soma may be necessary to contain synthetic machinery needed to maintain the larger neuritic processes, resulting in steroid effects on soma size. Finally, since steroid-accumulating neurons would undergo less cell death, this would increase the proportion of neurons in adulthood that accumulate steroids. This is, however, only one of several conceivable scenarios of the cascade theory. For example, it is possible that the enhancement of neurite outgrowth observed by Toran-Allerand is a result of a primary effect of steroids on viability of specific neuron types. If steroids differentially prevent the death of fast-growing neurons, this would result in the observed increase in neurite growth. Thus, prevention of cell death could be the initial crucial event. Finally, if the primary effect of neonatal steroids were to induce steroid receptors in specific neurons, this could in turn lead to enhanced neural growth and survival. It must be emphasized that these sequences of events are highly hypothetical at this point. To begin to discriminate among these theoretical possibilities, we need accurate descriptions of the sequences of developmental events: When do steroids act? When does cell death occur? What is the time course of development of steroid receptors? Now we can only point out that although steroids clearly modulate several developmental processes, these processes may be important enough to have pervasive secondary effects on the morphogenesis of neural circuits.

PRACTICAL IMPLICATIONS FOR NEUROSCIENCE

The use of steroid hormones in developmental neurobiology is comparatively recent, but there are indications that these hormones might become powerful tools for use in understanding basic processes such as cell death and survival. If it can be shown that steroids prevent cell death in certain neuronal populations but not in others, one can potentially compare the response of each population to steroids to help understand which cellular properties are potentially responsible for prevention of cell death. A second example of the potential use of steroids is to regulate cell number in a specific brain region, in order to study the effects of modification in size of a neuronal population on its afferents or efferents. This type of experiment might be accomplished by implanting functionally dimorphic male brain regions into females, an approach which has already been shown to modify behavior (Arendash & Gorski 1983) or by using local microimplants of steroids to masculinize selectively only neurons near the implant. These experiments and others are well within the range of techniques used routinely in neuroscience research, and we can expect exciting developments in the near future in the cellular analysis of steroid influences on neural development. At the same time, the application of relatively new techniques (such as the recombinant DNA technology) will increase our ability to recognize specific cell populations (with the advent of new cell markers), and to study precisely which gene products are regulated by steroids at various stages of development.

It is studies such as these that will constitute the fourth phase of the investigation of sexual differentiation. The discovery of discrete morphological sex differences should stimulate these studies, since we now have specific neuronal systems to investigate. We may suggest that the first three phases of the study of the sexual differentiation of the brain have produced results that have changed our concepts of the neuroendocrine function of the brain, but that the results of the studies to come, although further elucidating this important phenomenon, will also contribute to our understanding of neuronal development *per se*.

Acknowledgments

We thank Sue Einstein for expert secretarial assistance. Research in the authors' laboratories was supported by NSF grant 80–06798 (A.P.A.), NIH grants HD-15021 (A.P.A.), RR07009 (UCLA), and HD-01182 (R.A.G.), and by grants to R.A.G. from the Kroc and Grant Foundations.

Literature Cited

Altman, J., Bayer, S. A. 1978. Development of the diencephalon in the rat. I. Autoradiographic study of the time of origin and settling patterns of neurons of the hypothalmus. *J. Comp. Neurol.* 182:945–72

Anderson, C. H. 1978. Time of neuron origin in the anterior hypothalamus of the rat. *Brain Res.* 154:119–22

Arendash, G. W., Gorski, R. A. 1983. Effects of discrete lesions of the sexually dimorphic nucleus of the preoptic area or other medial preoptic regions on the sexual behavior of male rats. *Brain Res. Bull.* 10:147–54

Arnold, A. P. 1975. The effects of castration and androgen replacement on song, courtship, and aggression in zebra finches. *J. Exp. Zool.* 191:309–26

Arnold, A. P. 1980a. Quantitative analysis of sex differences in hormone accumulation in the zebra finch brain: Methodological and theoretical issues. *J. Comp. Neurol.* 189:421–36

Arnold, A. P. 1980b. Effects of androgens on volumes of sexually dimorphic brain regions in the zebra finch. *Brain Res.* 185:441–44

Arnold, A. P. 1982. Neural control of passerine song. In *Evolutionary and Ecological Aspects of Acoustic Communication in Birds*, ed. D. E. Kroodsma, E. H. Miller. New York: Academic. In press

Arnold, A. P., Saltiel, A. 1979. Sexual difference in pattern of hormone accumulation in

the brain of a song bird. *Science* 205:702–5

Barraclough, C. A., Gorski, R. A. 1961. Studies on mating behavior in the androgensterilized rat and their relation to the hypothalamic regulation of sexual behavior in the female rat. *J. Endocrinol.* 25:175–82

Baum, M. J., Erskine, M. A., Holbrook, P. G., Gallagher, C. A. 1982. Hormonal regulation of behavioral sexual differentiation in the male ferret. *Conf. Reproductive Behav. Abstr.*, Michigan State Univ., p. 8

Breedlove, S. M. 1983. The specificity of motoneuron to muscle innervation can be manipulated hormonally in the rat. *Soc. Neurosci. Abstr.* 9: In press

Breedlove, S. M., Arnold, A. P. 1980. Hormone accumulation in a sexually dimorphic motor nucleus of the rat spinal cord. *Science* 210:564–66

Breedlove, S. M., Arnold, A. P. 1981. Sexually dimorphic motor nucleus in rat spinal cord: Response to adult hormone manipulation, absence in androgen insensitive rats. *Brain Res.* 225:297–307

Breedlove, S. M., Arnold, A. P. 1983a. Hormonal control of a developing neuromuscular system: I. Complete demasculinization of the male rat spinal nucleus of the bulbocavernosus using the antiandrogen flutamide. *J. Neurosci.* 3:417–23

Breedlove, S. M., Arnold, A. P. 1983b. Hormonal control of a developing neuromuscular system: II. Sensitive periods for the androgen induced masculinization of the rat spinal nucleus of the bulbocavernosus. *J. Neurosci.* 3:424–32

Breedlove, S. M., Arnold, A. P. 1983c. Sex differences in the pattern of steroid accumulation by motoneurons of the rat lumbar spinal cord. *J. Comp. Neurol.* 215:211–16

Breedlove, S. M., Jacobson, C. D., Gorski, R. A., Arnold, A. P. 1982. Masculinization of the female rat spinal cord following a single neonatal injection of testosterone propionate but not estradiol benzoate. *Brain Res.* 237:173–81

Breedlove, S. M., Jordan, C. L., Arnold, A. P. 1983. Neurogenesis in the sexually dimorphic spinal nucleus of the bulbocavernosus of the rat. *Dev. Brain Res.* In press

Brown-Grant, K. 1975. On "critical periods" during the post-natal development of the rat. *Int. Symp. Sex. Endocrinol. Perinatal Period. INSERM* 32:357–76

Bubenik, G. A., Brown, G. M. 1973. Morphologic sex differences in primate brain areas involved in regulation of reproductive activity. *Experientia* 29:619–21

Christensen, L. W., Gorski, R. A. 1978. Independent masculinization of neuroendocrine systems by intracerebral implants of testosterone or estradiol in the neonatal female rat. *Brain Res.* 146:325–40

Cihak, R., Gutmann, E., Hanzlikova, V. 1970. Involution and hormone-induced persistence of the muscle sphincter (levator) ani in female rats. *J. Anat.* 106:93–110

Commins, D., Yahr, P. 1982. Morphology of the gerbil medial preoptic area (MPOA) is sexually dimorphic and influenced by adult gonadal steroids. *Conf. Reproductive Behav. Abstr.*, Michigan State Univ., p. 47

Cowan, W. M. 1978. Aspects of neural development. In *International Review of Physiology and Neurophysiology III*, ed. R. Porter, 17:149–91. Baltimore: University Park Press

DeVoogd, T. J., Nixdorf, B., Nottebohm, F. 1982. Recruitment of additional synapses into a brain network takes extra brain space. *Soc. Neurosci. Abstr.* 8:140

DeVoogd, T. J., Nottebohm, F. 1981. Sex differences in dendritic morphology of a song control nucleus in the canary: A quantitative Golgi study. *J. Comp. Neurol.* 196:309–16

DeVoogd, T., Nottebohm, F. 1981. Gonadal hormones induce dendritic growth in the adult avian brain. *Science* 214:202–4

Diamond, M. C., Dowling, G. A., Johnson, R. E. 1980. Morphologic cerebral cortical asymmetry in male and female rats. *Exp. Neurol.* 71:261–68

Döhler, K. D., Hines, M., Coquelin, A.,

Davis, F., Shryne, J. E., Gorski, R. A. 1982a. Pre- and postnatal influence of diethylstilboestrol on differentiation of the sexually dimorphic nucleus in the preoptic area of the female rat brain. *Neuroendocrinol. Lett.* 4:361–65

Döhler, K. D., Coquelin, A., Davis, F., Hines, M., Shryne, J. E., Gorski, R. A. 1982b. Differentiation of the sexually dimorphic nucleus in the preoptic area of the rat brain is determined by the perinatal hormone environment. *Neurosci. Lett.* 33:295–98

Dörner, G., Staudt, J. 1968. Structural changes in the preoptic anterior hypothalmic area of the male rat, following neonatal castration and androgen substitution. *Neuroendocrinology* 3:136–40

Dube, J. Y., Lesage, R., Tremblay, R. R. 1976. Androgen and estrogen binding in rat skeletal and perineal muscles. *Can. J. Biochem.* 54:50–55

Gorski, R. A. 1966. Localization and sexual differentiation of the nervous structures which regulate ovulation. *J. Reprod. Fertil. Suppl.* 1:67–88

Gorski, R. A. 1968. Influence of age on the response to paranatal administration of a low dose of androgen. *Endocrinology* 82:1001–4

Gorski, R. A. 1971. Gonadal hormones and the perinatal development of neuroendocrine function. In *Frontiers in Neuroendocrinology, 1971*, ed. L. Martini, W. F. Ganong, pp. 237–90. New York: Oxford Univ. Press

Gorski, R. A. 1983. Steroid-induced sexual characteristics in the brain. *Neuroendocrine Perspect.* 12: In press

Gorski, R. A., Gordon, J. H., Shryne, J. E., Southam, A. M. 1978. Evidence for a morphological sex difference within the medial preoptic area of the rat brain. *Brain Res.* 148:333–46

Gorski, R. A., Harlan, R. E., Jacobson, C. D., Shryne, J. E., Southam, A. M. 1980. Evidence for the existence of a sexually dimorphic nucleus in the preoptic area in the rat. *J. Comp. Neurol.* 193:529–39

Gorski, R. A., Csernus, V. J., Jacobson, C. D. 1981. Sexual dimorphism in the preoptic area. *Reproduction and Development. Adv. Physiol. Sci.* 15:121–30

Gorski, R. A., Wagner, J. W. 1965. Gonadal activity and sexual differentiation of the hypothalamus. *Endocrinology* 76:226–39

Goy, R. W., McEwen, B. S. 1980. *Sexual Differentiation of the Brain.* Cambridge, Mass: MIT Press

Greenough, W. T., Carter, C. S., Steerman, C., DeVoogd, T. J. 1977. Sex differences in dendritic patterns in hamster preoptic area. *Brain Res.* 126:63–72

Gregory, E. 1975. Comparison of postnatal

CNS development between male and female rats. *Brain Res.* 99:152–56

Gurney, M. 1980. *Sexual differentiation of brain and behavior in the zebra finch (Poephila guttata): A cellular analysis.* PhD thesis. Calif. Inst. Technol., Pasadena

Gurney, M. 1981. Hormonal control of cell form and number in the zebra finch song system. *J. Neurosci.* 1:658–73

Gurney, M. 1982. Behavioral correlates of sexual differentiation in the zebra finch song system. *Brain Res.* 231:153–72

Gurney, M., Konishi, M. 1980. Hormone induced sexual differentiation of brain and behavior in zebra finches. *Science* 208:1380–82

Hannigan, P. C., Kelley, D. B. 1981. Male and female laryngeal motoneurons in *Xenopus laevis. Soc. Neurosci. Abstr.* 7:269

Harlan, R. E., Gorski, R. A. 1978. Effects of postpubertal ovarian steroids on reproductive function and sexual differentiation of lightly androgenized rats. *Endocrinology* 102:1716–24

Harris, G. W., Levine, S. 1962. Sexual differentiation of the brain and its experimental control. *J. Physiol.* 163:42P–43P

Henry, J. L., Calaresu, F. R. 1972. Topography and numerical distribution of neurons of the thoraco-lumbar intermediolateral nucleus in the cat. *J. Comp. Neurol.* 144:205–13

Hines, M., Davis, F. C., Goy, R. W., Gorski, R. A. 1982. The existence of a sexually dimorphic nucleus in the preoptic area of the guinea pig brain. *Biol. Reprod.* 26 (Suppl. 1):49A

Ifft, J. D. 1972. An autoradiographic study of the time of final division of neurons in rat hypothalamic nuclei. *J. Comp. Neurol.* 144:193–204

Jacobson, C. D., Gorski, R. A. 1981. Neurogenesis of the sexually dimorphic nucleus of the preoptic area of the rat. *J. Comp. Neurol.* 196:519–29

Jacobson, C. D., Shryne, J. E., Shapiro, F., Gorski, R. A. 1980. Ontogeny of the sexually dimorphic nucleus of the preoptic area. *J. Comp. Neurol.* 193:541–48

Jacobson, C. D., Csernus, V. J., Shryne, J. E., Gorski, R. A. 1981a. The influence of gonadectomy, androgen exposure, or a gonadal graft in the neonatal rat on the volume of the sexually dimorphic nucleus of the preoptic area. *J. Neurosci.* 1:1142–47

Jacobson, C. D., Davis, F. C., Freiberg, E., Gorski, R. A. 1981b. Formation of the sexually dimorphic nucleus of the preoptic area of the male rat. *Soc. Neurosci. Abstr.* 7:286

Jacobson, C. D., Arnold, A. P., Gorski, R. A. 1982. Steroid accumulation in the sexually dimorphic nucleus of the preoptic area (SDN-POA). *Anat. Rec.* 202:88A

Jacobson, M. 1978. *Developmental Neurobiology.* New York: Plenum

Kawashima, S. 1960. Influence of continued injections of sex steroids on the estrous cycle in the adult rat. *Annot. Zool. Jpn.* 33:226–33

Lawrence, J. M., Raisman, G. 1980. Ontogeny of synapses in a sexually dimorphic part of the preoptic area in the rat. *Brain Res.* 183:466–71

Lieberburg, I., Wallach, G., McEwen, B. S. 1977. The effects of an inhibitor of aromatization (1,4,6-androstatriene-3,17-dione) and an antiestrogen (CI-628) on in vivo formed testosterone metabolites recovered from neonatal rat brain tissues and purified cell nuclei. Implications for sexual differentiation of the rat brain. *Brain Res.* 127:176–81

Lobl, R. T., Gorski, R. A. 1974. Neonatal intrahypothalamic androgen administration: The influence of dose and age on androgenization of female rats. *Endocrinology* 94:1325–30

McEwen, B. S. 1976. Gonadal steroid receptors in neuroendocrine tissues. In *Subcellular Mechanisms in Reproductive Neuroendocrinology,* ed. F. Naftolin, pp. 353–400. Amsterdam: Elsevier

Milner, T. A., Loy, R. 1982. Hormonal regulation of axonal sprouting in the hippocampus. *Brain Res.* 243:180–85

Naess, O., Haug, E., Attramadal, A., Aakvaag, A., Hansson, V., French, F. 1976. Androgen receptors in the anterior pituitary and central nervous system of the androgen "insensitive" *(Tfm)* rat: Correlation between receptor binding and effects of androgens on gonadotropin secretion. *Endocrinology* 99:1295–1303

Nottebohm, F. 1980a. Brain pathways for vocal learning in birds: A review of the first 10 years. *Prog. Psychobiol. Psychol.* 9:85–125

Nottebohm, F. 1980b. Testosterone triggers growth of brain vocal control nuclei in adult female canaries. *Brain Res.* 189:429–36

Nottebohm, F., Arnold, A. P. 1976. Sexual dimorphism in vocal control areas of the song bird brain. *Science* 194:211–13

Nottebohm, F., Stokes, T. M., Leonard, C. M. 1976. Central control of song in the canary *(Serinus canarius). J. Comp. Neurol.* 165:457–86

Nunez, E., Savu, L., Engelmann, F., Benassayag, C., Crepy, O., Jayle, M. F. 1971. Origine embryonnaire de la proteine serique fixant l'oestrone et l'oestradiol chex la Ratte impubere. *C. R. Acad. Sci. Paris* 273:242–45

Nurcombe, V., McGrath, P. A., Bennett, M. R. 1981. Postnatal death of motor neurons during the development of the brachial spinal cord of the rat. *Neurosci. Lett.* 27:249–54

Ojeda, S. R., Kalra, P. S., McCann, S. M. 1975. Further studies on the maturation of the estrogen negative feedback on gonadotropin release in the female rat. *Neuroendocrinology* 18:242–55

Oppenheim, R. W., 1981. Neuronal cell death and some related regressive phenomena during neurogenesis: A selective historical review and progress report. In *Studies in Developmental Neurobiology: Essays in Honor of Viktor Hamburger*, ed. W. M. Cowan, pp. 74–133. New York: Oxford Univ. Press

Pfaff, D. W. 1966. Morphological changes in the brains of adult male rats after neonatal castration. *J. Endocrinol.* 36:415–16

Pfeiffer, C. A. 1936. Sexual differences of the hyphophyses and their determination by the gonads. *Am. J. Anat.* 58:195–225

Phoenix, C. H., Goy, R. W., Gerall, A. A., Young, W. C. 1959. Organizing action of prenatally administered testosterone propionate on the tissues mediating mating behavior in the female guinea pig. *Endocrinology* 65:369–82

Plapinger, L., McEwen, B. S., Clemens, L. E. 1973. Ontogeny of estradiol-binding sites in rat brain. II. Characteristics of neonatal binding macromolecule. *Endocrinology* 93:1129–39

Raisman, G., Field, P. M. 1971. Sexual dimorphism in the preoptic area of the rat. *Science* 173:731–33

Raisman, G., Field, P. M. 1973. Sexual dimorphism in the neuropil of the preoptic area of the rat and its dependence on neonatal androgen. *Brain Res.* 54:1–29

Raynaud, J. P., Mercier-Bodard, C., Baulieu, E. E. 1971. Rat estradiol binding plasma protein (EBP). *Steroids* 18:767–88

Reier, P. J., Cullen, M. J., Froelich, J. S., Rothchild, I. 1977. The ultrastructure of the developing medial preoptic nucleus in the postnatal rat. *Brain Res.* 122:415–36

Ryan, S. M., Arnold, A. P. 1981. Evidence for cholinergic participation in the control of bird song: Acetylcholinesterase distribution and muscarinic receptor autoradiography in the zebra finch brain. *J. Comp. Neurol.* 202:211–19

Silverman, A. J., Krey, L. C., Zimmerman, E. A. 1979. A comparative study of the luteinizing hormone releasing hormone (LHRH) neuronal networks in mammals. *Biol. Reprod.* 20:98–110

Simerly, R. B., Swanson, L. W., Gorski, R. A. 1983. Demonstration of a sexual dimorphism in the distribution of serotonin immunoreactive fibers in the medial preoptic nucleus of the rat. *Anat. Record.* 205:185A–86A

Sutherland, S. D., Gorski, R. A. 1972. An evaluation of the inhibition of androgenization of the neonatal female rat brain by barbiturate. *Neuroendocrinology* 10:94–108

Toran-Allerand, C. D. 1976. Sex steroids and development of the newborn mouse hypothalamus and preoptic area in vitro: Implications for sexual differentiation. *Brain Res.* 106:407–12

Toran-Allerand, C. D., Gerlach, J. L., McEwen, B. S. 1980. Autoradiographic localization of 3H-estradiol related to steroid responsiveness in cultures of the hypothalamus and preoptic area. *Brain Res.* 184:517–22

Venable, J. H. 1966. Morphology of the cells of normal, testosterone-deprived and testosterone-stimulated levator ani muscles. *Am. J. Anat.* 119:271–302

Wainman, P., Shipounoff, G. C. 1941. The effects of castration and testosterone propionate on the striated perineal musculature of the rat. *Endocrinology* 29:955–78

Ward, I. 1969. Differential effect of pre- and postnatal androgen on the sexual behavior of intact and spayed rats. *Horm. Behav.* 1:25–36

Weisz, J., Gunsalus, P. 1973. Estrogen levels in immature female rats: True or spurious—ovarian or adrenal? *Endocrinology* 93:1057–65

Weisz, J., Ward, I. L. 1980. Plasma testosterone and progesterone titers of pregnant rat, their male and female fetuses, and neonatal offspring. *Endocrinology* 106:306–16

Whalen, R. E. 1981. Current issues in neurobiology of sexual differentiation. In *Hormones in Development and Aging*, ed. A. Vernadakis, P. Timiras, pp. 273–304. New York: Spectrum

Wieland, S. J., Fox, T. O. 1981. Androgen receptors from rat kidney and brain: DNA-binding properties of wild-type and tfm mutant. *J. Steroid Biochem.* 14:409–14

Wilson, J. G. 1943. Reproductive capacity of adult female rats treated prepuberally with estrogenic hormone. *Anat. Rec.* 86:341–59

Wilson, J. G., Hamilton, J. B., Young, W. C. 1941. Influence of age and presence of the ovaries on reproductive function in rats injected with androgens. *Endocrinology* 29:784–89

Wilson, J. G., Young, W. C., Hamilton, J. B. 1940. A technic suppressing development of reproductive function and sensitivity to estrogen in the female rat. *Yale J. Biol. Med.* 13:189–202

Wright, L. L., Smolen, A. J. 1982. Estradiol increases numbers of neurons and synapses in neonatal rat superior cervical ganglion. *Soc. Neurosci. Abstr.* 8:198

Young, W. C., Goy, R. W., Phoenix, C. H. 1964. Hormones and sexual behavior. *Science* 143:212–18

Ann. Rev. Neurosci. 1984. 7:443–78

MULTIPLE MECHANISMS OF WITHDRAWAL FROM OPIOID DRUGS

D. E. Redmond, Jr. and J. H. Krystal

Neurobehavior Laboratory, Yale University School of Medicine, New Haven, Connecticut 06510

INTRODUCTION AND OVERVIEW

The opiate withdrawal syndrome has attracted the interest of scientists and society since the realization in the 1880s that the medically useful opiate drugs induce a state of physiological and psychological dependency. Explanations for this syndrome of autonomic disturbance and psychic distress, which appears when drug intake ceases, have ranged from elaborate psychological and moral theories to increasingly precise biochemical theories. New compounds, introduced partly for their lack of withdrawal effects, have included both heroin and methadone; now both are known for their withdrawal syndromes and for substantial abuse. Nearly every treatment suspected to affect brain function has been tried without significant success against the withdrawal syndrome (Kleber & Riordan 1982), including electroconvulsive therapy, major and minor tranquilizers, insulin-induced hypoglycemic shock, and many others. No nonopiate treatment so far has been convincingly proven effective; and the problems of opiate dependency remain substantial for the individuals involved and costly to society due to lost productivity, treatment delivery, and crime associated with drug supply and procurement.

The scientific excitement about treatments for opiate withdrawal phenomena was increased by the discoveries that the opiates act at receptors that are targets for numerous endogenous opiate-like peptide neurotransmitters. The research into the mechanisms of opiate action, tolerance, and withdrawal suddenly appeared to offer insights into fundamental processes of central neural regulation and physiology. At the same time the increasing numbers of neurotrans-

443

0147-006X/84/0301-0443$02.00

mitters, receptors, intracellular modulators, and interacting neuronal systems make more complex models and explanations necessary. Unfortunately, a review that encompasses all of these aspects would require more than the space allotted. We focus, therefore, on particular aspects of the withdrawal syndrome that may help to organize and illuminate the data, which at first glance suggest that every known neurotransmitter and neuroregulatory system is involved. We emphasize the physiology of the classical opiate or true morphine withdrawal or abstinence syndrome, which appears to be physiologically distinct from analgesia and other opiate actions. This physiology suggests that neural processes that induce tolerance and withdrawal may converge in a final common biochemical and neuroanatomical pathway that is mediated substantially by central noradrenergic neuronal systems.

The opiate abstinence syndrome emerges after repeated administration of heroin, morphine, or methadone and lasts for hours to a few days, depending upon the specific drug and the duration and dose of prior administration. This syndrome commonly includes hot and cold flashes, goose-flesh, increases in body temperature, aching bones and muscles, perspiration, anorexia, abdominal cramps, diarrhea, nausea, emesis, lacrimation, rhinorrhea, anxiety, restlessness, yawning, pupillary dilation, tremor, insomnia, tachypnea, tachycardia, increased blood pressure, and craving for the drug (Kolb & Himmelsbach 1938, Himmelsbach 1942).

The time required for the establishment of tolerance and dependence (generally thought to be related) and the course and duration of the abstinence syndrome may be useful data with which to evaluate potential mechanisms underlying the phenomena. An important aim of reserachers is to separate acute and desirable pharmacologic effects of opioids from tolerance, dependence, and withdrawal phenomena. Days to weeks of opiate administration are usually required for dependence and opiate craving to become fully established, and the abstinence syndrome appears to become more severe with increasing dose and duration of opiate intake (Jaffe 1980); however, opiate abstinence phenomena can be produced by administration of opiate antagonists after a few hours of opiate administration (Wikler & Carter 1953, Martin & Eades 1961, Cochin & Kornetsky 1964, Cheney & Goldstein 1971b, Ritzmann 1981, Krystal & Redmond 1983) and a week after a single agonist/antagonist dose (Nutt & Jasinski 1973). Although acute administration of antagonists such as naloxone does not produce withdrawal signs in animals, chronic administration does produce them (Malin et al 1982).

Subtle but prolonged changes in the basal metabolic rate, sleeping patterns, and body temperature are noted in human subjects, following the acute withdrawal syndrome (Himmelsbach 1942). Two stages have been described in this "protracted abstinence" (Martin & Jasinski 1969). During an initial period of several weeks post-withdrawal, small changes are noted similar to the acute withdrawal phase: increase in blood pressure, pulse rate, pupil size, body temperature, and respiratory rate. For some time thereafter, body temperature, pulse, and respiratory rate drop below baseline values. These physiological changes occur during a period in which mild subjective distress, craving for opiates, insomnia, and behavioral disturbances also occur and when many formerly opiate-dependent individuals relapse to opiate use, although relationships between these events are difficult to ascertain.

In this review we focus on areas where overlap is apparent between different biochemical mechanisms of withdrawal phenomena; these areas may provide clues to possible "final common pathways." First, we briefly consider evidence for alterations of endogenous opiate receptors during opiate dependence. Second, we describe the involvement of some intracellular modulators. Third, we review evidence for alterations in six neurotransmitters. Fourth, we outline some behavioral and environmental factors in opioid withdrawal that may have implications for cellular mechanisms. And finally, we review the status of clinical pharmacotherapy for opiate dependence and suggest some issues for future research.[1]

OPIATE RECEPTORS

Characterization of Opiate Receptors

Opiates compete stereospecifically for saturable binding sites proportional to their physiological effects (Goldstein et al 1971, Pert & Snyder 1973, Simon et al 1973, Terenius 1973). Opiate receptors, therefore, are an appropriate starting point for considering mechanisms that might contribute to the development of tolerance and dependence. However, both endogenous and exogenous opioids show differences in relative potency when tested in a variety of brain and peripheral tissues (Lord et al 1977); some opioid compounds exhibit behavioral effects unlike those of morphine. These effects also are not antagonized by the opiate antagonist, naloxone (Martin et al 1976, Shearman & Herz 1982a). In addition, some opioids fail to develop cross-tolerance uniformly and show different binding patterns to the same tissue (Chang et al 1979, Robson &

---

[1]Abbreviations: ACh, acetylcholine; AMPT, alpha-methyl-*para*-tyrosine; AOAA, aminooxyacetic acid; BZD, benzodiazepine; cAMP, cyclic adenosine $3',5'$-monophosphate; cGMP, cyclic guanosine $3',5'$-monophosphate; DA, dopamine; DADLE, D-Ala2-D-Leu5-enkephalin; DHPG, dihydroxy phenylethylene; DHT, 5,6-dihydroxytryptamine; DPA, di-*n*-propylacetate; EKC, ethylketocyclazocine; GTP, guanosine triphosphate; GVG, gama-vinyl-GABA; H2, histamine-2; 5-HIAA, 5-hyroxyindolactic acid; IBMX, butyl-methyl xanthine; LC, locus coeruleus; MHPG, 3-methoxy-4-hydroxy phenethylene glycol; 6-OHDA, 6-hydroxydopamine; NE, norepinephrine; PCP, phencyclidine; PCPA, *para*-chlorophenylaline; PDE, phosphodiesterase; PGE1, prostaglandin-E1; QMAS, quasi-morphine-abstinence syndrome; TRH, thyrotropin releasing hormone.

Kosterlitz 1979, Wuster et al 1980, Zukin & Zukin 1981, Goodman & Snyder 1982). Multiple receptors designated mu, delta, kappa, sigma, and epsilon have been proposed to explain these discrepancies (Martin et al 1976, Lord et al 1977, Schulz et al 1979a, Zukin & Zukin 1981, Chang & Cuatrecasas 1981, Pfeiffer & Herz 1981). As investigators have attempted to relate these multiple receptors to the actions of the endogenous opioid peptides, a major hope has been that selectivity would be discovered for therapeutically useful receptors that were not associated with tolerance and dependence. We concentrate therefore on the evidence that associates these receptors with tolerance and dependence (see Akil et al 1984, this volume, for more detail).

The mu receptor is the classical morphine receptor and is highly sensitive to naloxone antagonism (Chang & Cuatrecasas 1981, Zukin & Zukin 1981). Endogenous ligands with selective affinity for the mu receptor include morphine, dihydromorphine, fentanyl, and (D-Ala2-MePhe4-Met(0)5-ol) enkephalin (Chang et al 1979). The delta receptor is distinguished from the mu receptor by its greater enkephalin potency and relative insensitivity to naloxone (Lord et al 1977). A highly selective agonist for the delta receptor is D-Ala2-D-Leu5-enkephalin (DADLE) (Robson & Kosterlitz 1979). The pharmacologic effects of mu and delta agonists are not easily distinguished (Shearman & Herz 1982a,b). Mu agonists produce meiosis, bradycardia, hypothermia, analgesia, decreased vigilance, and sedation with higher doses. After chronic administration of mu agonists, abrupt abstinence or mu antagonist administration produces the classical opiate withdrawal syndrome. Tolerance to mu and delta agonists develops separately at the two receptors without the development of cross-tolerance (Wuster et al 1980, Frederickson et al 1981, Herz et al 1982). In addition, mu agonists are more potent than delta agonists in producing physical dependence after intraventricular injection (Chang et al 1979, Chang & Cuatrecasas 1979). However, most mu and delta agonists have effects at both receptors, and receptor-selective actions remain an elusive goal.

The kappa receptor was suggested to explain the effects of drugs such as pentazocine or ethylketocyclazocine (EKC) (Martin et al 1976) that did not suppress the opiate withdrawal syndrome despite their analgesic activity in nontolerant individuals (Kosterlitz et al 1981). The kappa agonists constrict pupils, decrease flexor reflexes, and cause profound sedation without altering pulse or skin twitch (Cowan 1981a, Martin 1981, Iwamoto 1981), but do not generalize behaviorally to mu or delta agonists (Shearman & Herz 1982, Herling & Woods 1981). Kappa agonists do not alter delta receptor sensitivity (Chang et al 1982). Varying degrees of cross-tolerance between kappa agonists (Chavkin et al 1982) suggests heterogeneity in the classification, although the development of considerable tolerance has been demonstrated. Dynorphin (Goldstein et al 1979) was recently found to bind to the kappa receptor (Chavkin et al 1982).

The sigma receptor has been controversial because some data suggest that it might not be a true opioid receptor. It appears to be a selective phencyclidine (PCP) receptor as well as a binding site for the opioid agonists SKF-10,047 (N-allylnormetazocine) and cyclazocine (Zukin & Zukin 1979, Chang & Cuatrecasas 1981, Quirion et al 1981). The PCP-like effects of sigma agonists such as behavioral arousal and "bizarre behavior" in rodents are not naloxone reversible (Cowan 1981a, Byrd 1982). The effects of sigma agonists include mydriasis, tachycardia, and "canine-delirium." These effects do not generalize to the effects of mu, delta, or kappa agonists, but generalize to PCP (Herling & Woods 1981, Shannon 1982). Human subjects find cyclazocine to be opiate-like in low doses; but at higher doses, it is more often paradoxically described as an amphetamine or barbiturate (Martin et al 1965). Cyclazocine's abstinence syndrome, appearing three to four days after withdrawal, includes mydriasis, tachycardia, anorexia, weight loss, and increased body temperature, with lesser effects on blood pressure and respiratory rate (Martin et al 1965). Although the nonopiate-like effects of sigma agonists and PCP suggest the possibility that the sigma receptor is not an opiate receptor, the thermoregulatory effects of PCP are naloxone reversible (Glick & Guido 1982). Thus, the sigma receptor has both opioid and nonopioid characteristics.

Alterations in Receptor Number or Conformation

Collier (1965) hypothesized that alterations in receptor numbers or conversion from silent to active states might account for tolerance and dependence on opioids. However, many studies fail to demonstrate a change in opiate receptor numbers (Klee & Streaty 1973, Hollt et al 1975, Simon et al 1975, Creese & Sibley 1981). Two studies suggest that acute but not chronic opiate administration produces a temporary increase in receptors (Pert & Snyder 1975, Harris & Kazmierowski 1975), whereas others find evidence suggestive of down-regulation of opiate receptors (Hitzemann et al 1974, Davis et al 1975), which might occur selectively among the various receptor types (Chang et al 1982).

Conformational changes in the opiate receptor have also been postulated to explain changes in naloxone sensitivity found by some investigators after acute and chronic morphine administration (Takemori et al 1973, Tulanay et al 1979). But others did not find changes in receptor affinity (Klee & Streaty 1973, Pert & Snyder 1973, Hitzemann et al 1974, Simon et al 1975). In the "two conformation model" of opiate receptor function (Pert & Snyder 1974, Cheney et al 1982), the abstinence syndrome is postulated to arise from a decrease in the proportion of receptors held in the agonist conformation by the binding of agonists, consistent with the dissociation rates for morphine and endorphins (Pert & Bowie 1979). However, the two conformation model of the opiate receptor does not take multiple opiate receptor subclasses into account (Bowen et al 1982, Olgiati et al 1982), and a "three conformation model"

(Bowen et al 1982), which includes mu agonist, mu antagonist, and delta agonist, might account for limited interconvertibility of delta receptors and limited cross tolerance of the mu and delta receptors. Besides changes in conformation, conversion of receptors from a silent to an active state (Chavkin & Goldstein 1982) and clustering of receptors in the presence of enkephalin have also been noted experimentally (Chang et al 1979).

INTRACELLULAR MODULATORS OF OPIOID ACTION

Several intracellular modulators appear to be involved in opioid actions, tolerance, and dependency, as first suggested for enzymatic changes (Axelrod 1956, Shuster 1961, Goldstein & Goldstein 1961, 1968). The cyclic nucleotides may be directly related to synaptic transmission through the "receptor-second messenger" model, but also have other functions in nervous tissue, including regulation of microtubule function and neurotransmitter biosynthesis. Calcium and calmodulin have a number of roles in intracellular and receptor-coupled functions. Data relevant to specific synthetic or catabolic enzymes is described below in conjunction with the associated neurotransmitters.

Adenyl Nucleotides

Prostaglandin-induced elevation of intracellular cyclic adenosine 3',5'-monophosphate (cAMP) (Collier et al 1972) has proven valuable in evaluating the role of both adenylate cyclase and prostaglandin in opioid tolerance and dependency. Opioids inhibit the prostaglandin-E1 (PGE1) stimulation of adenylate cyclase, decreasing intracellular cAMP in cell lines, peripheral tissues, and brain homogenates (Collier & Roy 1974, Sharma et al 1975a,b, Mehta & Johnson 1975, Klee et al 1976, Brandt et al 1980, Goldstein et al 1977, Tsang et al 1978, Havemann & Kuschinsky 1978, Walczak et al 1979, Law et al 1982). These studies show that the action of morphine in these systems is naloxone reversible, stereospecific, and guanosine triphosphate (GTP) dependent. Tolerance to opioids is associated with an increase in adenylate cyclase activity and return of intracellular cAMP to baseline levels after two to three days of morphine treatment. Abrupt withdrawal of opioid after chronic administration produces adenylate cyclase hyperactivity associated with an acute flood of intracellular cAMP. Intravenous or intraventricular infusion of cAMP antagonizes morphine analgesia and accelerates the development of opioid tolerance and physical dependence (Ho et al 1973a,b).

Other evidence for cAMP changes in opioid dependence came from studies of isoproterenol-stimulated adenylate cyclase activity. Morphine dependent rodents (Llorens et al 1978) and monkeys (Nathanson & Redmond 1981) exhibit increases in this activity, which suggest increases in beta-adrenergic receptor sensitivity. Abrupt withdrawal of morphine as well as chronic administration of piperoxane [which increases norepinephrine (NE) release via alpha-2 adrenoceptors] decreases adenylate cyclase activity below normal. In view of the possible second messenger role of cAMP, more specific studies of transmitter-sensitive cAMP changes might also illuminate the involvement of histamine (H2), serotonin, octopamine, and peptide hormone receptors in addition to the beta-1-adrenoceptor and prostaglandin E1 effects described (Greengard 1979).

Methylxanthines such as theophylline and caffeine produce symptoms similar to opioid withdrawal in opioid-naive animals, an effect potentiated by naloxone (Collier et al 1972, 1974, 1981). The increases noted in cAMP were suggested as the common denominator between methylxanthine effects and naloxone-precipitated opiate withdrawal. This methylxanthine-induced syndrome was called the "quasi-morphine abstinence syndrome" (QMAS) and has been used to investigate other possible common mechanisms that might be relevant to opiate dependency. QMAS effects have now been attributed to a variety of drugs including caffeine, 3-iso butyl-methyl xanthine (IBMX), sodium valproate [or di-n-propylacetate (DPA)], thyrotropin releasing hormone (TRH), and others (Cowan 1981b).

Other systems implicated in QMAS include norepinephrine/ alpha-2 adrenoceptors, adenosine receptors, acetylcholine receptors, GABA-receptors, and benzodiazepine receptors. Alkylxanthines increase norepinephrine turnover and release (Berkowitz et al 1970, Karasawa et al 1976, Galloway & Roth 1983), and block reuptake (Cardinali 1977). IBMX increases noradrenergic neuronal activity while benzodiazepines inhibit it after microiontophoretic administration in the locus coeruleus (LC). This effect is enhanced by naloxone pretreatment (Grant et al 1980, Grant & Redmond 1982a,b). IBMX-induced increases in locus coeruleus activity and norepinephrine metabolism are reversed by the alpha-2 adrenoceptor agonist, clonidine, which also prevents the development of QMAS seen after administration of IBMX (Grant & Redmond 1982b, Galloway & Roth 1983). Thus, the methylxanthine-induced QMAS appears to involve alterations in the activity of central noradrenergic systems. A dopamine-dependent QMAS has also been suggested, based on effects of L-DOPA (Lal et al 1975a), which possibly acts via a beta-adrenergic receptor (Cramer & Kiessling 1976).

Adenosine receptors may also play a role in the actions of alkylxanthines in the QMAS. Caffeine and theophylline are weak phosphodiesterase (PDE) inhibitors at biologically relevant concentrations but are quite potent antagonists at adenosine receptors (Daly et al 1981). Caffeine, theophylline, and IBMX antagonize the "high affinity" inhibitory adenosine A-1 receptor in concentrations that produce the QMAS. The result of this antagonism is increased adenylate cyclase activity (Wachtel 1982). Extracellular cAMP is an agonist at a "low affinity" stimulatory A-2 receptor that increases adenylate cyclase activity (Fredholm et al 1982). Thus exogenous cAMP, used experimentally in QMAS, may increase intracellular cAMP by stimulating adenylate cyclase via a surface receptor or by passing through the cell membrane and directly increasing intracellular levels.

Actions at adenosine receptors, however, do not preclude the functional involvement of other neurotransmitters, since adenosine inhibits release of acetylcholine, norepinephrine, dopamine, and GABA. In addition, alkylxanthines may interact with benzodiazepine receptors directly, or perhaps indirectly through adenosine (Daly et al 1981). The relationship of the "true" opiate withdrawal syndrome to the various "quasi-opioid withdrawal" syndromes remains a challenging question, which now appears to involve multiple neural systems. QMAS research and the knowledge of multiple receptors are expanding the area of "drug withdrawal" research in general to include phenomena that are more relevant to withdrawal syndromes following nonopioid drugs.

Guanyl Nucleotides

Several studies show that opioid administration increases brain cyclic guanosine 3',5'-monophosphate (cGMP) levels as much as 200% while decreasing cAMP levels 80% (Gullis et al 1975, Minneman & Iversen 1976). In opiate dependent rats, cGMP levels are greater at 24 hours than at 72 hours after cessation of opiate administration (Volicer et al 1977). However, intraventricular administration of cAMP but not cGMP exacerbates the opioid withdrawal syndrome (Roy & Collier 1975). Although cAMP rather than cGMP appears to modulate opioid dependency, considering the possible role attributed to cGMP as a possible second messenger for alpha adrenoceptors, histamine H1, muscarinic acetylcholine, and glutamate receptors, these cGMP changes deserve further and more specific exploration of neurotransmitter sensitive effects (Greengard 1979).

Calcium and Calmodulin

Much evidence supports a relationship between calcium ions and the actions of opiates. Morphine alters intracellular calcium levels, and calcium-mediated changes modify opioid analgesia, tolerance, and dependence. Morphine decreases calcium uptake by synaptosomes; as tolerance develops to the morphine effect, synaptosomal calcium rises (Guerrero-Munoz et al 1979a). However, naloxone-precipitated withdrawal abolishes this increase in calcium uptake (Harris et al 1977). Although EDTA (which chelates both calcium and manganese) has no effect, EGTA (which selectively chelates calcium and makes it unavailable to the cell) facilitates morphine analgesia as well as the development of tolerance and opioid dependency (Guerrero-Munoz et al 1979b). In addition, lanthanum (which impairs calcium movement by blocking calcium channels) produces analgesia, which shows cross-tolerance to morphine and is partially reversible by naloxone (Harris et al 1976). Also, low intracellular levels of calcium stimulate adenylate cyclase, whereas high levels inhibit the enzyme (Rasmussen 1970). Calcium also affects alpha-1 vs alpha-2 adrenoceptor responses (McGrath 1983) and appears to regulate spontaneous activity and reactivity to orthodromic stimulation in the locus coeruleus (Aghajanian & VanderMaelen 1982, Aghajanian et al 1983). Thus, calcium could play a modulatory role in opioid abstinence via interactions with several neurotransmitters.

The crucial intermediary for calcium action within the neuron is a synaptosomal calcium-binding protein, calmodulin, which influences neurotransmitter synthesis and release (DeLorenzo 1981, 1982). Calmodulin is also an inhibitory subunit of a phosphodiesterase modulating hormone action and may modulate cGMP levels (Rasmussen 1970, Cheung 1982). Dynorphin and beta-endorphin, but not met- or leu-enkephalin, bind in an inhibitory fashion to a brain calmodulin, which stimulates a phosphodiesterase (Sellinger-Barnette & Weiss 1981). Acute morphine inhibits calmodulin in a dose-dependent, naloxone-reversible manner (Nehmed et al 1982, Simantov et al 1982). Chronic morphine administration produces a 28–37% increase in calmodulin activity of rat midbrain and thalamus, which correlates with an increase in calcium in end organ structures. Also, with chronic opioid administration, tolerance develops to the calcium-regulated phosphorylation of intracellular proteins seen after acute opiate administration (O'Callaghan et al 1982). Both calmodulin and adenylate cyclase are thought to exert their regulatory actions via alterations in protein phosphorylation (Cheung 1982, Greengard 1979). Thus calmodulin, like adenylate cyclase, appears to mediate the cellular actions of opioids in analgesia and opioid dependency. Consistent with this idea, inhibitors of protein synthesis inhibit the development of tolerance and dependence on opioids (Loh et al 1969).

ALTERATIONS IN NEUROTRANSMITTER SYSTEMS

Enkephalin

A number of auto-regulatory mechanisms have been suggested that would lead to functional deficits of endogenous opioids during chronic administration of exogenous opiates. These deficits might contribute to acute or protracted abstinence effects. On the other hand, if tolerance developed in autoreceptors, an increase in enkephalin release might result after ceasing opioid administration. Simantov & Snyder (1976) showed that acute morphine administration increases brain enkephalin 25%, and five days of administering morphine pellets doubles it. However, no increases (Childers et al 1977) or decreases (Shani et al 1979) in brain enkephalin after chronic morphine administration were found in later studies.

Changes in catabolic activity of endogenous opiates might interact with exogenous opioids to contribute to tolerance or dependence. Specific enkephalinases rapidly and selectively degrade enkephalin (Lane et al 1977, Dupont et al 1977). The enkephalinases with the highest affinity for the pentapeptides, A1 and A2, are endopeptidases, which cleave Met-enkephalin (Tyr-Gly-Gly-Phe-Met) and Leu-enkephalin (Tyr-Gly-Gly-Phe-Leu) at the Gly3-Phe4 bond (Malfroy et al 1978, Gorenstein & Snyder 1980). An additional enkephalinase, B, is an aminopeptidase, which cleaves enkephalins at the Gly2-Gly3 bond. Enkephalinase is selectively inhibited by enkephalin, its breakdown products, and the peptidase inhibitors bacitracin, aprotinin, and thiorphan (Miller et al 1977, Malfroy et al 1978, Gorenstein & Snyder 1980, Sullivan et al 1980) and by captopril (Greenberg et al 1981). Reuptake as a mechanism for terminating peptide neurotransmitter activity has not been demonstrated.

Opiate agonist-like effects result from decreases in enkephalinase activity (Simmons & Ritzmann 1976, Miller et al 1977, Roques et al 1980, Martin & Voight 1982, de la Baume et al 1982). Chronic morphine administration increases the high affinity enkephalinase 59% (Malfroy et al 1978). Aprotinin decreases naloxone-precipitated effects in opioid dependent rats, whereas an aprotinin-bacitracin mixture decreases the "total severity" of the precipitated abstinence syndrome (Pinsky et al 1982). However, it is still unclear whether alterations in enkephalinase activity or enkephalin levels play a direct role in the genesis of opioid dependence and the abstinence syndrome. The antiwithdrawal and analgesic effects of the enkephalinase inhibitors remain an important subject for clinical research, although some clinical data suggest that neither enkephalin nor enkephalinases play a major direct role in the acute methadone abstinence syndrome (see below).

Acetylcholine

Kolb & Himmelsbach (1938) pointed out that "the morphinist is one who has been accustomed to perverse vagal predominance and sympathetic depression. . . ." This link between acute morphine effects and cholinergic systems is supported by increased brain levels of acetylcholine (ACh) due to inhibition of its release and synthesis, with some tolerance to the effect with chronic opioid administration (Giarman & Pepeu 1962, Hano et al 1964, Howes et al 1970, Large & Milton 1970, Sharkawi 1972, Domino & Wilson 1973, Jhamandas & Sutak 1980). During spontaneous or naloxone-precipitated opioid withdrawal, ACh turnover increases. Cholinergic agonists exacerbate opioid withdrawal, whereas both muscarinic and nicotinic blockers attenuate some limited aspects of the syndrome (Martin & Eades 1967, Bhargava & Way 1972, Vasquez et al 1974).

Cholinergic systems appear to play a more central role in mediating a QMAS in rhesus monkeys (Swain & Seevers 1976) and in opioid-induced hyperthermia during chronic administration and tolerance (Oka et al 1972, Sharkawi

1972, Glick 1975, Rosow et al 1980). However, the acute hypothermic effects of morphine are antagonized by both the serotonin antagonist, *para*-chlorophenylalanine (PCPA), and the dopamine receptor blocker, haloperidol. The benzazocine compound, UM 1046 (*N*-cyclopropylmethyl-1,2,3,4,5,6-hexahydro-8-hydroxy-6-methyl-3-benzazocine), and the similar UM 1037 produce a QMAS, possibly via cholinergic mechanisms (Valentino et al 1979, 1981, Smith & Valentino 1980).

Gamma-Amino-Butyric Acid and Benzodiazepines

GABA and benzodiazepines (BZD) are also implicated in opioid mechanisms. Because many methodologies used to study them interact, as do their receptors (Tallman et al 1980), we discuss them together in this section. The GABA agonist, muscimol, potentiates morphine analgesia, although it has no analgesic effect by itself (Biggio et al 1977). Increases in brain GABA induced by gamma-vinyl-GABA (GVG) (Buckett 1980) and aminooxyacetic acid (AOAA) (Spaulding et al 1980) show opiate-like analgesic effects in rodents, although some controversy exists (Ho et al 1976). Both beta-endorphin and morphine administration increase GABA levels in the globus pallidus (Moroni et al 1978). This effect of morphine on GABA is not inhibited by haloperidol (van der Heyden et al 1980).

Morphine tolerance is associated with increased GABA binding in the medulla, diencephalon, striatum, and cerebellum, but naloxone-precipitated opiate withdrawal does not change the number of GABA receptors (Sivam et al 1981, 1982). GVG and bicuculline do not prevent the morphine withdrawal syndrome, but picrotoxin and diazepam ameliorate it in sedative doses (Hynes et al 1980). Clinically, both diazepam and baclofen, a GABA analogue that has both GABA and non-GABA effects, decrease the severity of the opiate abstinence syndrome in humans (Litt et al 1971, Jaffe et al 1982).

Some GABA or BZD effects may result from increases in enkephalin. Increasing GABA increases striatal met-enkephalin release (Harsing et al 1982, Sawynok & Labella 1981), and benzodiazepines decrease striatal and increase hypothalamic enkephalin (Duka et al 1979, 1980). However, the necessity of distinguishing pre- and postsynaptic GABA receptors complicates interpretation of these data (Arbilla & Langer 1979, de Boer et al 1980, Bourgoin et al 1982). The understanding of GABA-opiate interactions is further complicated by the observation that naloxone, particularly at high doses, acts as a GABA antagonist (Gumulka et al 1979, Dingledine et al 1978). Naloxone also antagonizes the anxiolytic action of benzodiazepines (Duka et al 1982).

A QMAS produced by di-*n*-propylacetate (DPA or valproate) has been linked to enhanced GABA-ergic activity (de Boer et al 1977, 1980, van der Laan & Bruinvels 1981, van der Laan et al 1980, 1982). DPA increases opioid withdrawal behaviors when injected into the nucleus center median parafascicularis of the thalamus and the central nucleus of the amygdala. Systemic DPA produces a more complete QMAS, including jumping, body shakes, chewing, and increased motor activity. The effects are inhibited by morphine, the GABA blockers bicuculline and picrotoxin, as well as a GABA synthesis inhibitor. Several lines of evidence suggest that noradrenergic mechanisms play an important modulatory role in this QMAS. Morphine injections into the noradrenergic nucleus, the locus coeruleus, 6-hydroxydopamine (6-OHDA) lesions of the dorsal noradrenergic bundle, a norepinephrine synthesis inhibitor (FLA-63), clonidine, and guanfacine, which all reduce NE activity, decrease the DPA-induced QMAS; but direct DPA injections into the locus coeruleus (where increased concentrations of GABA inhibit neuronal activity) do not produce the syndrome (van der Laan et al 1982).

Serotonin

A role for the monoamine neurotransmitters in the effects of opiates has been suggested for many years. Early studies using available biochemical techniques were often contradictory (see Eidelberg 1976) and, as with other neurotransmitter systems, accurate conclusions could not be drawn from biochemical concentrations of transmitters or metabolites alone. The discovery of specific neurotransmitter-containing systems, based on fluorescent staining techniques, has made it possible to study particular systems with greater neurophysiological and neurochemical specificity. Evidence for the involvement of serotonergic neurons in opiate actions includes increases in serotonin turnover and synthesis in acute and chronic morphine-treated animals, increases in brain's production of the serotonin metabolite 5-hydroxyindolactic acid (5-HIAA) acutely, and tolerance to morphine's analgesic action, corresponding with decreases in 5-HIAA levels to baseline (Way et al 1968, Ho et al 1972, Haubrich & Blake 1973, Laska & Fennessy 1976). Supersensitivity to serotonin (5-HT) has also been shown in morphine tolerant tissues (Schulz & Goldstein 1973). *Para*-chlorophenylalanine (PCPA), which inhibits serotonergic synthesis, and exogenous tryptophan, which increases brain serotonin levels, both have been reported to antagonize opiate analgesia (Ho et al 1975), but others found no effect of PCPA on acute effects of opiates (Cheney & Goldstein 1971a). Raphe lesions, which decrease brain serotonin, inhibit morphine analgesia (Proudfit 1980).

The results are inconclusive with regard to the effects of altering serotonin activity on the development of tolerance and dependence on opiates. Intraventricular administration of 5,6-dihydroxytryptamine (DHT), which destroys serotonergic neurons, decreases signs of morphine tolerance and dependence (Ho et al 1973c). PCPA and serotonin receptor antagonist administration produce similar results in some (Ho et al 1972, Cervo et al 1981) but not all studies (Cheney & Goldstein 1971a, Cheney et al 1971).

Alterations in some behavioral signs associated with opiate abstinence, in particular "wet dog" shakes (Bedard & Pycock 1977), sterotypy (Carter & Pycock 1981), and withdrawal jumping (Collier et al 1972), demonstrated interactions with dopamine systems (Dafny & Burks 1977, Waddington & Crow 1979); and the absence of effects on other prominent withdrawal signs has suggested the involvement of other neurotransmitters (Cervo et al 1981). Some alterations in serotonergic activity might be associated with changes in noradrenergic systems, suggested by anatomical and physiological links between the serotonergic raphe nuclei and the locus coeruleus (Svensson et al 1975, Gallager & Aghajanian 1976, Pickel et al 1977, Segal 1979, Baraban & Aghajanian 1980, 1981, McRae-Deguerce et al 1982).

Dopamine

Dopamine systems do not respond homogeneously to opioid administration. Opioids increase dopamine turnover in the mesolimbic and striato-nigral systems, while decreasing turnover in the tuberoinfundibular and cortical dopaminergic systems (Clouet & Ratner 1970, Kuschinsky & Hornykiewicz 1974, Moleman & Bruinvels 1976, Biggio et al 1978, Persson 1979, Deyo et al 1979). In the nigrostriatal system, neuronal firing rates and dopamine synthesis are similarly changed (Nowycky et al 1978). Prolactin release, which is inhibited by hypothalamic dopamine, increases during acute opioid administration, shows tolerance to the opioid effect, and declines during opiate withdrawal, although these effects may be directly mediated by opiate receptors (Tolis et al 1975, Deyo et al 1979). The striato-nigral system shows the opposite changes in response to opioids and during opiate withdrawal (Lal 1975, Blasig et al 1976).

L-DOPA, apmorphine, and amphetamine (which increase DA activity) exacerbate the opioid withdrawal syndrome, whereas antagonists such as haloperidol decrease the severity of the syndrome (Lal et al 1971, Gianutsos et al 1974, Hynes et al 1978). The withdrawal behaviors affected by these drugs include jumping, chewing, writhing, and aggression. That only selected aspects of opioid withdrawal such as jumping were affected suggested that dopamine plays a modulatory but not a primary role in opioid withdrawal. In addition, dopaminergic neurons are inhibited by drugs that produce the GABA-QMAS, whereas noradrenergic neurons are activated by these drugs (Yessian et al 1969, Biswas & Carlsson 1977, van der Laan et al 1982).

Many of the methods that have previously implicated dopamine (DA) mechanisms do not adequately distinguish between dopamine GABA, norepinephrine, and opioid systems. Opioids act on dopaminergic neurons via opiate receptors and not dopamine receptors (Schwartz et al 1978). Also, the butyrophenones, the effects of which have implicated dopaminergic systems in opiate withdrawal, bind to opioid receptors with affinity comparable to meperidine (Creese et al 1976, Clay & Brougham 1975), of which they are structural derivatives. The antiwithdrawal effects of haloperidol, which appear to support dopaminergic involvement in withdrawal in animals and humans, are blocked by naloxone (Las & Hynes 1978). Furthermore, dopamine and noradrenergic effects are not easily distinguished after 6-OHDA or alpha-methyl-*para*-tyrosine (AMPT) administration; and the systems may interact neurophysiologically. Several alpha-1 and alpha-2 adrenergic agonists and antagonists affect dopamine function secondarily to their effects at these receptors (Anden & Grabowska 1976, Svensson & Ahlenius 1982). In addition, a presynaptic dopaminergic receptor on posterior hypothalamic noradrenergic neurons has been noted (Galzin et al 1982). Thus, there is both behavioral and biochemical evidence of interaction between these two systems.

Norepinephrine

Evidence for central noradrenergic changes in opioid actions was first published by Vogt (1954) who found decreases in "sympathin" (NE) in cat brains after high doses of morphine. The studies that followed using biochemical techniques were often contradictory (see Clouet 1975, Eidelberg 1976 for reviews), but recent studies of norepinephrine-containing neuronal systems have clarified the role of central norepinephrine. Much recent evidence has been derived from studies of the nucleus locus coeruleus, the primary source of noradrenergic innervation of the limbic system, cerebral and cerebellar cortices, and a quantitatively smaller source of innervation of hypothalamic and other brainstem nuclei (Korf et al 1974, Moore & Bloom 1979, Grant & Redmond 1981). Its compact and homogeneous cells can be studied neurophysiologically and neuropharmacologically as representative of central NE nuclei, which may also be functionally connected. Neuroanatomically, the locus coeruleus receives afferents from primary pain and sensory systems and sends efferents to many areas of the brain associated with specific physiological changes described during the classical morphine withdrawal syndrome. Postsynaptic receptors for these efferents are alpha-1, alpha-2, beta-1, and beta-2 adrenoceptors, where NE release, in addition to some specific effects, seems to increase the "signal to noise ratio" and to modulate other neurotransmitter inputs (see Woodward et al 1979 for review). The locus coeruleus appears to contain presynaptic receptors on the perikarya or dendrites and on the nerve terminals that mediate modulatory responses. Receptors for substance P, muscarinic acetylcholine, and glutamate are excitatory, and inhibitory responses are seen following application of the alpha-2 adrenoceptor agonists norepinephrine and epinephrine, with evidence suggesting autoregulatory collateral inhibition. Clonidine is a potent agonist at these receptors, whereas antagonists yohimbine and piperoxane induce cell firing and increase NE release (Cedarbaum & Aghajanian 1976, Aghajanian et al 1977; see Moore & Bloom 1979,

Grant & Redmond 1981, Langer 1981 for reviews). A GABA receptor is inhibitory, and benzodiazepine compounds decrease locus coeruleus activity (Grant & Redmond 1981). Most relevant to interactions with opioids, the locus coeruleus receptors for endogenous opioid peptide ligands, including enkephalin, endorphin, and dynorphin (see Figure 1) (Atweh et al 1978, Miller & Cuatrecasas 1978, Miller & Pickel 1980, Watson et al 1980, Pickel 1982, Watson et al 1982); and opioids inhibit neuronal firing rates in the locus coeruleus (Korf et al 1974, Sasa et al 1975, Young et al 1977, Bird & Kuhar 1977, Aghajanian 1978). Paradoxically, acute doses of morphine rapidly increase the concentrations of the NE metabolites, 3-methoxy 4-hydroxy phenethylene glycol (MHPG) and dihydroxy phenylethylene glycol (DHPG), in rat brain (LoPachin & Reigle 1978, Huff & Reigle 1980), with rapid tolerance after a single dose (Reigle & Huff 1980).

Some evidence suggests that the locus coeruleus and perhaps similar brain NE nuclei may mediate much of the classical morphine withdrawal syndrome (and perhaps other similar syndromes). Stimulation of the LC produces several behavioral and physiological signs (Redmond et al 1976, Redmond & Huang 1979) that are similar to the opiate abstinence syndrome in opiate-naive monkeys (Redmond et al 1978, Redmond 1981). NE hyperactivity in opiate withdrawal therefore would be opposite to opioid effects. Gunne reported decreased norepinephrine concentrations in brain during withdrawal (Gunne 1959, 1963), which now appear to result from increased NE release. Tolerance develops to opioid inhibition, and abrupt withdrawal of opioids (or naloxone administration) produces hyperactivity of the locus coeruleus in opioid dependent animals as demonstrated by single unit activity (Aghajanian 1978) or measures of NE turnover (Crawley et al 1979, Laverty & Roth 1980, Roth et al 1982). These recent findings are all consistent with older data showing that alpha-methyl-p-tyrosine, an inhibitor of NE synthesis, substantially blocks the morphine withdrawal syndrome in animals (see Clouet 1975 for review).

Provocative evidence resulted from interactions of the alpha-2 agonist, clonidine, with LC associated effects. In low doses, clonidine decreases the firing of LC neurons (Svensson et al 1975), and blocks the effects of LC stimulation in monkeys (Redmond 1977). Although clonidine and morphine both inhibit the locus coeruleus, they do so at independent receptors. Effects of microiontophoretically applied clonidine are not reversed by naloxone but by alpha-2 antagonists, and morphine effects are reversed by opioid antagonists but not alpha-2 antagonists. Clonidine inhibits the LC hyperactivity elicited by naloxone in morphine tolerant rats (Aghajanian 1978). Administered together, clonidine and morphine have synergistic effects (Aceto & Harris 1981). Behavioral and physiological signs of opioid withdrawal in rodents and monkeys are reduced or blocked by clonidine (Tseng et al 1975, Meyer & Sparber 1976, Colelli et al 1976, Vetulani & Bednarczyk 1977, Fielding et al 1978, Crawley

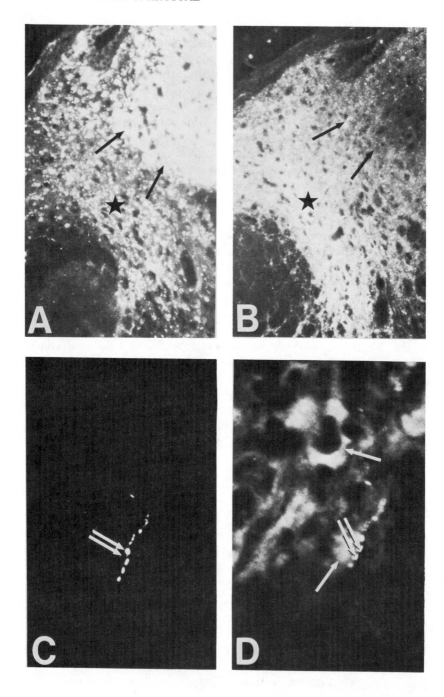

et al 1979, Laverty & Roth 1980, Roth et al 1982). The efficacy of clonidine in the alleviation of the opioid abstinence syndrome in human subjects is consistent with these preclinical effects (Gold et al 1978a,b) (discussed in more detail below). Other alpha-2 adrenoceptor agonists, such as lofexidine (Shearman et al 1980) and guanfacine, also inhibit the increased noradrenergic activity and withdrawal signs of opiate withdrawal in rats (Zigun et al 1981). Debrisoquin, which may decrease central NE activity by another route, also reduces withdrawal signs in rats proportionally with its effects on brain MHPG concentrations (Swann et al 1983).

Based on several measures of postsynaptic adrenoceptor sensitivity, chronic opioid administration produces a type of disuse hypersensitivity based on a variety of measures (Vasquez et al 1974, Montel et al 1975, Llorens et al 1978, Nathanson & Redmond 1981, Kuriyama 1982). The beta-receptor coupled changes are characterized by increased sensitivity of isoproterenal stimulated adenylate cyclase activity. Some effects of LC stimulation in monkeys are also blocked by the beta-1 antagonist propranolol; this is consistent with data cited above that propranolol has some anti-withdrawal properties in animal models and some effects in clinical trials (Grosz 1972, Resnick et al 1976). Failure of this treatment to block the syndrome more than partially may be due to remaining effects at alpha-1 or alpha-2 adrenoceptors that are left intact or increased. Opioid tolerance also increases the number of clonidine binding sites in the rat brain. This hypersensitivity and/or change in the size of the alpha-2 receptor population decreases with naloxone-precipitated withdrawal (Hamburg & Tallman 1981), and isoproterenol-stimulated adenylate cyclase becomes hypoactive after spontaneous withdrawal (Nathanson & Redmond 1981). The increased sensitivity resulting from chronic morphine treatment is similar to the effects of the neurotoxin 6-OHDA, which potentiates opiate

←———

Figure 1   Panels A and B: Thin serial sections were taken through the locus coeruleus. Panel A was stained with anti-DBH serum—*arrows* indicate the very heavily fluorescent NA cells of the locus coeruleus, the *star* is in the middle of a heavily fluorescent NA fiber field. Panel B was stained with anti-enkephalin serum. The *arrows* are in approximately the same position as in panel A, and are meant to indicate the NA cell region (now devoid of staining). The *star* indicates a dense enkephalin fiber field. The essence of this (and other) studies is that enkephalin fibers co-mingle with NA fibers, and surround the cells of the locus coeruleus.

Panels C and D: This is a black and white representation of an original two color fluorescent study. Panel C (*double arrows*) is a B-endorphin stained fiber (formerly green flourescent-FITC) in this locus coeruleus. Panel D (with different filters) shows the same fiber (*double arrows*) and the NA locus coeruleus cells (formerly red fluorescent-Rhodamine)—see *single arrows.* The implication of this type of study was of potential axo-somatic contacts between B-end/ACTH fibers and NA cells of the locus coeruleus—as well as potential axo-axonic contacts in the projection pathway of both systems (not shown). (Illustration courtesy of S. J. Watson et al 1980.)

analgesia. It does not prevent tolerance, however, and exacerbates the abstinence syndrome, as might be predicted if it contributed to further increases in postsynaptic sensitivity (Friedler et al 1972, Elchisak & Rosecrans 1979). For this reason 6-OHDA "lesions" may not be an interpretable test of the effect of noradrenergic deficits.

Previously noted data suggest that changes in other neurotransmitters or neuromodulators result in alterations of noradrenergic function, including GABA, dopamine, calcium, calmodulin, cAMP, and adenosine. The methylxanthine QMAS (Grant & Redmond 1982a,b, Galloway & Roth 1983), the DPA-GABA QMAS (van der Laan et al 1982), and the QMAS after UM 1046 (Valentino & Aston-Jones 1982) all may be substantially modulated by LC or clonidine sensitive mechanisms. On the other hand, clonidine and other relatively NE-specific compounds have effects at other receptors outside the locus coeruleus system, possibly at post-synaptic alpha-2 sites, on dopamine turnover, and outside the brain. These sites may be responsible for some clonidine effects (Hughes et al 1978, Franz et al 1982). Another highly selective alpha-2 adrenoceptor agonist, guanfacine, does not stimulate histamine H2 receptors, as clonidine does (Summers et al 1981), but still has anti-withdrawal potency (Zigun et al 1981). The widespread distribution of opioid receptors does support the probability of withdrawal effects mediated by non-noradrenergic systems, although the "rebound" withdrawal phenomena that these might produce would not necessarily contribute to the discomfort and dysphoria of the clinical withdrawal syndrome.

ENVIRONMENTAL AND LEARNING FACTORS IN OPIOID WITHDRAWAL

Animal and Human Studies

Environmental factors associated with opiates and with the production of opiate addiction have been extensively considered. Conditioned salivation in dogs using morphine as the unconditioned stimulus was one of the first examples of classical conditioning (Pavlov 1927). A naloxone-reversible attenuation of the morphine abstinence syndrome has also been conditioned (Drawbaugh & Lal 1974). Withdrawal-like responses have also been described in animals and man upon exposure to components of the drug-using experience or other drug-related stimuli (Wikler et al 1953, Irwin & Seevers 1956, Teasdale 1973, O'Brien et al 1975, Goldberg 1976, Goldberg & Schuster 1970, O'Brien et al 1977, Sideroff & Jarvik 1980). In addition to conditioned stimuli studied more formally, both stress (Whitehead 1974) and context have a profound effect on opioid effects and withdrawal (Siegel 1976, Siegel et al 1982).

A Possible Neural Substrate for Opiate Withdrawal, Conditioned Opiate Withdrawal, and Fear

If there were a primary neural substrate for fear or anxiety and if such a system were involved in the addiction process, abstinence should be sensitive to the same types of complex environmental, social, and internal cues as anxiety or fear. Both fear and opiate withdrawal should be phenomenologically similar, and both should be similarly affected by pharmacological changes in the relevant system. A growing body of evidence supports this hypothesis. Simultaneous suppression of fear and pain by endogenous opiatergic mechanisms (Scallet 1981) is supported by empirical evidence and, during severe "fight or flight" conditions or stress, would probably be adaptive. In the opposite direction, classical opiate withdrawal and fear are quite similar in 26 behavioral and physiological aspects in man and other primates (see Table 2, p. 153 in Redmond 1981), and in humans include subjective anxiety, restlessness, similar physiological changes, and, during severe withdrawal and panic states, fear of impending doom or death. Much of the opiate abstinence syndrome is so similar to anxiety, fear, and panic that it has been described as a "phobia," in which abstinence serves as the "phobic stimulus" (Hall 1979).

This same comparison between opiate withdrawal effects and fear or anxiety also reveals the extensive overlap between these phenomena and the effects of locus coeruleus activation or inhibition (Redmond 1981). The locus coeruleus and associated central noradrenergic systems therefore may be a major neural substrate for many opiate withdrawal signs and symptoms and for fear or anxiety. This hypothesis is based on studies of LC stimulation and lesions in monkeys (Redmond et al 1976, Redmond & Huang 1979), on the common effects of most anxiolytic drugs on the locus coeruleus, and on predicted effects of activation or inhibition of NE function in human subjects which have been reviewed elsewhere (Redmond 1982). Neuronal responses in the locus coeruleus to novel stimuli in awake animals (Aston-Jones & Bloom 1981, Foote et al 1980) and the stimulus enhancing and specific physiological effects of LC activation are consistent with this postulated function (Woodward et al 1979), although others have drawn different conclusions from these data (see van Dongen 1981). As might be predicted, the locus coeruleus system is involved in opioid-related interaction with environmental stimuli and opioid-seeking behavior in particular (Lewis et al 1976, Arnsten et al 1981, Smith et al 1982). The locus coerulus and its associated central noradrenergic systems and projection areas then may be a common neural substrate for aspects of fear, anxiety, "conditioned" opiate withdrawal, true morphine abstinence syndrome, and some quasi-morphine abstinence syndromes.

CLINICAL ISSUES IN PHARMACOTHERAPY FOR OPIATE DEPENDENCE

Opiate Agonists and Antagonists

Treatment for opiate addiction usually involves substitution of the long-acting opioid, methadone, which has the advantages of engaging subjects in psychosocial therapy, stabilizing the opioid level, preventing the withdrawal syndrome, and largely blocking the effects of illicit opiate drugs. The disadvantages of methadone treatment are that it is a substitute opioid agonist that is subject to abuse and that it produces a significant withdrawal syndrome in most subjects. The goal of obtaining a narcotic drug-free state is difficult to achieve. Twenty to 60% of patients are able to reach an opiate free state (Sorensen et al 1982), but of these 90%–95% eventually relapse to opioid use. From a biochemical point of view, the slow pharmacokinetics of methadone and the "stability," which is an advantage for social and psychological therapy for some individuals, may induce greater dependence across a wider spectrum of opioid systems than results from heroin use itself. The early assumption that methadone did not produce withdrawal (Isbell et al 1947) might have resulted from the more gradual withdrawal phase, although methadone-dependent individuals often feel that methadone withdrawal is "worse" than that from other opioids, do not like it, and frequently continue to take other opiates during treatment (Sutker & Allain 1974), although objective data are less negative (Goldstein 1972, Senay et al 1977). Several other opioid agonists (Jaffe et al 1972, Senay et al 1977, Jasinski et al 1977, Woody et al 1981, Marcovici et al 1981, Sorensen et al 1982) and opioid peptides [beta-endorphin, D-Ala2-MePhe4-Met-(0)-5-ol enkephalin, and dynorphin] have been tested in withdrawal studies and appear to share many of methadone's disadvantages (Su et al 1978, Wen 1982, Wen & Ho 1982).

According to behavioral paradigms, opiate-seeking behavior might be extinguished by blocking the reinforcing properties of opiates. Methadone achieves this blockade via cross tolerance, perhaps extinguishing the reinforcing properties of the injection process and facilitating the cessation of opiate use (Goldstein 1972). Some opiate antagonists, such as cyclazocine and nalorphine, block the action of heroin at the mu receptor but produce undesirable side effects such as hallucinations via their agonist activity at the kappa and sigma receptors, thereby limiting their clinical value (Martin & Gordetzky 1965). Naltrexone, a long-acting, relatively pure antagonist, decreases opiate self-administration (Meyer et al 1975, Mello et al 1981), but high dropout rates hinder its clinical use (O'Brien et al 1975). A psychological explanation is that antagonists decrease the perceived availability of opiates rather than extinguishing craving for opiates (Meyer et al 1975). A biological explanation might be that the antagonists block important endogenous opioids associated with feeling states, mood, and response to stresses—a deficiency of which might have contributed to the reinforcing properties of opiates in these individuals initially. An additional limitation of the use of the pure antagonists is the fact that opiate dependent individuals must be entirely opiate free to avoid severe withdrawal effects upon initial administration. One solution to these problems may be the use of mixed agonist-antagonists, such as buprenorphine, which produces acceptable subjective effects and few abstinence symptoms, perhaps at the expense of potential abuse (Jasinski et al 1978, Banks 1979, Mello et al 1981a).

Nonopiate Treatments

Nonopiate treatments for the opiate withdrawal syndrome have had limited success in human subjects (Grosz 1972, Resnick et al 1976, Drawbaugh & Lal 1974, Jaffe et al 1982). The discovery that clonidine attenuates opioid withdrawal in human subjects, therefore, is of theoretical as well as practical significance (Gold et al 1978a,b). Clonidine does not bind to opiate receptors, does not produce euphoria, and is discriminable from opiates, although it has some opiate-like properties (Young et al 1981). Five to 30 μg/kg orally in divided doses prevents or reverses nearly every sign or symptom of withdrawal from mu-type opioids, although not completely in every patient (Charney et al 1981). The percentage of patients who achieve a narcotic and symptom free state or begin antagonist administration in inpatient studies appears greater than with methadone schedules alone (Gold et al 1979a,b, 1980, Uhde et al 1980, Charney et al 1982). It is also successful, but less so, in outpatient studies where illicit drugs are more available; clonidine doses cannot be as precisely titrated; and higher doses may be prevented by hypotension or other side effects (Washton et al 1980a, Washton & Resnick 1980). In these different settings, success rates have been reported from 4/13 patients (Washton & Resnick 1981) to 99/100 (Gold et al 1981). A clondine withdrawal syndrome noted after long periods of administration to hypertensives (Whitsett et al 1978) or after high doses in monkeys (Woolverton et al 1981) has not been a clinical problem after opiate withdrawal, although most protocols discontinue clonidine over a two day period after seven or eight days of administration. Specifically, the opiate withdrawal syndrome is not merely suppressed and does not emerge when clonidine is completely discontinued, nor when subjects are challenged with naloxone or naltrexone. Lofexidine, another imidazoline alpha-2 adrenoceptor agonist, shows comparable antiwithdrawal activity (Gold et al 1982, Washton et al 1982) perhaps with fewer sedative or hypotensive effects (Washton & Resnick 1981).

Clonidine maintains anti-withdrawal effects in opiate-dependent humans administered naltrexone (Riordan & Kleber 1980, Charney et al 1982). A rapid treatment is therefore possible, using increasing doses of naltrexone, with abstinence signs and symptoms suppressed by clonidine. Subjects do experience some discomfort, but 45 subjects have completed the protocol and achieved full treatment doses of 50 mg naltrexone daily (D. S. Charney et al, unpublished). Although many of the inpatient studies have been placebo-controlled with both subjects and raters blinded to administration schedules, important comparisons with methadone withdrawal schedules within a single institutional study are difficult for many reasons, including "unblinding" by the powerful clonidine effect, and the necessity to compare symptom suppression, treatment retention, and reaching a drug free state under schedules of greatly differing lengths. One small outpatient study found no difference between abrupt discontinuation of methadone plus clonidine treatment and gradual methadone reduction 1 mg per day for 20 days, based on percentage of patients reaching 10 days opiate-free (Washton & Resnick 1981). Important controlled follow-up comparisons on relapse rates after various clonidine-withdrawal methods or traditional withdrawal schedules have also not been reported, although early data support the advantages of institution of naltrexone and comprehensive psychological and rehabilitative treatment (Gold et al 1981, Rawson et al 1981).

Since the acute abstinence syndrome is inhibited or reversed by clonidine, these clinical data have been interpreted as supporting a major role of clonidine-sensitive mechanisms in the classical morphine or methadone abstinence syndrome. The fact that the abstinence syndrome subsides during treatment with high doses of naltrexone can also be interpreted as evidence against the involvement of naltrexone-sensitive endogenous opioids in the recovery from the acute abstinence syndrome (Charney et al 1982). Instead, recent clinical data, combined with preclinical evidence reviewed above, continue to implicate central noradrenergic hyperactivity in the acute withdrawal syndrome (see Figure 2). Baseline to peak change in rated abstinence scores in methadone-dependent subjects who received naltrexone or placebo under controlled blinded conditions correlated significantly with baseline to peak change in plasma MHPG concentrations (Charney et al 1983), which may reflect brain MHPG and NE turnover (Elsworth et al 1982). The effects of naltrexone-precipitated withdrawal and clonidine on brain and plasma MHPG and the abstinence syndrome in monkeys (Roth et al 1982) and suppression of plasma MHPG and the abstinence syndrome in humans by clonidine (D. E. Redmond et al unpublished) are consistent with these data and the involvement of central noradrenergic hyperactivity in the acute opiate withdrawal syndrome.

SUMMARY

A selective review of biochemical changes and pharmacologic effects during the opiate abstinence syndrome reveals changes in several neurotransmitter,

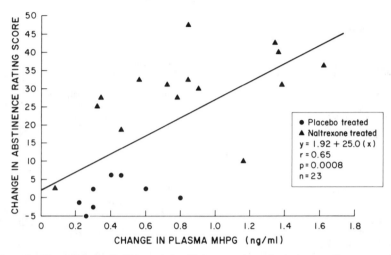

Figure 2   The statistically significant relationship between change in abstinence rating scores and change in concentration of plasma MHPG is shown in 23 methadone-dependent human subjects some of whom were given placebo and some were given naltrexone 60 to 90 minutes before sampling (see Charney et al 1983).

intracellular, and receptor-modulated activities. Considerable portions of this evidence, from endogenous opioid distributions and activities to the anti-withdrawal effects of clonidine in human subjects, suggest that central NE systems, including the nucleus locus coeruleus, may be important for further studies of the biochemistry of the classical opiate withdrawal syndrome. Furthermore, involvement of several known intracellular, receptor, and neurotransmitter mechanisms in opiate withdrawal may be traced to interactions with noradrenergic neurons.

ACKNOWLEDGMENTS

Supported in part by USPHS Grants DA02321, MH25642, MH31176, and by the Harry Frank Guggenheim Foundation, the St. Kitts Biomedical Research Foundation, and the State of Connecticut. D.E.R. is recipient of Research Scientist Career Development Award DA-00075 from NIDA. We thank R. H. Roth and M. P. Galloway for their helpful advice and comments, and Laura Fawcett, Marilyn Jones, and Joseph P. Fenerty for editorial and bibliographical help.

Literature Cited

Aceto, M. D., Harris, L. S. 1981. Antinociceptive mechanism and acute and chronic behavioral effects of clonidine. In *Psychopharmacology of Clonidine*, ed. H. Lal, S. Fielding, pp. 243–58. New York: Liss

Aghajanian, G. K. 1978. Tolerance of locus coeruleus neurones to morphine and suppression of withdrawal response by clonidine. *Nature* 276:186–88

Aghajanian, G. K., Cedarbaum, J. M., Wang, R. Y. 1977. Evidence for norepinephrine-mediated collateral inhibition of locus coeruleus neurons. *Brain Res.* 136:570–77

Aghajanian, G. K., VanderMaelen, C. P. 1982. Alpha$_2$-adrenoceptor-mediated hyperpolarization of locus coeruleus neurons: Intracellular studies in vivo. *Science* 215:1394–96

Aghajanian, G. K., VanderMaelen, C. P., Andrade, R. 1983. Intracellular studies on the role of calcium in regulating the activity and reactivity of locus coeruleus neurons in vivo. *Brain Res.* 273:237–43

Akil, H., Watson, S. J., Young, E., Lewis, M. E., Khachaturian, H., Walker, M. J. 1984. Endogenous opioids: Biology and function. *Ann. Rev. Neurosci.* 7:In press

Anden, N.-E., Grabowska, M. 1976. Pharmacological evidence for a stimulation of dopamine neurons by noradrenaline neurons in the brain. *Eur. J. Pharmacol.* 39:275–82

Arbilla, S., Langer, S. Z. 1979. Facilitation by GABA of the potassium-evoked release of ^{3}H-noradrenaline from the rat occipital cortex. *Naunyn-Schmiedebergs Arch. Pharmacol.* 306:161–68

Arnsten, A. F. T., Segal, D. S., Loughlin, S. E., Roberts, D. C. S. 1981. Evidence for an interaction of opioid and noradrenergic locus coeruleus systems in the regulation of environmental stimulus-directed behavior. *Brain Res.* 222:351–63

Aston-Jones, G., Bloom, F. E. 1981. Norepinephrine-containing locus coeruleus neurons in behaving rats exhibit pronounced responses to non-noxious environmental stimuli. *J. Neurosci.* 1(8):887–900

Atweh, S. F., Murrin, L. C. Kuhar, M. J. 1978. Presynaptic localization of opiate receptors in the vagal and accessory optic systems: An autoradiographic study. *Neuropharmacology* 17:65–71

Axelrod, J. 1956. Possible mechanism of tolerance to narcotic drugs. *Science* 124:263–64

Banks, C. D. 1979. Overdosage of buprenorphine: Case report. *N. Z. Med. J.* 89:255–56

Baraban, J. M., Aghajanian, G. K. 1980. Suppression of firing activity of 5-HT neurons in the dorsal raphe by alpha-adrenoceptor antagonists. *Neuropharmacology* 9:355–63

Baraban, J. M., Aghajanian, G. K. 1981. Noradrenergic innervation of serotonergic neurons in the dorsal raphe: Demonstration by electron microscopic autoradiography. *Brain Res.* 204:1–11

Bedard, P., Pycock, C. J. 1977. 'Wet-dog' shake behaviour in the rat: A possible quantitative model of central 5-hydroxytryptamine activity. *Neuropharmacology* 16:663–70

Berkowitz, B. A., Tarver, J. H., Spector, S. 1970. Release of norepinephrine in the central nervous system by theophylline and caffeine. *Eur. J. Pharmacol.* 10:64–71

Bhargava, H. N., Way, E. L. 1972. Acetylcholinesterase inhibition and morphine effects in morphine tolerant and dependent mice. *J. Pharmacol. Exp. Ther.* 183(1):31–40

Biggio, G., Casu, M., Corda, M. G., DiBello, C., Gessa, G. L. 1978. Stimulation of dopamine synthesis in caudate nucleus by intrastriatal enkephalins and antagonism by naloxone. *Science* 200:552–54

Biggio, G., Della Bella, D., Frigeni, V., Guidotti, A. 1977. Potentiation of morphine analgesia by muscimol. *Neuropharmacology* 16:149–50

Biswas, B., Carlsson, A. 1977. The effect of intracerebroventricularly administered GABA on brain monoamine metabolism. *Naunyn-Schmiedebergs Arch. Pharmacol.* 299:41–46

Bird, S. J., Kuhar, M. J. 1977. Iontophoretic application of opiates to the locus coeruleus. *Brain Res.* 122:523–33

Blasig, J., Gramsch, C., Laschka, E., Herz, A. 1976. The role of dopamine in withdrawal jumping in morphine dependent rats. *Arzneim. Forsch.* 26(6):1104–06

Bourgoin, S., Cesselin, F., Artaud, F., Glowinski, J., Hamon, M. 1982. *In vivo* modulations by GABA-related drugs of met-enkephalin release in basal ganglia of the cat brain. *Brain Res.* 248:321–30

Bowen, W. D., Pert, C. B., Pert, A. 1982. Nigral 6-hydroxydopamine lesions equally decrease mu and delta opiate binding to striatal patches: Further evidence for a conformationally malleable type 1 opiate receptor. *Life Sci.* 31:1679–82

Brandt, M., Buchen, C., Hamprecht, B. 1980. Relationship between the actions of calcium ions, opioids, and prostaglandin E$_1$ on the level of cyclic AMP in neuroblastoma ×

glioma hybrid cells. *J. Neurochem.* 34(3):643–51

Buckett, W. R. 1980. Irreversible inhibitors of GABA transaminase induce antinociceptive effects and potentiate morphine. *Neuropharmacology* 19:715–22

Byrd, L. D. 1982. Comparison of the behavioural effects of phencyclidine, ketamine, *d*-amphetamine and morphine in the squirrel monkey. *J. Pharmacol. Exp. Ther.* 220(1):139–44

Cardinali, D. P. 1977. Effects of pentoxifylline and theophylline on neurotransmitter uptake and release by synaptosome-rich homogenates of the rat hypothalamus. *Neuropharmacology* 16:785–90

Carter, C. J., Pycock, C. J. 1981. The role of 5-hydroxytryptamine in dopamine-dependent stereotyped behaviour. *Neuropharmacology* 20:261–65

Cedarbaum, J. M., Aghajanian, G. K. 1976. Noradrenergic neurons of the locus coeruleus: Inhibition by epinephrine and activation by the alpha-antagonist piperoxane. *Brain Res.* 112:413–19

Cervo, L., Rochat, C., Romandini, S., Samanin, R. 1981. Evidence of a preferential role of brain serotonin in the mechanisms leading to naloxone-precipitated compulsive jumping in morphine-dependent rats. *Psychopharmacology* 74:271–74

Chang, K.-J., Cooper, B. R., Hazum, E., Cuatrecasas, P. 1979. Multiple opiate receptors: Different regional distribution in the brain and differential binding of opiates and opioid peptides. *Mol. Pharmacol.* 16:91–104

Chang, K.-J., Cuatrecasas, P. 1979. Multiple opiate receptors: Enkephalins and morphine bind to receptors of different specificity. *J. Biol. Chem.* 254(8):2610–18

Chang, K.-J., Cuatrecasas, P. 1981. Heterogeneity and properties of opiate receptors. *Fed. Proc.* 40(13):2729–34

Chang, K.-J., Eckel, R. W., Blanchard, S. G. 1982. Opiate receptor down regulation: An opioid peptide specific effect. *Fed. Proc.* 41:1076 (Abstr.)

Charney, D. C., Riordan, C. E., Kleber, H. D., Murburg, M., Braverman, P., et al. 1982. A safe, effective, and rapid treatment of abrupt withdrawal from methadone therapy. *Arch. Gen. Psychiat.* 39:1327–32

Charney, D. S., Sternberg, D. E., Kleber, H. D., Heninger, G. R., Redmond, D. E. Jr. 1981. The clinical use of clonidine in abrupt withdrawal from methadone. *Arch. Gen. Psychiat.* 38:1273–77

Charney, D. S., Murburg, M., Galloway, M., Heninger, G. R., Redmond, D. E. Jr. 1983. Plasma MHPG changes in proportion to naltrexone-precipitated methadone withdrawal signs and symptoms: Evidence for noradrenergic hyperactivity. Submitted for publication

Chavkin, C., Goldstein, A. 1982. Reduction in opiate receptor reserve in morphine tolerant guinea pig ilea. *Life Sci.* 31:1687–90

Chavkin, C., James, I. F., Goldstein, A. 1982. Dynorphin is a specific endogenous ligand of the kappa opioid receptor. *Science* 215:413–15

Cheney, B. V., Lahti, R. A., Barsuhn, C. 1982. Drug affinities for the agonist and antagonist states of the opioid receptor. *Eur. J. Pharmacol.* 77:259–63

Cheney, D. L., Goldstein, A. 1971a. The effect of *p*-chlorophenylalanine on opiate-induced running, analgesia, tolerance and physical dependence in mice. *J. Pharmacol. Exp. Ther.* 177:309–15

Cheney, D. L., Goldstein, A. 1971b. Tolerance to opioid narcotics: Time course and reversibility of physical dependence in mice. *Nature* 232:477–79

Cheney, D. L., Goldstein, A., Algeri, S., Costa, E. 1971. Narcotic tolerance and dependence: Lack of relationship with serotonin turnover in the brain. *Science* 171:1169–71

Cheung, W. Y. 1982. Calmodulin. *Sci. Am.* 246(6):62–71

Childers, S. R., Mantov, S., Snyder, S. H. 1977. Enkephalin: Radioimmuno-assay and radioreceptor assay in morphine dependent rats. *Eur. J. Pharmacol.* 46:289–93

Clay, G. A., Brougham, L. R. 1975. Haloperidol binding to an opiate receptor site. *Biochem. Pharmacol.* 24:1363–67

Clouet, D. 1975. Possible roles of catecholamines in the action of narcotic drugs. In *Catecholamines and Behavior*, ed. A. J. Friedhoff, pp. 103–7. New York: Plenum

Clouet, D. H., Ratner, M. 1970. Catecholamine biosynthesis in brains of rats treated with morphine. *Science* 168:854–56

Cochin, J., Kornetsky, C. 1964. Development and loss of tolerance to morphine in the rat after single and multiple injections. *J. Pharmacol. Exp. Ther.* 145(1):1–10

Colelli, B., Meyer, D. R., Sparber, S. B. 1976. Clonidine antagonizes disruption of fixed ratio operant behavior in morphine pelleted rats given naloxone. *Pharmacologist* 18(672):236 (Abstr.)

Collier, H. O. J. 1965. A general theory of the genesis of drug dependence by induction of receptors. *Nature* 205(4967):181–82

Collier, H. O. J., Cuthbert, N. J., Francis, D. L. 1981. Character and meaning of quasi-morphine withdrawal phenomena elicited by methylxanthines. *Fed. Proc.* 40(5):1513–18

Collier, H. O. J., Francis, D. L., Henderson, G., Schneider, C. 1974. Quasi morphine-abstinence syndrome. *Nature* 249:471–73

Collier, H. O. J., Francis, D. L., Schneider, C. 1972. Modification of morphine withdrawal by drugs interacting with humoral mechanisms: Some contradictions and their interpretation. *Nature* 237:220–23

Collier, H. O. J., Roy, A. C. 1974. Morphine-like drugs inhibit the stimulation by E prostaglandins of cyclic AMP formation by rat brain homogenate. *Nature* 248:24–27

Cowan, A. 1981a. Simple *in vivo* tests that differentiate prototype agonists at opiate receptors. *Life Sci.* 28:1559–70

Cowan, A. 1981b. Quasi-morphine withdrawal syndrome: Recent developments. *Fed. Proc.* 40(5):1489–90

Cowan, A. 1981c. RX 336-M, a new chemical tool in the analysis of the quasi-morphine withdrawal syndrome. *Fed. Proc.* 40:1497–1501

Cramer, V. H., Kiessling, M. 1976. Zur dopaminempfindlichen adenylzyklase des gehirns. *Arzneim. Forsch.* 26(6):1106–7

Crawley, J. N., Laverty, R., Roth, R. H. 1979. Clonidine reversal of increased norepinephrine metabolite levels during morphine withdrawal. *Eur. J. Pharmacol.* 57:247–50

Creese, I., Feinberg, A. P., Snyder, S. H. 1976. Butyrophenone influences on the opiate receptor. *Eur. J. Pharmacol.* 36:231–35

Creese, I., Sibley, D. R. 1981. Receptor adaptations to centrally acting drugs. *Ann. Rev. Pharmacol. Toxicol.* 21:357–91

Dafny, N., Burks, T. F. 1977. 5-HT and morphine interaction, effects on sensory input in caudate nucleus and substantia nigra. *Neuropharmacology* 16:577–85

Daly, J. W., Bruns, R. F., Snyder, S. H. 1981. Adenosine receptors in the central nervous system: Relationship to the central actions of methylxanthines. *Life Sci.* 28:2083–97

Davis, M. E., Akera, T., Brody, T. M. 1975. Saturable binding of morphine to rat brainstem slices and the effect of chronic morphine treatment. *Res. Commun. Chem. Pathol. Pharmacol.* 12(3):409–18

de Boer, T., Bartels, K., Metserlaar, H. J., Bruinvels, J. 1980. Di-n-propylacetate-induced abstinence behaviour as a possible correlate of increased GABA-ergic activity in the rat. *Psychopharmacology* 71:257–67

de Boer, T., Metselaar, H. J., Bruinvels, J. 1977. Suppression of GABA-induced abstinence behaviour in naive rats by morphine and bicuculine. *Life Sci.* 20:933–42

de la Baume, S., Gros, C., Yi, C. C., Chaillet, P., Marcais-Collado, H., et al. 1982. Selective participation of both "enkephalinase" and aminopeptidase activities in the metabolism of endogenous enkephalins. *Life Sci.* 31:1753–56

DeLorenzo, R. J. 1981. The calmodulin hypothesis of neurotransmission. *Cell Calcium* 2:365–85

DeLorenzo, R. J. 1982. Calmodulin in neurotransmitter release and synaptic function. *Fed. Proc.* 41(7):2265–72

Deyo, S. N., Swift, R. M., Miller, R. J. 1979. Morphine and endorphins modulate dopamine turnover in rat median eminence. *Proc. Natl. Acad. Sci. USA* 76(6):3006–9

Dingledine, R., Iversen, L. L., Breuker, E. 1978. Naloxone as a GABA antagonist: Evidence from iontophoretic, receptor binding and convulsant studies. *Eur. J. Pharmacol.* 47:19–27

Domino, E. F., Wilson, A. E. 1973. Enhanced utilization of brain acetylcholine during morphine withdrawal in the rat. *Nature* 243:285–86

Drawbaugh, R., Lal, H. 1974. Reversal by narcotic antagonist of a narcotic action elicited by a conditional stimulus. *Nature* 247:65–67

Duka, T., Millan, M. J., Ulsamer, B., Doenicke, A. 1982. Naloxone attenuates the anxiolytic action of diazepam in man. *Life Sci.* 31:1833–36

Duka, T., Wuster, M., Herz, A. 1979. Rapid changes in enkephalin levels in rat striatum and hypothalamus induced by diazepam. *Naunyn-Schmiedebergs Arch. Pharmacol.* 309:1–5

Duka, T., Wuster, M., Herz, A. 1980. Benzodiazepines modulate striatal enkephalin levels via a GABA-ergic mechanism. *Life Sci.* 26:771–76

Dupont, A., Cusan, L., Garon, M., Alvarado-Urbina, G., Labrie, F. 1977. Extremely rapid degradation of [³H] methionine-enkephalin by various rat tissues *in vivo* and *in vitro*. *Life Sci.* 21:907–14

Eidelberg, E. 1976. Possible actions of opiates upon synapses. *Prog. Neurobiol.* 6:81–102

Elchisak, M. A., Rosecrans, J. A. 1979. Development of morphine tolerance and physical dependence in rats depleted of brain catecholamines by 6-hydroxydopamine. *Neuropharmacology* 18:175–82

Elsworth, J. D., Redmond, D. E. Jr., Roth, R. H. 1982. Plasma and cerebrospinal fluid 3-methoxy 4-hydroxyphenyl ethylene glycol (MHPG) as indices of brain norepinephrine metabolites in primates. *Brain Res.* 235:115–24

Fielding, S., Wikler, J., Hynes, M., Szewczak, M., Novick, W. J., Lal, H. 1978. A comparison of clonidine with morphine for antinociceptive and antiwithdrawal actions. *J. Pharmacol. Exp. Ther.* 207(3):899–905

Foote, S. L., Aston-Jones, G., Bloom, F. E. 1980. Impulse activity of locus coeruleus neurons in awake rats and monkeys is a function of sensory stimulation and arousal.

Proc. Natl. Acad. Sci. USA 77(5):3033–37

Franz, D. N., Hare, B. D., McCloskey, K. L. 1982. Spinal sympathetic neurons: Possible sites of opiate-withdrawal suppression by clonidine. *Science* 215:1643–45

Frederickson, R. C. A., Smithwick, E. L., Shuman, R., Bemis, K. G. 1981. Metkephamid, a systemically active analog of methionine enkephalin with potent opioid delta-receptor activity. *Science* 211:603–5

Fredholm, B. B., Jonzon, B., Lindgren, E., Lindstrom, K. 1982. Adenosine receptors mediating cyclic AMP production in the rat hippocampus. *J. Neurochem.* 39:165–75

Friedler, G., Bhargava, H. N., Quock, R., Way, E. L. 1972. The effect of 6-hydroxydopamine on morphine tolerance and physical dependence. *J. Pharmacol. Exp. Ther.* 183(1):49–55

Gallager, D. W., Aghajanian, G. K. 1976. Effect of antipsychotic drugs on the firing of dorsal raphe cells. I. Role of adrenergic system. *Eur. J. Pharmacol.* 39:341–55

Galloway, M. P., Roth, R. H. 1983. Clonidine prevents methylxanthine stimulation of norepinephrine metabolism in rat brain. *J. Neurochem.* 40(1):246–51

Galzin, A. M., Dubocovich, M. L., Langer, S. Z. 1982. Presynaptic inhibition by dopamine receptor agonists of noradrenergic neurotransmission in the rabbit hypothalamus. *J. Pharmacol. Exp. Ther.* 221(2):461–71

Gianutsos, G., Haynes, M. D., Puri, S. K., Drawbaugh, R. B., Lal, H. 1974. Effect of apomorphine and nigrostriatal lesions on aggression and striatal dopamine turnover during morphine withdrawal: Evidence for dopaminergic supersensitivity in protracted abstinence. *Psychopharmacologia* 34:37–44

Giarman, N. J., Pepeu, G. 1962. Drug-induced changes in brain acetylcholine. *Br. J. Pharmacol.* 19:226–34

Glick, S. D. 1975. Hyperthermic and hypothermic effects of morphine in mice: Interactions with apomorphine and pilocarpine and changes in sensitivity after caudate nucleus lesions. *Arch. Int. Pharmacodyn.* 213:264–71

Glick, S. D., Guido, R. A. 1982. Naloxone antagonism of the thermoregulatory effects of phencyclidine. *Science* 217:1272–73

Gold, M. S., Pottash, A. C., Sweeney, D. R., Kleber, H. D. 1980. Opiate withdrawal using clonidine: A safe, effective, and rapid nonopiate treatment. *J. Am. Med. Assoc.* 243(4):343–46

Gold, M. S., Pottash, A. L. C., Extein, I., Stoll, A. 1981. Clinical utility of clonidine in opiate withdrawal. In *Problems of Drug Dependence, 1980*, ed. L. S. Harris, pp. 95–100. NIDA Res. Monogr. 34, Rockville, Md.

Gold, M. S., Pottash, A. L. C., Sweeney, D. R., Kleber, H. D., Redmond, D. E. Jr. 1979a. Rapid opiate detoxification: clinical evidence of antidepressant and antipanic effects of opiates. *Am. J. Psychiat.* 136(7):982–83

Gold, M. S., Pottash, A. C., Sweeney, D. R., Extein, I., Annitto, W. J. 1982. Lofexidine blocks acute opiate withdrawal. In *Problems of Drug Dependence, 1981*, ed. L. S. Harris, pp. 264–68. NIDA Res. Monogr. 41, Rockville, Md.

Gold, M. S., Redmond, D. E. Jr., Donabedian, R. K. 1978a. Prolactin secretion, a measurable central effect of opiate-receptor antagonists. *Lancet* 1(8059):323–24

Gold, M. S., Redmond, D. E. Jr., Kleber, H. D. 1978b. Clonidine blocks acute opiate withdrawal symptoms. *Lancet* 2:599–602

Gold, M. S., Redmond, D. E. Jr., Kleber, H. D. 1979b. Noradrenergic hyperactivity in opiate withdrawal supported by clonidine reversal of opiate withdrawal. *Am. J. Psychiat.* 136(1):100–2

Goldberg, S. R. 1976. Conditioned behavioral and physiological changes associated with injections of a narcotic antagonist in morphine-dependent monkeys. *Pavlov. J. Biol. Sci.* 4:203–21

Goldberg, S. R., Schuster, C. R. 1970. Conditioned nalorphine-induced abstinence changes: Persistence in post morphine-dependent monkeys. *J. Exp. Anal. Behav.* 14:33–46

Goldstein, A. 1972. Heroin addiction and the role of methadone in its treatment. *Arch. Gen. Psychiat.* 26:291–97

Goldstein, A., Cox, B. M., Klee, W. A., Nirenberg, M. 1977. Endorphin from pituitary inhibits cyclic AMP formation in homogenates of neuroblastoma X glioma hybrid cells. *Nature* 265(5592):362–63

Goldstein, D. B., Goldstein, A. 1961. Possible role of enzyme inhibition and repression in drug tolerance and addiction. *Biochem. Pharmacol.* 8:48

Goldstein, A., Goldstein, D. B. 1968. Enzyme expansion theory of drug tolerance and physical dependence. *Nervous Ment. Dis.* 44:265–67

Goldstein, A., Lowney, L. I., Pal, B. K. 1971. Stereospecific and nonspecific interactions of the morphine congener levorphanol in subcellular fractions of mouse brain. *Proc. Natl. Acad. Sci. USA* 68(8):1742–47

Goldstein, A., Tachibana, S., Lowney, L. I., Hunkapiller, M., Hood, L. 1979. Dynorphin-(1–13), an extraordinarily potent opioid peptide. *Proc. Natl. Acad. Sci. USA* 76(12):6666–70

Goodman, R. R., Snyder, S. H. 1982. Kappa opiate receptors localized by autoradiography to deep layers of cerebral cortex: Relation to sedative effects. *Proc. Natl. Acad. Sci. USA* 79:5703–7

Gorenstein, C., Snyder, S. H. 1980. Enkephalinases. *Proc. R. Soc. London* 210:123–32

Grant, S. J., Huang, Y. H., Redmond, D. E. Jr. 1980. Benzodiazepines attenuate single unit activity in the locus coeruleus. *Life Sci.* 27:2231–36

Grant, S. J., Redmond, D. E. Jr. 1981. The neuroanatomy and pharmacology of the nucleus locus coeruleus. In *Psychopharmacology of Clonidine*, ed. H. Lal, S. Fielding, pp. 5–27. New York: Liss

Grant, S. J., Redmond, D. E. Jr. 1982a. Methylxanthine activation of noradrenergic unit activity and reversal by clonidine. *Eur. J. Pharmacol.* 85:105–9

Grant, S. J., Redmond, D. E. Jr. 1982b. Clonidine suppresses methylxanthine-induced quasi-morphine withdrawal syndrome. *J. Pharmacol. Biochem. Behav.* 17:655–58

Greenberg, R., O'Keefe, E. H., Antonaccio, M. J. 1981. The effect of captopril and ascorbic acid on nonenzymatic methionine enkephalin inactivation in the guinea-pig ileum. *J. Pharmacol. Exp. Ther.* 217:750–56

Greengard, P. 1979. Cyclic nucleotides, phosphorylated proteins, and the nervous system. *Fed. Proc.* 38:2208–17

Grosz, H. J. 1972. Narcotic withdrawal symptoms in heroin users treated with propranolol. *Lancet* 2:564–66

Guerrero-Munoz, F., Cerreta, K. V., Guerrero, M. L., Way, E. L. 1979a. Effect of morphine on synaptosomal Ca^{++} uptake. *J. Pharmacol. Exp. Ther.* 209(1):132–36

Guerrero-Munoz, F., Guerrero, M. D. L., Way, E. L. 1979b. Effect of morphine on calcium uptake by lysed synaptosomes. *J. Pharmacol. Exp. Ther.* 211(2):370–74

Gullis, R., Traber, J., Hamprecht, B. 1975. Morphine elevates levels of cyclic GMP in a neuroblastoma X glioma hybrid cell line. *Nature* 256:57–59

Gumulka, S. W., Dinnendahl, V., Schonhofer, P. S. 1979. The effect of naloxone on cerebellar cGMP content. *Naunyn-Schmiedebergs Arch. Pharmakol.* 306:169–72

Gunne, L. M. 1959. Noradrenaline and adrenaline in the rat brain during acute and chronic morphine administration and during withdrawal. *Nature* 184:1950–51

Gunne, L. M. 1963. Catecholamines and 5-hydroxytryptamine in morphine tolerance and withdrawal. *Acta. Physiol. Scand.* 58(Suppl. 204):1

Hall, S. M. 1979. The abstinence phobia. In *Behavioral Analysis and Treatment of Substance Abuse*, ed. N. A. Krasnegor, pp. 55–67. NIDA Res. Monogr. 25, Rockville, Md.

Hamburg, M., Tallman, J. F. 1981. Chronic morphine administration increases the apparent number of alpha$_2$-adrenergic receptors in rat brain. *Nature* 291:493–95

Hano, K., Kaneto, H., Kakunaga, T., Moribayashi, N. 1964. The administration of morphine and changes in acetylcholine metabolism in mouse brain. *Biochem. Pharm.* 13:441–47

Harris, J., Kazmierowski, D. T. 1975. Morphine tolerance and naloxone receptor binding. *Life Sci.* 16:1831–36

Harris, R. A., Loh, H. H., Way, E. L. 1976. Antinociceptive effects of lanthanum and cerium in nontolerant and morphine tolerant-dependent animals. *J. Pharmacol. Exp. Ther.* 196(2):288–97

Harris, R. H., Yamamoto, H., Loh, H. H., Way, E. L. 1977. Discrete changes in brain calcium with morphine analgesia, tolerance-dependence, and abstinence. *Life Sci.* 20:501–6

Harsing, L. G. Jr., Yang, H.-Y. T., Costa, E. 1982. Evidence for a gamma-aminobutyric acid (GABA) mediation in the benzodiazepine inhibition of the release of met^5-enkephalin elicited by depolarization. *J. Pharmacol. Exp. Ther.* 220:616–20

Haubrich, D. R., Blake, D. E. 1973. Modification of serotonin metabolism in rat brain after acute or chronic administration of morphine. *Biochem. Pharmacol.* 22:2753–59

Havemann, U., Kuschinsky, K. 1978. Effects of morphine on prostaglandin E$_2$ (PGE$_2$)-sensitive adenylate cyclase in corpus striatum of rats and its cellular localization by using kainic acid. *Brain Res.* 150:441–45

Herling, S., Woods, J. H. 1981. Discriminative stimulus effects of narcotics: Evidence for multiple receptor-mediated actions. *Life Sci.* 28:1571–84

Herz, A., Schulz, R., Wuster, M. 1982. Development of selective tolerance to particular types of opiate receptors. In *Problems of Drug Dependence, 1981*, ed. L. S. Harris, pp. 215–22. NIDA Res. Monogr. 41, Rockville, Md.

Himmelsbach, C. K. 1942. Clinical studies of drug addiction. *Arch. Intern. Med.* 69:766–72

Hitzemann, R. J., Hitzemann, B. A., Loh, H. H. 1974. Binding of ^{3}H-naloxone in the mouse brain: Effect of ions and tolerance development. *Life Sci.* 14(12):2393–404

Ho, I. K., Brase, D. A., Loh, H. H. 1975. Influence of *l*-tryptophan on morphine analgesia, tolerance, and physical dependence. *J. Pharmacol. Exp. Ther.* 193:35–43

Ho, I. K., Loh, H. H., Way, E. L. 1973a.

Cyclic adenosine monophosphate antagonism of morphine analgesia. *J. Pharmacol. Exp. Ther.* 185(2):336–46

Ho, I. K., Loh, H. H., Way, E. L. 1973b. Effects of cyclic 3', 5'-adenosine monophosphate on morphine tolerance and physical dependence. *J. Pharmacol. Exp. Ther.* 185(2):347–57

Ho, I. K., Loh, H. H., Way, E. L. 1973c. Influence of 5,6-dihydroxytryptamine on morphine tolerance and physical dependence. *Eur. J. Pharmacol.* 21:331–36

Ho, I. K., Loh, H. H., Way, E. L. 1976. Pharmacological manipulation of gamma-aminobutyric acid (GABA) in morphine analgesia, tolerance and physical dependence. *Life Sci.* 18:1111–24

Ho, I. K., Lu, S. E., Stolman, S., Loh, H. H., Way, E. L. 1972. Influence of p-chlorophenylalanine on morphine tolerance and physical dependence and regional brain serotonin turnover studies in morphine tolerant-dependent mice. *J. Pharmacol. Exp. Ther.* 182(1):155–65

Hollt, V., Dum, J., Blasig, J., Schubert, P., Herz, A. 1975. Comparison of *in vivo* and *in vitro* parameters of opiate receptor binding in naive and tolerant/dependent rodents. *Life Sci.* 16(12):1823–28

Howes, J. F., Harris, L. S., Dewey, W. L. 1970. The effect of morphine, nalorphine, naloxone, pentazocine, cyclazocine and oxotremorine on the synthesis and release of acetylcholine by mouse cerebral cortex slices in vitro. *Arch. Int. Pharmacodyn.* 184:267–76

Huff, J. W., Reigle, T. G. 1980. Effects of morphine and other centrally acting drugs on 3-4-dihydroxyphenylethylene glycol sulfate (DOPEG-SO$_4$) in rat brain. *Life Sci.* 27:1483–88

Hughes, J., Kosterlitz, H. W., Robson, L. E., Waterfield, A. A. 1978. The inhibitory effects of clonidine on the contractions of the guinea-pig ileum in the morphine-dependent and withdrawal states. *Br. J. Pharmacol.* 62:388–89

Hynes, M. D., McCarten, M. D., Shearman, G., Lal, H. 1978. Differential reduction of morphine-withdrawal body shakes by buta-clamol enantiomers. *Life Sci.* 22:133–36

Hynes, M. D., Shearman, G. T., Lal, H. 1980. Alterations in brain GABA fail to influence morphine withdrawal body shakes. *Brain Res. Bull.* 5(2):805–8

Irwin, S., Seevers, M. H. 1956. Altered response to drugs in the post addict (*Macaca mulatta*). *J. Pharmacol. Exp. Ther.* 116:31–32

Isbell, H., Wikler, A., Eddy, N. B., Wilson, J. L., Moran, C. F. 1947. Tolerance and addic-

tion liability of 6-dimethylamino 4-4-diphenylheptanone 3-(methadon). *J. Am. Med. Assoc.* 135(14):888–94

Iwamoto, E. T. 1981. Locomotor activity and antinociception after putative mu, kappa and sigma opioid receptor agonists in the rat: Influence of dopaminergic agonists and antagonists. *J. Pharmacol Exp. Ther.* 217(2):451–60

Jaffe, J. H. 1980. Drug addiction and drug abuse. In *Goodman and Gilman's, The Pharmacological Basis of Therapeutics*, ed. A. G. Gilman, L. S. Goodman, A. Gilman, pp. 535–84. New York: Macmillan. 6th ed.

Jaffe, J. H., Kanzler, M., Brady, R., Friedman, L. 1982. Methodology for assessing agents that suppress methadone withdrawal: A study of baclofen. In *Problems of Drug Dependence, 1981*, ed. L. S. Harris, pp. 269–75. NIDA Res. Monogr. 41, Rockville, Md.

Jaffe, J. H., Senay, E. C., Shuster, C. R., Renault, P. R., Smith, B., Dimenza, S. 1972. Methadyl acetate vs. methadone, a double blind study in heroin users. *J. Am. Med. Assoc.* 222:437–43

Jasinski, D. R., Pevnick, J. S., Clark, S. C., Griffith, J. D. 1977. Therapeutic usefulness of propoxyphene napsylate in narcotic addiction. *Arch. Gen. Psychiat.* 34:227–33

Jasinski, D. R., Pevnick, J. S., Griffith, J. D. 1978. Human pharmacology and abuse potential of the analgesic buprenorphine. *Arch. Gen. Psychiat.* 35:501–16

Jhamandas, K., Elliott, J. 1980. Investigation of action of enkephalin on the spontaneous and evoked release of acetylcholine from rat cortical and striatal slices. *Br. J. Pharmacol.* 71:211–17

Jhamandas, K., Sutak, M. 1980. Action of enkephalin analogues and morphine on brain acetylcholine release: Differential reversal by naloxone and an opiate pentapeptide. *Br. J. Pharmacol.* 71:201–10

Karasawa, T., Furukawa, K., Yoshida, K., Shimizu, M. 1976. Effect of theophylline on monoamine metabolism in the rat brain. *Eur. J. Pharmacol.* 37:97–104

Kleber, H. D., Riordan, C. E. 1982. The treatment of narcotic withdrawal: A historical review. *J. Clin. Psychiat.* 43:30–34

Klee, W. A., Lampert, A., Nirenberg, M. 1976. Dual regulation of adenylate cyclase by endogenous opiate peptides. In *Opiates and Endogenous Opioid Peptides*, ed. H. W. Kosterlitz, pp. 153–59. Amsterdam: Elsevier

Klee, W. A., Streaty, R. A. 1973. Narcotic receptor sites in morphine-dependent rats. *Nature* 248:61–63

Kolb, L., Himmelsbach, C. K. 1938. Clinical

studies of drug addiction, III. A critical review of the withdrawal treatments with method of evaluating abstinence syndromes. *Am. J. Psychiat.* 94:759–99

Korf, J., Bunney, B. S., Aghajanian, G. K. 1974. Noradrenergic neurons: Morphine inhibition of spontaneous activity. *Eur. J. Pharmacol.* 25:165–69

Kosterlitz, H. W., Paterson, S. J., Robson, L. E. 1981. Characterization of the kappa-subtype of the opiate receptor in the guinea-pig brain. *Br. J. Pharmacol.* 73:939–49

Krystal, J. H., Redmond, D. E. Jr. 1983. A preliminary description of acute physical dependence on morphine in the vervet monkey. *Pharmacol. Biochem. Behav.* 18:289–91

Kuriyama, K. 1982. Central beta-adrenergic receptor-adenylate cyclase system and formation of morphine withdrawal syndrome. *Trends Pharmacol. Sci.* 3:473–6

Kuschinsky, K., Hornykiewicz, O. 1974. Effects of morphine on striatal dopamine metabolism: Possible mechanism of its opposite effect on locomotor activity in rats and mice. *Eur. J. Pharmacol.* 26:41–50

Lal, H. 1975. Narcotic dependence, narcotic action and dopamine receptors. *Life Sci.* 17(4):483–96

Lal, H., Colpaert, F. C., Laduron, P. 1975a. Narcotic withdrawal like mouse jumping produced by amphetamine and L-dopa. *Eur. J. Pharmacol.* 30:113–16

Lal, H., Gianutsos, G., Puri, S. K. 1975b. A comparison of narcotic analgesics with neuroleptics on behavioral measures of dopaminergic activity. *Life Sci.* 17(1):29–34

Lal, H., Hynes, M. D. 1978. Effectiveness of butyrophenones and related drugs in narcotic withdrawal. In *Neuro-psychopharmacology,* ed. P. Deniker, C. Radouco-Thomas, A. Villeneuve, D. Baronet-Lacroix, F. Garcin, 1:289–95. Toronto; Pergamon

Lal, H., Puri, S. K., Karkalas, Y. 1971. Blockage of opioid-withdrawal symptoms by haloperidol in rats and humans. *Pharmacologist* 13:263

Lane, A. C., Rance, M. J., Walter, D. S. 1977. Subcellular localization of leucine-enkephalin-hydrolysing activity in rat brain. *Nature* 269:75–76

Langer, S. Z. 1981. Presynaptic regulation of the release of catecholamines. *Pharmacol. Rev.* 32(4):337–62

Large, W. A., Milton, A. S. 1970. The effect of acute and chronic morphine administration on brain acetylcholine levels in the rat. *Br. J. Pharmacol.* 38:451–52

Laska, F. J., Fennessy, M. R. 1976. Physical dependence in the rat induced by slow release morphine: Dose-response, time course and brain biogenic amines. *Clin. Exp. Pharmacol. Physiol.* 3:587–98

Laverty, R., Roth, R. H. 1980. Clonidine reverses the increased norepinephrine turnover during morphine withdrawal in rats. *Brain Res.* 182:482–85

Law, P. Y., Koehler, J. E., Loh, H. H. 1982. Comparison of opiate inhibition of adenylate cyclase activity in neuroblastoma N18TG2 and neuroblastoma × glioma NG108-15 hybrid cell lines. *Mol. Pharmacol.* 21:483–91

Lewis, M. J., Costa, J. L., Jacobowitz, D. M., Margules, D. L. 1976. Tolerance, physical dependence and opioid-seeking behavior: Dependence on diencephalic norepinephrine. *Brain Res.* 107:156–65

Litt, I. F., Colli, A. S., Cohen, M. I. 1971. Diazepam in the management of heroin withdrawal in adolescents: Preliminary report. *J. Pediatr.* 78(4):692–96

Llorens, C., Martres, M. P., Schwartz, J. C. 1978. Hypersensitivity to noradrenaline in cortex after chronic morphine: Relevance to tolerance and dependence. *Nature* 274:603–5

Loh, H. H., Shen, F.-H., Way, E. L. 1969. Inhibition of morphine tolerance and physical dependence development and brain serotonin synthesis by cycloheximide. *Biochem. Pharmacol.* 18:2711–21

LoPachin, R. M., Reigle, T. G. 1978. The effects of several narcotic analgesics on brain levels of 3-methoxy 4-hydroxy phenylethylene glycol sulfate in the rat. *J. Pharmacol. Exp. Ther.* 207:151–58

Lord, J. A. H., Waterfield, A. A., Hughes, J., Kosterlitz, H. W. 1977. Endogenous opioid peptides: Multiple agonists and receptors. *Nature* 267:495–99

McGrath, J. C. 1983. The variety of vascular alpha-adrenoceptors. *Trends Pharmacol. Sci.* 4:14–8

McRae-Degueurce, A., Berod, A., Mermet, A., Keller, A., Chouvet, G., et al. 1982. Alterations in tyrosine hydroxylase activity elicited by raphe nuclei lesions in the rat locus coeruleus: Evidence for the involvement of serotonin afferents. *Brain Res.* 235:285–301

Malfroy, B., Swerts, J. P., Guyon, A., Roques, B. P., Schwartz, J. C. 1978. High-affinity enkephalin-degrading peptidase in brain is increased after morphine. *Nature* 276:523–26

Malin, D. H., Layng, M. P., Swank, P., Baker, M. J., Hood, J. L. 1982. Behavioral alterations produced by chronic naloxone injections. *Pharmacol. Biochem. Behav.* 17:389–392

Marcovici, M., O'Brien, C. P., McLellan, A. T., Kacian, J. 1981. A clinical, controlled study of L-alpha-acetylmethadol in the treatment of narcotic addiction. *Am. J. Psychiat.* 138(2):234–36

Martin, R., Voigt, K. H. 1982. Leucine-enkephalin-like immunoreactivity in vasopressin terminals is enhanced by treatment with peptidases. *Life Sci.* 31:1729–32

Martin, W. R., Eades, C. G. 1961. Demonstration of tolerance and physical dependence in the dog following a short-term infusion of morphine. *J. Pharmacol. Exp. Ther.* 133:262–70

Martin, W. R., Eades, C. G. 1967. Pharmacological studies of spinal cord adrenergic and cholinergic mechanisms and their relation to physical dependence on morphine. *Psychopharmacologia* 11:195–223

Martin, W. R., Eades, C. G., Thompson, J. A., Huppler, R. E., Gilbert, P. E. 1976. The effects of morphine- and nalorphine-like drugs in the nondependent and morphine-dependent chronic spinal dog. *J. Pharmacol Exp. Ther.* 197(3):517–32

Martin, W. R., Fraser, H. F., Gorodetzky, C. W., Rosenberg, D. E. 1965. Studies of the dependence-producing potential of the narcotic antagonist 2-cyclopropylmethyl-2'-hydroxy-5,9-dimethyl-6,7-benzomorphan (cyclazocine). *J. Pharmacol. Exp. Ther.* 150:426–36

Martin, W. R., Gordetzky, C. W. 1965. Demonstration of tolerance to and physical dependence on *n*-allylnormorphine (nalorphine). *J. Pharmacol. Exp. Ther.* 150(3):437–42

Martin, W. R., Jasinski, D. R. 1969. Physiological parameters of morphine dependence in man-tolerance, early abstinence, protracted abstinence. *J. Psychiat. Res.* 7:9–17

Mehta, C. S., Johnson, W. E. 1975. Possible role of cyclic AMP and dopamine in morphine tolerance and physical dependence. *Life Sci.* 16:1883–88

Mello, N. K., Bree, M. P., Mendelson, J. H. 1981a. Comparison of the effects of buprenorphine and methadone on opiate self-administration in primates. In *Problems of Drug Dependence, 1980,* ed. L. S. Harris, pp. 67–73. NIDA Res. Monogr. 34, Rockville, Md.

Mello, N. K., Mendelson, J. H., Kuehnle, J. C., Sellers, M. S. 1981b. Operant analysis of human heroin self-administration and the effects of naltrexone. *J. Pharmacol. Exp. Ther.* 216:45–54

Meyer, D. R., Sparber, S. B. 1976. Clonidine antagonizes body weight loss and other symptoms used to measure withdrawal in morphine pelleted rats given naloxone. *Pharmacologist* 18(673):236 (Abstr.)

Meyer, R. E., Mirin, S. M., Altman, J. L. 1975. The clinical usefulness of narcotic antagonists: Implications of behavioral research. *Am. J. Drug Alcohol Abuse* 2(3–4):417–32

Miller, R. J., Chang, K.-J., Cuatrecasas, P. 1977. The metabolic stability of the enkephalins. *Biochem. Biophys. Res. Commun.* 74(4):1311–17

Miller, R. J., Cuatrecasas, P. 1978. Enkephalins and endorphins. *Vitam. Horm.* 36:297–381

Miller, R. J., Pickel, V. M. 1980. Immunohistochemical distribution of enkephalins: Interactions with catecholamine-containing systems. In *Histochemistry and Cell Biology of Autonomic Neurons, SIF Cells, and Paraneurons,* ed. O. Eranko, pp. 349–59. New York: Raven

Minneman, K. P., Iversen, L. L. 1976. Enkephalin and opiate narcotics increase cyclic GMP accumulation in slices of rat neostriatum. *Nature* 262:313–14

Moleman, P., Bruinvels, J. 1976. Differential effect of morphine on dopaminergic neurons in frontal cortex and striatum of the rat. *Life Sci.* 19:1277–82

Montel, H., Starke, K., Taube, H. D. 1975. Morphine tolerance and dependence in noradrenaline neurons of the rat cortex/ *Naunyn-Schmiedergs Arch. Pharmacol.* 288:415–26

Moore, R. Y., Bloom, F. E. 1979. Central catecholamine neuron system: Anatomy and physiology of the norepinephrine and epinephrine system. *Ann. Rev. Neurosci.* 2:1113–68

Moroni, F., Cheney, D. L., Peralta, E., Costa, E. 1978. Opiate receptor agonists as modulators of gamma-aminobutyric acid turnover in the nucleus caudatus, globus pallidus and substantia nigra of the rat. *J. Pharmacol. Exp. Ther.* 207:870–77

Nathanson, J. A., Redmond, D. E. Jr. 1981. Morphine withdrawal causes subsensitivity of adrenergic receptor response. *Life Sci.* 28:1353–60

Nehmed, R., Nadler, H., Simantov, R. 1982. Effects of acute and chronic morphine treatment on calmodulin activity of rat brain. *Mol. Pharmacol.* 22:389–94

Nowycky, M. C., Walters, J. R., Roth, R. H. 1978. Dopaminergic neurons: Effect of acute and chronic morphine administration on single cell activity and transmitter metabolism. *J. Neural Transm.* 42:99–116

Nutt, J. G., Jasinski, D. R. 1973. Methadone-naloxone mixtures for use in methadone maintenance programs. I. An evaluation in man of their pharmacological feasibility. II. Demonstration of acute physical dependence. *Clin. Pharmacol. Ther.* 15(2):156–66

O'Brien, C. P., O'Brien, T. J., Mintz, J., Brady, J. P. 1975/76. Conditioning of narcotic abstinence symptoms in human subjects. *Drug Alcohol Depend.* 1:115–23

O'Brien, C. P., Testa, T., O'Brien, T. J., Brady, J. P., Wells, B. 1977. Conditioned narcotic withdrawal in humans. *Science* 195:1000–2

O'Callaghan, J. P., Juskevich, J. C., Lovenberg, W. 1982. The effects of morphine on calcium-regulated phosphorylation of synaptosomal cytosolic proteins from rat striatum. *J. Pharmacol. Exp. Ther.* 220:696–702

Oka, T., Nozaki, M., Hosoya, E. 1972. Effects of *p*-chlorophenylalanine and cholinergic antagonists on body temperature changes induced by the administration of morphine to nontolerant and morphine-tolerant rats. *J. Pharmacol. Exp. Ther.* 180(1):136–43

Olgiati, V., Quirion, R., Bowen, W. D., Pert, C. B. 1982. Characterization of type 2 opiate receptors. *Life Sci.* 31:1675–78

Pavlov, I. P. 1927. *Conditioned Reflexes.* Transl. G. V. Anrep, pp. 35–38. London: Oxford Univ. Press

Persson, S.-A. 1979. Effect of morphine on the accumulation of dopa after decarboxylase inhibition in the rat. *Eur. J. Pharmacol.* 55:121–28

Pert, C. B., Bowie, D. L. 1979. Behavioral manipulation of rats causes alterations in opiate receptor occupancy. In *Endorphins in Mental Health Research,* ed. E. Usdin, W. E. Bunney, Jr., N. S. Kline, pp. 93–104. New York: Oxford Univ. Press

Pert, C. B., Snyder, S. H. 1973. Properties of opiate-receptor binding in rat brain. *Proc. Natl. Acad. Sci. USA* 70(8):2243–47

Pert, C. B., Snyder, S. H. 1974. Opiate receptor binding of angonists and antagonists affected differentially by sodium. *Mol. Pharmacol.* 10:868–79

Pert, C. B., Snyder, S. H. 1975. Opiate receptor binding-enhancement by opiate administration *in vivo. Biochem. Pharmacol.* 25:847–53

Pfeiffer, A., Herz, A. 1981. Discrimination of three opiate receptor binding sites with the use of a computerized curve-fitting technique. *Mol. Pharmacol.* 21:266–71

Pickel, V. 1982. Central noradrenergic neurons: Identification, distribution, and synaptic interactions with axons containing morphine-like peptides. *J. Clin. Psychiatry* 43:13–16

Pickel, V. M., Joh, T. H., Reis, D. J. 1977. A serotonergic innervation of noradrenergic neurons in nucleus locus coeruleus: Demonstration by immunocytochemical localization of the transmitter specific enzymes tyrosine and tryptophan hydroxylase. *Brain Res.* 131:197–214

Pinsky, C., Dua, A. K., LaBella, F. S. 1982. Peptidase inhibitors reduce opiate narcotic withdrawal signs, including seizure activity, in the rat. *Brain Res.* 243:301–7

Proudfit, H. K. 1980. Effects of raphe magnus and raphe pallidus lesions on morphine-induced analgesia and spinal cord monoamines. *Pharmacol. Biochem. Behav.* 13(5):705–14

Quirion, R., Hammer, R. P. Jr., Herkenham, M., Pert, C. B. 1981. Phencyclidine (angel dust)/sigma "opiate" receptor: Visualization by tritium-sensitive film. *Proc. Natl. Acad. Sci. USA* 78(9):5881–85

Rasmussen, H. 1970. Cell communication, calcium ion, and cyclic adenosine monophosphate. *Science* 170:404–12

Rawson, R. A., Washton, A. M., Resnick, R. B., Tennant, F. S. Jr. 1981. Clonidine hydrochloride detoxification from methadone treatments: The value of naltrexone aftercare. In *Problems of Drug Dependence, 1980,* ed. L. S. Harris, pp. 101–8. NIDA Res. Monogr. 34, Rockville, Md.

Redmond, D. E. Jr. 1977. Alterations in the function of the nucleus: A possible model for studies of anxiety. In *Animal Models in Psychiatry and Neurology,* ed. E. Usdin, I. Manin, pp. 293–305. New York: Pergammon

Redmond, D. E. Jr. 1981. Clonidine and the primate locus coeruleus: Evidence suggesting anxiolytic and anti-withdrawal effects. In *Psychopharmacology of Clonidine,* ed. H. Lal, S. Fielding, pp. 147–53. New York: Liss

Redmond, D. E. Jr. 1982. Central mechanisms and alpha-adrenergic receptors in opiate withdrawal and other psychiatric syndromes: New studies with clonidine. *J. Clin. Psychiat.* 43(Suppl. 6)

Redmond, D. E. Jr., Gold, M. D., Huang, Y. H. 1978. Enkephalin acts to inhibit locus coeruleus mediated behaviors. *Neurosci. Abstr.* 4:413(1306)

Redmond, D. E. Jr., Huang, Y. H. 1979. Current concepts II. New evidence for a locus coeruleus-norepinephrine connection with anxiety. *Life Sci.* 25(26):2149–62

Redmond, D. E. Jr., Huang, Y. H., Snyder, D. R., Maas, J. W. 1976. Behavioral effects of stimulation of the nucleus locus coeruleus in the stump-tailed monkey (*Macaca arctoides*). *Brain Res.* 116:502–10

Reigle, T. G., Huff, J. W. 1980. Single-dose tolerance to the effects of morphine on brain 3-methyoxy 4-hydroxy phenylethylene glycol sulfate. *Biochem. Pharmacol.* 29:2249–51

Resnick, R. B., Kestenbaum, R. S., Schwartz, L. K., Smith, A. 1976. Evaluation of propranolol in opiate dependence. *Arch. Gen. Psychiat.* 33:993–97

Riordan, C. E., Kleber, H. D. 1980. Rapid opiate detoxification with clonidine and naloxone. *Lancet* 1:1079–80

Ritzmann, R. F. 1981. Opiate dependence following acute injections of morphine and naloxone: The assessment of various withdrawal signs. *Pharmacol. Biochem. Behav.* 14:575–77

Robson, L. E., Kosterlitz, H. W. 1979. Specific protection of the binding sites of D-ala²-d-leu⁵-enkephalin (delta-receptors) and dihydromorphine (mu-receptors). *Proc. R. Soc. London* 205:425–32

Roques, B. P., Fournie-Zaluski, M. C., Soroca, E., Lecomte, J. M., Malfroy, B., et al. 1980. The enkephalinase inhibitor thiorphan shows antinociceptive activity in mice. *Nature* 288:286–88

Rosow, C. E., Miller, J. M., Pelikan, E. W., Cochin, J. 1980. Opiates and thermoregulation in mice. I. Agonists. *J. Pharmacol. Exp. Ther.* 211(2):275–83

Roth, R. H., Elsworth, J. D., Redmond, D. E. Jr. 1982. Clonidine suppression of noradrenergic hyperactivity during morphine withdrawal by clonidine: Biochemical studies in rodents and primates. *J. Clin. Psychiat.* 43(6):[Sec. 2]42–46

Roy, A. C., Collier, H. O. J. 1975. Prostaglandins, cyclic AMP and the biochemical mechanism of opiate agonist action. *Life Sci.* 16:1857–62

Sasa, M., Munekiyo, K., Takaori, S. 1975. Morphine interference with noradrenaline-mediated inhibition from the locus coeruleus. *Life Sci.* 17(9):1373–80

Sawynok, J., Labella, F. S. 1981. GABA and baclofen potentiate the K⁺-evoked release of methionine-enkephalin from rat striatal slices. *Eur. J. Pharmacol.* 70:103–10

Scallet, A. C. 1981. Effects of conditioned fear and environmental novelty on plasma beta-endorphin in the rat. *Peptides* 1:203–6

Schulz, R., Faase, E., Wuster, M., Herz, A. 1979a. Selective receptors for beta-endorphin on the rat vas deferens. *Life Sci.* 24:843–50

Schulz, R., Goldstein, A. 1973. Morphine tolerance and supersensitivity to 5-hydroxytryptamine in the myenteric plexus of the guinea-pig. *Nature* 244:168–70

Schulz, R., Wuster, M., Herz, A. 1979b. Supersensitivity to opioids following the chronic blockade of endorphin action by naloxone. *Naunyn-Schmiedebergs Arch. Pharmacol.* 306:93–96

Schwartz, J. C., Pollard, H., Llorens, C., Malfroy, B., Gros, C., et al. 1978. Endorphins and endorphin receptors in striatum: Relationships with dopaminergic neurons. *Adv. Biochem. Psychopharmacol.* 18:245–64

Segal, M. 1979. Serotonergic innervation of the locus coeruleus from the dorsal raphe and its action on responses to noxious stimuli. *J. Physiol.* 286:401–15

Sellinger-Barnette, M., Weiss, B. 1982. Interaction of beta-endorphin and other opioid peptides with calmodulin. *Mol. Pharmacol.* 21:86–91

Senay, E. C., Dorus, W., Goldberg, F., Thornton, W. 1977. Withdrawal from methadone maintenance. *Arch. Gen. Psychiat.* 34:361–67

Shani, J., Azov, R., Weissman, B. A. 1979. Enkephalin levels in rat brain after various regimens of morphine administration. *Neurosci. Lett.* 12:319–22

Shannon, H. E. 1982. Pharmacological analysis of the phencyclidine-like discriminative stimulus properties of narcotic derivatives in rats. *J. Pharmacol. Exp. Ther.* 222:146–51

Sharkawi, M. 1972. Morphine hyperthermia in the rat: Its attenuation by physostigmine. *Br. J. Pharmacol.* 44:544–48

Sharma, S. K., Klee, W. A., Nirenberg, M. 1975a. Dual regulation of adenylate cyclase accounts for narcotic dependence and tolerance. *Proc. Natl. Acad. Sci. USA* 72(8):3092–96

Sharma, S. K., Nirenberg, M., Klee, W. A. 1975b. Morphine receptors as regulators of adenylate cyclase activity. *Proc. Natl. Acad. Sci. USA* 72(2):590–94

Shearman, G. T., Herz, A. 1982b. D-ala²-, D-leu⁵-enkephalin generalizes to a discriminative stimulus produced by fentanyl but not ethylketocyclazocine. *Pharmacol. Biochem. Behav.* 16:249–52

Shearman, G. T., Herz, A. 1982a. Evidence that the discriminative stimulus properties of fentanyl and ethylketocyclazocine are mediated by an interaction with different opiate receptors. *J. Pharmacol. Exp. Ther.* 221:735–39

Shearman, G. T., Lal, H., Ursillo, R. C. 1980. Effectiveness of lofexidine in blocking morphine-withdrawal signs in the rat. *Pharmacol. Biochem. Behav.* 12(4):573–75

Shuster, L. 1961. Repression and de-repression of enzyme synthesis as a possible explanation of some aspects of drug action. *Nature* 189:314–15

Sideroff, S. I., Jarvik, M. E. 1980. Conditioned responses to a videotape showing herion-related stimuli. *Int. J. Addict.* 15(4):529–36

Siegel, S. 1976. Morphine analgesic tolerance: Its situation specificity supports a Pavlovian conditioning model. *Science* 23:323–25

Siegel, S., Hinson, R. E., Krank, M. D., McCully, J. 1982. Heroin "overdose" death: Contribution of drug-associated environmental cues. *Science* 216:436–37

Simantov, R., Baram, D., Dornay, M. 1982. Receptor and post-receptor alterations induced by opiates and solubilization of opiate receptors. In *From Molecular Biology to Function,* ed. E. Costa, M. Trabucchi, pp. 291–300. New York: Raven

Simantov, R., Snyder, S. H. 1976. Elevated levels of enkephalin in morphine-dependent rats. *Nature* 262:505–7

Simmons, W. H., Ritzmann, R. F. 1980. An inhibitor of opioid peptide degradation produces analgesia in mice. *Pharmacol. Biochem. Behav.* 13(5):715–18

Simon, E. J., Hiller, J. M., Edelman, I. 1973. Stereospecific binding of the potent narcotic analgesic [³H]etorphine to rat-brain homogenate. *Proc. Natl. Acad. Sci. USA* 70(7):1947–49

Simon, E. J., Hiller, J. M., Groth, J., Edelman, I. 1975. Further properties of stereospecific opiate binding sites in rat brain: On the nature of the sodium effect. *J. Pharmacol. Exp. Ther.* 192:531–37

Sivam, S. P., Nabeshima, T., Ho, I. K. 1981. Alterations of regional gamma-aminobutyric acid receptors in morphine tolerant mice. *Biochem. Pharmacol.* 30(15):2187–90

Sivam, S. P., Nabeshima, T., Ho, I. K. 1982. An analysis of GABA receptor changes in the discrete regions of mouse brain after acute and chronic treatments with morphine. *J. Neurochem.* 39:933–39

Smith, J. E., Co, C., Freeman, M. E., Lane, J. D. 1982. Brain neurotransmitter turnover correlated with morphine-seeking behavior of rats. *Pharmacol. Biochem. Behav.* 16:509–19

Smith, C. B., Valentino, R. J. 1980. Interactions between clonidine and a withdrawal-inducing benzazocine upon the isolated mouse vas deferens and guinea pig ileum. In *Endogenous and Exogenous Opioid Agonists and Antagonists*, ed. E. L. Way, pp. 513–16. New York: Pergamon

Sorensen, J. L., Hargreaves, W. A., Weinberg, J. A. 1982. Withdrawal from heroin in three or six weeks. *Arch. Gen. Psychiat.* 39:167–71

Spaulding, T. C., Little, J., McCormack, K., Fielding, S. 1980. The antinociceptive effects of GABA antagonists in mice. *Brain Res. Bull.* 5(2):415–19

Su, C.-Y., Lin, S.-H., Wang, Y.-T., Li, C.-H., Huang, L. H., et al. 1978. Effects of beta-endorphin on narcotic abstinence syndrome in man. *J. Formosan Med. Assoc.* 77(2):133–41

Sullivan, S., Akil, H., Blacker, D., Barchas, J. D. 1980. Enkephalinase: Selective inhibitors and partial characterization. *Peptides* 1:31–35

Summers, R. J., Jarrett, B., Louis, W. J. 1981. Comparison of (³H) clonidine and (³H) guanfasine binding to adrenoreceptors in membrane from rat cerebral cortex. *Neurosci. Lett.* 25:31–36

Sutker, P. B., Allain, A. N. 1974. Addict attitudes toward methadone maintenance: A preliminary report. *Int. J. Addict.* 9:237–43

Svensson, L., Ahlenius, S. 1982. Functional importance of nucleus accumbens noradrenaline in the rat. *Acta Pharmacol. Toxicol.* 50:22–24

Svensson, T. H., Bunney, B. S., Aghajanian, G. K. 1975. Inhibition of both noradrenergic and serotonergic neurons in brain by the alpha-adrenergic agonist clonidine. *Brain Res.* 92:291–306

Swain, H. H., Seevers, M. H. 1976. Evaluation of new compounds for morphine-like physical dependence in the rhesus monkey. In *Problems of Drug Dependence, 1975*, ed. L. S. Harris, pp. 773–95. NIDA Res. Monogr., Rockville, Md.

Swann, A. C., Charney, D. S., Elsworth, J. D., Jablons, D. M., Roth, R. H., et al. 1983. Catecholamine metabolites and behavior in morphine withdrawal: Partial reversal by debrisoquin sulphate. *Eur. J. Pharmacol.* 86:167–75

Takemori, A. E., Oka, T., Nishiyama, N. 1973. Alteration of analgesic receptor-antagonist interaction induced by morphine. *J. Pharmacol. Exp. Ther.* 186(2):261–65

Tallman, J. F., Paul, S. M., Skolnick, P., Gallager, D. W. 1980. Receptors for the age of anxiety: Pharmacology of the benzodiazepines. *Science* 207:274–81

Teasdale, J. D. 1973. Conditioned abstinence in narcotic addicts. *Int. J. Addict.* 8(2):273–92

Terenius, L. 1973. Stereospecific interaction between narcotic analgesics and a synaptic plasma membrane fraction of rat cerebral cortex. *Acta Pharmacol. Toxicol.* 32:317–20

Tolis, G., Hickey, J., Guyda, H. 1975. Effects of morphine on serum growth hormone, cortisol, prolactin, and thyroid stimulating hormone in man. *J. Clin. Endocrinol. Metab.* 41:797–800

Tsang, D., Tan, A. T., Henry, J. L., Lal, S. 1978. Effect of opioid peptides on L-noradrenaline-stimulated cyclic AMP formation in homogenates of rat cerebral cortex and hypothalamus. *Brain Res.* 152:521–27

Tseng, L-F., Loh, H. H., Wei, E. T. 1975. Effects of clonidine on morphine withdrawal signs in the rat. *Eur. J. Pharmacol.* 30:93–99

Tulunay, F. C., Yano, I., Takemori, A. E. 1979. Enhanced naloxone potency and the development of narcotic tolerance. *Eur. J. Pharmacol.* 53:247–53

Uhde, T. W., Redmond, D. E. Jr., Kleber, H. D. 1980. Clonidine suppresses the opioid abstinence syndrome without clonidine-withdrawal symptoms: A blind inpatient study. *Psychiat. Res.* 2(1):37–47

Valentino, R. J., Aston-Jones, G. 1982.

Activation of locus coeruleus neurons in the rat by UM 1046, a drug that mimics opiate withdrawal in normal animals. *Neurosci. Abstr.* 8:228

Valentino, R. J., Smith, C. B., Woods, J. H. 1979. An unusual benzazocine elicits acetylcholine release in the isolated guinea pig ileum. *Nature* 281:370–72

Valentino, R. J., Smith, C. B., Woods, J. H. 1981. Physiological and behavioral approaches to the study of the quasimorphine withdrawal syndrome. *Fed. Proc.* 40:1502–7

van der Heyden, J. A. M., Venema, K., Korf, J. 1980. In vivo release of endogenous gamma-aminobutyric acid from rat striatum: Effects of muscimol, oxotremorine, and morphine. *J. Neurochem.* 34(6):1648–53

van der Laan, J. W., Bruinvels, J. 1981. Dipropylacetate-induced quasi-morphine abstinence behaviour in the rat: Involvement of amygdaloid and thalamic structures. *Brain Res.* 229:133–46

van der Laan, J. W., Bruinvels, J., Cools, A. R. 1982. Dipropylacetate-induced quasimorphine abstinence behaviour in the rat: Participation of the locus coeruleus system. *Brain Res.* 247:309–14

van der Laan, J. W., Jacobs, A. W. C. M., Bruinvels, J. 1980. Effects of banched-chain fatty acids on GABA-degradation and behavior: Further evidence for a role of GABA in quasi-morphine abstinence behavior. *Pharmacol. Biochem. Behav.* 13(6):843–49

van Dongen, P. A. M. 1981. The human locus coeruleus in neurology and psychiatry. *Prog. Neurobiol.* 17:97–139

Vasquez, B. J., Overstreet, D. H., Russell, R. W. 1974. Psychopharmacological evidence for increase in receptor sensitivity following chronic morphine treatment. *Psychopharmacologia* 38:287–302

Vetulani, J., Bednarczyk, B. 1977. Depression by clonidine of shaking behavior elicited by nalorphine in morphine-dependent rats. *J. Pharm. Pharmacol.* 29:567–69

Vogt, M. 1954. The concentrations of sympathin in different parts of the central nervous system under normal conditions and after the administration of drugs. *J. Physiol.* 123:451–81

Volicer, L., Puri, S. K., Choma, P. 1977. Cyclic GMP and GABA levels in rat striatum and cerebellum during morphine withdrawal: Effect of apomorphine. *Neuropharmacology* 16:791–94

Wachtel, H. 1982. Characteristic behavioral alterations in rats induced by rolipram and other selective adenosine cyclic 3',5'-monophosphate phosphodiesterase inhibitors. *Psychopharmacology* 77:309–16

Waddington, J. L., Crow, T. J. 1979. Rota-

tional responses to serotonergic and dopaminergic agonists after unilateral dihydroxytryptamine lesions of the medial forebrain bundle: Co-operative interactions of serotonin and dopamine in neostriatum. *Life Sci.* 25(15):1307–14

Walczak, S. A., Wilkening, D., Makman, M. H. 1979. Interaction of morphine, etorphine and enkephalins with dopamine-stimulated adenylate cyclase of monkey amygdala. *Brain Res.* 160:105–16

Washton, A. M., Resnick, R. B., Perzel, J. F., Garwood, J. 1982. Opiate detoxification using lofexidine. In *Problems of Drug Dependence, 1981*, ed. L. S. Harris, pp. 261–63. NIDA Res. Monogr. 41, Rockville, Md.

Washton, A. M., Resnick, R. B. 1980. Clonidine for opiate detoxification outpatient clinical trials. *Am. J. Psychiat.* 137:1121–22

Washton, A. M., Resnick, R. B. 1981. Clonidine vs. methadone for opiate detoxification: Double-blind outpatient trials. In *Problems of Drug Dependence, 1980*, ed. L. S. Harris, pp. 89–94. NIDA Res. Monogr. 34, Rockville, Md.

Washton, A. M., Resnick, R. B., Rawson, R. A. 1980. Clonidine for outpatient opiate detoxification. *Lancet* 1:1078–79

Watson, S. J., Khachaturian, H., Coy, D., Taylor, L., Akil, H. 1982. Dynorphin is located throughout the CNS and is often co-localized with alpha-neo-endorphin. *Life Sci.* 31:1773–76

Watson, S. J., Richard, C. W. III, Ciaranello, R. D., Barchas, J. D. 1980. Interaction of opiate peptide and noradrenalin systems: Light microscopic studies. *Peptides* 1(1):23–30

Way, E. L., Loh, H. H., Shein, F. 1968. Morphine tolerance, physical dependence, and synthesis of brain 5-hydroxytryptamine. *Science* 162:1290–92

Wen, H. L. 1982. Clinical trial of met-enkephalin analogue (damme, fk 33–824) on heroin abusers. In *Regulatory Peptides: From Molecular Biology to Function*, ed. E. Costa, M. Trabucchi, pp. 397–403. New York: Raven

Wen, H. L., Ho, W. K. K. 1982. Suppression of withdrawal symptoms by dynorphin in heroin addicts. *Eur. J. Pharmacol.* 82:183–86

Whitehead, C. C. 1974. Methadone pseudo-withdrawal syndrome: Paradigm for a psychopharmacological model of opiate addiction. *Psychosom. Med.* 36(3):189–98

Whitsett, T. L., Chrysant, S. G., Dillard, B. L., Anton, A. H. 1978. Abrupt cessation of clonidine administration: A prospective study. *Am. J. Cardiol.* 41:1285–90

Wikler, A., Carter, R. L. 1953. Effects of single doses of n-allylnormorphine on hindlimb reflexes of chronic spinal dogs during cycles of morphine addiction. *J. Pharmacol. Exp. Ther.* 109:92–101

Wikler, A., Fraser, H. F., Isbell, H. 1953. *N*-allylnormorphine: Effects of single doses and precipitation of acute "abstinence syndromes" during addiction to morphine, methadone or heroin in man (post-addicts). *J. Pharmacol. Exp. Ther.* 109:1625–32

Woodward, D. J., Moises, H. C., Waterhouse, B. D., Hoffer, B. J., Freedman, R. 1979. Modulatory actions of norepinephrine in the central nervous system. *Fed. Proc.* 38:2109–16

Woody, G. E., Mintz, J., Tennant, F., O'Brien, C. P., McLellan, A. T., et al. 1981. Propoxyphene for maintenance treatment of narcotic addiction. *Arch. Gen. Psychiat.* 38:898–900

Woolverton, W. L., Wessinger, W. D., Balster, R. L., Harris, L. S. 1981. Intravenous clonidine self-administration by rhesus monkeys. In *Problems of Drug Dependence, 1980,* ed. L. S. Harris, pp. 166–73. NIDA Res. Monogr. 34, Rockville, Md.

Wuster, M., Schulz, R., Herz, A. 1980. Highly specific opiate receptors for dynorphin-(1–13) in the mouse vas deferens. *Eur. J. Pharmacol.* 62:235–36

Yessaian, N. H., Armenian, A. R., Buniatian, H. Ch. 1969. Effect of gamma-aminobutyric acid on brain serotonin and catecholamines. *J. Neurochem.* 16:1425–33

Young, W. S. III, Bird, S. J., Kuhar, M. J. 1977. Iontophoresis of methionine-enkephalin in the locus coeruleus area. *Brain Res.* 129:366–70

Young, W. S. III, Kuhar, M. J. 1981. Anatomical mapping of clonidine (alpha-2 noradrenergic) receptors in rat brain: Relationship to function. In *Psychopharmacology of Clonidine,* ed. H. Lal, S. Fielding, pp. 41–52. New York: Liss

Zigun, J. R., Bannon, M. J., Roth, R. H. 1981. Comparison of two alpha-noradrenergic agonists (clonidine and guanfacine) on norepinephrine turnover in the cortex of rats during morphine abstinence. *Eur. J. Pharmacol.* 70:565–70

Zukin, R. S., Zukin, S. R. 1981. Multiple opiate receptors: Emerging concepts. *Life Sci.* 29:2681–90

Zukin, S. R., Zukin, R. S. 1979. Specific [³H]phencyclidine binding in rat central nervous system. *Proc. Natl. Acad. Sci. USA* 76(10):5372–76

SUBJECT INDEX

CUMULATIVE INDEXES

CONTRIBUTING AUTHORS, VOLUMES 3–7

489

CHAPTER TITLES, VOLUMES 3–7

491

ORDER FORM

A NONPROFIT SCIENTIFIC PUBLISHER

Annual Reviews Inc.

4139 EL CAMINO WAY • PALO ALTO, CA 94306 USA • (415) 493-4400

ease list the volumes you wish to order by volume number. If you wish a standing order (the latest volume
nt to you automatically each year), indicate volume number to begin order. Volumes not yet published will
shipped in month and year indicated. All prices subject to change without notice. Prepayment required
m individuals. Telephone orders charged to VISA, MasterCard, American Express, welcomed.

ANNUAL REVIEW SERIES

		Prices Postpaid per volume USA/elsewhere	Regular Order Please send: Vol. number	Standing Order Begin with: Vol. number
nnual Review of ANTHROPOLOGY				
Vols. 1-10	(1972-1981)	$20.00/$21.00		
Vol. 11	(1982)	$22.00/$25.00		
Vol. 12	(1983)	$27.00/$30.00		
Vol. 13	(avail. Oct. 1984)	$27.00/$30.00	Vol(s). _____	Vol. _____
nnual Review of ASTRONOMY AND ASTROPHYSICS				
Vols. 1-19	(1963-1981)	$20.00/$21.00		
Vol. 20	(1982)	$22.00/$25.00		
Vol. 21	(1983)	$44.00/$47.00		
Vol. 22	(avail. Sept. 1984)	$44.00/$47.00	Vol(s). _____	Vol. _____
nnual Review of BIOCHEMISTRY				
Vols. 29-50	(1960-1981)	$21.00/$22.00		
Vol. 51	(1982)	$23.00/$26.00		
Vol. 52	(1983)	$29.00/$32.00		
Vol. 53	(avail. July 1984)	$29.00/$32.00	Vol(s). _____	Vol. _____
nnual Review of BIOPHYSICS AND BIOENGINEERING				
Vols. 1-10	(1972-1981)	$20.00/$21.00		
Vol. 11	(1982)	$22.00/$25.00		
Vol. 12	(1983)	$47.00/$50.00		
Vol. 13	(avail. June 1984)	$47.00/$50.00	Vol(s). _____	Vol. _____
nnual Review of EARTH AND PLANETARY SCIENCES				
Vols. 1-9	(1973-1981)	$20.00/$21.00		
Vol. 10	(1982)	$22.00/$25.00		
Vol. 11	(1983)	$44.00/$47.00		
Vol. 12	(avail. May 1984)	$44.00/$47.00	Vol(s). _____	Vol. _____
nnual Review of ECOLOGY AND SYSTEMATICS				
Vols. 1-12	(1970-1981)	$20.00/$21.00		
Vol. 13	(1982)	$22.00/$25.00		
Vol. 14	(1983)	$27.00/$30.00		
Vol. 15	(avail. Nov. 1984)	$27.00/$30.00	Vol(s). _____	Vol. _____

1

E ORDERING INFORMATION ON PAGE 4.

		Prices Postpaid per volume USA/elsewhere	Regular Order Please send:	Standing Order Begin with:
			Vol. number	Vol. number

Annual Review of ENERGY

Vols. 1-6	(1976-1981)	$20.00/$21.00		
Vol. 7	(1982)	$22.00/$25.00		
Vol. 8	(1983)	$56.00/$59.00		
Vol. 9	(avail. Oct. 1984)	$56.00/$59.00	Vol(s). _____	Vol. _____

Annual Review of ENTOMOLOGY

Vols. 7-16, 18-26	(1962-1971; 1973-1981)	$20.00/$21.00		
Vol. 27	(1982)	$22.00/$25.00		
Vol. 28	(1983)	$27.00/$30.00		
Vol. 29	(avail. Jan. 1984)	$27.00/$30.00	Vol(s). _____	Vol. _____

Annual Review of FLUID MECHANICS

Vols. 1-13	(1969-1981)	$20.00/$21.00		
Vol. 14	(1982)	$22.00/$25.00		
Vol. 15	(1983)	$28.00/$31.00		
Vol. 16	(avail. Jan. 1984)	$28.00/$31.00	Vol(s). _____	Vol. _____

Annual Review of GENETICS

Vols. 1-15	(1967-1981)	$20.00/$21.00		
Vol. 16	(1982)	$22.00/$25.00		
Vol. 17	(1983)	$27.00/$30.00		
Vol. 18	(avail. Dec. 1984)	$27.00/$30.00	Vol(s). _____	Vol. _____

Annual Review of IMMUNOLOGY

Vol. 1	(1983)	$27.00/$30.00		
Vol. 2	(avail. April 1984)	$27.00/$30.00	Vol(s). _____	Vol. _____

Annual Review of MATERIALS SCIENCE

Vols. 1-11	(1971-1981)	$20.00/$21.00		
Vol. 12	(1982)	$22.00/$25.00		
Vol. 13	(1983)	$64.00/$67.00		
Vol. 14	(avail. Aug. 1984)	$64.00/$67.00	Vol(s). _____	Vol. _____

Annual Review of MEDICINE: Selected Topics in the Clinical Sciences

Vols. 1-3, 5-15	(1950-1952; 1954-1964)	$20.00/$21.00		
Vols. 17-32	(1966-1981)	$20.00/$21.00		
Vol. 33	(1982)	$22.00/$25.00		
Vol. 34	(1983)	$27.00/$30.00		
Vol. 35	(avail. April 1984)	$27.00/$30.00	Vol(s). _____	Vol. _____

Annual Review of MICROBIOLOGY

Vols. 17-35	(1963-1981)	$20.00/$21.00		
Vol. 36	(1982)	$22.00/$25.00		
Vol. 37	(1983)	$27.00/$30.00		
Vol. 38	(avail. Oct. 1984)	$27.00/$30.00	Vol(s). _____	Vol. _____

Annual Review of NEUROSCIENCE

Vols. 1-4	(1978-1981)	$20.00/$21.00		
Vol. 5	(1982)	$22.00/$25.00		
Vol. 6	(1983)	$27.00/$30.00		
Vol. 7	(avail. March 1984)	$27.00/$30.00	Vol(s). _____	Vol. _____

Annual Review of NUCLEAR AND PARTICLE SCIENCE

Vols. 12-31	(1962-1981)	$22.50/$23.50		
Vol. 32	(1982)	$25.00/$28.00		
Vol. 33	(1983)	$30.00/$33.00		
Vol. 34	(avail. Dec. 1984)	$30.00/$33.00	Vol(s). _____	Vol. _____